Geodynamics of Lithosphere & Earth's Mantle

Seismic Anisotropy as a Record of the Past and Present Dynamic Processes

Edited by
Jaroslava Plomerová
Robert C. Liebermann
Vladislav Babuška

1998

Springer Basel AG

Reprint from Pageoph
(PAGEOPH), Volume 151 (1998), No. 2/3/4

Editors:

Jaroslava Plomerová
Geophysical Institute
Czech Academy of Sciences
Bočni II, 1401
141 31 Praha 4
Czech Republic

Robert C. Liebermann
Department of Geosciences
ESS Building
State University of New York
Stony Brook, NY 11794-2100
USA

Vladislav Babuška
UNESCO
Div. Earth Sciences
1, rue Miollis
75732 Paris
France

A CIP catalogue record for this book is available from the Library of Congress, Washington D.C., USA

Deutsche Bibliothek Cataloging-in-Publication Data

Geodynamics of lithosphere & earth's mantle : seismic anisotropy as a record of the past and present dynamic processes / ed. by Jaroslava Plomerová ... - Basel ; Boston ; Berlin : Birkhäuser 1998
  (Pageoph topical volumes)
  Aus: Pure and applied geophysics ; Vol. 151, No. 2/4. –
  ISBN 978-3-0348-9770-9   ISBN 978-3-0348-8777-9 (eBook)
  DOI 10.1007/978-3-0348-8777-9

© 1998 Springer Basel AG
Originally published by Birkhäuser Verlag in 1998
Softcover reprint of the hardcover 1st edition 1998

Printed on acid-free paper produced of chlorine-free pulp

ISBN 978-3-0348-9770-9

9 8 7 6 5 4 3 2 1

Contents

IV. Mathematical Aspects of Complex Wave Propagation and their Applications

Pure appl. geophys. 151 (1998) 213–219
0033–4553/98/040213–07 $ 1.50 + 0.20/0

❘Pure and Applied Geophysics

Geodynamics of Lithosphere and Earth's Mantle: Seismic Anisotropy as a Record of the Past and Present Dynamic Processes

Preface

Plate tectonics has significantly broadened our view of the dynamics of continental evolution, involving both the processes currently active at the surface and those extending deep into the interior of the Earth. Seismic anisotropy provides some of the most diagnostic evidence for mapping past and present deformation of the entire crust-mantle system.

This Topical Issue is derived mainly from papers presented in an International Workshop on "Geodynamics of the Lithosphere and the Earth's Mantle—Seismic Anisotropy as a Record of the Past and Present Dynamic Processes" held on July 8–13, 1996 at the Chateau of Třešt' in the Czech Republic. The workshop was organized by the Geophysical Institute of the Academy of Sciences of the Czech Republic in Prague and the Center for High Pressure Research (CHiPR) at Stony Brook. The workshop was linked in spirit to two previous international meetings in the Czech Republic in 1976 (Liblice) and 1986 (Bechyně) and brought together geophysicists and geologists who work in the field of observational and theoretical seismology, mineral and rock physics, gravity studies and geodynamic modelling.

The bulk of this Topical Issue is devoted to seismic anisotropy which provides a new dimension in the investigation of processes of our dynamic Earth. Various observations of seismic anisotropy and their implications for geology and geodynamic processes are presented in these papers, demonstrating the enormous, and heretofore largely unexploited, potential of seismic anisotropy to reveal the deformation of the Earth's interior. The papers are organized into four groups; we summarize below the major themes of these papers.

I. Large-scale Anisotropy of the Earth's Mantle

Montagner reviews in his paper global seismic data sets which provide insight into the depth location of large-scale anisotropy, from the D''-layer to the lithosphere. In addition to the generally well-documented seismic anisotropy in the lithosphere and asthenosphere, there is new evidence of anisotropy in the upper

(400–600 km) and lower (660–900 km) portions of the transition zone and in the D''-layer (just above the core-mantle boundary), whereas the bulk of the lower mantle appears to be isotropic. Montagner addresses the ability of seismic anisotropy to define continental roots, to discriminate between competing convective as well as petrological models of the mantle. Babuška *et al.* analyze systematic variations of the seismic radial anisotropy to about 250 km deep and relate them to age of continental provinces. An interpretation of the observed seismic anisotropy by the preferred orientation of olivine crystals results in a model of the mantle lithosphere characterized by anisotropic structures plunging steeply beneath old shields and platforms, as compared to less inclined anisotropies beneath Phanerozoic regions. Such "surface wave models" are compatible with anisotropic models derived from P and S observations. Kubo and Hiramatsu examine the global correlation between the fast directions of anisotropy derived from SKS splitting and absolute plate motions, as defined by three different reference models. They demonstrate conclusively that the selection of the reference frame is important to describe shear deformation in the asthenosphere beneath continents due to plate motion; a fact that is often omitted in discussions of results of shear-wave splitting analyses. The authors found a critical plate velocity for the formation of anisotropy caused by the dislocation-diffusion transition as a function of strain rate on a deformation mechanism of upper mantle olivine.

Regional studies of the upper mantle anisotropy attempt to model and interpret its observed lateral variations. Several papers deal with the upper mantle anisotropy beneath continents, employing various seismic waves and methods. Brechner *et al.* interpret standard analyses of the horizontal components of SKS phases recorded at the German Regional Seismic Network in central Europe and backazimuthal variations of the splitting parameters at some stations, using a model composed of two anisotropic layers mainly with horizontal symmetry axes. A similar study by Granet *et al.* focuses on two geodynamically important areas in western Europe— the Rhinegraben-Urach and the French Massif Central volcanic field. The results suggest that the complex anisotropic pattern is a consequence of a mixture of older deformation inherited from Hercynian structures and recent deformation associated with the rifting process. A significant part of the anisotropy is ascribed to the subcrustal lithosphere in the southern Rhinegraben area and related to recent deformation. In the French Massif Central area, two domains are observed with the main mechanisms produced by the deformation along a strike-slip transformlike fault and by the volcanic flow. The complexity of the results suggests two main sources of seismic anisotropy in the upper mantle lithospheric and asthenospheric components—both of which must be taken into account, as was found recently in several other regions as well. Similar results have been obtained beneath the Saxothuringicum-Moldanubicum contact in central Europe by Plomerová *et al.*, who report spatial variations of P-wave velocities and lateral variations of the particle motion of split shear waves which reflect lateral changes of structure and anisotropy within the deep lithosphere and asthenosphere. A joint interpretation of P-residual spheres and shear-wave splitting parameters results in a self-consistent 3-D anisotropic model of the lithosphere with high velocities plunging divergently from the contact of both tectonic units and southward thickening of the lithosphere beneath the Moldanubicum.

Four more papers pertain to anisotropy in different regions. Diaz *et al.* analyze the *SKS* splitting beneath the Iberian peninsula, for which they found significant variations in the inferred fast velocity directions in different domains. Although large-scale mechanisms such as the absolute plate motion of Eurasia can be invoked to explain the origin of anisotropic features in many sites, the regional variations observed in some domains imply that different origins of the anisotropy must be considered (for example, particular tectonics or present-day flow in the mantle of a smaller scale than the absolute plate motion). There are also significant variations in the delay time measurements at single stations, which cannot be explained by the classical hypotheses of anisotropy and may be attributed to the existence of a more complex anisotropic pattern. A tight array of seismographs spanning a 500 km traverse of southern Tibet resolved anisotropy from *S* and *SKS* as well as spatial variations of its direction and the magnitude of the splitting delay in the paper by Hirn *et al.* Anisotropy with a slow vertical axis is proposed to account for the observations with its origin related to horizontal shear or flow in the low-velocity layer. Average fast shear-wave polarization directions and splitting times of about 1 second were found at stations of two temporary seismic arrays in the Victoria Land region, Antarctica (paper by Pondrelli and Azzara). The fast polarization direction is quite parallel to the opening of the Ross Sea, an active rift system, but also to the absolute plate motion direction. Contrary to the preceding papers, Schutt *et al.* found in the Yellowstone Hot Spot wake, eastern Snake River Plane, Idaho, that the *SKS*-splitting parameters are independent of backazimuth, suggesting that anisotropy is constant beneath the region, although different from nearby Colorado and Nevada: the anisotropy is attributed to simple shear strain caused by the absolute motion of North America. They hypothesize that a fossil anisotropy created in past orogens and continent building events in the Snake River Plain have been reset or erased by the passage of the hot spot, and that subsequent strain of the hot spot-related asthenospheric wake created a uniformly oriented fast axis. Fischer *et al.* use both local *S* phases and teleseismic core phases such as *SKS* to constrain strength and orientation of anisotropy in the mantle beneath the western Pacific subduction zones. Their observations are consistent with a model in which the lower transition zone (520–660 km) and lower mantle are isotropic. Weak anisotropy appears to be contained in the upper transition zone (410–520 km) beneath the southern Kuriles. Significant anisotropy occurs in the backarc upper mantle. The geometry of backarc strain in the upper mantle varies systematically across the western Pacific rim.

II. Mantle Heterogeneity vs. Anisotropy—3-D Velocity and Density Structures and Inferences on Mantle Dynamics

High-quality teleseismic data of Amato *et al.* allow them to refine tomographic images of the lithosphere-asthenosphere structure with an improved resolution in the Northern and Central Apennines, and to study the deformation of the upper mantle by observing seismic anisotropy through shear-wave splitting analysis. They find clear evidence of seismic anisotropy in the uppermost mantle related to the main tectonic processes: either NE–SW compressional deformation of the lithosphere beneath the mountain belt, or arc-parallel asthenospheric flow, and successive extensional deformation in the backarc basin of northern Tyrrhenian and Tuscany. Mele uses P_n travel times to image the azimuthal variations of seismic velocity in the upper mantle beneath the northern part of the Apennines. She found that about 5% seismic anisotropy characterizes the uppermost mantle with the fastest direction of P_n following the arcuate trend of the chain.

Two papers treat modelling of mantle convection utilizing recent global high-resolution tomographic models. Čadek *et al.* study mantle viscosity structure and geodynamics processes. Models associated with the global geoid data indicate clearly the plausible existence of strong vertical viscosity stratification in the top 1000 km of the mantle. Lateral variations in the transition zone suggest interaction between lower-mantle plumes and the region from 600 to 1000 km. Čížková *et al.* present a regional correlation analysis between the seismic structure of the lower mantle and the reconstructions of subduction sites in the past 180 Myr, with the aim of estimating individual styles of slab motion over different parts of the Earth. They found that the correlation patterns obtained for individual branches are remarkably different. The authors conclude that even if no slabs penetrate to the lower mantle, the fast seismic regions will exist there and some of them will correlate with subduction sites at the surface.

Density heterogeneities are incorporated into the tomographic studies in two papers: Sabitova *et al.* focus on velocity and density heterogeneities of the Tien-Shan lithosphere and Yegorova *et al.* analyze large-scale 3-D gravity inhomogeneities in the European-Mediterranean upper mantle.

III. Mineral and Rock Physics Studies

Karato explores the geodynamical significance of seismic anisotropy in the deep mantle on the basis of mineral physics. His synthesis appears to provide a plausible interpretation of seismic anisotropy in terms of deep mantle dynamics. The results provide constraints on the style of mantle convection assuming: (i) regional variations in the magnitude of anisotropy as deformation mechanisms change from dislocation to diffusion creep; and (ii) changes in the elastic anisotropy with pressure for major minerals. A simple shear deformation geometry identifies the controlling factors of microstructural evolution and large strains at high pressures and temperatures, and thus provides a unique opportunity to investigate the "structural geology of the mantle." A brief outline of experimental studies of microstructural evolution and the rheology of mantle mineral aggregates is presented in the paper by Karato *et al.* An optimization algorithm based on the method of simulated annealing is of utility in calculating equilibrium phase assemblages as functions of pressure, temperature and chemical composition; with this method Bina reproduces phase diagrams and illustrates the expected thermal deflection of phase transitions in thermal models of subducting lithospheric slabs and buoyant mantle plumes. In two papers Pros *et al.* describe their laboratory approach to studying elastic anisotropy of rocks on spherical samples under hydrostatic pressure reaching 400 MPa and the elastic parameters they obtain for West-Bohemian granites.

IV. Mathematical Aspects of Complex Wave Propagation and their Applications

The remaining three papers address mathematical aspects of complex wave propagation. Friederich infers a general relation linking the elasticity tensor of an anisotropic medium with that of the constitutive single crystals and the function describing the orientation distribution of crystals. The theoretical results are applied to the inversion of a surface wave dispersion curve for an anisotropic 1-D model. Levin and Park advance a CookBook for *P-SH* conversions in layered media with hexagonally symmetric anisotropy. This type of study can be used to constrain candidate models of crustal anisotropy; e.g., the thicknesses and depths of anisotropic layers as well as the tilt of the anisotropic symmetry axis. Pšenčík and Vavryčuk present approximate *PP* plane wave displacement coefficients of reflection and transmission for weak contrast interfaces separating weakly but arbitrarily anisotropic elastic media. While the reflection coefficient depends only on 8 of the complete set of the weak anisotropy parameters describing *P*-wave phase velocity in weakly anisotropic media, the transmission coefficient depends on their complete set.

Acknowledgements

We thank all the authors for considering this Topical Issue of Pure and Applied Geophysics as the venue for publication of their results. Critical and constructive comments of all the reviewers helped to substantially improve many of the manuscripts, and are greatly appreciated. Our thanks are extended to namely Achauer U., Amato A., Anderson D. L., Babuška V., Barruol G., Blackman D., Bock G., Čadek O., Diaz J., Farra V., Friederich W., Furlong K., Gaherty J. B., Gajewski D., Girardin N., Granet M., Herquel G., Hirn A., Holt W. E., Karato

S.-I., Kendall M., Kern H., Klinge K., Kubo A., Leveque J.-J., Li B., Lillie R. J., Margheriti L., Molnar P., Montagner J.-P., Nataf H.-C., Nicolas A., Park J., Perchuc E., Plenefisch T., Plomerová J., Pondrelli S., Pšenčík I., Reichmann, H. J., Ricard Y., Romanowicz B., Savage M. K., Saxena S. K., Schutt D., Silver P., Sobolev S., Tommasi A., Vauchez A., Vavryčuk V. and Zeyen H.

We also thank institutions which sponsored the workshop—UNESCO, IASPEI, ICL-ILP and the Committee on Mineral and Rock Physics of the American Geophysical Union. Editorial work of JP was partly done within the framework of a grant of the Grant Agency of the Czech Academy of Sciences No. A3012604.

Jaroslava Plomerová   Robert C. Liebermann   Vladislav Babuška

Geophysical Institute   Department of Geosciences   UNESCO, Division
Czech Academy of   ESS Building   of Earth Sciences
Sciences; Boční II   State University of New York   1, rue Miollis
141 31 Praha 4   Stony Brook, NY 11794-2100   75732 Paris
Czech Republic   U.S.A.   France

 To access this journal online:
http://www.birkhauser.ch

Participants of the International Workshop on Geodynamics of the Lithosphere and the Earth's Mantle, July 8–13, 1996, in Třešt' in the Czech Republic.

I. Large-scale Anisotropy of the Earth's Mantle

Pure appl. geophys. 151 (1998) 223–256
0033–4553/98/040223–34 $ 1.50 + 0.20/0

⌐ Pure and Applied Geophysics

Where Can Seismic Anisotropy Be Detected in the Earth's Mantle? In Boundary Layers . . .

JEAN-PAUL MONTAGNER[1]

Abstract—During the last 30 years, considerable evidence of seismic anisotropy has accumulated demonstrating that it is present at all scales, but not in all depth ranges. We detail which conditions are necessary to detect large-scale seismic anisotropy. Firstly, minerals must display a strong anisotropy at the microscopic scale, and/or the medium must be finely layered. Secondly, the relative orientations of symmetry axes in the different crystals must not counteract in destroying the intrinsic anisotropy of each mineral, and there must be efficient mechanisms of orientation of minerals and aggregates. Finally, the strain field must be coherent at large scale in order to preserve long wavelength anisotropy. Part of shallow anisotropy can be related to the past strain field (frozen-in anisotropy), however the deep anisotropy is due to the present strain field. All these conditions are fulfilled only in boundary layers of convective mantle.

We review in this paper, the seismic data sets which provide insight into the location at depth of large-scale anisotropy from the D''-layer up to the lithosphere. In addition to the well-documented seismic anisotropy in the lithosphere and asthenosphere, there is new evidence of seismic anisotropy in the upper (400–660 km) and lower (660–900 km) transition zones and in the D''-layer. Nonetheless the bulk of the lower mantle seems close to isotropy. If we assume the hypothesis that seismic anisotropy is associated with boundary layers in convective systems, these observations strongly suggest that the transition zone is a boundary layer which makes the pasage of matter between the upper and the lower mantle difficult. However, this general statement does not rule out flow circulation between the upper and lower mantles. Finally, the geophysical, mineral physics and geological applications are briefly reviewed. An intercomparison between surface wave anisotropy and body-wave anisotropy data sets is presented. We discuss the scientific potential of seismic anisotropy and how it makes it possible to gain more insight into continental root, deformation and geodynamics processes.

Key words: Seismic anisotropy, mantle convection, boundary layers.

1. Introduction

For theoretical and practical reasons, the earth was long considered as composed of isotropic and laterally homogeneous layers. When an isotropic elastic medium can be described by two independent elastic parameters (λ and μ Lamé parameters), the simplest anisotropic medium (transverse isotropy with vertical symmetry axis) necessitates five independent parameters (LOVE, 1927; ANDERSON,

---

[1] Seismological Laboratory, CNRS URA 195, Institut Universitaire de France, Institut de Physique du Globe, Paris, France.

1961). To date, seismic observations have been explained in terms of isotropic lateral heterogeneities instead of manifestations of anisotropy. However, since, the 1960s, it was recognized that most parts of the earth are not only laterally heterogeneous but also anisotropic. Though the lateral heterogeneities of seismic velocities were long used for geodynamical applications, the importance of anisotropy for understanding geodynamic processes is only recently recognized. The early evidence of seismic anisotropy was the discrepancy between Rayleigh and Love wave dispersions (ANDERSON, 1961; AKI and KAMINUMA, 1963) and the azimuthal dependence of P_n velocities (HESS, 1964). These first observations were followed by many in the 1970s and 1980s (see CRAMPIN, 1984; MONTAGNER and NATAF, 1986; BABUŠKA and CARA, 1991 for references).

Different geophysical fields are involved in the investigation of the manifestations of anisotropy of earth materials, mineral physics and geology for the study of the microscopic scale, and seismologists for scales larger than typically one kilometer. It is evident that it is present in most depth ranges of the earth. But the origin of seismic anisotropy is nonunique. In the crust, the crack distribution seems to play a major role (CRAMPIN and BOOTH, 1985). In the lithosphere, it is usually explained by the preferred orientation of olivine (NICOLAS and CHRISTENSEN, 1987) and is related to plate tectonics processes. The intrinsic anisotropy of minerals (olivine and to a less extent orthopyroxene and clinopyroxene) associated with lattice-preferred orientation induces large-scale observable and unambiguous effects, either on body waves (S-wave splitting observed on SKS (VINNIK et al., 1984); P-wave anisotropy (BABUŠKA et al., 1984) or surface waves through the azimuthal anisotropy (FORSYTH, 1975) and the 'polarization' anisotropy (SCHLUE and KNOPOFF, 1977). Both kinds of observable anisotropy can be simultaneously explained by the theoretical developments of MONTAGNER and NATAF (1986). Most of the lower mantle seems to be isotropic (MEADE et al., 1995) except the D''-layer (VINNIK et al., 1989; MAUPIN, 1994). It recurs in the inner core, where anisotropy was discovered about 10 years ago from free oscillations (WOODHOUSE et al., 1986) and from the P-wave travel times reported in the ISC bulletins (MORELLI et al., 1986). However the origin and mechanisms creating the anisotropy in the core are still the subject of controversy.

Since these early observations of seismic anisotropy, numerous studies have confirmed the existence of anisotropy in different depth ranges of the earth and will not be reviewed in this paper. From the global geodynamics point of view, seismic anisotropy has many applications. We will show that it makes it possible to define the root of continents and to investigate the coupling between the lithosphere and the rest of the mantle (MONTAGNER and TANIMOTO, 1991; SILVER, 1996), and more generally to gain insight into mantle convection. Due to the high Rayleigh number, mantle convection is highly chaotic and numerical modeling demonstrates that most of the deformation takes place in boundary layers. Since seismic anisotropy is closely related to large-scale deformation (NICOLAS and CHRIS-

TENSEN, 1987; KARATO, 1989), we suggest that, conversely, boundary layers can be detected by the existence of seismic anisotropy. We will present in this paper, the different depth ranges in the mantle, where seismic anisotropy was detected, i.e., D''-layer, transition zone and uppermost mantle. In this top boundary layer, seismic anisotropy can be directly compared to geological results, and it provides fundamental information on processes involved in mountain building and collision between continents such as in Central Asia. The robust features of these different investigations in different depth ranges are presented in this paper and the enormous scientific potential of seismic anisotropy is underlined.

2. Evidence of Anisotropy in the Different Layers of the Mantle

2.1 Conditions for Observing Seismic Anisotropy

The observations of seismic anisotropy usually involve a very large spatial range from the microscopic wavelength extending to the thousands of kilometers. Different geophysical fields are involved in the investigation of the manifestations of anisotropy of earth materials, mineral physics and geology for the study of the microscopic scale, and seismology for scales larger than typically one kilometer. At least four conditions are necessary to detect large-scale mantle seismic anisotropy and they are quite severe.

Intrinsic anisotropy of minerals. The different minerals present in the upper mantle are anisotropic (PESELNICK et al., 1974). The main constituent, olivine, is strongly anisotropic; the difference of velocity between the fast axis and the slow axis is larger than 20%. Other important minerals such as orthopyroxene or clinopyroxene are anisotropic as well ($>10\%$) (see for example CHRISTENSEN and LUNDQVIST, 1982 and ANDERSON, 1989, for a review). Other constituents such as garnet display a cubic crystallographic structure which presents a smaller anisotropy.

As depth increases, the minerals undergo phase transformations. There is some tendency (though not systematic) that with increasing pressure and temperature, the crystallographic structure evolves towards a more closely packed structure, more isotropic, such as cubic structure. For example, olivine transforms towards β-spinel and then γ-spinel in the upper transition zone (410–660 km of depth) and towards perovskite and magnesiowüstite in the lower mantle. The ideal structure of perovskite is cubic, but it can display some important distortion which can induce large anisotropy at least in the uppermost lower mantle. Therefore, at microscopic scales, earth materials in the upper mantle are strongly anisotropic, although the anisotropy tends to decrease as depth is increasing, with some exceptions such as MgO (KARATO, 1997).

Efficient mechanisms of orientation of crystals. In order to observe seismic anisotropy the crystals must be sensitive to the strain field, slip systems must be activated, and a lattice-preferred orientation can develop from dislocation creep (KARATO, 1989). Through the mechanisms of lattice-preferred orientation, it is found that the anisotropy of many minerals (NICOLAS and CHRISTENSEN, 1987) can be very large.

Anisotropy of assemblages of minerals. Mantle rocks are assemblages of different minerals. All of them are less anisotropic than pure olivine. The amount of anisotropy is largely dependent on the percentage of these different minerals. The relative orientations of symmetry axes in the different minerals must not counteract in destroying the intrinsic anisotropy of each mineral. Consequently, the resulting anisotropy will depend on the mechanisms which align the crystallographic axes of different minerals. For example, the anisotropy of the pyrolitic model, mainly composed of olivine and orthopyroxene (RINGWOOD, 1975), is affected by the relative orientation of their crystallographic axes (CHRISTENSEN and LUNDQUIST, 1982). According to their results, the fast axis of olivine is parallel to the intermediate axis of orthopyroxene in the shear plane and parallel to the flow direction, however the fast axis of orthopyroxene is orthogonal to the shear plane. The resulting anisotropy, however, will be larger than 10% (ESTEY and DOUGLAS, 1986; MONTAGNER and ANDERSON, 1989a). For competing petrological models such as piclogite (ANDERSON and BASS, 1984, 1986), where the percentage of olivine is smaller, and of garnet larger than in pyrolite, the amount of anisotropy will be smaller (about 5%). At slightly larger scales, the scale of rock samples, several studies of anisotropy were undertaken (NICOLAS *et al.*, 1973). Dunite, which is also pure olivine, displays a large anisotropy (PESELNICK and NICOLAS, 1978). Moreover, this anisotropy is coherent in whole massifs of ophiolites over several tens of kilometers (NICOLAS, 1993; VAUCHEZ and NICOLAS, 1991).

Coherent strain field. At large scale, the deformation due to mantle convection must be large enough (MCKENZIE, 1979), and the principal directions of the strain tensor must be parallel or subparallel in order to preserve long wavelength anisotropy (RIBE, 1989). From P_n studies and models of the formation of the oceanic lithosphere, it is possible to infer that anisotropy remains homogeneous on horizontal length sales in excess of 1000 km.

However, it must be noted that there is one case in which anisotropy is observed without microscopic anisotropic minerals: If a medium is finely layered, with an average thickness of layer smaller than the wavelength, the equivalent medium behaves like a transversely isotropic solid (BACKUS, 1962). This condition can be met for example in a laminated structure, such as marble cake (ALLÈGRE and TURCOTTE, 1986), and this kind of anisotropy is often referred to as SPO (shape-preferred orientation) anisotropy (KARATO, 1997). Therefore, though the medium is effectively equivalent to an anisotropic medium from the seismic point of view, it can lead to invalid interpretations (LEVSHIN and RATNIKOVA, 1984). That is a very important statement: the observation of large-scale anisotropy does not lead to a unique interpretation, and necessitates additional information in order to discriminate between the different origins of anisotropy.

If one of these conditions is not satisfied, the overall seismic anisotropy effect will be close to zero. Fortunately, the necessary conditions for observing seismic anisotropy are satisfied in many geological and geodynamical contexts. Let us now consider the evidence of anisotropy at very large wavelengths (>100 km), which have been investigated from seismic observations. Different and independent seismic data sets evidence that the effect of anisotropy is not negligible for explaining the propagation of seismic waves inside the earth. Most of this anisotropy is thus far attributed to the upper mantle but we will see that there is new evidence that seismic anisotropy is not only confined to the upper mantle but that it is present in other depth ranges of the mantle as well.

2.2 Reference Earth Models

Before considering measurements of anisotropy at global, regional or local scales, let us consider the laterally averaged earth or equivalently, the spherically symmetric reference earth models. The last ten years have seen the rapid development of seismic tomographic models from the surface down to the center of the earth. It might therefore be considered that there is no need for inversion for radial reference earth models. However, tomographic models are derived from a reference model by linearized inversion schemes applying first-order perturbation theory. The quality of the reference model will strongly condition the outcome of inversions for 3-D models. As pointed out by ANDERSON and DZIEWONSKI (1982), an inappropriate theory might lead to biased models and such a statement motivated the introduction of anisotropy down to 220 km depth in the derivation of the Preliminary Reference Earth Model (hereafter referred as PREM, DZIEWONSKI and ANDERSON, 1981).

The most general case of anisotropy for a spherically symmetric earth is the transverse isotropy with vertical symmetry axis (also termed radial anisotropy). Such a medium can be described using six functions of radius r, the density ρ, the wave velocities, $V_{PH} = \sqrt{A/\rho}$, $V_{SV} = \sqrt{L/\rho}$, and the anisotropic parameters $\xi = N/L$, $\phi = C/A$, $\eta = F/(A - 2L)$, where A, C, F, L, N are the five elastic moduli needed to describe the transversely isotropic medium (LOVE, 1927). In the case of isotropy, V_{PH} and V_{SV} are the P- and S-wave velocities and ξ, ϕ, η are unity.

PREM introduced for the first time the radial anisotropy in the uppermost 220 km of the mantle. However, some aspects of the normal mode data are not well explained by PREM (MONTAGNER and ANDERSON, 1989a), for example fundamental toroidal mode. A modified version of PREM was derived ACY (MONTAGNER and ANDERSON, 1989b), with a smaller anisotropy in the layer 24–220 km and with evidence of small anisotropy down to 400 km discontinuity. Recently, WIDMER

(1991) proposed a new model CORE11 which incudes radial anisotropy in the entire mantle. MONTAGNER and KENNETT (1996) showed that these different reference earth models do not explain body-wave observations and conversely that the eigenperiods calculated from reference models derived from body wave IASP91 (KENNETT and ENGDAHL, 1991); SP6 (MORELLI and DZIEWONSKI, 1993); AK135 and AK303 (KENNETT et al., 1996) differ significantly from the observed eigenperiods. They proposed to reconcile body-wave and normal-mode approaches by inverting for free parameters which cannot be extracted from body waves, in the context of a radially anisotropic medium, but which can be derived from eigenperiods, i.e., the density ρ, the quality factor Q_μ and the anisotropy parameters η, ϕ, η.

Several robust features have been found from the inversions, based on the different body-wave models, regarding the radial anisotropy. First of all, anisotropy is significant in the whole upper mantle with a minimum value in the depth range 300–500 km. It is very small in the whole lower mantle except in the lower transition zone (between 660 km discontinuity and 1000 km depth) and in the D''-layer. The corresponding models for ξ, ϕ, η are plotted in Figure 1. An interesting feature of these models is the existence of radial anisotropy in the upper (410–600 km) and lower (660–1000 km) transition zones with opposite signs. KARATO (1997) proposed an explanation to this reversal by estimating the lattice-preferred oreintations of the main constituents of the upper and lower transition zones. The behavior of anisotropy in the D''-layer seems robust as well, but is more complex with a small S-wave anisotropy but large P wave and η anisotropies.

Therefore, these new reference earth models provide some indication of the existence of anisotropy in two boundary layers; the first one in D''-layer at the core-mantle boundary and the second one in upper and lower transition zones at the ULM (upper-lower mantle boundary around the 660 km discontinuity). Independent seismological studies tend to corroborate these findings in the D''-layer and in the transition zone, as will be seen in the next section.

2.3 Anisotropy in the D''-Layer

During recent years, the structure of the mysterious D''-layer has been extensively investigated. It is not the goal of this paper to review these different studies which make use of different kinds of body waves but only to draw attention to recent studies which make evident the presence of anisotropy in the D''-layer (VINNIK et al., 1989; LAY and YOUNG, 1991). By studying S_d waves (S-diffracted waves at the core-mantle boundary), VINNIK et al. (1995) found that SV_d waves are delayed relative to SH_d waves by 3 s (Fig. 2). That is an unambiguous observation in the sense that the authors present a careful comparison of their waveforms with SKS and $SKKS$ waveforms. Their observations are characteristic of a transversely isotropic medium with vertical symmetry axis with $V_{SH} > V_{SV}$. Other seismic

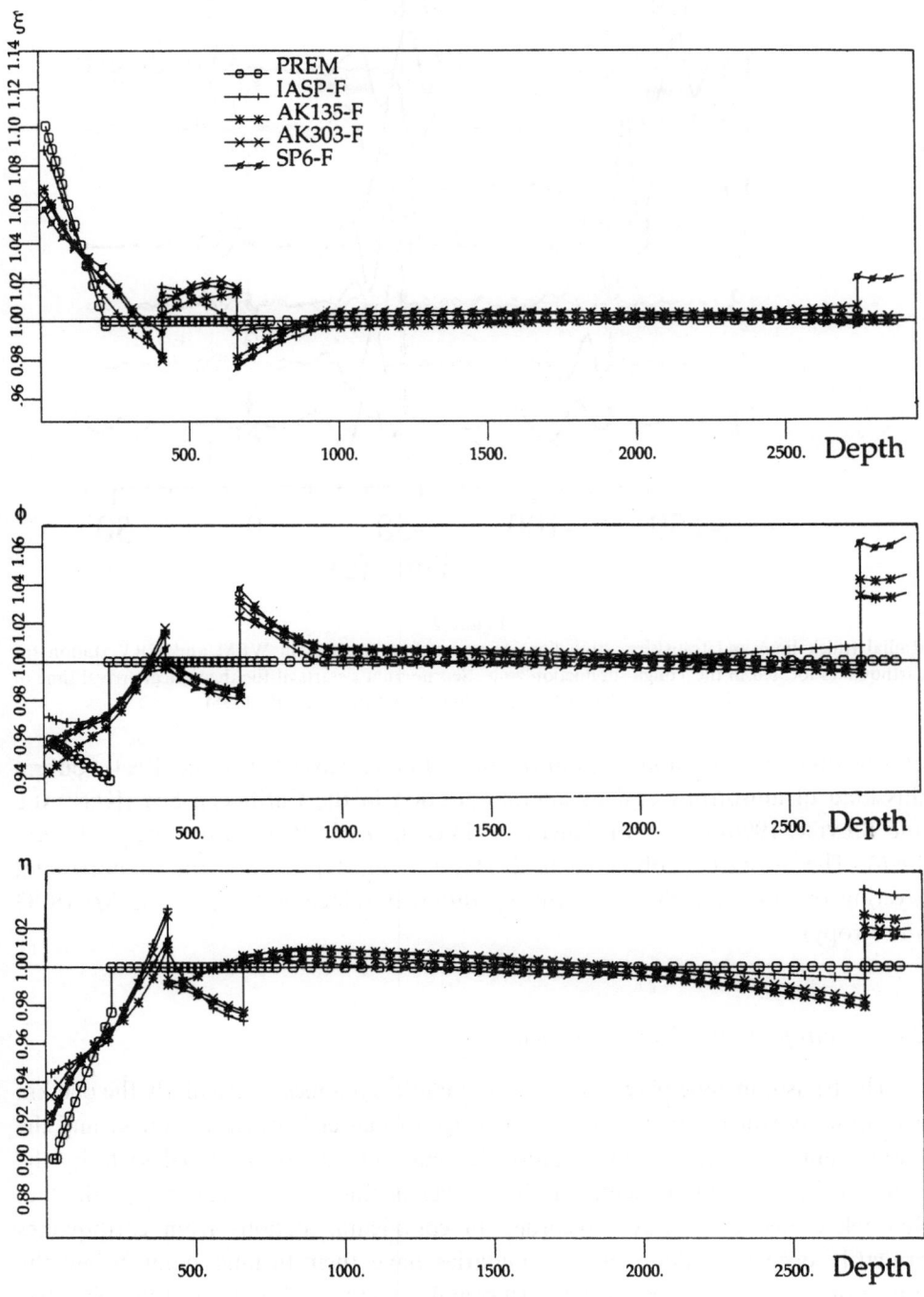

Figure 1
Anisotropic parameters after inversion of normal mode eigenfrequencies, starting from body-wave models IASP91, AK135, AK303, SP6. (a) ξ: S-wave anisotropy; (b) ϕ: P-wave anisotropy; (c) anisotropic parameter η (adapted from MONTAGNER and KENNETT, 1996).

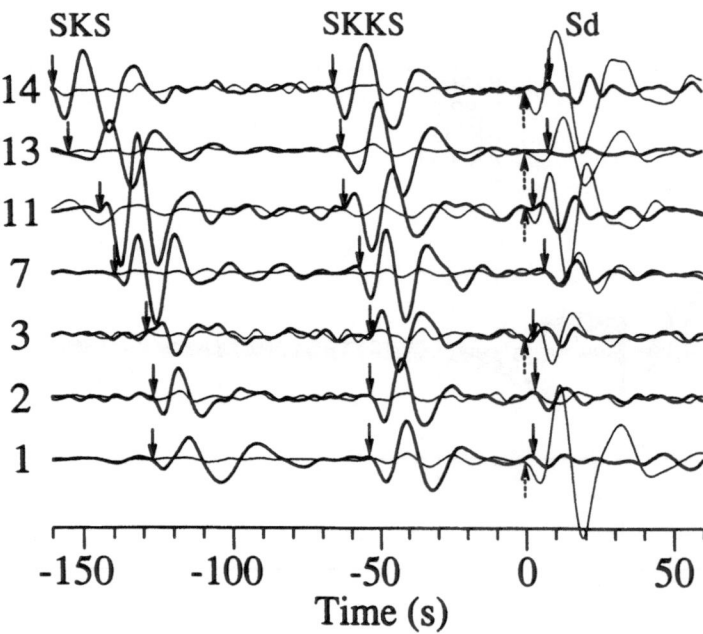

Figure 2
Radial (thick line) and transverse (thin line) components of records in WFM and HRV station for earthquakes located in the Tonga subduction zone. Seismogram *s* starts at the theoretical arrival time of S_{diff} for model IASP91 (after Fig. 2 of VINNIK *et al.*, 1995).

observations such as anomalous diffraction of body waves (MAUPIN, 1994) confirm this kind of anisotropy and its lateral variation in the Caribbean Sea (KENDALL and SILVER, 1996) or North America (MATZEL *et al.*, 1996; GARNERO and LAY, 1996). The anisotropy observed in D''-layer is most probably due to horizontal layering or aligned inclusions inducing different velocities for *SV* and *SH* (SPO anisotropy).

2.4 Anisotropy in the Transition Zone

The transition zone plays a key role in mantle dynamics, particularly the 660 km discontinuity which might inhibit the passage of matter between the upper and the lower mantle. Its seismic investigation is made difficult on a global scale by the poor sensitivity of fundamental surface waves in this depth range and by the fact that teleseismic body waves recorded in continental stations from earthquakes primarily occurring along plate boundaries have their turning point below the transition zone. However, MONTAGNER and KENNETT (1996), by using eigenfrequency data, display evidence of radial anisotropy in the upper and lower transition zones. Another important feature of transition zone is that, contrary to the rest of the upper mantle, the upper transition zone is characterized by a large degree 2 pattern (MASTERS *et al.*, 1982). MONTAGNER and ROMANOWICZ (1993) explained this degree 2 pattern by the predominance of a simple large-scale flow pattern characterized by two upwellings in the Central Pacific Ocean and Eastern Africa and two downwellings in the Western and Eastern Pacific Oceans. This scheme is corroborated by the existence in the upper transition zone of a slight but significant radial anisotropy displayed by MONTAGNER and TANIMOTO (1991) and ROULT *et al.* (1990). The distribution of radial anisotropy at a depth of 470 km is presented in Figure 3. We will explain in section 3.1 why the pattern of radial anisotropy is dominated by degree 4 in agreement with the prediction by this model.

The existence of anisotropy close to the 660 km discontinuity is also confirmed by VINNIK and MONTAGNER (1996) below Germany. By studying *P*-to-*S* converted waves at GRF network, they observed that part of the initial *P* wave is converted into *SH* wave. This signal can be observed on the transverse component in Figure 4. The amplitude of this wave cannot be explained by a dipping 660 km discontinuity and this constitutes sound evidence for the existence of anisotropy just above this discontinuity. However, there is evidence of lateral variation of anisotropy in the transition zone as found by the investigation of several subduction zones (FISCHER and YANG, 1994; FISCHER and WIENS, 1996). FOUCH and FISCHER (1996) present a synthesis of these different studies and demonstrate that some subduction zones such as Sakhalin Islands require deep anisotropy in the transition zone, whereas others such as Tonga need no anisotropy. However, they conclude that their data might be reconciled by considering the upper transition

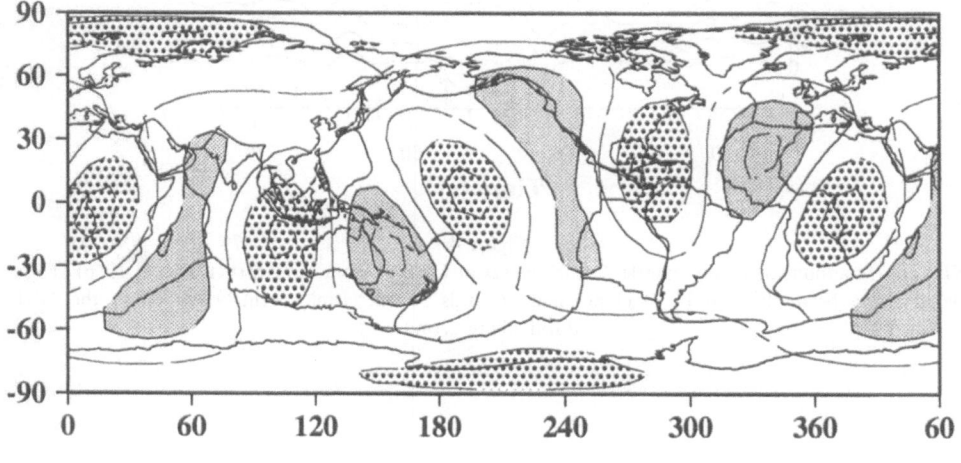

RADIAL ANIS. - DEGREE 4 - DEPTH= 467km

Figure 3

Radial ξ-anisotropy map for degree 4 at a depth of 470 km. Isolines 0.25%. Four zones of dominant radial flow can be observed and associated with Central Pacific and Africa for upgoing flow, Western Pacific and America for downgoing flow. They are therefore in agreement with the degree 2 pattern of the transition zone.

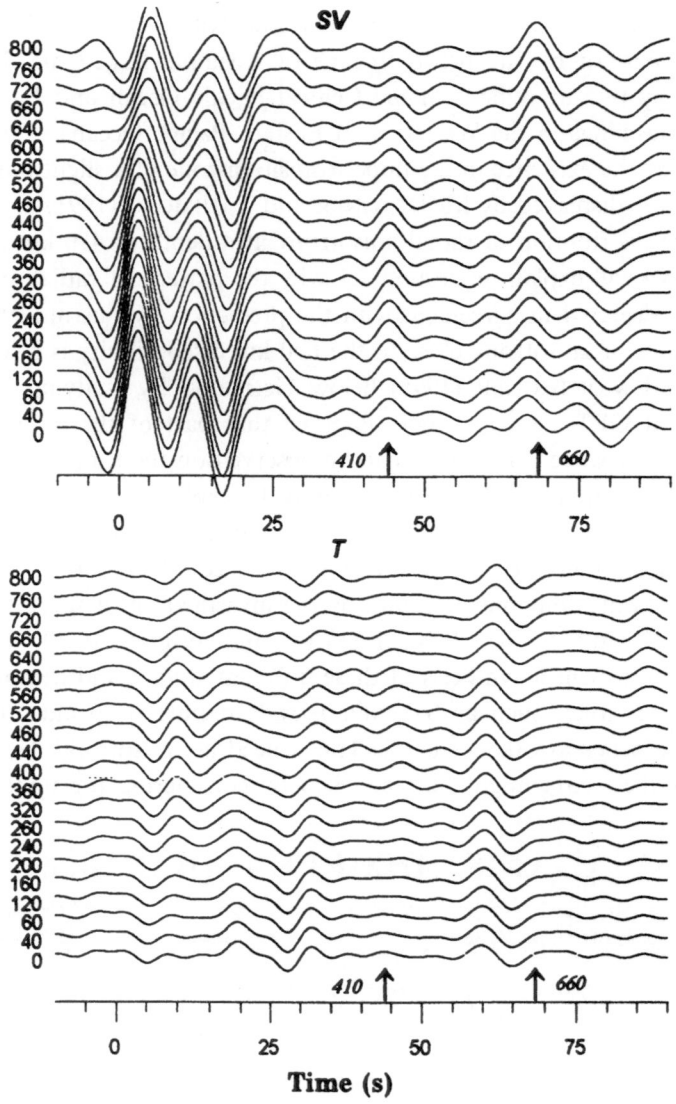

Figure 4

SV and *T* components of the mantle *P*-to-*S* phases at GRF (VINNIK and MONTAGNER, 1996). These signals have been obtained by stacking many records of teleseismic events recorded in the GRF (Gräfenberg array).

zone (410–520 km) intermittently anisotropic, and the rest of the transition zone isotropic. Moreover, they make a very strong assumption that the orientation and strength of anisotropy does not vary with depth, which is not supported by surface wave studies (MONTAGNER and TANIMOTO, 1991).

2.5 Evidence of Anisotropy in the Upper 410 km of the Mantle

The upper 410 km of the mantle is the depth range where the existence of seismic anisotropy is now widely recognized and well documented. The early evidence was the discrepancy between Rayleigh and Love wave dispersion (ANDERSON, 1961; AKI and KAMINUMA, 1963) and the azimuthal dependence of P_n velocities (HESS, 1964). Azimuthal variations have been found for different areas in the world as well as for body waves and surface waves.

Body waves. For body waves the evidence of anisotropy results from the investigation of the splitting in teleseismic shear waves such as *SKS* (VINNIK *et al.*, 1984, 1989a,b, 1992; SILVER and CHAN, 1988; ANSEL and NATAF, 1989), *ScS* (ANDO, 1984; FUKAO, 1984) and *S* (ANDO and ISHIKAWA, 1983; BOWMAN and ANDO, 1987; FISCHER and WIENS, 1996; GAHERTY and JORDAN, 1995). There is also evidence of *P*-wave anisotropy (BABUŠKA *et al.*, 1984, 1993). These waves are shown to provide an excellent lateral resolution. Among these different observations, the splitting information derived from *SKS* is probably the less ambiguous and has been extensively used in teleseismic anisotropy investigations (see SILVER, 1996 for a review). The drawbacks of this technique are that it is almost impossible to locate at depth the anisotropic area, that it cannot take account for a dipping symmetry axis, and that only continental areas can be extensively investigated. The rapid variation of directions of fast velocity which can be observed in some continental regions on a short spatial scale (VINNIK *et al.*, 1989a; HIRN *et al.*, 1995), cannot be explained by a very deep anisotropy and the origin of anisotropy at depth is confined to the first 410 km, either in the lithosphere or in the top of the asthenosphere.

Surface waves. Surface waves are also well suited for investigating upper mantle anisotropy. Two kinds of observable anisotropy can be considered. The first one results from the well-known discrepancy between Love and Rayleigh waves, often referred to as "polarization" anisotropy (SCHLUE and KNOPOFF, 1977) or radial anisotropy. In order to remove this discrepancy, it is necessary to consider a transversely isotropic medium with a vertical symmetry axis (also termed radial anisotropic medium). This kind of anisotropy is characterized by five anisotropic parameters plus density (ANDERSON, 1961). On a global scale, NATAF *et al.* (1984, 1986) derived by the simultaneous inversion of Rayleigh and Love wave dispersion, the geographical distributions of *S*-wave anisotropy at different depths, assuming radial anisotropy.

The second kind of observable anisotropy is the azimuthal anisotropy, directly derived from the azimuthal variation of phase velocity of surface waves. It was observed initially on Rayleigh waves by FORSYTH (1975) in Nazca plate. Since these pioneering studies, global and regional models have been derived for both kinds of anisotropy (MITCHELL and YU, 1980; MONTAGNER, 1985). TANIMOTO and ANDERSON (1985) obtained a global distribution of the Rayleigh wave azimuthal anisotropy at different periods. On a regional scale, several tomographic investigations report the existence of azimuthal anisotropy in the Indian Ocean (MONTAGNER, 1986a), in the Pacific Ocean (SUETSUGU and NAKANISHI, 1987; NISHIMURA and FORSYTH, 1987, 1988) and in Africa (HADIOUCHE *et al.*, 1988). LÉVÊQUE and CARA (1985), CARA and LÉVÊQUE (1988) used higher mode data to display radial anisotropy under the Pacific Ocean and North America to at least 300 km.

The radial anisotropy (or "polarization" anisotropy) and the azimuthal anisotropy are two different manifestations of the same phenomenon, the anisotropy of the upper mantle. MONTAGNER and NATAF (1986) derived a technique which makes it possible to simultaneously explain these two forms of seismically observable anisotropy. The principles of this technique will be only briefly described, for the most general case of anisotropy (on the condition that it is small). A complete description of the whole procedure can be found in MONTAGNER (1996). The number of independent anisotropic parameters is 13 for surface waves. This high number of independent parameters explains the difficulty in implementing such a technique, from a practical point of view. However, the method can be slightly simplified by assuming that the medium displays a symmetry axis with any orientation. In that case, 7 independent anisotropic parameters are necessary, 5 related to the transverse isotropy plus 2 angles to express the orientation in space of the symmetry axis. This simplified technique was coined "vectorial tomography" (MONTAGNER and NATAF, 1988), and was applied to the 3-D anisotropic investigation of the Indian Ocean (MONTAGNER and JOBERT, 1988). Moreover, this latter study showed that, paradoxically, in order to explain their data on the Indian Ocean, a parameterization with anisotropy requires less parameters than a parameterization with only isotropic terms. This can be explained by the fact that the increase in the number of physical parameters in the case of anisotropy is largely compensated by a reduced number of spatial parameters. A first global 3-D anisotropic tomography (with 13 anisotropic parameters) was derived by MONTAGNER and TANIMOTO (1990, 1991). Therefore, contrarily to body waves, surface waves enable location of anisotropy at depth but, so far, its lateral resolution (several thousands of kilometers) is very poor.

From a theoretical point of view, a general slight elastic anisotropy in a plane-layered medium gives rise to an azimuthal dependence of the local phase or group velocities of Love and Rayleigh waves of the form (SMITH and DAHLEN, 1973, 1975):

$$v(\omega, \Psi) - v_0(\omega, \Psi) = \alpha_0(\omega) + \alpha_1(\omega) \cos 2\Psi + \alpha_2(\omega) \sin 2\Psi$$

$$+ \alpha_3(\omega) \cos 4\Psi + \alpha_4(\omega) \sin 4\Psi \qquad (1)$$

where ω is the frequency of the wave and Ψ is the azimuth along the path.

MONTAGNER and NATAF (1986), following the same approach, displayed the simple linear combinations of the elastic tensor components C_{ij} sufficient to describe the two seismically observable effects of anisotropy, the "polarization" anisotropy and the azimuthal anisotropy. The 0–Ψ term corresponds to the average over all azimuths and involves five independent parameters, A, C, F, L, N, which express the equivalent transversely isotropic medium with vertical symmetry axis. The other azimuthal terms (2–Ψ and 4–Ψ) depend on 4 groups of 2 parameters, B, G, H, respectively describing the 2–Ψ azimuthal variation of A, L, F, and E describing the 4–Ψ azimuthal variation of A and N. Therefore, the different azimuthal terms α_0, α_1, α_2, α_3, α_4, depend on 13 three-dimensional parameters, which are assumed independent:

Constant term (0 Ψ-azimuthal term)

(α_0)

$$A = \rho v_{PH}^2 = \tfrac{3}{8}(C_{11} + C_{22}) + \tfrac{1}{4}C_{12} + \tfrac{1}{2}C_{66}$$

$$C = \rho v_{PV}^2 = C_{33}$$

$$F = \tfrac{1}{2}(C_{13} + C_{23})$$

$$L = \rho v_{SV}^2 = \tfrac{1}{2}(C_{44} + C_{55})$$

$$N = \rho v_{SH}^2 = \tfrac{1}{8}(C_{11} + C_{22}) - \tfrac{1}{4}C_{12} + \tfrac{1}{2}C_{66}$$

2 Ψ-azimuthal term:

$(\alpha_1)\cos 2\Psi$ $\qquad\qquad\qquad$ $(\alpha_2)\sin 2\Psi$

$$B_c = \tfrac{1}{2}(C_{11} - C_{22}) \qquad\qquad B_s = C_{16} + C_{26}$$

$$G_c = \tfrac{1}{2}(C_{55} - C_{44}) \qquad\qquad G_s = C_{54}$$

$$H_c = \tfrac{1}{2}(C_{13} - C_{23}) \qquad\qquad H_s = C_{36}$$

4 Ψ-azimuthal term:

$(\alpha_3)\cos 4\Psi$ $\qquad\qquad\qquad\qquad$ $(\alpha_4)\sin 4\Psi$

$$E_c = \tfrac{1}{8}(C_{11} + C_{22}) - \tfrac{1}{4}C_{12} - \tfrac{1}{2}C_{66} \qquad E_s = \tfrac{1}{2}(C_{16} - C_{26})$$

where indices 1 and 2 refer to horizontal coordinates (1: North; 2: East) and index 3 refers to vertical coordinates. ρ is the density, v_{PH}, v_{PV} are respectively horizontal and vertical P-wave velocities, v_{SH}, v_{SV} horizontal and vertical S-wave velocities. Therefore, the different parameters present in the different azimuthal terms are simply related to elastic moduli C_{ij} and the corresponding kernels are detailed and their variation at depth is plotted in MONTAGNER and NATAF (1986). Therefore, in the most general case for a slight anisotropy, thirteen combinations of elastic moduli are necessary to describe the total effect of anisotropy on seismic surface waves. That means that, from a theoretical point of view, seismic surface waves have the ability to provide 13 tomographic models. However, from a practical point of view, data does not have the resolving power to invert for so many parameters. MONTAGNER and ANDERSON (1989a) proposed the usage of constraints from petrology in order to reduce the parameter space. Actually, they ascertained that some of these parameters display large correlations independent of the petrological model used. Two extreme models were used to derive these correlations: the pyrolite model (RINGWOOD, 1975) and the piclogite model (ANDERSON and BASS, 1984, 1986; BASS and ANDERSON, 1984). In the depth inversion process, the slightest correlations between parameters of both models are kept. This approach was already followed by MONTAGNER and ANDERSON (1989b) to derive an average reference earth model, and by MONTAGNER and TANIMOTO (1991) for the first global 3-D anisotropic model. The complete tomographic technique (regionalization + inversion at depth) has been applied to investigate either regional structures of the Indian Ocean (MONTAGNER and JOBERT, 1988), of the Atlantic Ocean (MOCQUET et al., 1989; SILVEIRA et al., 1997), of Africa (HADIOUCHE et al., 1989), of the Pacific Ocean (NISHIMURA and FORSYTH, 1988), of Antarctica (ROULT et al., 1994) and Central Asia (GRIOT et al., 1996) or global structure (MONTAGNER and TANIMOTO, 1990, 1991).

Figure 5 presents some maps which illustrate the two kinds of anisotropy which can be retrieved by simultaneous inversion of Rayleigh and Love waves constant and azimuthal terms of equation (1). The map of Figure 5a is the distribution of the G parameter which is related to the azimuthal variation of SV-wave velocity. The

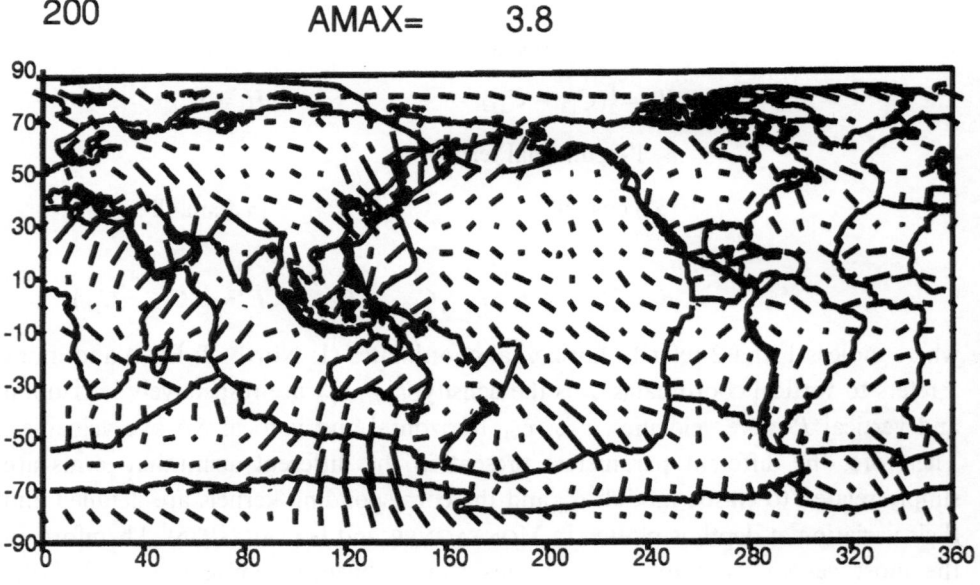

Figure 5a

AUM model for =100 km

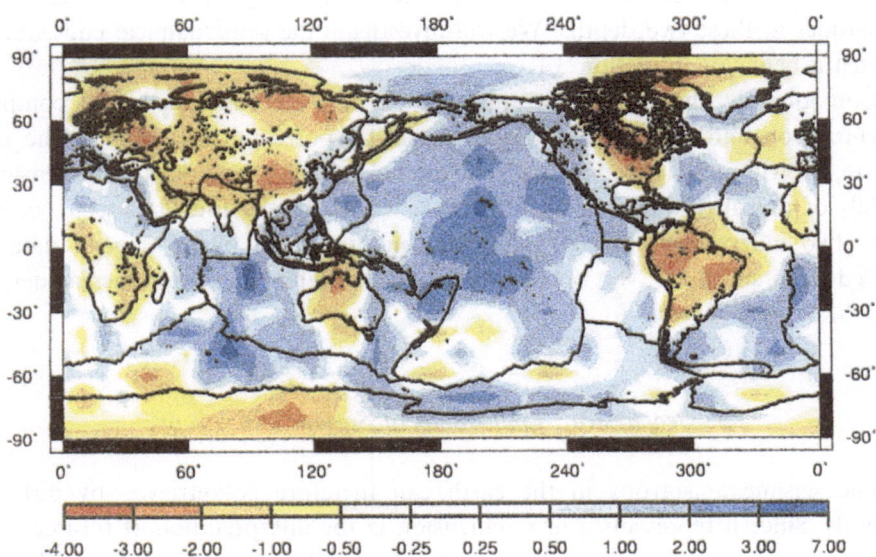

AUM model for =310 km

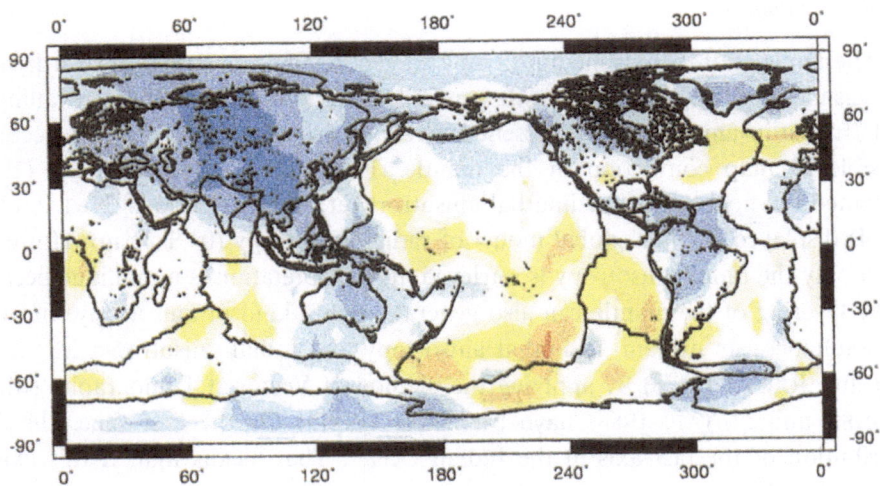

Figure 5

Result of the simultaneous inversion of Rayleigh and Love waves dispersion and their azimuthal variations. (a) Distribution of the *G* parameter at 200 km (adapted from MONTAGNER and TANIMOTO, 1991). *G* is related to the azimuthal variation of v_{SV} velocity. (b) ξ distributions at two depths (100 km and 300 km) in % with respect to ACY400 (MONTAGNER and ANDERSON, 1989b). Be aware that ξ anomalies are plotted at the 2 depths with respect to a reference value different from 0.

maximum amplitude of G is around 2% and rapidly decreasing as depth is increasing. On Figure 5b, the equivalent radial anisotropy of the medium for S wave expressed by the ξ parameter is displayed for 2 different depths, 100 km and 310 km. The distribution of anisotropy has completely different patterns and amplitudes at these two depths. We will investigate the geodynamical consequence of such a behavior in section 3.1.

Consequently, the evidence of anisotropy is now quite robust, but its complete interpretation and utilization for geophysical purposes are ongoing. In the next section, we will present examples of geodynamical applications of three-dimensional models of anisotropy derived from surface waves. In the fourth section, we will relate two independent seismic data sets, on one hand, phase velocity data set derived from surface waves, on the other hand, differential travel times as derived from SKS body waves.

3. Geophysical Applications of Seismic Anisotropy

The seismic anisotropy in the earth can therefore be retrieved by different methods. Since it represents a new dimension in the interpretation of seismic data, its scientific potential is enormous, and still largely unexploited. We will only present examples of interesting applications of anisotropy in geodynamics and tectonics.

3.1 Geodynamics

The application of seismic anisotropy to geodynamics is straightforward, since the seismic anisotropy in the mantle generally reflects the strain field prevailing in past (frozen-in anisotropy) or present convective processes. Therefore, it becomes possible to map convection in the mantle. TANIMOTO and ANDERSON (1985) presented the first maps of azimuthal anisotropy at different periods. NATAF et al. (1984) displayed the first global model of radial anisotropy (0–Ψ term). However, when only the radial anisotropy is retrieved, its interpretation is nonunique because a fine layering of the mantle can also generate such a kind of anisotropy. In order to simultaneously explain azimuthal anisotropy and radial anisotropy, MONTAGNER and JOBERT (1988) by applying the method of Vectorial Tomography (MONTAGNER and NATAF, 1988) have been able to plot for the first time, the 3-D distribution of the fast axis in the Indian Ocean. That means that, if this axis is effectively related to the strain induced by plate motion and convective processes, it is possible to visualize convection cells. A quite similar technique has been applied on a global scale by MONTAGNER and TANIMOTO (1990, 1991). When both kinds of observable anisotropy (azimuthal and radial anisotropies) are retrieved, the most likely interpretation is the presence of large-scale flow which can align the symmetry axes of the different anisotropic minerals of the mantle; primarily olivine and to a less extent the ortho- and clino-pyroxenes.

For the upper mantle, the $\xi = (N - L)/_L$ parameter (radial S anisotropy) can be simply interpreted as a tendency for a horizontal or subhorizontal flow when ξ is positive, and a radial (or very steep) flow when it is negative. The simultaneous existence of azimuthal anisotropy for example through the G parameter (SV-wave azimuthal anisotropy) will provide in addition the horizontal orientation of the flow, but not its vectorial direction because surface waves are insensitive to the sense of propagation: In the expression of the phase velocity (equation 1), the transformation of the azimuth Ψ into $\Psi + \pi$ does not change the value of phase velocity. In order to define the actual direction of flow, it is necessary to add information such as plate velocity directions or numerical modeling strain principal directions. Though surface wave tomography enables simultaneous retrieval of thirteen anisotropic parameters, we will only focus on the best resolved anisotropic parameters, the G parameter (related to azimuthal variations of SV-wave velocity) and the ξ parameter which expresses the relative importance of the azimuthally averages of v_{SH} and v_{SV} or in other words the radiality of the flow.

The distribution of the G parameter is plotted on Figure 5a at 200 km depth. It shows that the agreement of directions of maximum velocity with plate tectonics is reasonable in the depth range 100–300 km (MONTAGNER, 1994). However, the azimuth of G parameter can largely vary as a function of depth (MONTAGNER and TANIMOTO, 1991). For instance, at shallow depths (to 60 km), the maximum velocity is very often parallel to mountain belts (VINNIK et al., 1992; SILVER, 1996; BABUŠKA et al., this issue). This means that, at a given place, the orientation of fast axis is a function of depth. Contrary to seismic anisotropy derived from SKS, the anisotropy inferred from surface waves can be located at detph, and both measurements must be combined in order to provide a complete understanding of the processes prevailing in a given tectonic context.

The simultaneous inversion of Rayleigh, Love wave dispersion curves and their azimuthal variations provides better estimates of the second kind of observable seismic anisotropy, i.e., the radial anisotropy expressed through the parameter $\xi = (N - L)/_L$. The ξ parameter is retrieved through the inversion of the 0–Ψ azimuthally averaged term of the expansion of the local phase velocity and thus is not biased by an imperfect azimuthal coverage of the criss-crossing paths in the area under investigation. It results that a close inspection of the three-dimensional distribution of ξ displays different patterns below continents and oceans (Fig. 5b). Whereas ξ is primarily positive in most oceanic areas (except near ridges) at 100 km depth, the pattern is reversed at large depths (310 km), where positive ξ is now associated with continents. However, the amplitude of ξ is rapidly decreasing in average at depth below 200 km and is less resolved.

Continents. ξ is usually very heterogeneous below continents in the first 150–200 km of depth with positive or negative areas according to geology. However, it seems to display a systematic tendency to be positive at larger depth (down to 300–400 km), whereas it is very large in the oceanic lithosphere in the depth range 50–200 km and decreases rapidly at larger depths (Fig. 6). Conversely, radial anisotropy displays a maximum (though smaller than in oceanic lithosphere) below very old continents (such as the Siberian and Canadian Shields) in the depth range 200–400 km. Contrary to the assertion of GAHERTY and JORDAN (1995), seismic anisotropy below continents is not necessarily limited to the upper 220 km. A more quantitative comparison of radial anisotropy between different continental provinces is presented in this issue by BABUŠKA *et al.* (1997), and it demonstrates systematic differences according to the tectonic context. The existence of positive large-scale radial anisotropy below continents at depth might be a good indicator of the continental root which has been largely debated since the presentation of the controversial model of tectosphere by JORDAN (1978, 1981). If we assume that this maximum of anisotropy is related to an intense strain field in this depth range, it might be characteristic of the boundary between continents and "normal" upper mantle material. Our results show that the root of continents is located between 200 and 300 km.

It must be emphasized that the anisotropy near the surface is probably different from the deep one. Part of the observed anisotropy might be related to the fossile strain field prevailing during the setting of materials and the other deeper part is related to the present strain field. If we bear in mind that the anisotropy displayed from surface waves is the long wavelength filtered anisotropy (approximately 1500 km), it can be easily understood that the average anisotropy displayed from surface waves in the first 200 km might be very different from the one found from body waves. The tectonics of continents is the result of a long and complex history. The characteristic length scale under continents is related to the size of blocks successively accreted to existing initial cores and probably smaller than 1500 km. This statement is supported by different studies of *SKS* body waves which demonstrate that the direction of maximum velocity can change on scales smaller than 100 km (VINNIK *et al.*, 1992; SILVER, 1996). This difference of characteristic scales can explain the apparent contradictions between surface wave and body wave anisotropy measurements. The fact that we do not observe a systematic behavior in the first hundreds of kilometers for similar continental geological zones, does not mean that anisotropy is not present but only that its characteristic scale is different from one region to another. Due to the low-pass filtering effect of our technique, its long wavelength signature is diluted.

The fossile shallow anisotropy (first hundred of kilometers) that reflects the past strain field, should be very useful for understanding the processes involved in surficial tectonics. If we were able to determine the age of this shallow anisotropy, the measurement of this kind of anisotropy should open wide a new field in Earth

Sciences: the Paleo-Seismology which might provide fundamental information of Structural Geology. We will return to the scientific applications of shallow anisotropy in the next section.

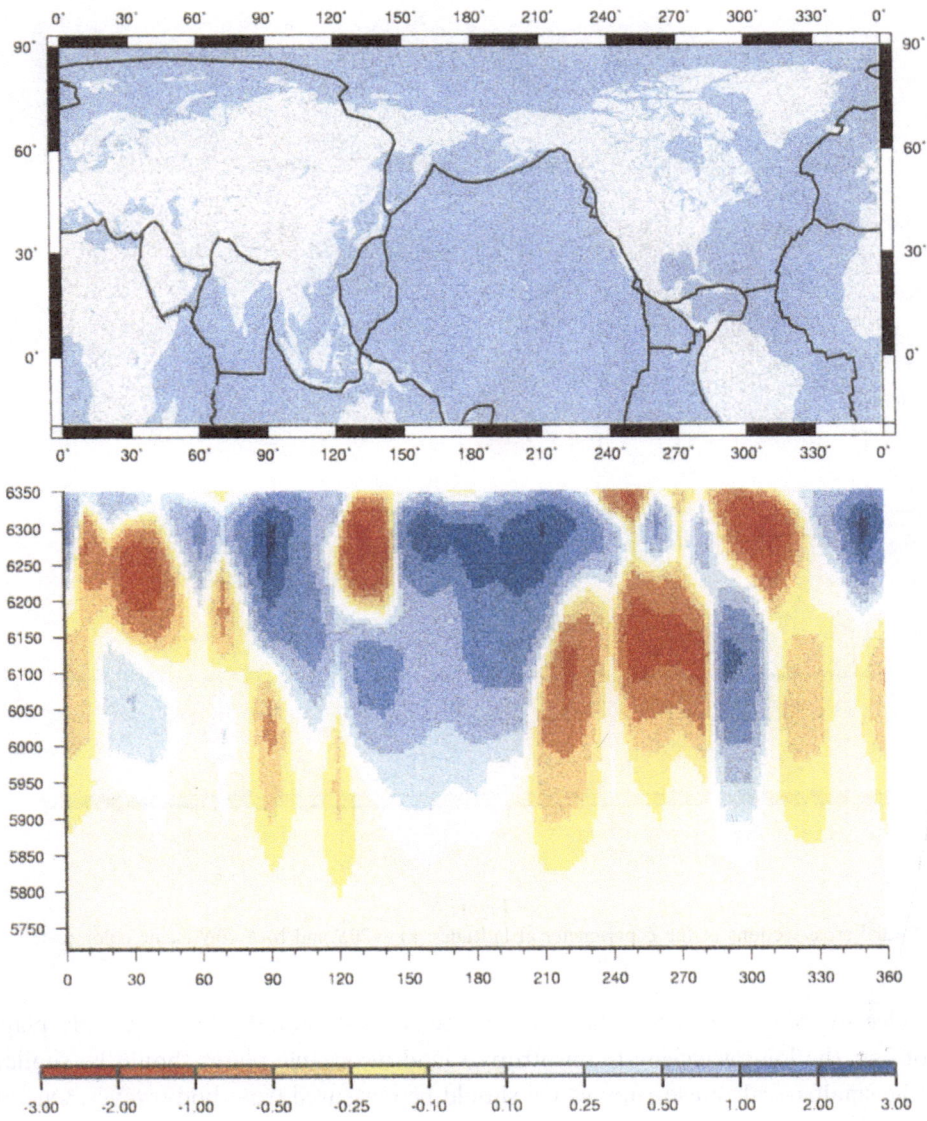

Figure 6a

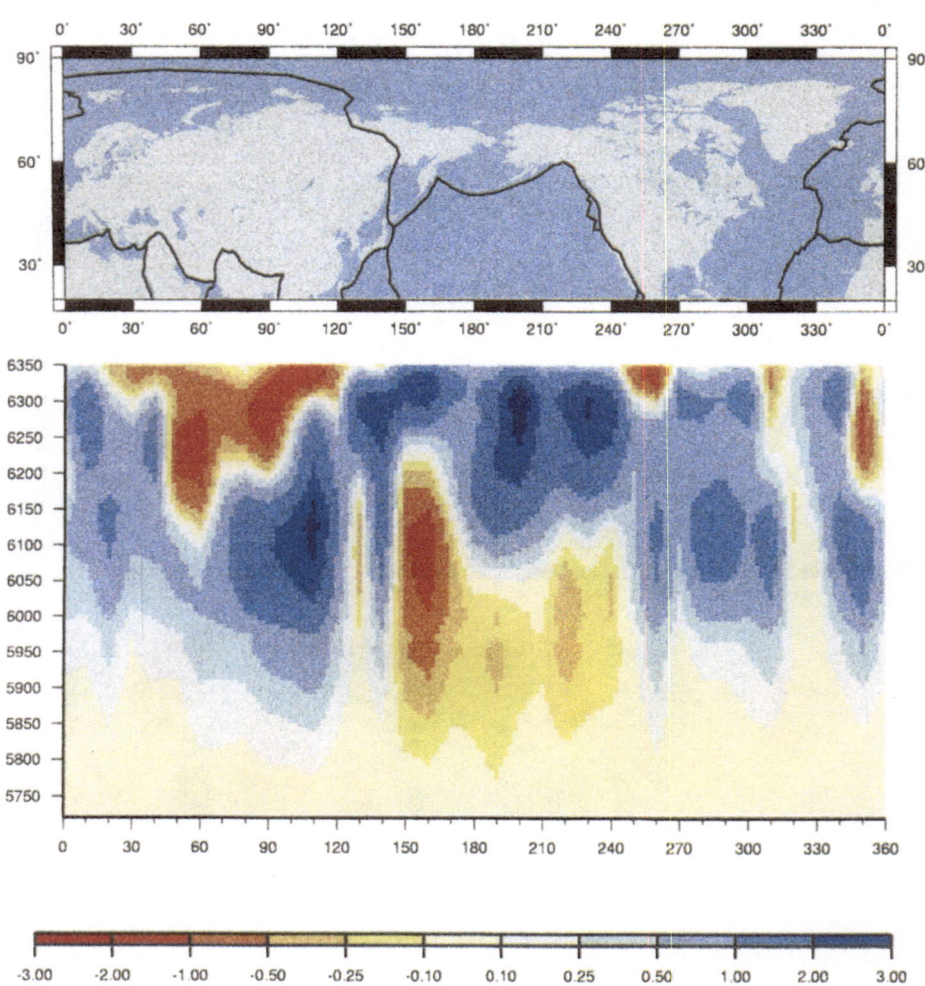

Figure 6
Radial cross-sections of the ξ parameter at latitudes: a) $-20S$ and b) $+20N$ (same color scale).

Oceans. Since convective flow below oceans is dominated by large-scale plate motions, the long wavelength anisotropy found in oceanic plates should be similar to the smaller-scale anisotropy which should be measured from body waves. One of the first evidence of anisotropy was found in the Pacific Ocean by HESS (1964) for P_n waves. Since that time there have been many measurements of the subcrustal anisotropy (see BABUŠKA and CARA, 1991 for a review). Unfortunately, to date, there are very few measurements of deep anisotropy by *SKS* splitting in the oceans. Due to the absence of a seismic station on the sea floor, the only measurements available for *SKS* were performed in stations located on ocean islands (ANSEL and NATAF, 1989; KUO and FORSYTH, 1992; RUSSO and OKAL, 1997), which are by nature anomalous objects, such as hotspots where the strain field is perturbed by the ascending material and not necessarily representative of the main flow field. There were other attempts to determine anisotropy for other kinds of body waves such as *ScS* and multiple *ScS* (ANDO, 1984; FARRA and VINNIK, 1994; GAHERTY *et al.*, 1997), *PS* waves (SU and PARK, 1994), or differential times of *sS-S*, or *SS-S* waves (KUO *et al.*, 1987; SHEEHAN and SOLOMON, 1991; FISCHER and YANG, 1994; GAHERTY *et al.*, 1996; FOUCH and FISCHER, 1996; YANG and FISCHER, 1994). However, the two cross-sections of Figure 6 illustrate that the ξ parameter is negative and small, where flow is primarily radial (mid-ocean ridges and subducting zones). Between plate boundaries, oceans display vast areas with a positive radial anisotropy, characteristic of an overall horizontal flow field. Oceanic plates are zones where the comparison between directions of plate velocities (MINSTER and JORDAN, 1978) and directions of *G* parameter is the most successful in the entire lithosphere and asthenosphere beneath 250–300 km (MONTAGNER, 1994). Conversely, such a comparison is more controversial below plates bearing a large proportion of continents, such as the European-Asian plate, characterized by a very small absolute motion in the hotspot coordinate system. Another interesting application of the ξ parameter can be found in the process of understanding the convection pattern in the transition zone. Recent tomographic models of seismic velocity, anisotropy and anelasticity obtained from GEOSCOPE and IRIS data demonstrate that at depth larger than 410 km $\pm$ 100 km, the degree 2 and to a less extent the degree 6 arise as the most important features. These models also display a large degree 4 for radial anisotropy at large depths (ROULT *et al.*, 1990; MONTAGNER and TANIMOTO, 1991). A simple flow pattern with two upgoing flows below central Pacific and Africa and two downgoing flows below western Pacific and America can explain the predominance of these different degrees 2 and 6. MONTAGNER and ROMANOWICZ (1993) demonstrate that the degree-2 pattern correctly predicts the predominance of degree 4 of ξ-anisotropy because ξ is not affected by the sense of flow but only by its radiality or horizontality. Moreover, the geographical distribution of radial anisotropy is in agreement with the degree-2 pattern as it is displayed in Figure 3. Therefore, the observations of the geographical distributions of degrees 2, 4, 6 in the transition zone are coherent and spatially dependent. MONTAGNER (1994) compared these different degrees to the corresponding degrees of the hotspot and slab distribution. In this simple framework the distribution of plumes and slabs is merely a byproduct of the large-scale simple flow in the transition zone.

3.2 Tectonics

The measurement of seismic anisotropy can provide fundamental information for the understanding of tectonic processes. SILVER (1996) presents a review of the information provided by shear-wave splitting beneath continents. The poor lateral resolution of global scale anisotropic tomography can be considered as a strong limitation in continental areas. This technique can only be efficiently applied in areas where large-scale tectonic forces are implied. The best candidate where this condition is fulfilled in continents is the collision zone between India and Asia. GRIOT et al. (1996) undertook such an investigation in Central Asia. The primary goal of this study is the discrimination between two competing extreme models of deformations, the heterogeneous model of AVOUAC and TAPPONNIER (1993) and the homogeneous model of ENGLAND and HOUSEMAN (1986). It was necessary to use shorter wavelength surface waves (40–200 s) in order to obtain a lateral resolution of 350 km. Synthetic models of seismic anisotropy can be inferred from the heterogeneous and homogeneous models. In order to perform correct and quantitative comparisons between observed seismic anisotropy and the deformation models, the short wavelengths of the synthetic models (spatial scale smaller than 350 km) were filtered out (Fig. 7).

The statistical comparison between observed and synthetic azimuthal anisotropies for both models enables a determination in different depth ranges of which the deformation model dominates. GRIOT et al. (1997) show that the heterogeneous model is in better agreement with observations in the first 200 km, whereas the homogeneous model better fits the deep anisotropy below 200 km. We must be aware that such a comparison is only valid from a statistical point of view and that a comparison at a more local scale (the scale of body-wave measurements) might display some differences with the observed SKS anisotropy. This kind of investigation only underlines large-scale ongoing and prevailing active processes and is not devoted to a precise measurement of anisotropy at any specific place.

4. Discussion

In the previous sections we have illustrated by different examples that the effect of anisotropy (though small) is significant and must be taken into account to correctly explain independent data sets related to the propagation of seismic waves inside the earth. Nonetheless, we must wonder whether these data sets are consistent, and which general statement regarding mantle convection can be put forward when considering the different depth ranges in which seismic anisotropy has been observed.

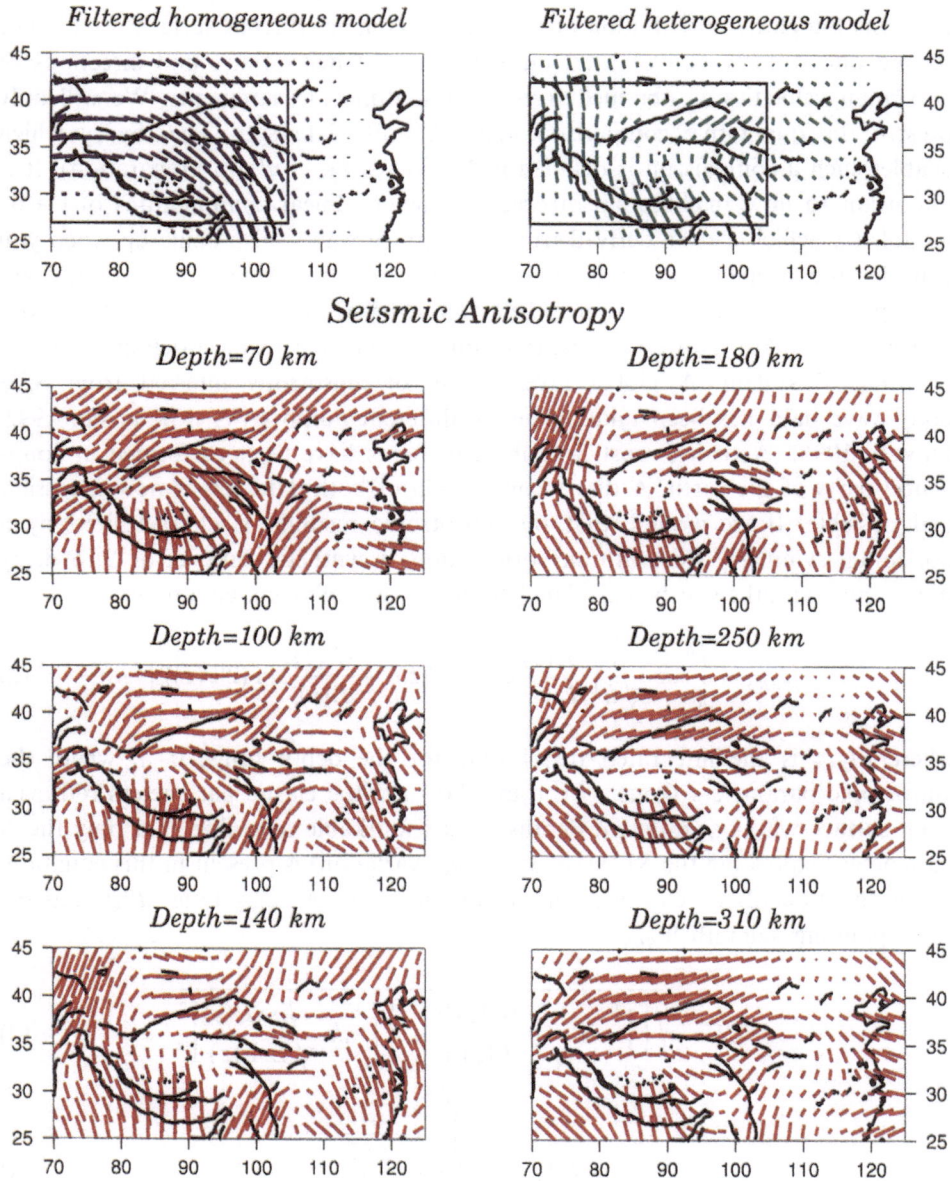

Figure 7
Distributions of azimuthal anisotropy at different depths (after GRIOT *et al.*, 1997). Top—Synthetic anisotropic models. Left: Heterogeneous model. Right: Homogeneous model. Bottom figures: *G* parameter at different depths derived from surface-wave data.

4.1 Comparison between Surface-wave Anisotropy and SKS Delay Times

For simplicity only two data sets are considered in this section. The first one concerns the global distribution of anisotropy as inferred from surface waves. The second one is composed of local measurements of delay times and directions of maximum velocities as obtained from SKS-splitting measurements. We will only present the qualitative comparison between both data sets. The theory which enables such a comparison is detailed in MONTAGNER et al. (in preparation). It is important to note that the anisotropic parameters, linear combinations of elastic moduli C_{ij}, which can be derived from the surface waves, also come up when you consider the propagation of body waves in symmetry planes for a slightly anisotropic medium (see for instance, BACKUS, 1965; CRAMPIN, 1984; MONTAGNER and NATAF, 1986). However, attention must be directed to the orientation of the coordinates' system. A global investigation of anisotropy inferred from SKS body-wave data has been undertaken by different authors (VINNIK et al., 1992; SILVER, 1996). Actually, most SKS measurements have been performed in continental parts of the earth. A direct comparison of both data sets is now necessary and possible. If we assume that the anisotropic medium is characterized by a horizontal symmetry axis with any orientation, a synthetic data set of SKS delay times and azimuths can be calculated by using the following equations:

$$\delta t_{SKS} = \int_0^h dz \, \sqrt{\frac{\rho}{h}} \left(\frac{G_c(z)}{L(z)} \cos(2\Psi(z)) + \frac{G_s(z)}{L(z)} \sin(2\Psi(z)) \right) \tag{2}$$

where δt_{SKS} is the integrated travel time for the depth range $0-h$, where the anisotropic parameters $G_c(z)$, $G_s(z)$ and $L(z)$ are the elastic parameters retrieved from surface waves at different depths. It is remarkable to realize that only the G parameter (expressing the SV-wave azimuthal variation) is present in this equation. From equation (2) we can infer the maximum value of delay time δt_{SKS}^{max} and the corresponding azimuth Ψ_{SKS}:

$$\delta t_{SKS}^{max} = \sqrt{ \left\{ \int_0^h dz \, \sqrt{\frac{\rho}{L}} \frac{G_c(z)}{L(z)} \right\}^2 + \left\{ \int_0^h dz \, \sqrt{\frac{\rho}{L}} \frac{G_s(z)}{L(z)} \right\}^2 } \tag{3}$$

$$\tan(2\Psi_{SKS}) = \frac{\int_0^h dz \, \frac{G_s(z)}{L(z)}}{\int_0^h dz \, \frac{G_c(z)}{L(z)}}. \tag{4}$$

However, equation (2) is approximate and only valid when the wavelength is much larger than the thickness of layers. It should be possible to make more precise calculations by using the technique derived for two layers by SILVER and SAVAGE (1994). The resulting map is presented in Figure 8.

Initially the map illustrates that both data sets are compatible in magnitude although not necessarily in directions. Some contradictions between measurements derived from surface waves and from body waves can be noted. The agreement of directions is correct in tectonically active areas but not in old cratonic zones. The discrepancy in these areas results from the rapid change of directions of anisotropy at a small scale and that the hypothesis of horizontal symmetry axis is not valid (PLOMEROVÁ et al., 1996). These changes stem from the complex history of these areas, which have been built by successive collages of continental pieces. The positive consequence of this discrepancy is that a small-scale mapping of anisotropy in such areas might provide clues to understand the processes of growth of continents and mountain building.

Contrary to surface waves, body waves have a good lateral resolution but no vertical resolution. They are sensitive to the short wavelength anisotropy just below and around the stations. On the other hand, global anisotropy tomography derived from surface waves only provides long wavelength anisotropy (poor lateral resolution) although it enables location at depth of anisotropy. Therefore the long wavelength anisotropy derived from surface waves will display the same direction as the short wavelength anisotropy inferred from body waves only when large-scale geodynamic processes are dominating. In some continental areas, short-scale anisotropy, the result of a complex history, might be important and even mask the large-scale anisotropy more related to present convective processes.

4.2 Location of Anisotropy in Boundary Layers

The review of the presence of anisotropy in different layers of the earth demonstrated that the anisotropy is a very general feature but that it is not present in whole depth ranges nor at all scales. As discussed in section 2.1, the observation of seismic anisotropy at large scales requires several redhibitory conditions, starting with the presence of anisotropic crystals extending to the existence of an efficient large-scale present or past strain field. Theoretical studies suggest that, when anisotropic minerals such as olivine or pyroxene are subjected to strain, their crystallographic axes develop a systematic relationship to the principal axes of finite strain (MCKENZIE, 1979; RIBE, 1989). Many numerical modelings of the convective mantle show that in a convective system, the strain field is not spatially uniform. Streamlines are substantially more concentrated in boundary layers than centrally in the cells. Consequently, the amplitude of the strain field is very heterogeneous and the largest in boundary layers. Conversely, we can assume that the observation of mantle seismic anisotropy is an indication of a strong present-day strain field (excepting crust, topmost oceanic and continental lithospheres where fossil anisotropy may be present), associated with boundary layers. In the previous sections we noted sound evidence of the presence of seismic anisotropy in the D''-layer, in

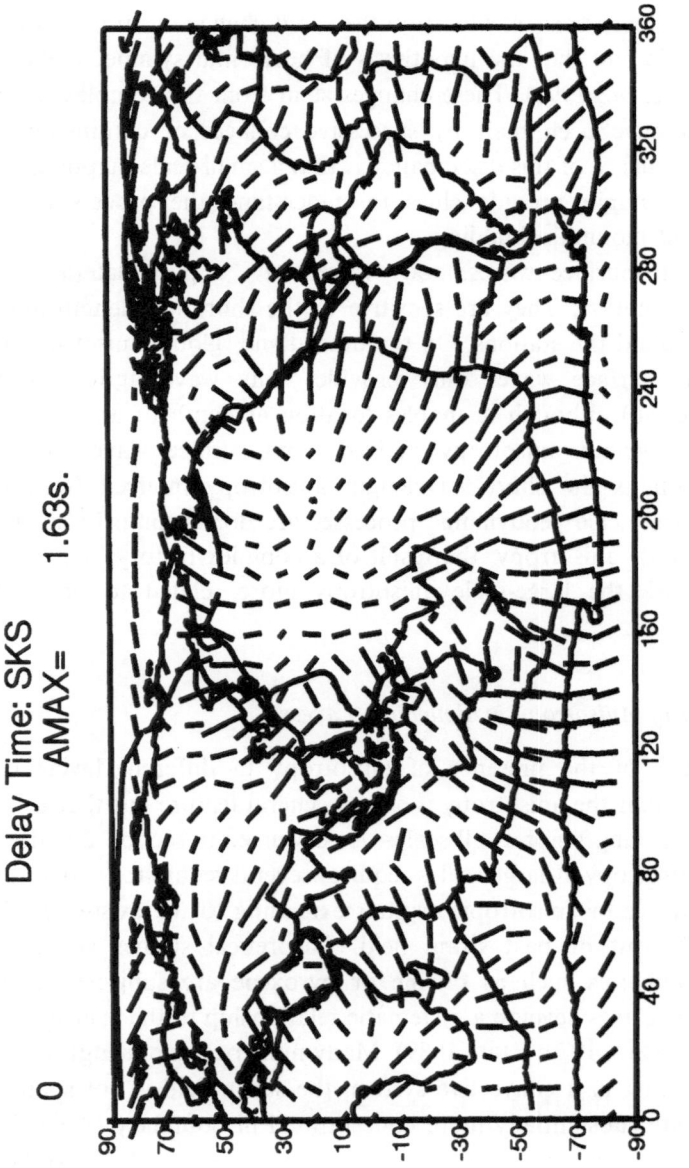

Figure 8a

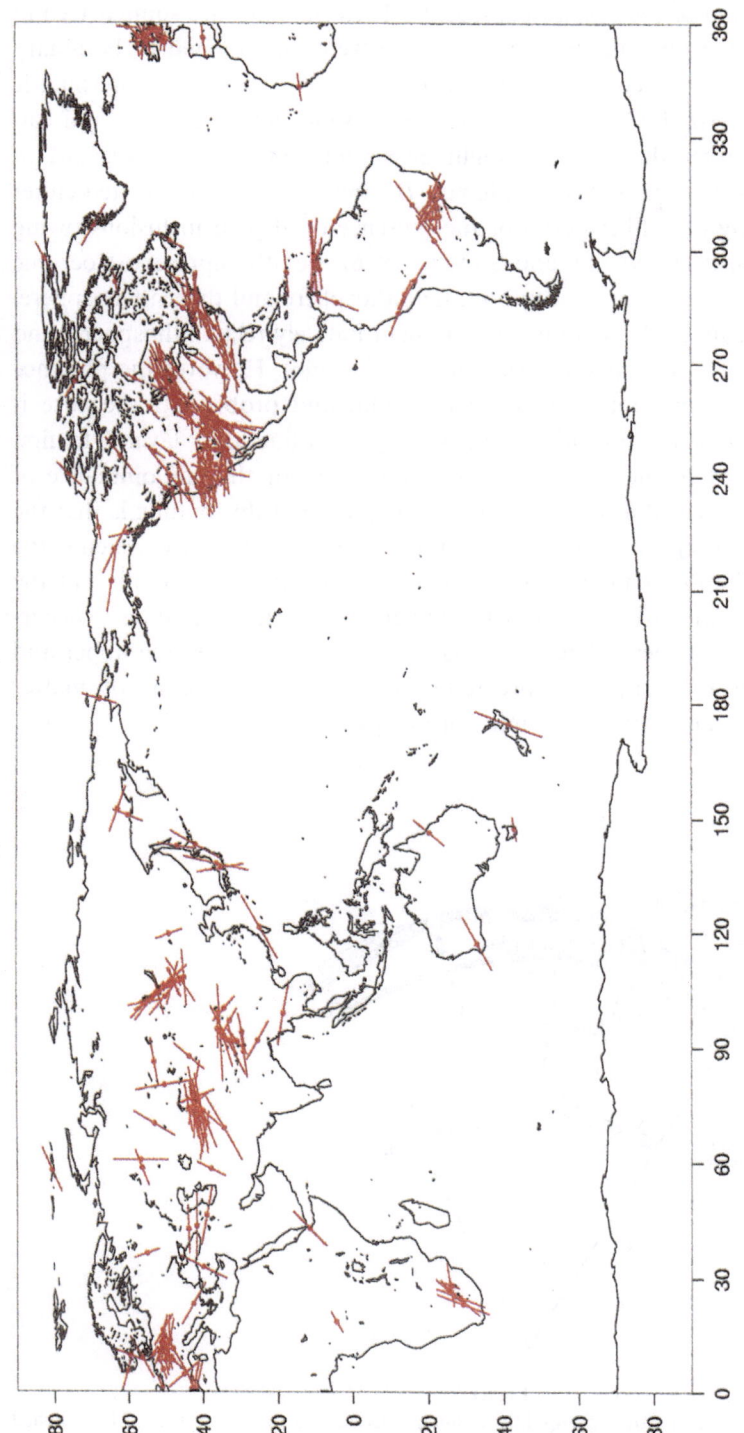

Silver [1996]

Figure 8

Top: Distributions of synthetic delay time δt^{max}_{SKS} and azimuth Ψ_{SKS} at the surface of the earth, such as derived from AUM anisotropic tomographic model (MONTAGNER and TANIMOTO, 1991). Bottom: Map of distribution of observed SKS maximum directions and delay times (SILVER, 1996). On both figures the length of lines is proportional to δt_{SKS}.

the transition zone and in the uppermost mantle. These findings are summarized in Figure 9. The D''-layer and the uppermost mantle were long related to boundary layers of the mantle convective system. The D''-layer above the core-mantle boundary is characterized by a large degree of seismic heterogeneities and anisotropy with v_{SH} larger than v_{SV}. It might be simultaneously the graveyard of subducted slabs and the source of megaplumes. D'' anisotropy can be related either to horizontal layering of cold material or the presence of aligned inclusions owing to the presence of melt (KENDALL and SILVER, 1996). For the uppermost oceanic mantle, seismic anisotropy is present in both the lithosphere and the asthenosphere, and for oceans some finite-element models can quantitatively relate lithospheric and asthenospheric strain to anisotropy (TOMMASI et al., 1996). However the presence of anisotropy in the transition zone is fundamental and problematic because it provides a new clue that the transition zone is acting as a boundary layer. The first evidence of anisotropy in the transition zone tends to favor the predominance of horizontal flow over vertical flow. The major consequence of this finding is that the transition zone is dividing the mantle, on average, into two convective systems, the upper mantle and the lower mantle. This general statement does not rule out the possibility that flow circulation between the upper and the lower mantle is occurring. Nonetheless it does mean that the exchange of matter between the upper and the lower mantle is difficult. It is premature to assess the amount of matter circulating from measurements of seismic anisotropy.

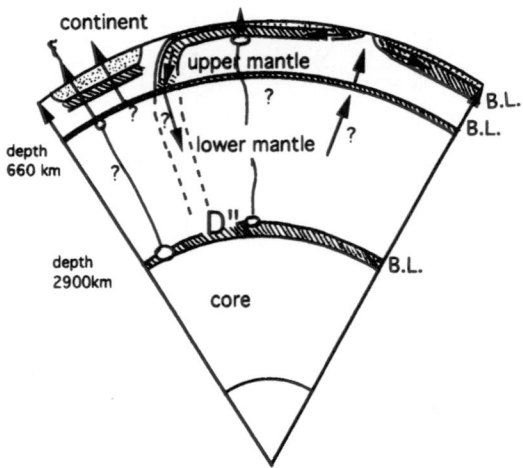

Figure 9

Cross-section of the earth from the core-mantle boundary extending to the surface. The hatched (respectively dotted) areas show where there is robust evidence of (resp. fossile) anisotropy. B.L. stands for Boundary Layer.

5. Conclusions

We have presented in this paper different observations of seismic anisotropy and their applications in geology and geodynamics. Seismic anisotropy defines continental roots so as to discriminate different competing convective models. Three boundary layers as yet have been detected by seismic anisotropy: the uppermost mantle, the transition zone (though new work is necessary to confirm it) and the D''-layer. Other applications of seismic anisotropy can be easily found. For example, MONTAGNER and ANDERSON (1989a) demonstrate that different anisotropic parameters might be used to discriminate competing petrological models such as pyrolite or piclogite. Some seismologists claim that the temporal variation of anisotropy in the crust might be an efficient tool to investigate the earthquake cycle. It is now used routinely in seismic exploration. Therefore the scientific potential of seismic anisotropy is enormous and largely unexploited. In conclusion, seismic anisotropy provides a new dimension in the investigation of processes of our dynamic earth.

Acknowledgments

I am grateful to Lev Vinnik, Alessandro Forte, Vlada Babuška, Jarka Plomerová, and Bob Liebermann for fruitful discussions. I thank Shun-Ichiro Karato and an anonymous reviewer for their critical review of the paper, and Daphné-Anne Griot for providing some of the figures.

REFERENCES

AKI, K., and KAMINUMA, K. (1963), *Phase Velocity of Love Waves in Japan (Part 1): Love Waves from the Aleutian Shock of March 1957*, Bull. Earthq. Res. Inst. *41*, 243–259.

ALLÈGRE, C. J., and TURCOTTE, D. L. (1986), *Implications of a Two-component Marble-cake Mantle*, Nature *323*, 123–127.

ANDERSON, D. L. (1961), *Elastic Wave Propagation in Layered Anisotropic Media*, J. Geophys. Res. *66*, 2953–2963.

ANDERSON, D. L., *Theory of the Earth* (Blackwell Scientific Publications, Oxford 1989).

ANDERSON, D. L., and BASS, J. D. (1984), *Mineralogy and Composition of the Upper Mantle*, Geophys. Res. Lett. *11*, 637–640.

ANDERSON, D. L., and BASS, J. D. (1986), *Transition Region of the Earth's Upper Mantle*, Nature *320*, 321–328.

ANDERSON, D. L., and DZIEWONSKI, A. M. (1982), *Upper Mantle Anisotropy: Evidence from Free Oscillations*, Geophys. J. R. Astron. Soc. *69*, 383–404.

ANDERSON, D. L., and REGAN, J. (1983), *Upper Mantle Anisotropy and the Oceanic Lithosphere*, Geophys. Res. Lett. *10*, 841–844.

ANDO, M. (1984), *ScS Polarization Anisotropy around the Pacific Ocean*, J. Phys. Earth *32*, 179–196.

ANDO, M., ISHIKAWA, Y., and YAMAZAKI, F. (1983), *Shear-wave Polarization Anisotropy in the Upper Mantle beneath Honshu, Japan*, J. Geophys. Res. *88*, 5850–5864.

ANSEL, V., and NATAF, H.-C. (1989), *Anisotropy beneath 9 Stations of the Geoscope Broadband Network as Deduced from Shear-wave Splitting*, Geophys. Res. Lett. *16*, 409–412.

AVOUAC, J.-P., and TAPPONNIER, P. (1993), *Kinematic Model of Active Deformation in Central Asia*, Geophys. Res. Lett. *20*, 895–898.

BABUŠKA, V., and CARA, M., *Seismic Anisotropy in the Earth* (Kluwer Academic Press, Dordrecht, The Netherlands 1991), 217 pp.

BABUŠKA, V., PLOMEROVÁ, J., and SILENY, J. (1984), *Large-scale Oriented Structures in the Subcrustal Lithosphere of Central Europe*, Ann. Geophys. *2*, 649–662.

BABUŠKA, V., PLOMEROVÁ, J., and SILENY, J. (1993) *Models of Seismic Anisotropy in the Deep Continental Lithosphere*, Phys. Earth Planet. Int. *78*, 167–191.

BABUŠKA, V., MONTAGNER, J.-P., PLOMEROVÁ, J., and GIRARDIN, N. (1997), *Age-dependent Large-scale Fabric of the Mantle Lithosphere as Derived from Surface-wave Velocity Anisotropy*, Pure appl. geophys. *151*, 257–280.

BACKUS, G. E. (1962), *Long-wave Elastic Anisotropy Produced by Horizontal Layering*, J. Geophys. Res. *67*, 4427–4440.

BACKUS, G. E. (1965), *Possible Forms of Seismic Anisotropy of the Upper Mantle under Oceans*, J. Geophys. Res. *70*, 3249–3439.

BARRUOL, G., and MAINPRICE, D. (1993), *A Quantitative Evaluation of the Contribution of Crustal Rocks to the Shear-wave Splitting of Teleseismic SKS Waves*, Phys. Earth Planet. Int. *78*, 281–300.

BARRUOL, G., SILVER, P. G., and VAUCHEZ, A. (1997), *Shear-wave Splitting in the Eastern U.S.: Deep Structure of a Complex Continental Plate*, J. Geophys. Res. *102*, 8329–8348.

BASS, J., and ANDERSON, D. L. (1984), *Composition of the Upper Mantle: Geophysical Tests of 2 Petrological Models*, Geophys. Res. Lett. *11*, 237–240.

BOWMAN, J. R., and ANDO, M. (1987), *Shear-wave Splitting in the Upper Mantle Wedge above the Tonga Subduction Zone*, Geophys. J. R. Astron. Soc. *88*, 25–41.

CARA, M., and LÉVÊQUE, J.-J. (1988), *Anisotropy of the Asthenosphere: The Higher Mode Data of the Pacific Revisited*, Geophys. Res. Lett. *15*, 205–208.

CHRISTENSEN, N. I., and LUNDQUIST, S. (1982), *Pyroxene Orientation within the Upper Mantle*, Bull. Geol. Soc. Am. *93*, 279–288.

CRAMPIN, S. (1984), *An Introduction to Wave Propagation in Anisotropic Media*, Geophys. J. R. Astron. Soc. *76*, 17–28.

CRAMPIN, S., and BOOTH, D. C. (1985), *Shear-wave Polarizations near the North Anatolian Fault, II. Interpretation in Terms of Crack-induced Anisotropy*, Geophys. J. R. Astron. Soc. *83*, 75–92.

DZIEWONSKI, A. M., and ANDERSON, D. L. (1981), *Preliminary Reference Earth Model*, Phys. Earth Planet. Int. *25*, 297–365.

ENGLAND, P., and HOUSEMAN, G. (1986), *Finite Strain Calculations of Continental Deformation, 2. Comparison with the India-Asia Collision Zone*, J. Geophys. Res. *91*, 3664–3676.

ESTEY, L. H., and DOUGLAS, D. J. (1986), *Upper Mantle Anisotropy: A Preliminary Model*, J. Geophys. Res *91*, 11,393–11,406.

FARRA, V., and VINNIK, L. (1994), *Shear-wave Splitting in the Mantle of the Pacific*, Geophys. J. Int. *119*, 195–218.

FISCHER, K. M., and YANG, X. (1994), *Anisotropy in Kuril-Kamtchatka Subduction Zone Structure*, Geophys. Res. Lett. *21*, 5–8.

FISCHER, K. M., and WIENS, D. A. (1996), *The Depth Distribution of Mantle Anisotropy beneath the Tonga Subduction Zone*, Earth Planet. Sci. Lett. *142*, 253–260.

FORSYTH, D. W. (1975), *The Early Structural Evolution and Anisotropy of the Oceanic Upper Mantle*, Geophys. J. R. Astron. Soc. *43*, 103–162.

FOUCH, M. J., and FISCHER, K. M. (1996), *Mantle Anisotropy beneath Northwest Pacific Subduction Zones*, J. Geophys. Res. *101*, 15,987–16,002.

FUKAO, Y. (1984), *Evidence from Core-reflected Shear Waves for Anisotropy in the Earth's Mantle*, Nature *309*, 695–698.

GAHERTY, J. B., and JORDAN, T. H. (1995), *Lehmann Discontinuity as the Base of an Anisotropic Layer beneath Continents*, Science *268*, 1468–1471.

GAHERTY, J. B., JORDAN, T. H., and GEE, L. S. (1997), *Seismic Structure of the Upper Mantle in a Central Pacific Corridor*, J. Geophys. Res., in press.

GARNERO, E. J., and LAY, T. (1996), *Lateral Variations in Lowermost Mantle Shear-wave Anisotropy beneath the North Pacific and Alaska*, J. Geophys. Res., *102*, 8121–8135.

GRIOT, D.-A., MONTAGNER, J.-P., and TAPPONNIER, P. (1996), *Surface Wave Phase Velocity and Azimuthal Anisotropy in Central Asia*, J. Geophys. Res.

GRIOT, D.-A., MONTAGNER, J.-P., and TAPPONNIER, P. (1997), *Heterogeneous versus Homogeneous Strain in Central Asia*, Geophys. Res. Lett.

HADIOUCHE, O., JOBERT, N., and MONTAGNER, J.-P. (1989), *Anisotropy of the African Continent Inferred from Surface Waves*, Phys. Earth Planet. Int. *58*, 61–81.

HESS, H. (1964), *Seismic Anisotropy of the Uppermost Mantle under the Oceans*, Nature *203*, 629–631.

HIRN, A. *et al.* (1995), *Seismic Anisotropy as an Indicator of Mantle Flow beneath the Himalayas and Tibet*, Nature *375*, 571–574.

JORDAN, T. H. (1978), *Composition and Development of the Continental Tectonosphere*, Nature *274*, 544–548.

JORDAN, T. H. (1981), *Continents as a Chemical Boundary Layer*, Philos. Trans. R. Soc. London, Ser. A, *301*, 359–373.

KANESHIMA, S., and SILVER, P. G. (1992), *A Search for Source-side Anisotropy*, Geophys. Res. Lett. *19*, 1049–1052.

KARATO, S.-I., *Seismic anisotropy: mechanisms and tectonic implications*. In *Rheology of Solids and of the Earth* (eds. S. Karato, and M. Toriumi) (Oxford University Press, Oxford 1989) pp. 393–342.

KARATO, S.-I. (1997), *Seismic Anisotropy in the Deep Mantle, Boundary Layers and the Geometry of Mantle Convection*, Pure appl. geophys. *151*, 565–587.

KENDALL, J.-M., and SILVER, P. G. (1996), *Constraints from Seismic Anisotropy on the Nature of the Lowermost Mantle*, Nature *381*, 409–412.

KENNETT, B. L. N., and ENGDAHL, E. R. (1991), *Travel Times for Global Earthquake Location and Phase Identification*, Geophys. J. Int. *105*, 429–465.

KENNETT, B. L. N., ENGDAHL, E. R., and BULAND, R. (1995), *Constraints on Seismic Velocities in the Earth from Travel Times*, Geophys. J. Int., *122*, 108–124.

KUO, B.-Y., and FORSYTH, D. W. (1992), *A Search for Split SKS Waveforms in the North Atlantic*, Geophys. J. Int. *108*, 557–574.

KUO, B.-Y., FORSYTH, D. W., and WYSESSION, M. (1987), *Lateral Heterogeneity and Azimuthal Anisotropy in the North Atlantic Determined from SS-S Differential Travel Times*, J. Geophys. Res. *92*, 6421–6436.

LAY, T., and YOUNG, C. J. (1991), *Analysis of Seismic SV Waves in the Core's Penumbra*, Geophys. Res. Lett. *18*, 1373–1376.

LÉVÊQUE, J. J., and CARA, M. (1985), *Inversion of Multimode Surface Wave Data: Evidence for Sub-lithospheric Anisotropy*, Geophys. J. R. Astron. Soc. *83*, 753–773.

LEVSHIN, A., and RATNIKOVA, L. (1984), *Apparent Anisotropy in Inhomogeneous Media*, Geophys. J. R. Astron. Soc. *76*, 65–69.

LOVE, A. E. H., *A Treatise on the Theory of Elasticity*, 4th ed. (Cambridge University Press 1927) 643 pp.

MASTERS, G., JORDAN, T. H., SILVER, P. G., and GILBERT, F. (1982), *Aspherical Earth Structure from Fundamental Spheroidal-mode Data*, Nature *298*, 609–613.

MATZEL, E., SEN, M. K., and GRAND, S. P. (1996), *Evidence for Anisotropy in the Deep Mantle beneath Alaska*, Geophys. Res. Lett. *23*, 2417–2420.

MAUPIN, V. (1994), *On the Possibility of Anisotropy in the D″ Layer as Inferred from the Polarization of Diffracted S Waves*, Phys. Earth Planet. Int. *87*, 1–32.

McKENZIE, D. (1979), *Finite Deformation during Fluid Flow*, Geophys. J. R. Astron. Soc. *58*, 689–715.

MEADE, C., SILVER, P. G., and KANESHIMA, S. (1995), *Laboratory and Seismological Observations of Lower Mantle Anisotropy*, Geophys. Res. Lett. *22*, 1293–1296.

MINSTER, J. B., and JORDAN, T. H. (1978), *Present-day Plate Motions*, J. Geophys. Res. *83*, 5331–5354.

MITCHELL, B. J., and YU, G.-K. (1980), *Surface Wave Dispersion, Regionalized Velocity Models and Anisotropy of the Pacific Crust and Upper Mantle*, Geophys. J. R. Astron. Soc. *63*, 497–514.

MONTAGNER, J.-P. (1985), *Seismic Anisotropy of the Pacific Ocean Inferred from Long-period Surface Wave Dispersion*, Phys. Earth Planet. Int. *38*, 28–50.

MONTAGNER, J.-P. (1986a), *First Results on the Three-dimensional Structure of the Indian Ocean Inferred from Long-period Surface Waves*, Geophys. Res. Lett. *13*, 315–318.

MONTAGNER, J. P. (1986b), *Regional Three-dimensional Structures Using Long-period Surface Waves*, Ann. Geophys. *4*, B3, 283–294.

MOCQUET, A., ROMANOWICZ, B., and MONTAGNER, J.-P. (1989), *Three D Structure of the Upper Mantle beneath the Atlantic Ocean From Long-period Rayleigh Waves. I: Group and Phase Velocity Distributions*, J. Geophys. Res. *94*, 7449–7468.

MONTAGNER, J. P. (1994), *What Can Seismology Tell us about Mantle Convection?* Rev. Geophys. *32*, 2, 115–137.

MONTAGNER, J. P., (1996), *Surface waves on a global scale—Influence of anisotropy and anelasticity*, Summer School of Erice, *Seismic Modeling of the Earth's Structure* (eds. E. Boschi, G. Ekström, A. Morelli), pp. 81–148.

MONTAGNER, J. P., and ANDERSON, D. L. (1989a), *Constraints on Elastic Combinations Inferred from Petrological Models*, Phys. Earth Planet. Int. *54*, 82–105.

MONTAGNER, J. P., and ANDERSON, D. L. (1989b), *Constrained Reference Mantle Model*, Phys. Earth Planet. Int. *58*, 205–227.

MONTAGNER, J.-P., and JOBERT, N. (1988), *Vectorial Tomography II: Application to the Indian Ocean*, Geophys. J. R. Astron. Soc. *94*, 309–344.

MONTAGNER, J.-P., and NATAF, H.-C. (1986), *On the Inversion of the Azimuthal Anisotropy of Surface Waves*, J. Geophys. Res. *91*, 511–520.

MONTAGNER, J.-P., and NATAF, H.-C. (1988), *Vectorial Tomography. I: Theory*, Geophys. J. R. Astron. Soc. *94*, 295–307.

MONTAGNER, J.-P., and TANIMOTO, T. (1990), *Global Anisotropy in the Upper Mantle Inferred from the Regionalization of Phase Velocities*, J. Geophys. Res. *95*, 4797–4819.

MONTAGNER, J.-P., and TANIMOTO, T. (1991), *Global Upper Mantle Tomography of Seismic Velocities and Anisotropies*, J. Geophys. Res. *96*, 20,337–20,351.

MONTAGNER, J.-P., and ROMANOWICZ, B. (1993), *Degrees 2, 4, 6 Inferred from Seismic Tomography*, Geophys. Res. Lett. *20*, 631–634.

MORELLI, A., DZIEWONSKI, A. M., and WOODHOUSE, J. H. (1986), *Anisotropy of Inner Core Inferred PKIKP Travel Times*, Geophys. Res. Lett. *13*, 1545, 1548.

MORELLI, A., and DZIEWONSKI, A. M. (1993), *Body Wave Travel Times and a Spherically Symmetric P- and S-wave Velocity Model*, Geophys. J. Int. *112*, 178–194.

NATAF, H.-C., NAKANISHI, I., and ANDERSON, D. L. (1984), *Anisotropy and Shear Velocity Heterogeneities in the Upper Mantle*, Geophys. Res. Lett. *11*, 109–112.

NATAF, H.-C., NAKANISHI, I., and ANDERSON, D. L. (1986), *Measurement of Mantle Wave Velocities and Inversion for Lateral Heterogeneity and Anisotropy, III. Inversion*, J. Geophys. Res. *91*, 7261–7307.

NICOLAS, A. (1993), *Why Fast Polarization Directions of SKS Seismic Waves are Parallel to Mountain Belts?* Phys. Earth Planet. Int. *78*, 337–342.

NICOLAS, A., BOUDIER, F., and BOULLIER, A. M. (1973), *Mechanisms of Flow in Naturally and Experimentally Deformed Peridotites*, Am. J. Sci. *273*, 853–876.

NICOLAS, A., and CHRISTENSEN, N. I., *Formation of anisotropy in upper mantle peridotites: A review*. In *Composition, Structure and Dynamics of the Lithosphere/Asthenosphere System* (eds K. Fuchs, and C. Froidevaux) (American Geophysical Union, Washington, D.C. 1987) pp. 111–123.

NISHIMURA, C. E., and FORSYTH, D. W. (1987), *Rayleigh Wave Phase Velocities in the Pacific with Implications for Azimuthal Anisotropy and Lateral Heterogeneities*, Geophys. J. R. astr. Soc.

NISHIMURA, C. E., and FORSYTH, D. W. (1989), *The Anisotropic Structure of the Upper Mantle in the Pacific*, Geophys. J. *96*, 203–229.

PESELNICK, L., NICOLAS, A., and STEVENSON, P. R. (1974), *Velocity Anisotropy in a Mantle Peridotite from Ivrea Zone: Application to Upper Mantle Anisotropy*, J. Geophys. Res. *79*, 1175–1182.

PESELNICK, L., and NICOLAS, A. (1978), *Seismic Anisotropy in an Ophiolite Peridotite. Application to Oceanic Upper Mantle*, J. Geophys. Res. *83*, 1227–1235.

PLOMEROVÁ, J., SILENY, J., and BABUŠKA, V. (1996), *Joint Interpretation of Upper-Mantle Anisotropy Based on Teleseismic P-travel Time Delay and Inversion of Shear-wave Splitting Parameters*, Phys. Earth Planet. Int. *95*, 293–309.

RIBE, N. M. (1989), *Seismic Anisotropy and Mantle Flow*, J. Geophys. Res. *94*, 4213–4223.

RICARD, Y., NATAF, H.-C., and MONTAGNER, J.-P. (1996), *The 3S-Mac Model: Confrontation with Seismic Data*, J. Geophys. Res. *101*, 8457–8472.

RINGWOOD, A. E., *Composition and Petrology of the Earth's Mantle* (McGraw-Hill, New York 1975) 618 pp.

ROULT, G., ROULAND, D., and MONTAGNER, J.-P. (1994), *Antarctica II: Upper Mantle Structure from Velocity and Anisotropy*, Phys. Earth Planet. Int. *84*, 33.

ROULT, G., ROMANOWICZ, B., and MONTAGNER, J.-P. (1990), *3D Upper Mantle Shear Velocity and Attenuation from Fundamental Mode-free Oscillation Data*, Geophys. J. Int. *101*, 61–80.

RUSSO, R. M., and OKAL, E. A. (1997), *Shear-wave Splitting in French Polynesia*, Geophys. J. Int., in press.

SCHLUE, J. W., and KNOPOFF, L. (1977), *Shear-wave Polarization Anisotropy in the Pacific Ocean*, Geophys. J. R. Astron. Soc. *49*, 145–165.

SHEEHAN, A. F., and SOLOMON, S. C. (1991), *Joint Inversion of Shear-wave Travel-time Residuals and Geoid and Depth Anomalies for Long-wavelength Variations in Upper Mantle Temperature and Composition along the Mid-Atlantic Ridge*, J. Geophys. Res. *96*, 19,981–20,351.

SILVER, P. G. (1996), *Seismic Anisotropy beneath the Continents: Probing the Depths of Geology*, Ann. Rev. Earth Planet. Sci. *24*, 385–432.

SILVER, P. G., and CHAN, W. W. (1988), *Implications for Continental Structure and Evolution from Seismic Anisotropy*, Nature *335*, 34–39.

SILVER, P. G., and CHAN, W. W. (1991), *Shear-wave Splitting and Subcontinental Mantle Deformation*, J. Geophys. Res. *96*, 16,429–16,454.

SILVER, P. G., and SAVAGE, M. K. (1994), *The Interpretation of Shear-wave Splitting Parameters in the Presence of Two Anisotropic Layers*, Geophys. J. Int. *119*, 959–963.

SILVEIRA, G., STUTZMANN, E., MONTAGNER, J.-P., and MENDES-VICTOR, L. (1997), *Anisotropic Tomography of the Atlantic Ocean from Rayleigh Surface Waves*, Phys. Earth Planet. Int., submitted.

SMITH, M. L., and DAHLEN, F. A. (1973), *The Azimuthal Dependence of Love and Rayleigh Wave Propagation in a Slightly Anisotropic Medium*, J. Geophys. Res. *78*, 3321–3333.

SMITH, M. L., and DAHLEN, F. A. (1975), *Correction to 'The Azimuthal Dependence of Love and Rayleigh Wave Propagation in a Slightly Anisotropic Medium'*, J. Geophys. Res. *80*, 1923.

SU, L., and PARK, J. (1994), *Anisotropy and the Splitting of PS Waves*, Phys. Earth Planet. Int. *86*, 263–276.

SUETSUGU, D., and NAKANISHI, I. (1987), *Regional and Azimuthal Dependence of Phase Velocities of Mantle Reyleigh Waves in the Pacific Ocean*, Phys. Earth Planet. Int. *47*, 230–245.

TANIMOTO, T., and ANDERSON, D. L. (1985), *Lateral Heterogeneity and Azimuthal Anisotropy of the Upper Mantle: Love and Rayleigh Waves 100–250 s*, J. Geophys. Res. *90*, 1842–1858.

TARANTOLA, A., and VALETTE, B. (1982), *Generalized Non-linear Inverse Problems Solved Using the Least-squares Criterion*, Rev. Geophys. Space Phys. *20*, 219–232.

TOMMASI, A., VAUCHEZ, A., and RUSSO, R. (1996), *Seismic Anisotropy in Oceanic Basins: Resistive Drag of the Sublithospheric Mantle*, Geophys. Res. Lett. *23*, 2991–2994.

TRAMPERT, J., and WOODHOUSE, J. H. (1996), *Global Phase Velocity Maps of Love and Rayleigh Waves between 40 and 150 Seconds*, Geophys. J. Int. *212*, 675–690.

VAUCHEZ, A., and NICOLAS, A. (1991), *Mountain Building: Strike-parallel Motion and Mantle Anisotropy*, Tectonophys. *185*, 183–191.

VINNIK, L., FARRA, F., and ROMANOWICZ, B. (1989), *Observational Evidence for Diffracted SV in the Shadow, of the Earth's Core*, Geophys. Res. Lett. *16*, 519–522.

VINNIK, L. P., KOSAREV, G. L., and MAKEYEVA, L. I. (1984), *Anisotropiya litosfery po nablyudeniyam voln SKS and SKKS*, Doklady Akademii Nauk USSR *278*, 1335–1339.

VINNIK, L. P., KIND, R., KOSAREV, G. L., and MAKEYEVA, L. I. (1989a), *Azimuthal Anisotropy in the Lithosphere from Observations of Long-period S Waves*, Geophys. J. Int. *99*, 549–559.

VINNIK, L. P., FARRA, V., and ROMANOWICZ, B. (1989b), *Azimuthal Anisotropy in the Earth from Observations of SKS at GEOSCOPE and NARS Broadband Stations*, Bull. Seismol. Soc. Am. *79*, 1542–1558.

VINNIK, L., FARRA, V., and ROMANOWICZ, B. (1989c), *Observational Evidence for Diffracted SV in the Shadow of the Earth's Core*, Geophys. Res. Lett. *16*, 519–522.

VINNIK, L., MAKEYEVA, L. I., MILEV, A., USENKO, A. Y. (1992), *Global Patterns of Azimuthal Anisotropy and Deformations in the Continental Mantle*, Geophys. J. Int. *111*, 433–447.

VINNIK, L., ROMANOWICZ, B., LE STUNFF, Y., and MAKEYEVA, L. I. (1995), *Seismic Anisotropy in D''-layer*, Geophys. Res. Lett. *22*, 1657–1660.

VINNIK, L., and MONTAGNER, J.-P. (1996), *Shear-wave Splitting in the Mantle from Ps Phases*, Geophys. Res. Lett. *23*, 2449–2452.

WIDMER, R. (1991), *The Large-scale Structure of the Deep Earth as Constrained by Free Oscillations Observations*, Ph.D. Thesis, 149 pp., University of California, San Diego.

WIDMER, R., MASTERS, G., and GILBERT, F. (1993), *Spherically Symmetric Attenuation within the Earth from Normal Mode Data*, Geophys. J. Int. *104*, 541–553.

WOODHOUSE, J. H., GIARDINI, D., and LI, X. D. (1986), *Evidence for Inner Core Anisotropy from Free Oscillations*, Geophys. Res. Lett. *13*, 1549–1552.

YANG, X., and FISCHER, K. M. (1994), *Constraints on North Atlantic Upper Mantle Anisotropy from S and SS Phase*, Geophys. Res. Lett. *21*, 309–312.

(Received December 30, 1996, revised/accepted October 8, 1997)

Pure appl. geophys. 151 (1998) 257–280
0033–4553/98/040257–24 $ 1.50 + 0.20/0

❚ Pure and Applied Geophysics

Age-dependent Large-scale Fabric of the Mantle Lithosphere as Derived from Surface-wave Velocity Anisotropy

V. Babuška,[1] J.-P. Montagner,[2] J. Plomerová[3] and N. Girardin[2]

Abstract—Systematic variations of the seismic radial anisotropy ξ to depths of 200–250 km in North America and Eurasia and their surroundings are related to the age of continental provinces, and typical depth dependences of ξ_R are determined. The relative radial anisotropy ξ_R in the mantle lithosphere of Phanerozoic orogenic belts is characterized by $v_{SH} > v_{SV}$, with its maximum depth of about 70 km, on the average, while beneath old shields and platforms, it exhibits a maximum deviation from ACY400 model (Montagner and Anderson, 1989) at depths of about 100 km with $v_{SV} \geq v_{SH}$ signature. An interpretation of the observed seismic anisotropy by the preferred orientation of olivine crystals results in a model of the mantle lithosphere characterized by anisotropic structures plunging steeply beneath old shields and platforms, compared to less inclined anisotropies beneath Phanerozoic regions. This observation supports the idea derived from petrological and geochemical observations that a mode of continental lithosphere generation may have changed throughout earth's history.

Key words: Mantle lithosphere, radial seismic anisotropy.

1. Introduction

Even though the continents form one of the more accessible regions of the earth, considerable research remains to be undertaken to understand how they were formed and subsequently deformed. The continental lithosphere, which preserves about 95 percent of the earth's history, consists of a silica-rich, low-density crust and an ultramafic, high-density mantle lithosphere. Although a variety of geologic and geophysical studies contributed substantially to our understanding of the continental crust, serious deficiencies remain in our knowledge of the structure of the subcrustal lithosphere. Plate tectonic and global geodynamic models however underline the relevance of upper mantle structure and dynamics for the evolution of near-surface structures.

Compared to highly deformed continental crust of a very complicated petrology, the composition of the subcrustal lithosphere is relatively simple, comprising about 70% of olivine (see Section 4). Thanks to a systematic preferred orientation of olivine crystals and to their high elastic anisotropy, both the oceanic and the

[1] UNESCO, Division Earth Sciences, 1, rue Miollis, 75 732 Paris, France.
[2] I.P.G.P., 4, Place Jussieu, 72 252 Paris, France.
[3] Geophysical Institute, Czech Academy of Sciences, 141 31 Praha 4, Czech Republic.

continental mantle lithospheres exhibit a large-scale elastic anisotropy as a ubiqui-
tous property observable by both body- and surface-wave velocities. It is the seismic
anisotropy which contains information regarding the fabric and the deformation of
deeper parts of the earth.

In this paper we discuss variations of anisotropic parameters derived from
observations of Rayleigh and Love wave velocities. According to a depth variation
of one of the parameters—the radial anisotropy ξ, we resolved six types of
depth-dependences and we compare the radial anisotropy in the subcrustal lithosphere with the age of the crust. For such an analysis we use data of North America
and Eurasia, the two continents which are sufficiently covered by surface-wave
observations. The laterally integral effect of the surface waves is compensated by
their vertical resolution. The surface-wave anisotropy thus provides information on
the large-scale fabric which reflects plastic flow at different depth levels, either
present-day or frozen flow, depending on whether the waves penetrate the asthenosphere or remain in the lithosphere.

2. Seismic Anisotropy of the Upper Mantle

Anisotropy of physical properties is inherent to minerals, and laboratory studies
of rock samples show that many mineral assemblages exhibit anisotropic elastic
properties. The fabric which is detected by anisotropic propagation of elastic waves
may have many causes: stacking of isotropic layers with different elastic properties
(e.g., RIZNICHENKO, 1949) and fluid-filled cracks (e.g., CRAMPIN and BOOTH,
1985) is typical for the upper continental crust. The lattice preferred orientation of
minerals, such as mica in the lower crust or olivine in the upper mantle, is a
dominant cause of elastic anisotropy in the lithosphere-asthenosphere system.
Though the crust contains strongly deformed and oriented anisotropic minerals,
due to its heterogeneity and petrological complexity the overall crustal anisotropy
is small (BARRUOL and MAINPRICE, 1993) and the main source of the large-scale
anisotropy observed by different types of seismic waves must be sought in the upper
mantle. Therefore, in this paper which is oriented towards the long wavelength
anisotropy derived from surface-wave observations, we will limit our discussion to
the mantle lithosphere and the underlying asthenosphere.

To orientate anisotropic upper mantle minerals by flow or by other types of
deformation, an ordering process that generates coherent strains over a sufficiently
large region is needed in order to be detected macroscopically by seismic waves with
wavelengths of tens of kilometers. The use of anisotropy as a measure of deformation thus requires knowledge of its seismological manifestations, its relationship to
the properties of crystalline aggregates, and ultimately to the style of tectonic
deformation (SILVER, 1996).

2.1 Body-wave Anisotropy

Short wavelength information is derived from body-wave velocities. Many observations showed azimuthal variations in seismic velocities of the *Pn* phase that propagates subhorizontally beneath the *M* discontinuity. HESS (1964) first suggested that the anisotropy of about 7%, observed in the Northwest Pacific (RAITT, 1963; SHOR and POLLARD, 1964), was due to a preferred orientation of olivine crystals. The fastest *Pn* velocities, i.e., preferably orientated *a* axes of olivine, were approximately parallel to trends of fracture zones which were presumably fossil traces of the paleospreading of the Pacific plate. The azimuthal anisotropy of *Pn* velocities in the oceanic lithosphere can amount to 10%, although it generally ranges from 3 to 8% (CHRISTENSEN, 1984).

Probably the first irrefutable finding of *Pn* anisotropy in the continental mantle lithosphere was presented by BAMFORD (1973, 1977); that of long refraction profiles with a wide variety of propagation directions. Using a time-term analysis exceeding 80 seismic profiles in western Germany, the author determined the anisotropy of 7–8% with an azimuth of the maximum velocity N20E, i.e., oblique to the orientation of the Variscan orogenic belt in Central Europe. To interpret the observations, FUCHS (1983) introduced a model with two different olivine orientations, assuming that orientation in the topmost mantle is caused by a "leakage" of the present crustal stress field into the mantle. Another example of the *Pn* anisotropy originates in the western United States where BAMFORD *et al.* (1979) found a high-velocity direction between N70E and N80E, consistent with the offshore refraction measurements of RAITT *et al.* (1969) for the Pacific upper mantle.

The azimuthal variations of *Pn* velocities in seismic refraction studies are obtained from the travel times of waves that have travelled more or less horizontally through the uppermost mantle. Thus the information is incomplete, describing variations in a horizontal plane and giving very little indication of velocities in other directions inclined to the horizontal plane, as well as the vertical extent of anisotropy.

More complete information regarding *P*-wave anisotropy might be obtained with the use of teleseismic waves which arrive at stations from different azimuths and under different angles of incidence. Ray paths of teleseismic body waves (*P* and *S*) with periods between 1 and 12 s cover a broader fan of propagations, from subvertically propagating SKS waves to obliquely incident *P* waves, and thus they allow us to infer a more realistic 3-D model of the upper mantle anisotropy and map it with a better lateral resolution (BABUŠKA *et al.*, 1993). Lateral changes of *P*-velocity anisotropy within the subcrustal lithosphere are mapped by teleseismic *P*-residual spheres which display that part of the relative travel-time residuals which depends on azimuth and incidence angles. Many regions manifest a systematic pattern of residual spheres throughout large tectonic units. Often, negative residuals are concentrated prevailingly on one side of the lower-hemisphere projection whereas positive ones are on the opposite side. Such P-residual pattern, characterized by 2π periodicity, can be interpreted by P-wave propagation within anisotropic structures with plunging symmetry axes (BABUŠKA et al., 1984, 1993; PLOMEROVÁ et al., 1996).

While it is challenging to prove that the directional dependence of P-wave velocities (propagational anisotropy) is solely caused by an anisotropic propagation, the shear-wave splitting (polarization anisotropy), is generally accepted as direct proof of an anisotropic medium. Studies of polarization anisotropy, and more specifically, of shear-wave splitting in teleseismic shear phases such as SKS and S, were recently reviewed by SILVER (1996). Because of the excellent lateral resolution afforded by these phases, small geologic domains and their boundaries can be carefully sampled. However, the vertical resolution is low due to the integrating effect as a result of subvertical propagation of the SKS phases. Thus in case of SKS, the anisotropy could reside anywhere between the core-mantle boundary and the earth's surface. Recent findings of lateral variations of the observed shear-wave anisotropy are related to prominent geological sutures and support the generally accepted idea that the dominant source of anisotropy in shear-wave splitting appears to be in the mantle lithosphere (SILVER, 1996) and partly in the asthenosphere (VINNIK et al., 1992).

2.2 Surface-wave Anisotropy

Many indications of anisotropic properties of the oceanic and continental lithosphere derive from surface-wave observations, mainly from Love/Rayleigh-wave incompatibilities (e.g., CARA et al., 1980) or from the inversion of regional dispersion data (e.g., MONTAGNER, 1986). From an observational point of view, the anisotropy of surface-wave velocities induces two kinds of observable effects: the azimuthal anisotropy (FORSYTH, 1975) and the radial anisotropy which can be inferred from the simultaneous inversion of Rayleigh and Love wave data (ANDERSON, 1961; MONTAGNER and TANIMOTO, 1991) in order to remove the Rayleigh-Love wave dispersion discrepancy.

A weak elastic anisotropy gives rise to an azimuthal dependence of Rayleigh and Love wave velocities (SMITH and DAHLEN, 1973). Five mutually independent parameters, independent of azimuth ψ along the surface wave path, $A = \rho v_{PH}^2 = 3/8(C_{11} + C_{22}) + 1/4(C_{12}) + 1/2(C_{66})$, $C = \rho v_{PV}^2 = C_{33}$, $F = 1/2(C_{13} + C_{33})$, $L = \rho v_{SV}^2 = 1/2(C_{44} + C_{55})$ and $N = \rho v_{SH}^2 = 1/8(C_{11} + C_{22}) - 1/4(C_{12}) + 1/2(C_{66})$ express the equivalent transverse isotropic medium with vertical symmetry axis (C_{ij}, ρ and v are elastic moduli, density and velocity). Azimuthal terms with 2ψ and 4ψ periodicity describe the azimuthal variation of Rayleigh and Love waves, respectively (MONTAGNER and NATAF, 1986, 1988). Two parameters relate to S-wave anisotropy: the radial anisotropy ζ defined as $\zeta = (N - L)/L \cong 2(v_{SH} - v_{SV})/v_{SV}$, and the azimuthal anisotropy $\mathbf{G}(G, \psi_G)$. The parameter $\mathbf{G}$ expresses the azimuthal dependence of v_{SV} and provides the most important contribution to the Rayleigh wave azimuthal variation (MONTAGNER, 1994). MONTAGNER and NATAF (1988) present inferences pertaining to convection and mineral composition of the earth's mantle from the vectorial tomography. In this paper we focus on lateral changes and depth dependences of the radial anisotropy ξ.

MONTAGNER and ANDERSON (1989) used a dataset of updated normal mode eigenperiods to develop a new average anisotropic mantle model ACY400. In the model, the anisotropy is required down to at least 200 km and preferably to 400 km, although with a smaller amplitude than in the PREM model (DZIEWONSKI and ANDERSON, 1981). For azimuthal anisotropy, a good correlation between absolute plate motion (MINSTER and JORDAN, 1978) and fast v_{SV} directions at 200 km depth was noted below oceanic plates (MONTAGNER, 1994). However, this kind of correlation was not found for continental plates which underwent a more complex development.

3. Surface-wave Radial Anisotropy

We use data concerning the 3-D variations of parameters of the anisotropic upper mantle model (AUM) by MONTAGNER and TANIMOTO (1991). These authors inverted a large set of Love- and Rayleigh-wave dispersion curves, applying fundamental mode data in the period range 70–250 s. The data allow the authors to acquire information about anisotropy of the upper mantle. Due to a multitude of epicenter-station and great-circle path dispersion curves, a superior coverage of the earth's surface is obtained. The anisotropic parameters of the AUM model are given in $10° \times 10°$ grid to 500 km, with a vertical resolution of about 50 km for the first 200 km and an error less than 1% (MONTAGNER and JOBERT, 1988). Depth resolution decreases with increasing depth. In this section we discuss a depth dependence of the radial anisotropy ξ at different tectonic environments and the magnitude of the anisotropy within the subcrustal lithosphere in relation to the geologic age of the overlying crust.

3.1 Depth Distribution of Anisotropy

To examine regional changes of the anisotropy and its possible correlation with the age of major tectonic units, we have analyzed a depth distribution of the relative radial anisotropy ξ_R and the azimuthal anisotropy $\mathbf{G}$. The relative radial anisotropy ξ_R relates to model ACY400 (ξ_0), Figure 1:

$$\xi_R = \xi - \xi_0.$$

Thus $\xi_R = 0$ means a positive radial anisotropy of about 2% which continues to a depth of 200 km. The signature $v_{SH} > v_{SV}$ continues in model ACY400 to a depth of 310 km (zero radial anisotropy). At a depth of 400 km the model radial anisotropy attains -1.5% ($v_{SH} < v_{SV}$). The relative radial anisotropy ξ_R thus allows us to trace deviations from the average anisotropic earth's model ACY400 in various tectonic provinces. As ξ values we use the original data file by MONTAGNER and TANIMOTO (1991) and limit our discussions to the two continents mentioned above and surrounding oceans.

Figure 2 displays an example of a systematic variation of the relative radial anisotropy ξ_R with depth in the lithosphere-asthenosphere system of a part of the North American continent. Similar $\xi_R(h)$ curves form groups related to large tectonic provinces. While the oceanic lithosphere (left-low corner of Fig. 2) is characterized by positive values of $\xi_R (v_{SH} > v_{SV})$ with their maximum amplitude at a depth approaching 70 km, i.e., near, the lithosphere-asthenosphere transition, the Archean lithosphere of the Canadian shield (e.g., intersections 50–100, 50–90) shows a completely different pattern. Negative ξ_R values are typical for this "shield" type radial anisotropy ($v_{SH} < v_{SV}$) with their maximum amplitude at a depth about 100 km, i.e., within the Archean subcrustal lithosphere. In the mantle lithosphere of the Trans-Hudson Orogen and other tectonic units towards the West

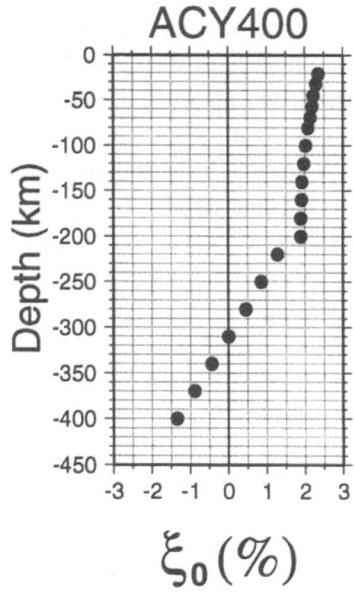

Figure 1
Reference model ACY400 of the radial anisotropy ξ_0 (MONTAGNER and ANDERSON, 1989).

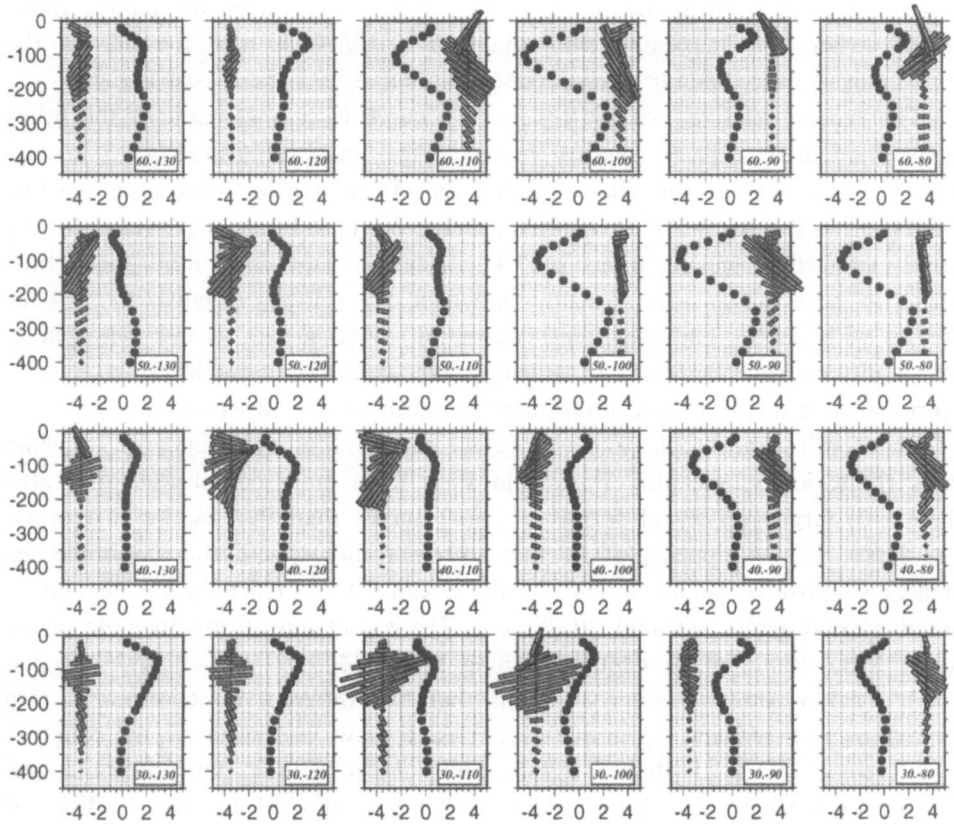

Figure 2
Lateral variation of the depth dependence (depth in km) of two anisotropic parameters: relative radial anisotropy ξ_R (dots; scale in percents) and azimuthal anisotropy $G(G, \psi_G)$—oriented bars—in $10° \times 10°$ intersections in North America (see Fig. 5a). North is oriented in a positive depth direction (upwards), East is oriented in a direction of positive ξ_R (to the right). Length of G is proportional to G.

Coast of North America, the amplitudes of ξ_R are smaller and the $\xi_R(h)$ dependence is more complex. A prominent change in the $\xi_R(h)$ dependence, which seems to be related to the transition between the Superior Province and the Trans-Hudson Orogen, is also marked by a change in azimuths of parameter G (e.g., between intersections 50–110 and 50–100 in Fig. 2).

When performing the $\xi_R(h)$ type analysis, we examined 210 curves corresponding to $10° \times 10°$ intersections in both continents and their surroundings. Three main types of curves, expressing a depth dependence of the relative radial anisotropy $\xi_R(h)$, are depicted in Figure 3. In stable parts of both continents, such as in the Canadian (3a) and Baltic Shields (3b), and the Siberian Platform (3c), ξ_R is negative to a depth of about 200 km. Maximum negative amplitudes of ξ_R (-2.5% to

-4.5%) are observed at depths of 70 to 120 km, i.e., within the mantle lithosphere, where the maximum azimuthal anisotropy **G** is also located. In the type analysis of $\xi_R(h)$ curves we denote this curve as Type 1 (see Fig. 4).

The second relatively simple type of ξ_R curve is characteristic of the lithosphere and asthenosphere beneath oceans (Figs. 3a,c; see also Type 3 in Fig. 4). The amplitude of the relative radial anisotropy reaches maximum positive values of $+1\%$ to $+3\%$ at depths around 60 to 80 km, i.e., close to the lithosphere-asthenosphere transition. The oceanic subcrustal lithosphere thus seems to be a single-layer structure in which *SH* velocity (relative to *SV*) increases with depth. The consistent orientation of **G** in the oceanic lithosphere, which differs from that in the asthenosphere, confirms a one-layer, large-scale fabric as seen in the mantle lithosphere by surface-wave observations.

The depth dependence of ξ_R in Paleozoic and Mesozoic-Cenozoic orogenic belts (Figs. 3a,b,c; see also Type 2 in Fig. 4), is more complicated and also there is a larger variety of forms of $\xi_R(h)$ curves than that in the two previous types. In the upper part of the subcrustal lithosphere the amplitude increases to positive values similar to the oceanic type, reaching the maximum amplitudes at depths around 70–80 km. The radial anisotropy then decreases to its local minimum of depths reaching about 140–180 km. Below 200–250 km a local maximum occurs. It seems that two different directions of **G** are related to two extreme ξ_R amplitudes. Due to low resolution of the long-period surface waves at shallow depth, only upper mantle data on both anisotropic parameters can be considered reliable and used for the analysis of $\xi_R(h)$ curves.

Besides the three basic types described above, i.e., Type 1 observed mainly in shields and platforms, Type 2 in the Mesozoic-Cenozoic and Paleozoic orogenic belts, and Type 3 beneath oceans, we found curves showing a more or less monotonous increase of $\xi_R(h)$ from negative to positive values to 200 km (Type 4 in Fig. 4). For less than 4% of the curves it was difficult to decide if they belong to the Types 2 or 3, due to less developed local minima at depth around 160 km. They are considered as a transitional Type 5. Only less than 10% of the $\xi_R(h)$ curves show no deviation from the ACY400 model within the upper 200 km ($\xi_R(h)$ is close to zero). Figure 4 summarizes six types of $\xi_R(h)$ curves and presents average $\xi_R(h)$ computed for corresponding types along with the standard deviations (SD) of individual values. The error bars mark a narrow band along the average curves. Regions with $\xi_R(h)$ characterized as Type 1 require as much as -1.8% reduction, on average, of the model radial anisotropy within the lithosphere at a depth of 100 km. On the other hand, regions with $\xi_R(h)$ of Types 2 and 3 need a positive correction of the $\xi_0(h)$ of about 1.3% and 1.8%, on average, at depths of 70 km. The $\xi_R(h)$ curve of Type 2 attains the model $\xi_0(h)$ value at depths of 140–200 km, on average, while at greater depths the curve marks slight positive

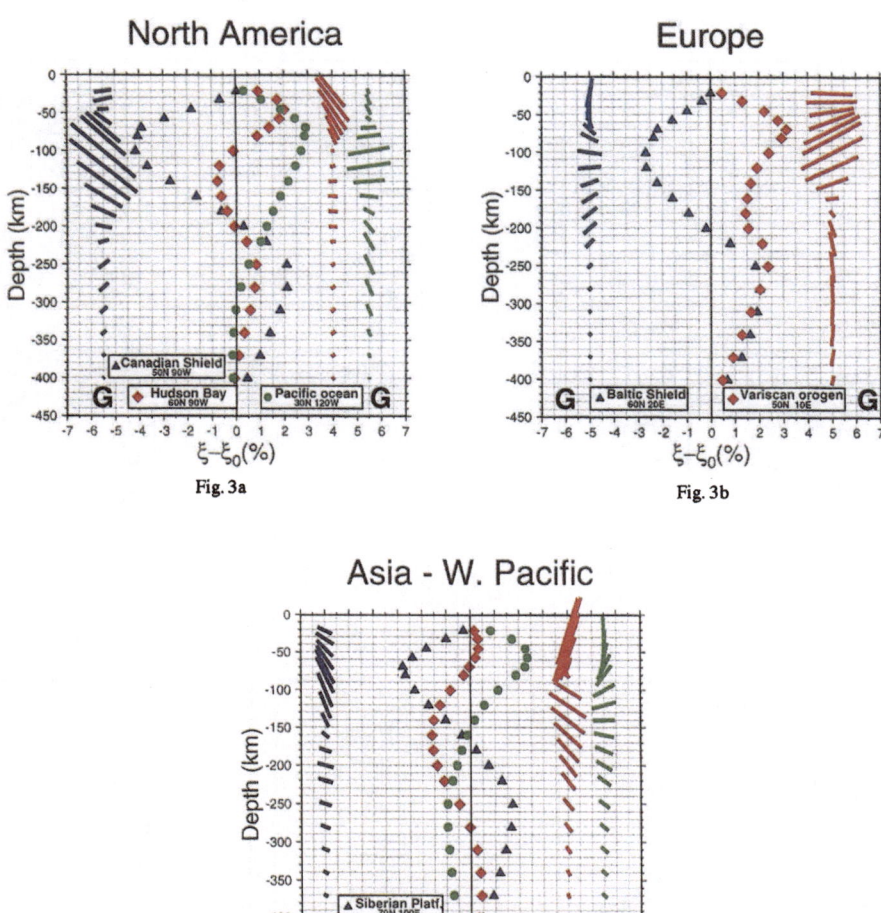

Figure 3

Examples of the depth dependence of the relative radial anisotropy ξ_R and the azimuthal anisotropy **G** for three types of tectonic environments in three geographic regions.

corrections. The average $\xi_R(h)$ curve of Type 3 returns to the model values at greater depths -250 km. The largest deviations in the asthenospheric part of the mantle are observed for curves of Type 1 and Type 4, about 1.4%–1.8%, respec-

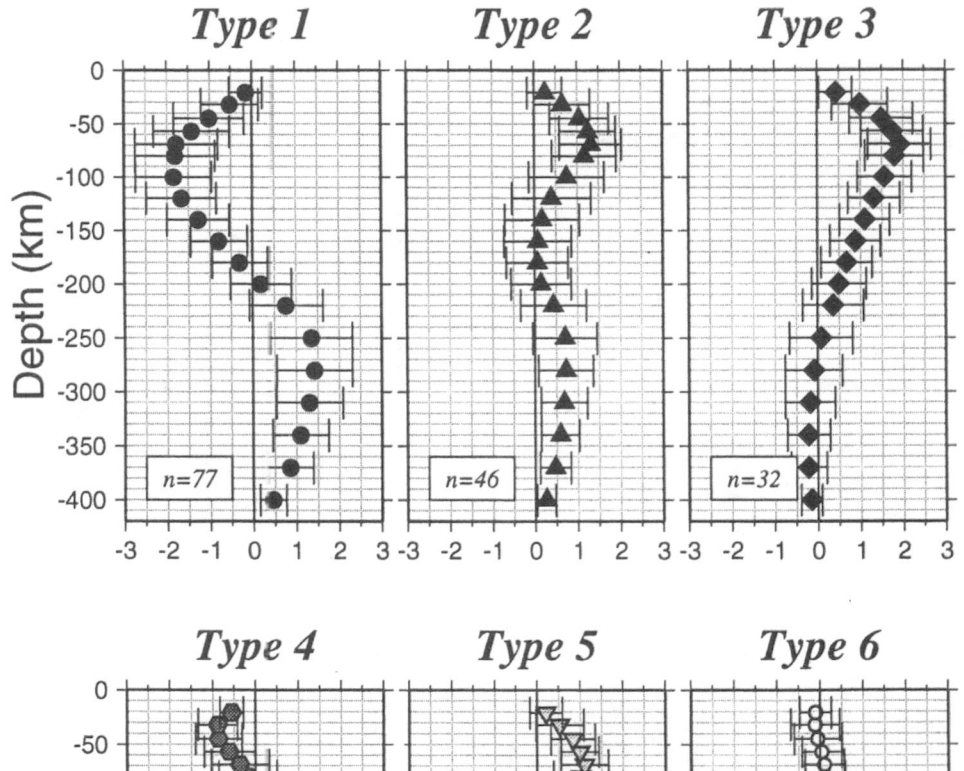

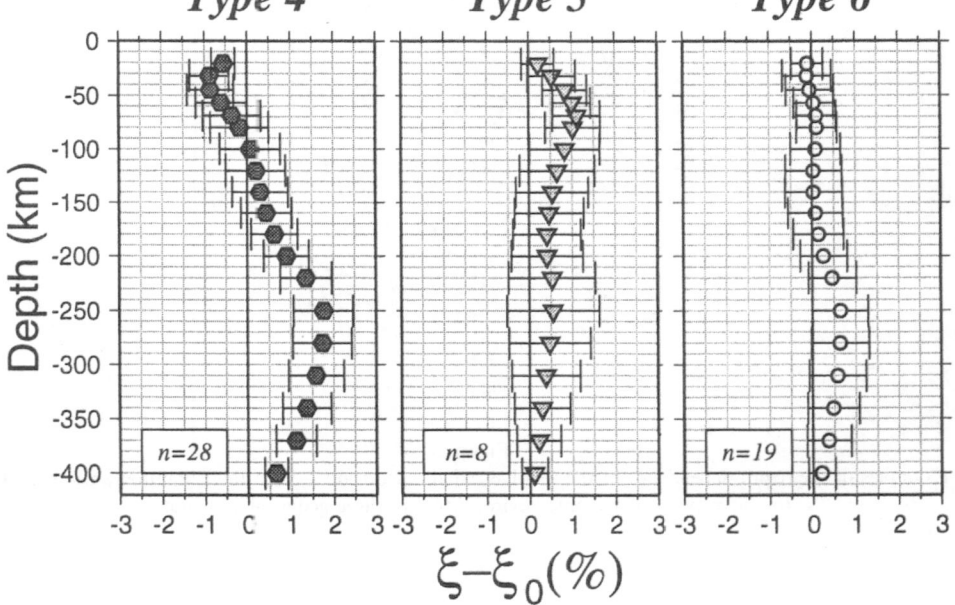

$$\xi-\xi_0(\%)$$

Figure 4
Six types of depth dependence of the relative radial anisotropy ξ_R represented by average values
combined with the standard deviations. For a description see Section 3.1.

tively. In general, the standard deviations are larger in the sublithospheric part of the mantle.

3.2 Lateral Changes of Anisotropy

The distribution of the individual types of $\xi_R(h)$ curves on both continents (Figs. 5a,b) marks distinctly regions with a similar relative radial anisotropy. Though the types are ascribed to $10° \times 10°$ intersections, due to the long wavelength of the surface waves, the curves represent a relatively large volume of the upper mantle which extends to a depth with increasing wavelengths. Inevitably, in some of the intersections the volume comprises tectonic units of different origin and age.

The distribution of Type 1 of $\xi_R(h)$ corresponds to large parts of tectonically stable parts of continents—shields and platforms. In North America, Type 1 extends further to the southeast of the USA, beyond the boundary of the shield (Fig. 5a) and, on the other hand, the Hudson Bay is characterized by Type 2. In Europe, the boundary between the East-European Platform and the Paleozoic and younger units is well reflected in $\xi_R(h)$ curves (see intersections 50.10 and 50.20 in Fig. 5b). The signature $v_{SH} > v_{SV}$, which we have observed in Paleozoic Central Europe (see Fig. 3b) is in accord with the findings of FRIEDRICH and HUANG (1996), who applied a different set of data.

The maps in Figures 6a,b delineate the maximum amplitude of the relative radial anisotropy within the subcrustal lithosphere, derived from dependence on the type of $\xi_R(h)$. In the case of the curve Types 1 and 4, the minimum of the $\xi_R(h)$ curves between depths 32 and 200 km is plotted. For Types 2, 3 and 5 the maximum of the $\xi_R(h)$ curves is shown. Values ξ_R at depths of 100 km are used for Type 6. Positive values and also partly values between 0 and -1% are characteristic of Phanerozoic units, while negative ones are typical of Archean and Proterozoic provinces (cf., Figs. 6a,b and 7a,b). Within Europe, the Trans-European Suture Zone (TESZ) is the most prominent geological boundary. It separates mobile Phanerozoic terranes, characterized by positive extremal values of the $\xi_R(h)$ curves (Fig. 6b), from the Precambrian East-European Platform, with negative extremal values of the $\xi_R(h)$ curves. It can be observed (Figs. 2, 4 and 5a,b) that within the continental lithosphere extremes reside deeper beneath the shields and platforms (100 km on the average) than beneath the Phanerozoic regions (70 km), where also a local minimum exists deeper in the lithosphere. Beneath the oceans the maximum amplitude of the radial anisotropy is also observed at a depth of about 70 km, although in this case it means near the lithosphere-asthenosphere transition.

We examine the types of radial anisotropy in the lithosphere on the basis of crustal types and their approximate ages, namely shields, platforms, Paleozoic

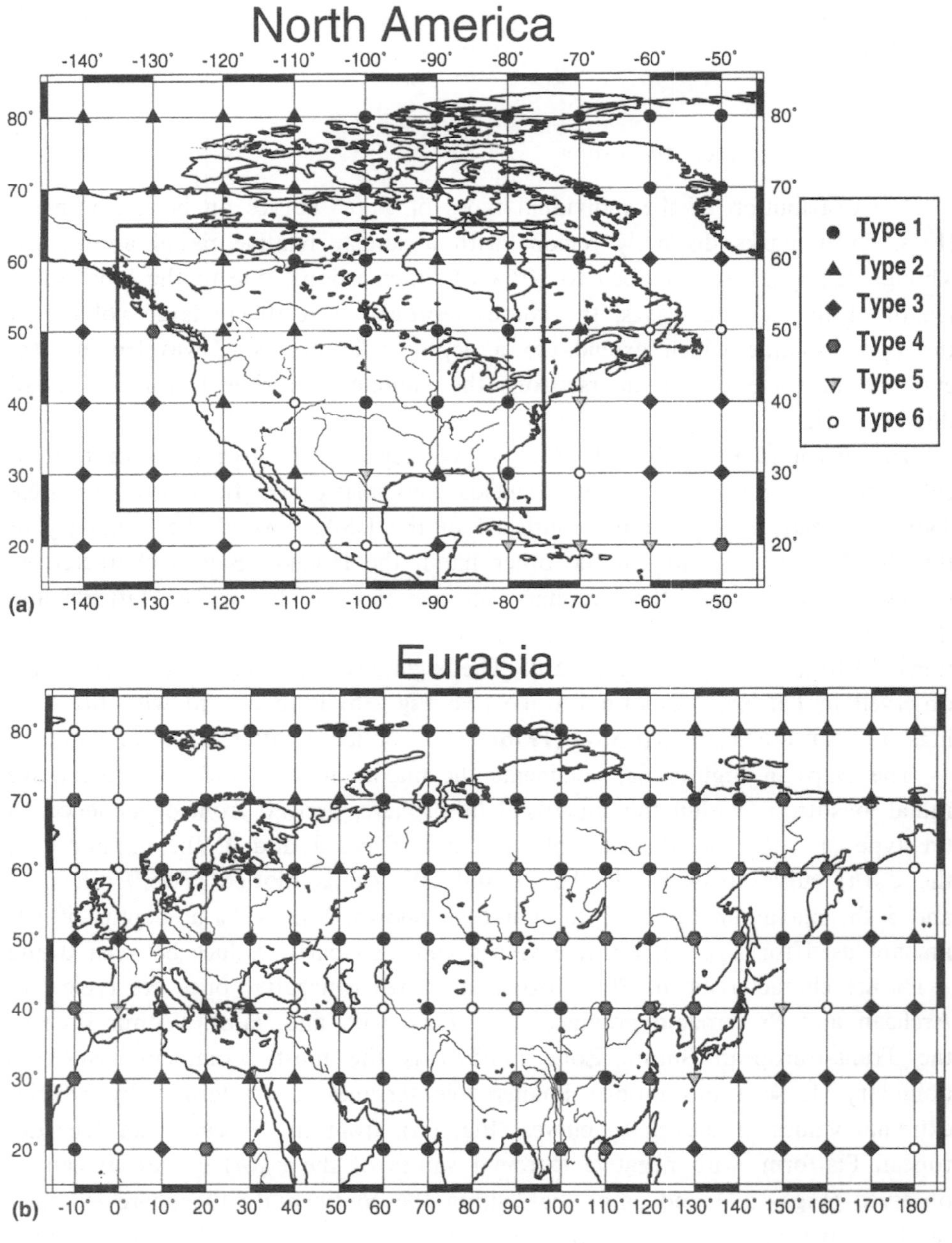

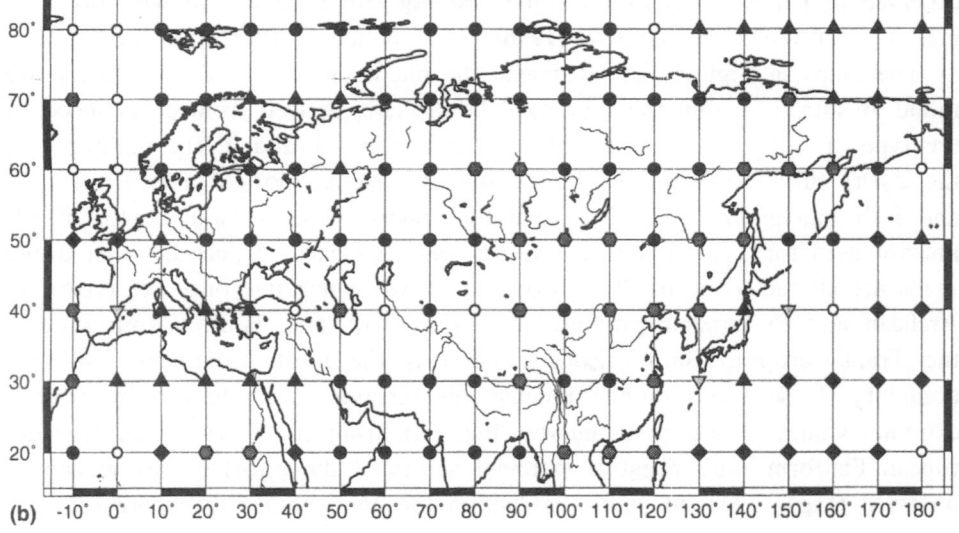

Figure 5
Geographical distribution of the six types of $\xi_R(h)$ defined in Figure 4, shown for $10° \times 10°$ intersections at both investigated continents: (a) North America and (b) Eurasia.

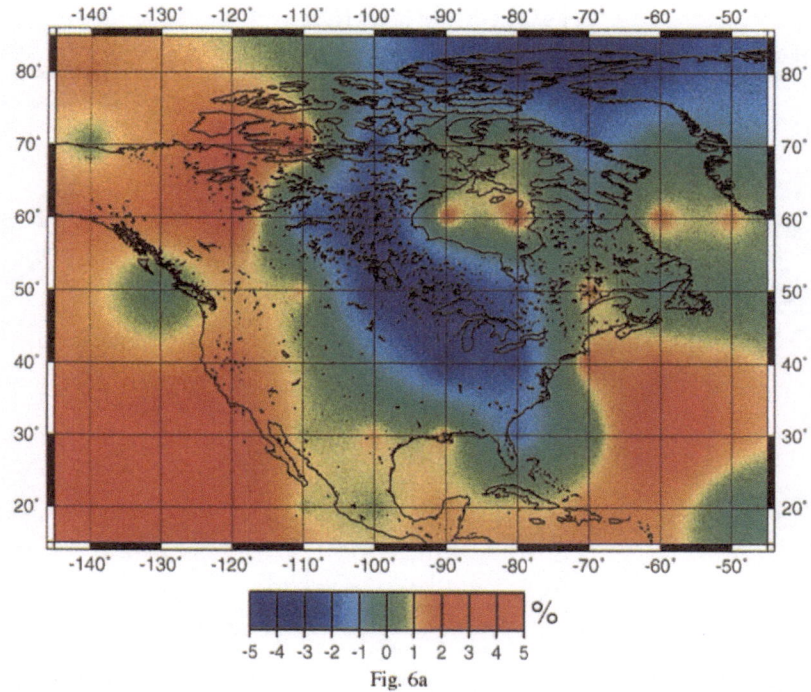

Fig. 6a

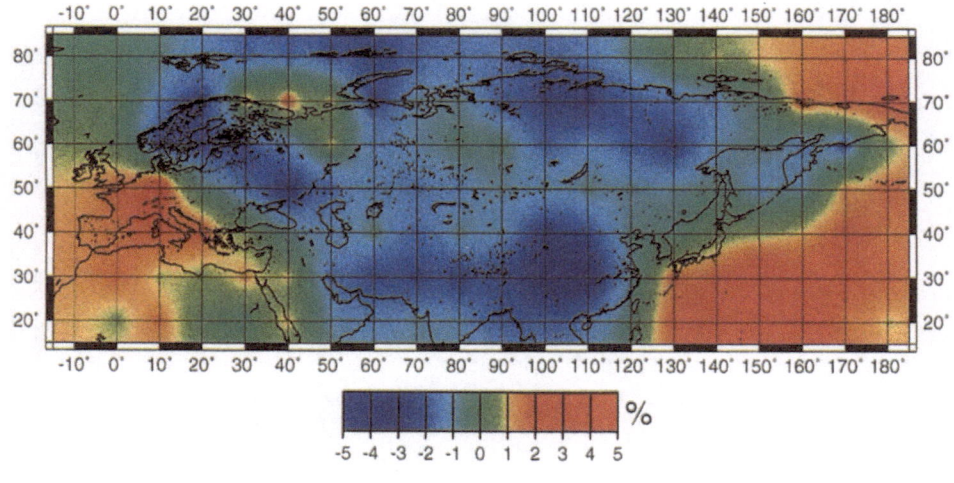

Fig. 6b

Figure 6

Lateral variations of the maximum amplitude of the relative radial anisotropy ξ_R in the subcrustal lithosphere of North America (6a) and Eurasia (6b). For a description see Section 3.2.

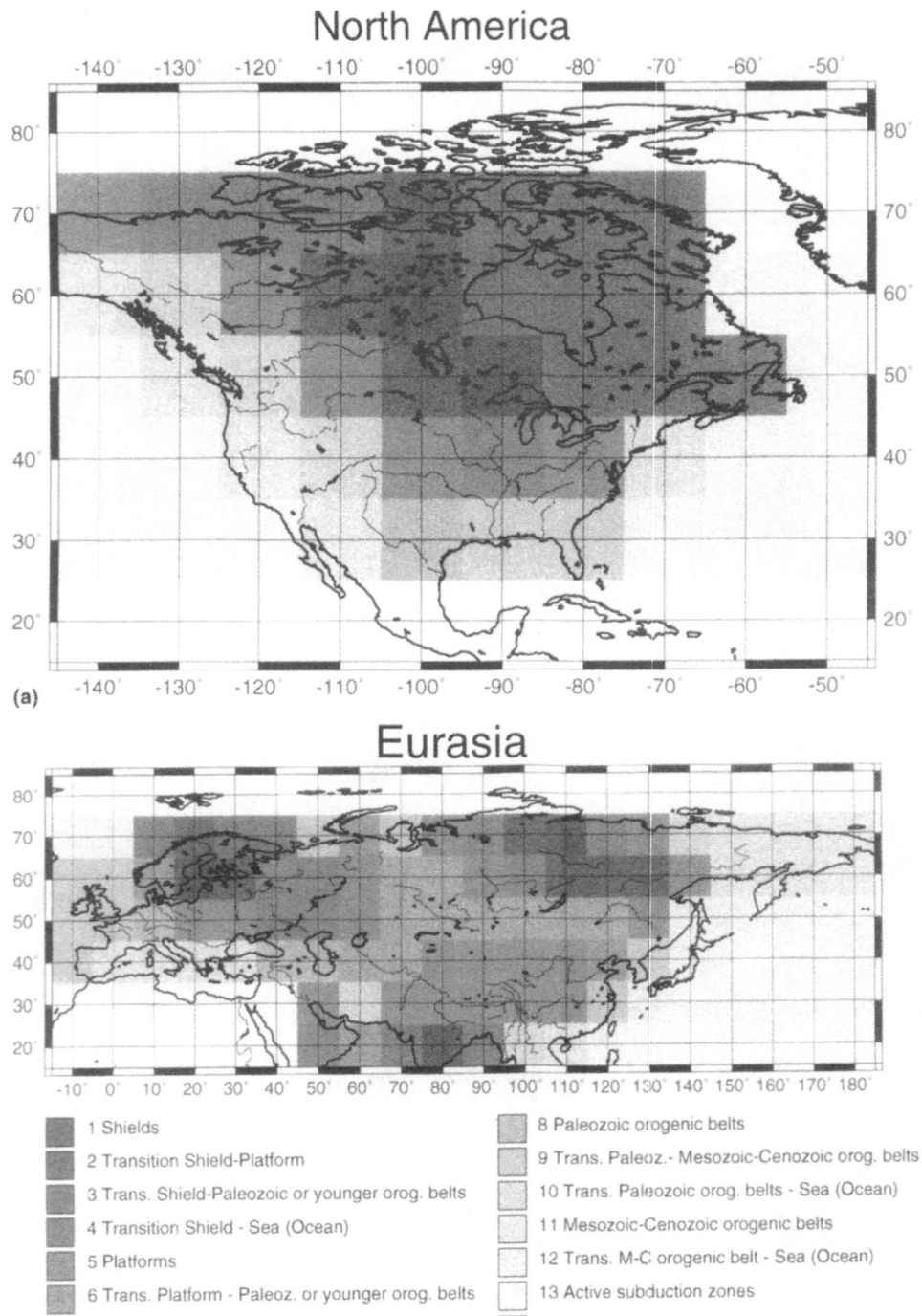

Figure 7(a and b).

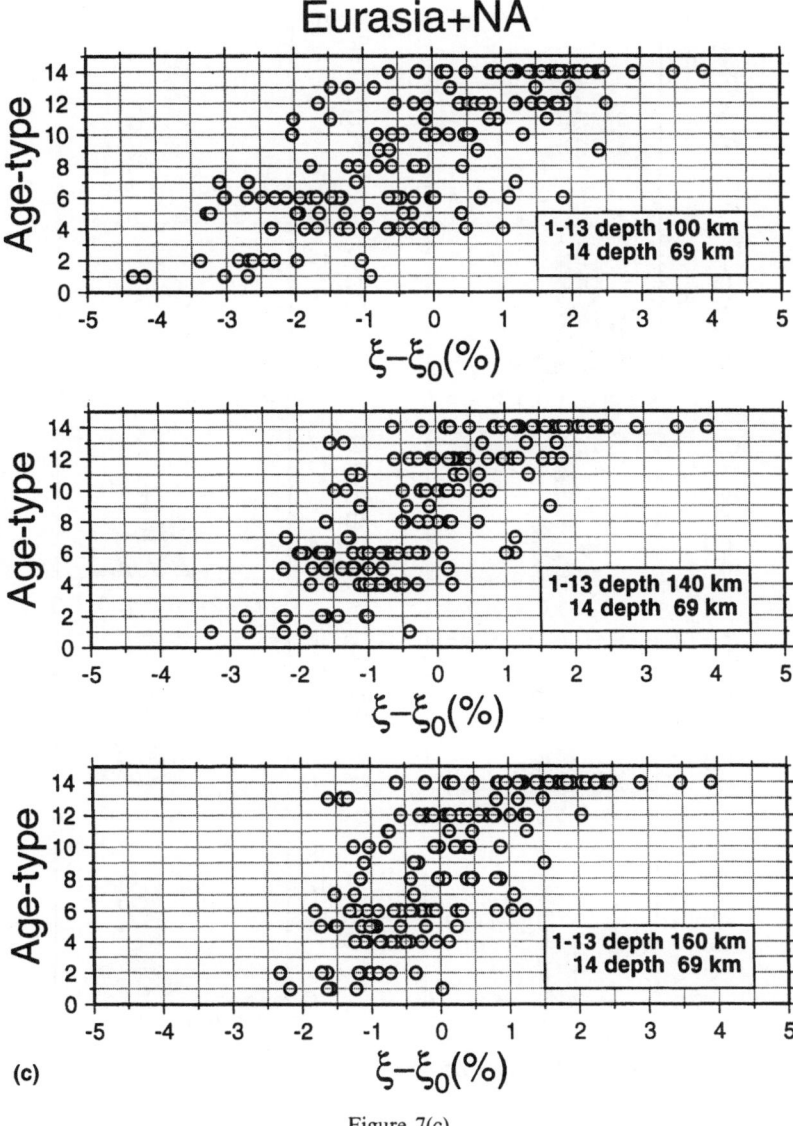

Figure 7(c)

orogenic belts, Mesozoic-Cenozoic orogenic belts, and ocean island arcs, as compiled globally by ZANDT and AMMON (1995) with modifications in Paleozoic Central Europe (Figs. 7a,b). In comparing the maximum amplitude of the relative radial anisotropy ξ_R with the crustal age, we assume that the entire lithosphere beneath the individual intersections is of the same age, which, however, is not always the case. This may be the case, e.g., in the Himalayas and Tibet, where the

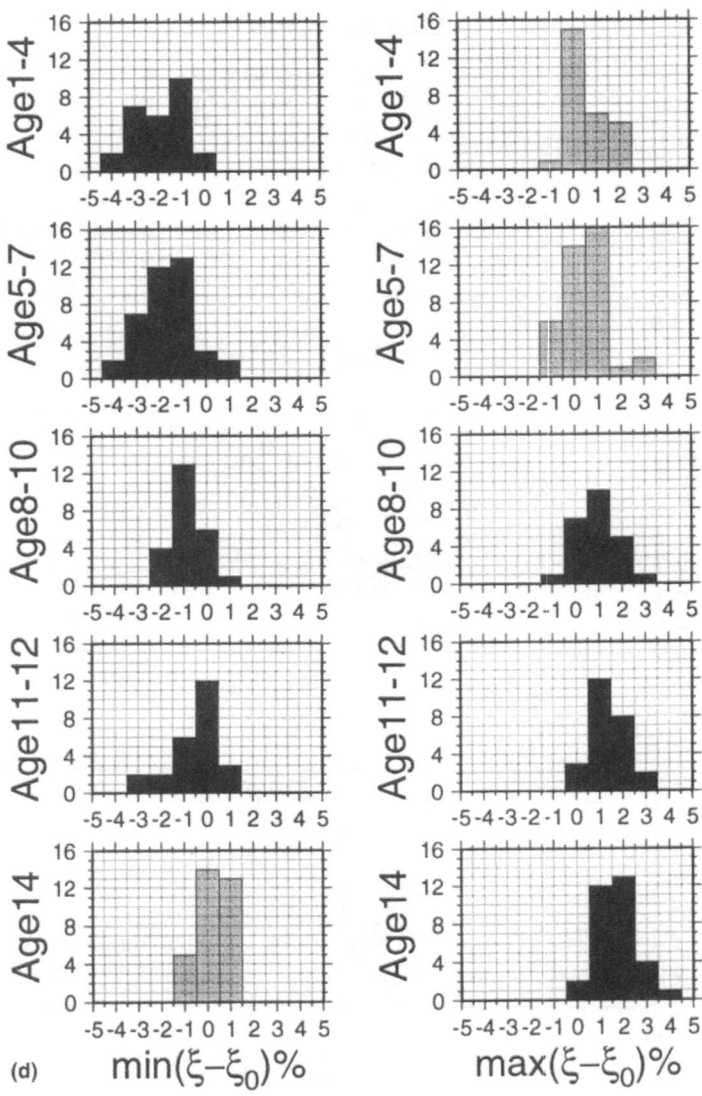

Figure 7
Crustal age types in North America (a) and in Eurasia (b), based on a global compilation by ZANDT and AMMON (1995). Shields are areas of exposed Precambrian-age crust. Platforms represent shield-like crust with sedimentary cover 1–5 km thick. Orogenic belts experienced one or more Phanerozoic-age tectonic events. Paleozoic (Pal.) orogenic belts are relatively stable with eroded mountains. Mesozoic-Cenozoic (M-C) belts are often still tectonically active, with considerable surface relief and crustal thickness variation. (c)—Relative radial anisotropy ξ_R at depths 100, 140 and 160 km beneath continents for crustal types 1–12, and at a depth of 69 km beneath the oceans (crustal age type 14); (d)—Histograms of minimum and maximum values of $\xi_R(h)$ to 200 km according to the age types, for details see Section 3.2.

entire mountain belt is characterized by $v_{SV} \geq v_{SH}$ signature, typical for the cratonic type of the subcrustal lithosphere. ROYDEN et al. (1997) suggest that during the continental convergence in eastern Tibet, the upper crustal deformation is decoupled from the motion of the underlying mantle. While field observations and satellite geodesy indicate that little crustal shortening has occurred in that region since about 4 million years ago, the authors estimate the northward motion of the underlying mantle at about 50 mm/year. The depth-dependent rheology can thus explain a model in which a major part of the Himalayan belt is underlaid by a cratonic northward moving Indian mantle lithosphere. A detailed tomographic study of central Asia by surface waves (GRIOT et al., 1998) also indicates that the mantle below 100 km is cold. A distinct crustal thickening, typical for this region, can also affect the Love-wave propagation as well as the radial anisotropy ξ_R.

In Figures 7c,d we present the maximum amplitude (or values at defined depths) of $\xi_R(h)$ in the subcrustal lithosphere in relation to the age of crustal units. Figure 7c shows $\xi_R(h)$ at three depth levels beneath continents (100, 140 and 160 km) and at one level beneath the oceans (69 km). Vertical scale is arbitrary and represents an approximate ordering of crustal units according to their age (Figs. 7a,b), from the oldest crust (Archean -1) to the youngest (oceanic -14). Although there is a large variance of ξ_R at individual "crustal age" levels, a general tendency of increasing ξ_R with decreasing age of crust is evident. The effect decreases with increasing depth. As mentioned above, the maximum amplitude of deviations from the reference ξ_0 occurs at different depths (Figs. 4 and 6).

Figure 7d illustrates histograms of minimum and maximum values of $\xi_R(h)$, curves in the subcrustal lithosphere, to 200 km. In groups with crustal age characterized as 1–4 and 5–7, the $\xi_R(h)$ curve of Type 1 prevails and thus the minimum of $\xi_R(h)$ represents the distribution of the maximum deviation (histograms with the black fill), while the maximum only reflects the values at 200 km, i.e., values close to the ACY400 model (histograms with the gray fill). Type 2 is dominant for groups with crustal ages of 8–10 and 11–12, with a real local minimum and maximum (see Fig. 4). For oceans (age 14), where we observe mainly Type 3, only the maxima of $\xi_R(h)$ reflect the distribution of the deviations from the model (histograms with the black fill). Similar to maxima of $\xi_R(h)$ for crustal ages 1–7, minima of $\xi_R(h)$ (the histogram with a gray fill) for age 14 express no deviation from the model at a depth of 200 km. Both Figures 7c and 7d reflect that there is, in general, a correlation between the age and the relative radial anisotropy. The most negative values of $\xi_R(h)$ belong to the mantle lithosphere of shields and platforms, and the most positive values are characteristic of the lithosphere of Phanerozoic regions.

4. Discussion and Conclusion

4.1 Petrofabrics and Seismic Anisotropy

The existence of seismic anisotropy discussed in this paper indicates an ordered medium in the mantle lithosphere and the asthenosphere. In the upper mantle this order is created primarily by a lattice preferred orientation of anisotropic crystals in response to finite strain. If the relation between lattice preferred orientation and strain fabric (such as the orientation of the foliation plane and lineation direction) is known, then seismic anisotropy can be used as a measure of mantle fabric and an indicator of different processes that lead to lithosphere formation.

Geophysical and petrological constraints restrict the dominant mineralogical composition of the subcrustal lithosphere and asthenosphere to olivine and pyroxene. As pyroxenes have a diluting or negative effect on the anisotropy (NICOLAS and CHRISTENSEN, 1987) and other minerals of upper-mantle peridotites (garnet, spinel) have simply a diluting effect, it has been generally recognized that the alignment of olivine crystals is a dominant cause of seismic anisotropy. This is also due to the fact that olivine is the most abundant phase (about 70%) in the upper mantle. It forms about 75% of the volume of oceanic peridotites (DICK, 1987) and, e.g., between 56 and 80% of the continental mantle lithosphere estimated from the composition of garnet lherzolite and harzburgite nodules from South Africa (MAINPRICE and SILVER, 1993).

A large elastic anisotropy of olivine single crystal (25% for P velocities and 14% for S velocities, BABUŠKA and CARA, 1991) is another favorable condition for seismic anisotropy in the upper mantle. The seismic anisotropy increases with fabric strength and finite strain. MAINPRICE and SILVER (1993) surveyed 25 naturally deformed peridotites of oceanic origin and found an average maximum S-wave anisotropy of 9% for the olivine component and about 8% when the orthopyroxene component was taken into account, twice the fabric strength and the resulting anisotropy of the kimberlite nodules (3.7%). The authors conluded that all the olivine fabrics had the same general characteristics, a maximum of preferentially oriented a axes parallel to the grain shape lineation and c axes maxima oriented in the foliation plane and normal to the lineation. Similarly the b axes are preferentially oriented perpendicularly to the foliation. The overall seismic anisotropy is thus dominated by the orientation of the $a-c$ foliation plane along which we observe the highest P velocities as well as the largest S-wave splitting.

In the following paragraphs we will discuss the radial anisotropy in relation to the olivine fabric and to published results on the anisotropy determined from shear-wave splitting. Our most robust observations are negative values of ξ_R, showing that SV velocities are either higher or close to SH within the subcrustal lithosphere of shields and platforms on one hand, and positive values of ξ_R in the lithosphere of Phanerozoic regions and oceans, representing $v_{SH} > v_{SV}$, on the other. Positive ξ values are compatible either with a horizontal layering or with a horizontal flow which would lead to a subhorizontal preferred orientation of the "fast" a axes of olivine and to a subhorizontal foliation plane.

4.2 Large-scale Fabric of Oceanic Lithosphere

It is generally accepted that the oceanic lithosphere has a relatively simple large-scale anisotropy. It follows both from the observed Pn velocity anisotropy (see Section 2) and the observed Love/Rayleigh-wave discrepancy displayed, e.g., in the Pacific Plate (CARA and LEVEQUE, 1988), that seismic anisotropy of the oceanic lithosphere is due to a preferred orientation of the olivine a axes in the horizontal plane. Such orientation is also confirmed by positive values of the radial anisotropy which attains the highest amplitude near the lithosphere-asthenosphere transition (Figs. 2, 3 and 4), where the shear strain accumulates with plate displacement (TOMMASI et al., 1996).

For high-temperature conditions typical for the asthenosphere, NICOLAS and CHRISTENSEN (1987) demonstrated that the a axis of olivine (high P-velocity direction) was oriented in the main shearing direction prevailing parallel to the flow lines, while axes b and c, the lowest and the intermediate velocity directions, respectively, formed a girdle in the plane perpendicular to the flow. Such a hexagonal symmetry, with the symmetry axis oriented parallel to the flow direction, might be the main cause of the observed seismic anisotropy in the depth range of 100–200 km below oceans. This model can also explain azimuthal variations of both Rayleigh- and Love-wave velocities as related to the present-day flow in the asthenosphere.

Though many surface-wave observations suggest a transversely isotropic medium with the vertical symmetry axis for the oceanic subcrustal lithosphere, Pn-anisotropy observations demonstrated that its fabric is more complicated. The frozen-in orientations of olivine crystals result from a plastic flow beneath the base of the newly formed and gradually thickening oceanic lithosphere. Maximum P-wave velocities are normal to mid-oceanic ridges, thus reflecting orientations of olivine a axes along the mantle flow (NICOLAS and CHRISTENSEN, 1987). The anisotropy is probably a typical property of the entire thickness of the subcrustal lithosphere (ASADA, 1984), though the orientation of velocity extremes changes both laterally and vertically (e.g., SHIMAMURA et al., 1983) depending on the orientation of the flow in the asthenosphere in different stages of the lithosphere formation (the anisotropic plate thickening model of ISHIKAWA, 1984).

4.3 Radial Seismic Anisotropy vs. Age of Continents

While the oceanic lithosphere is characterized by positive values of the radial anisotropy, in the lithosphere of continental regions we observe both positive and negative values of ξ_R. The most prominent radial anisotropy with a negative amplitude of ξ_R has been recorded in the Superior Province of the Canadian Shield (intersections 50–100 and 50–90 in Figs. 2 and 6). In that region, a large shear-wave anisotropy was observed for subvertically propagating SKS waves with delay times between 1 and 2 s (SILVER *et al.*, 1993). The extreme negative values of ξ_R (about -4%, representing absolute $\xi = -2\%$) are observed at a depth close to 100 km, where also the largest **G** values reside. An abrupt decrease of **G** and a change of its orientation at a depth about 200 km seem to mark the lithosphere base, which was quoted at that depth beneath the Canadian Shield also by SILVER and CHAN (1991). Negative absolute ξ values can be observed in regions above anisotropic media with steeply dipping foliation planes containing the *a* axes. Such orientation is in accord with the observations of large SKS delay times (SILVER, 1996).

As mentioned above, we have found a negative amplitude of $\xi_R(h)$ and a similar $\xi_R(h)$ dependence also in the Fennoscandia and in other Precambrian shields. Though the maximum amplitudes of the negative ξ_R reach -4%, the average maximum amplitude of ξ_R related to its maximum amplitude in the subcrustal lithosphere of shields is about -2% (Type 1 in Fig. 4).

The steep orientations of olivine foliation planes mean that a major part of the subcrustal lithosphere was not formed by the conductive cooling mechanism which would produce a subhorizontal orientation of the *a* axes and the foliation planes. Such a mechanism would be acceptable only for the lowermost part of the lithosphere of shields (deeper than approximately 150 km) where the signature $v_{SV} < v_{SH}$ dominates (as to the absolute velocities, see Fig. 4, Type 1). Our observations also indicate that $v_{SV} \geq v_{SH}$ signature, characteristic for a major part of the mantle lithosphere of shields, is not compatible with models derived from long refraction profiles which are characterized by a sequence of alternating high- and low-velocity layers in the uppermost mantle (e.g., GUGGISBERG, 1986), as such a medium would be expressed by long wave-length *SH* velocities higher than *SV* (e.g., BACKUS, 1965). Existence of the large shear-wave splitting and azimuthal anisotropy of surface waves rules out an anisotropic approximation by a hexagonal model with vertical symmetry axis for such regions.

Keeping in mind the long-term stability of shields indicated by a number of independent petrological, geochemical (e.g., REISBERG and LORAND, 1995) and seismological observations (e.g., SNYDER *et al.*, 1996), we can assume that the $v_{SV} > v_{SH}$ signature reflects an old fabric.

Which processes could produce a large-scale fabric characterized by steeply inclined foliation planes containing the "fast" *a*-olivine axes? Could it be perhaps an accretion of pieces of a hot and relatively thin Archean oceanic lithosphere? Such steep anisotropic structures can also represent remnants of systems of paleosubductions of ancient oceanic lithosphere (BABUŠKA *et al.*, 1984), as also indicated by deep mantle reflectors occurring predominantly beneath outcrops of Archean or Proterozoic continental crust (CALVERT *et al.*, 1995).

Another question might be, to what depth such an old fabric can be preserved so that it can be observed today? Survival of a fossil fabric requires the mantle lithosphere to cool below a critical blocking temperature (800–900°C) for olivine creep to become ineffective over geologically significant time scales (ESTEY and DOUGLAS, 1986). The lithosphere-asthenosphere transition being determined by 1300°C isotherm, fossil fabric would survive in the upper two thirds of the lithosphere (ABALOS and DIAZ, 1995). Therefore, we can assume that $v_{SV} \geq v_{SH}$ signature, as well as the significant azimuthal dependence of v_{SV} in the upper and central part of the mantle lithosphere beneath shields (Figs. 2 and 3a,b) are caused by a fossil preferred orientation of olivine crystals.

Compared to the "shield" type radial anisotropy, $\xi_R(h)$ in the mantle lithosphere of Phanerozoic regions is different (Type 2 in Fig. 4). It reaches its maximum amplitude at a shallower depth (about 70 km) and it is positive. This means $v_{SH} > v_{SV}$ and in terms of the olivine orientation that either similar to the oceanic lithosphere, the foliation planes containing the a axes of olivine are oriented subhorizontally, or they plunge moderately. If the high P-velocity directions within the subcrustal lithosphere are inclined between 30° and 60°, then such anisotropic structures within the subcrustal lithosphere produce the "bipolar" pattern of teleseismic P-residual spheres mentioned above (BABUŠKA et al., 1993) and a directional dependence of shear-wave polarization parameters (PLOMEROVÁ et al., 1996).

By inverting shear-wave splitting parameters evaluated in 3 D, ŠÍLENÝ and PLOMEROVÁ (1996) were able to retrieve 3-D orientation of symmetry axes of an anisotropic medium with hexagonal or orthorhombic symmetry. Applying this method to the Baltic shield, the best solution was found for the hexagonal model with the steeply dipping fast velocity plane (a', c')—plunging at 49° from horizontal. On the other hand, using shear-wave splitting parameters by SAVAGE and SILVER (1993) for the Phanerozoic western U.S., PLOMEROVÁ et al. (1996) found there shallow inclinations of the fast velocity directions, which is in accord with findings of this paper. The following self-consistent anisotropic inferences were found: the orthorhombic a' axis plunges at 23° and 36° at Mojave Desert (station LAC) and Great Central Valey (MHC), respectively, and the (a', c') plane of the hexagonal model inclines at about 20° at San Francisco Bay (station BKS). The results obtained from the inversion of the shear-wave splitting parameters with delay times exceeding 1 s, are compatible with results of lateral variations of the depth distribution of the radial anisotropy $\xi_R(h)$, leading us to conclusions about different fabrics within the lithosphere of substantially different age.

The different olivine orientations can result from different processes that lead to the formation of the continental lithosphere in the Archean, compared to the Phanerozoic, and that probably changed considerably during the Proterozoic. It has been shown by different authors that the continental crust has an andesitic bulk composition, which could not have been produced by the basaltic magmatism that dominates sites of present-day crustal growth. Andesite production seems to be limited to hot and wet conditions that may have been more common in the Archean than today (see RUDNICK, 1995 for a review). It is thus possible that under high-heat flow conditions of the Archean times, the oceanic lithosphere was thinner than in Phanerozoic times and that also the mode of growth of the entire continental lithosphere from oceanic sources was different.

Acknowledgements

We thank the reviewers, B. Romanowicz and A. Vauchez, for their comments and suggestions which significantly improved the manuscript. This study was partly supported by grant No. A3012604 of the Grant Agency of the Czech Academy of Sciences.

REFERENCES

ABALOS, B., and DIAZ, J. (1995), *Correlation between Seismic Anisotropy and Major Geological Structures in SW Iberia: A Case Study on Continental Lithosphere Deformation*, Tectonics *14*, 1021–1040.

ANDERSON, D. L. (1961), *Elastic Wave Propagation in Layered Anisotropic Media*, J. Geophys. Res. *66*, 2953–2963.

ASADA, T. (1984), *Seismic Anisotropy beneath the Ocean*, J. Phys. Earth *32*, 177–178.

BABUŠKA, V., and CARA, M., *Seismic Anisotropy in the Earth* (Kluwer, Dordrecht, 1991).

BABUŠKA, V., PLOMEROVÁ, J., and ŠÍLENÝ, J. (1984), *Large-scale Oriented Structures in the Subcrustal Lithosphere of Central Europe*, Ann. Geophys. *2*, 649–662.

BABUŠKA, V., PLOMEROVÁ, J., and ŠÍLENÝ, J. (1993), *Models of Seismic Anisotropy in the Deep Continental Lithosphere*, Phys. Earth Planet. Int. *78*, 167–191.

BACKUS, G. E. (1965), *Possible Forms of Seismic Anisotropy of the Upper Mantle under Oceans*, J. Geophys. Res. *70*, 3429–3439.

BAMFORD, D. (1973), *Refraction Data in Western Germany—A Time-term Interpretation*, Z. Geophys. *39*, 907–927.

BAMFORD, D. (1977), *Pn Velocity Anisotropy in a Continental Upper Mantle*, Geophys. J. R. Astr. Soc. *49*, 29–48.

BAMFORD, D., JENTSCH, M., and PRODEHL, C. (1979), *Pn Anisotropy Studies in Northern Britain and the Eastern and Western United States*, Geophys. J. R. Astr. Soc. *57*, 397–429.

BARRUOL, G., and MAINPRICE, D. (1993), *A Quantitative Evaluation of the Contribution of Crustal Rocks to the Shear-wave Splitting of Teleseismic SKS Waves*, Phys. Earth. Planet. Int. *78*, 281–300.

CALVERT, A. J., SAWYER, E. W., DAWIS, W. J., and LUDDEN, J. N. (1995), *Archean Subduction Inferred from Seismic Images of a Mantle Suture in the Superior Province*, Nature *375*, 670–674.

CARA, M., and LEVEQUE, J. J. (1988), *Anisotropy of the Asthenosphere: The Higher Mode Data of the Pacific Revisited*, Geophys. Res. Lett. *15*, 205–208.

CARA, M., NERCESSIAN, A., and NOLET, G. (1980), *New Inferences from Higher Mode Data in Western Europe and Northern Eurasia*, Geophys. J. R. Astr. Soc. *61*, 459–478.

CHRISTENSEN, N. I. (1984), *The Magnitude, Symmetry and Origin of Upper Mantle Anisotropy Based on Fabric Analyses of Ultramafic Tectonites*, Geophys. J. R. Astr. Soc. *76*, 89–111.

CRAMPIN, S., and BOOTH, D. C. (1985), *Shear-wave Polarizations near North Anatolian Fault, II, Interpretation in Terms of Crack-induced Anisotropy*, Geophys. J. R. Astr. Soc. *83*, 75–92.

DICK, H. J. B., *Petrologic variability of the oceanic uppermost mantle*. In *Geophysics and Petrology of the Deep Crust and Upper Mantle* (eds. Noller, J. S., Kirby, S. H., and Nielson-Pike, J. E.) (U.S. Geol. Survey Circular 956, 1987) pp. 17–20.

DZIEWONSKI, A. M., and ANDERSON, D. L. (1981), *Preliminary Reference Earth Model*, Phys. Earth Planet. Int. *25*, 297–356.

ESTEY, L. H., and DOUGLAS, B. J. (1986), *Upper-mantle Anisotropy: A Preliminary Model*, J. Geophys. Res. *91*, 11393–11406.

FORSYTH, D. W. (1975), *The Early Structural Evolution and Anisotropy of the Oceanic Upper Mantle*, Geophys. J. R. Astr. Soc. *43*, 103–162.

FRIEDRICH, W., and HUANG, Z.-X. (1996), *Evidence for Upper Mantle Anisotropy beneath Southern Germany from Love and Rayleigh-wave Dispersion*, Geophys. Res. Lett. *23*, 1135–1138.

FUCHS, K. (1983), *Recently Formed Elastic Anisotropy and Petrological Models for the Continental Subcrustal Lithosphere in Southern Germany*, Phys. Earth Planet. Int. *31*, 93–118.

GRIOT, D. A., MONTAGNER, J.-P., and TAPPONIER, P. (1998), *Surface Wave Phase Velocity Tomography and Azimuthal Anisotropy in Central Asia*, J. Geophys. Res., in press.

GUGGISBERG, B., *Eine zweidimensionale refraktionsseismische Interpretation der Geschwindigkeits-Tiefen-Struktur des oberen Erdmantels unter dem fennoskandischen Schild (Projekt Fennolora)*, Diss. ETH 7945 (Zürich, 1986) 199 pp.

HESS, H. (1964), *Seismic Anisotropy of the Uppermost Mantle under Oceans*, Nature *203*, 629–631.

ISHIKAWA (1984), *Anisotropic Plate Thickening Model*, J. Phys. Earth *32*, 219–228.

MAINPRICE, D., and SILVER, P. (1993), *Interpretation of SKS-waves Using Samples from the Subcontinental Lithosphere*, Phys. Earth Planet. Int. *78*, 257–280.

MINSTER, J. B., and JORDAN, T. H. (1978), *Present-day Plate Motions*, J. Geophys. Res. *83*, 5331–5354.

MONTAGNER, J.-P. (1986), *Regional Three-dimensional Structures Using Long-period Surface Waves*, Ann. Geophys. B *4*, 283–294.

MONTAGNER, J.-P. (1994), *Can Seismology Tell us Anything about Convection in the Mantle?*, Reviews of Geophysics *32*, 115–137.

MONTAGNER, J. P., and ANDERSON, D. L. (1989), *Constrained Reference Mantle Model*, Phys. Earth Planet. Int. *58*, 205–227.

MONTAGNER, J.-P., and NATAF, H. C. (1986), *A Simple Method for Inverting the Azimuthal Anisotropy of Surface Waves*, J. Geophys. Res. *91*, 511–520.

MONTAGNER, J.-P., and NATAF, H. C. (1988), *Vectorial Tomography—I. Theory*, Geophys. J. R. Astr. Soc. *94*, 295–307.

MONTAGNER, J.-P., and JOBERT, N. (1988), *Vectorial Tomography—II. Application to the Indian Ocean*, Geophys. J. *94*, 309–344.

MONTAGNER, J.-P., and TANIMOTO, T. (1991), *Global Upper Mantle Tomography of Seismic Velocities and Anisotropies*, J. Geophys. Res. *96*, 20,337–20,351.

NICOLAS, A., and CHRISTENSEN, N. I., *Formation of anisotropy in upper mantle peridotites—A review*. In *Composition, Structure and Dynamics of the Lithosphere-Asthenosphere System* (eds. Fuchs, K., and Froidevaux, C.) (Geodyn. Ser. AGU 16, 1987) pp. 111–123.

PLOMEROVÁ, J., ŠÍLENÝ, J., and BABUŠKA, V. (1996), *Joint Interpretation of Upper-mantle Anisotropy Based on Teleseismic P-travel Time Delay and Inversion of Shear-wave Splitting Parameters*, Phys. Earth Planet. Int. *95*, 293–309.

RAITT, R. W., *Seismic Refraction Studies of the Mendocino Fracture Zone* (Rep. MPL-U-23/63, Mar. Phys. Lab. Scripps Inst. of Oceanography, Univ. of Calif., San Diego, 1963).

RAITT, R. W., SHOR, G. G., JR., FRANCIS, T. J. G., and MORRIS, G. B. (1969), *Anisotropy of the Pacific Upper Mantle*, J. Geophys. Res. *74*, 3095–3109.

REISBERG, L., and LORAND, J.-P. (1995), *Longevity of Subcontinental Mantle Lithosphere from Osmium Isotope Systematics in Orogenic Peridotite Massifs*, Nature *376*, 159–162.

RIZNICHENKO, Y. V. (1949), *Seismic Quasi-anisotropy*, Izv. Akad. Nauk SSSR *13*, 518–544 (in Russian).

ROYDEN, L. H., BURCHFIELD, B. C., KING, R. W., WANG, E., CHEN, Z., SHEN, F., and LIU, Y. (1997), *Surface Deformation and Lower Crustal Flow in Eastern Tibet*, Science *276*, 788–790.

RUDNICK, R. L. (1995), *Making Continental Crust*, Nature *378*, 571–578.

SAVAGE, M. K., and SILVER, P. G. (1993), *Mantle Deformation and Tectonics: Constraints from Seismic Anisotropy in Western United States*, Phys. Earth Planet. Int. *78*, 207–227.

SHIMAMURA, H., ASADA, T., SUYEHIRO, K., YAMADA, T., and INATANI, H. (1983), *Longshot Experiments to Study Velocity Anisotropy in the Oceanic Lithosphere of the Northwestern Pacific*, Phys. Earth Planet. Int. *31*, 348–362.

SHOR, G. G., JR., and POLLARD, D. D. (1964), *Mohole Site Selection Studies North of Mani*, J. Geophys. Res. *69*, 1627–1637.

ŠÍLENÝ, J., and PLOMEROVÁ, J. (1996), *Inversion of Shear-wave Splitting Parameters to Retrieve Three-dimensional Orientation of Anisotropy in Continental Lithosphere*, Phys. Earth Planet. Int. *95*, 277–292.

SILVER, P. G. (1996), *Seismic Anisotropy beneath the Continents: Probing the Depths of Geology*, Annu. Rev. Earth Planet. Sci. *24*, 385–432.

SILVER, P. G., and CHAN, W. W. (1991), *Shear-wave Splitting and Subcontinental Mantle Deformation*, J. Geophys. Res. *96*, 429–454.

SILVER, P. G., MEYER, R. P., and JAMES, D. E. (1993), *Intermediate-scale Observations of the Earth's Deep Interior from APT89 Transportable Teleseismic Experiment*, Geophys. Res. Lett. *20*, 1123–1126.

SMITH, M. L., and DAHLEN, F. A. (1973), *The Azimuthal Dependence of Love- and Rayleigh-wave Propagation in a Slightly Anisotropic Medium*, J. Geophys. Res. *78*, 3321–3333.

SNYDER, D. B., LUCAS, S. B., and MCBRIDE, J. H., *Crustal and mantle reflectors from Palaeoproterozoic orogens and their relation to arc-continent collisions*. In *Precambrian Crustal Evolution in the North Atlantic Region* (ed. Brewer, T. S.) (Geol. Soc. Spec. Publ. No. 112, 1996) pp. 1–23.

TOMMASI, A., VAUCHEZ, A., and RUSSO, R. (1996), *Seismic Anisotropy in Ocean Basins: Resistive Drag of the Sublithopsheric Mantle?*, Geophys. Res. Lett. *23*, 2991–2994.

VINNIK, L. P., MAKEEVA, L. I., MILEV, A., and USENKO, Y. (1992), *Global Patterns of Azimuthal Anisotropy and Deformation in the Continental Mantle*, Geophys. J. Int. *111*, 433–447.

ZANDT, G., and AMMON, CH. J. (1995), *Continental Crust Composition Constrained by Measurements of Crustal Poisson's Ratio*, Nature *374*, 152–154.

(Received January 18, 1997, revised July 1, 1997, accepted August 10, 1997)

To access this journal online:
http://www.birkhauser.ch

Pure appl. geophys. 151 (1998) 281–303
0033–4553/98/040281–23 $ 1.50 + 0.20/0

Pure and Applied Geophysics

On Presence of Seismic Anisotropy in the Asthenosphere beneath Continents and its Dependence on Plate Velocity: Significance of Reference Frame Selection

ATSUKI KUBO[1] and YOSHIHIRO HIRAMATSU[2]

Abstract—We examine the possibility of seismic anisotropy in the asthenosphere due to present plate motion using SKS splitting results. The fast directions of anisotropy correlate weakly with the directions of the absolute plate motion (APM) for all APM models. Weak correlation indicates the possibility of asthenospheric anisotropy as well as frozen anisotropy in the lithosphere. Detection of strain rate dependence of anisotropy is helpful to further conclusion of the problem. The selection of reference frame is important to describe shear deformation in the asthenosphere beneath continent due to plate motion. The behavior of hot spots to the mesosphere, fixed or drifted by mantle return flow, is a key of the selection of the reference frame. For the NNR-NUVEL1 model, APM correlated anisotropy appears only at plate velocity faster than 1.4 cm/yr. It suggests the new possibility of the formation of asthenospheric anisotropy in addition to frozen anisotropy in the lithosphere. A critical plate velocity for the formation of anisotropy can be caused by the dislocation-diffusion transition as a function of strain rate on a deformation mechanism map of the upper mantle olivine.

Key words: SKS splitting, asthenosphere, plate motion, dislocation-diffusion transition.

Introduction

Seismic anisotropy is recognized as able to draw flow line in the upper mantle. This interpretation has been derived mainly from studies of the ocean floor (e.g., HESS, 1964; FORSYTH, 1975; MONTAGNER and TANIMOTO, 1991). CHRISTENSEN and SALISBERY (1979) support this interpretation of seismic observation from fabric analysis of mantle material for ophiolite complex. A simple shear experiment in high temperature and high pressure environment directly supports this interpretation (ZHANG and KARATO, 1995). A axes of olivine align parallel to the flow direction of the mantle.

The analysis of SKS wave splitting reveals seismic azimuthal anisotropy with higher horizontal resolution than those of long-range refraction studies of P waves and tomographic studies of surface waves. The disadvantage of this method is poor resolution of anisotropy with depth. Although SKS splitting samples seismic

[1] National Institute of Polar Research, Kaga, Itabashi, Tokyo, 173, Japan.
[2] Department of Earth Sciences, Kanazawa University, Kakuma-machi, Kanazawa, 920–11, Japan.

anisotropy between surface to core-mantle boundary, the dominant seismic anisotropy may be restricted in the upper mantle by other techniques (MONTAGNER and TANIMOTO, 1991; MONTAGNER and KENNET, 1996). The dominant frozen anisotropy in the lithosphere was proposed as the cause of SKS wave splitting from a world-wide review of SKS splitting studies (e.g., SILVER, 1996). The recognition of the asthenospheric anisotropy, as well as the dominant anisotropy in the lithosphere in frozen form, should be important for applications to geodynamic studies. Precise separation of asthenospheric anisotropy is also important for lithospheric mantle dynamics of past orogenic study. The present status of advanced analysis of shear-wave splitting extends to the treatment of two-layer anisotropy (SAVAGE and SILVER, 1993; Özaleybey and SAVAGE, 1995). Such analysis could potentially give splitting parameters separated into both in the lithosphere and asthenosphere. The results of two-layer anisotropy are insufficient for world-wide discussion because sufficient coverage of backazimuth is required for two-layer anisotropy analysis. We discuss the possibility of asthenospheric anisotropy using accumulated results derived from single-layer analysis of SKS splitting.

The correlation between the fast polarization direction (FPD) and the direction of the absolute plate motion (APM) has been discussed (Fig. 1) to separate the contribution of anisotropy formed by the present process in the asthenosphere from that in the lithosphere by fossil deformation (VINNIK et al., 1989; VINNIK et al., 1992; SILVER and CHAN, 1991; SILVER, 1996). Recent space geodesy (SOUDARIN and CAZENAVE, 1995; LARSON et al., 1997) revealed that the plate motions of the continents were similar to those predicted by a plate motion model on no net rotation (NNR) frame (ARGUS and GORDON, 1991). Shear deformation in the asthenosphere due to present plate motion is one of the candidates to explain observed anisotropy (hereafter we call this deformation simple asthenospheric flow: SAF).

The correlation between FPD and APM was discussed and two extreme conclusions were obtained. VINNIK et al. (1992) concluded that the present mantle flow forms anisotropy in the asthenosphere at stations around the world from the correlation between FPD and APM. On the other hand, many studies (e.g., SILVER and CHAN, 1988, 1991; SILVER, 1996) found that the dominant anisotropy is in the lithosphere because of good correlation between local geologic trends and FPDs. However, continents in fact show plate motions with respect to the deep mantle as revealed by space geodesy, although continents generally behave as anchors to the plate motions (STODDARD and ABBOTT, 1996). Thus, a shear zone due to present SAF should be defined beneath the continental lithosphere.

In previous studies, APM models are used for comparison with seismic anisotropy without a positive reason for the model selection. For example, AM1-2 (MINSTER and JORDAN, 1978) is used by VINNIK et al. (1992) and SILVER (1996); HS2-NUVEL1 (GRIPP and GORDON, 1990) is used by KANESHIMA and SILVER

(1994), and JAMES and ASSUMPÇÃO (1996); NNR-NUVEL1 (ARGUS and GORDON, 1991) is used by BORMANN *et al.* (1996). In this study we also discuss the effect of model selection on the shear deformation and focus on strain-rate dependence of APM correlated anisotropy.

Data

We prepared a data set of the SKS splitting analysis, adding results tabulated in Table 1 to those listed in the appendix of SILVER (1996). The total number of stations is 343. Each of these results is presented by two splitting parameters ϕ: the FPD and δt: the delay time of two split waves. In Table 1, we adopt results of single-layer analysis of ÖZALEYBEY and SAVAGE (1995) rather than two-layer anisotropy tabulated in the appendix of SILVER (1996) for North America, to avoid duplicated counting for frequency distribution of FPD. To confirm the quality of the SKS splitting analysis, we compare the splitting parameters at the stations studied by several researchers. Figures 2a,b show histograms of standard deviations of splitting parameters at 44 stations where numerous results are obtained. Clear reduction of frequency is recognized at standard deviations larger than 20° and 0.3 s, respectively. SKS splitting results have sufficient accuracy to discuss the formation of anisotropy in spite of differences of data selection and methods by each

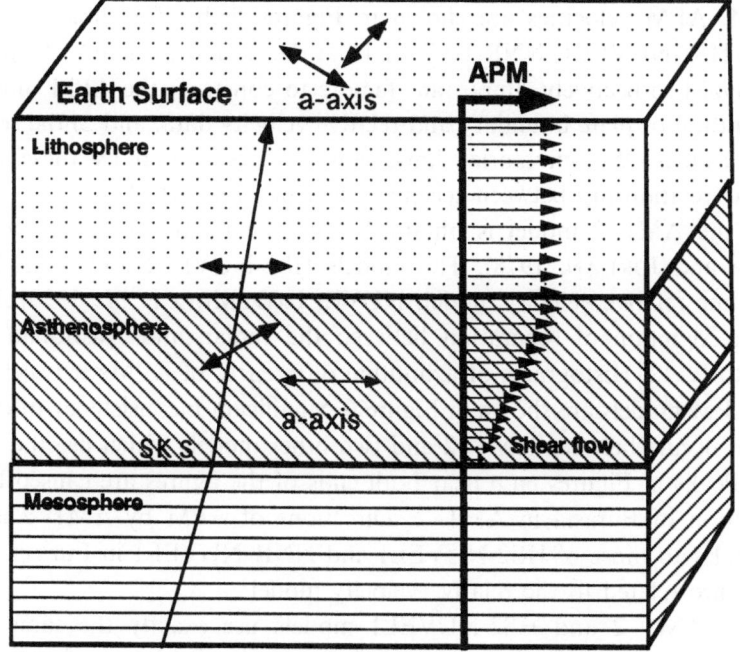

Figure 1
Cartoon of formation of anisotropy both in the asthenosphere due to present plate motion and in the lithosphere due to fossil frozen deformation, and setting of SKS splitting.

researcher. We exclude data which show standard deviation larger than 20° or 0.3 s. All results used in this study are shown in Figure 3. To avoid the effect of inhomogeneous station distribution, we use the spatially averaged splitting parameters at the grid points of equal area spacing (5° × 5° spacing at the equator). Spatially averaged splitting parameters within 2.5° distance from each grid point are shown in Figure 4. For averaging ϕ, we use the following equation:

$$\bar{\phi} = 1/2 \tan^{-1}\left(\frac{\displaystyle\sum_{i=1}^{n} \sin 2\phi_i}{\displaystyle\sum_{i=1}^{n} \cos 2\phi_i}\right),$$

where ϕ_i are measurements of FPD at the i-th station, and n is the number of data. By using the spatially averaged splitting parameters, the histogram of angle between FPD and APM exhibits a meaningful correlation between seismic anisotropy and plate motion. We exclude averaged splitting parameters which show standard deviation greater than 45°, because these results strongly reflect the lateral variation of FPDs.

APM Models and Motions for Comparison with Seismic Anisotropy

We use three APM models to discuss the directional deviation between FPD and APM. APM velocities are shown in Figures 5a,b,c for three APM models (AM1–2; HS2-NUVEL1; NNR-NUVEL1), at the stations, where SKS splitting has been analyzed respectively. Important factors to construct the APM model are 1) the recognition of plate boundary and relative velocity data, and 2) the selection of reference frames. We summarize important features caused by these factors to discuss the relationship between seismic anisotropy and APM.

First, the AM1–2 model is obtained from the relative velocity model RM2 (MINSTER and JORDAN, 1978). The models HS2-NUVEL1 and NNR-NUVEL1 use the same relative velocity model, NUVEL1 (DEMETZ et al., 1990). NUVEL1 includes recognition of separation between the Indian and Australian plates in addition to RM2. Histograms of azimuthal differences of APMs between different pairs of APM models are shown in Figure 6. Significant inconsistency of directional estimation occurs between AM1–2 and two other models which use the NUVEL1 data set. Peaks in Figures 6a,b near both ends of the figures are caused by a poor determination of Eurasia in AM1–2. Directions of APM by AM1–2 are nearly perpendicular to those of HS2-NUVEL1 and NNR-NUVEL1 in Eurasia. This is a major feature related to the relative velocity model.

Second, AM1–2 and HS2-NUVEL1 models use exactly the same hot spot reference frame. Six of nine hot spots used to construct a hot spot frame are in the Pacific (MINSTER and JORDAN, 1978). In contrast, the NNR-NUVEL1 model is constructed on the frame of no net rotation (LLIBOUTRY, 1974). It is derived from

Table 1

Data set of SKS splitting used in this study, except data already tabulated in SILVER *(1996)*

Stn.	Lat. (°)	Lon. (°)	δt (s)	ϕ (°)	Plate	Ref.
MOX	50.7	11.6	0.7	100	EUR	VI92, BOR
NE04	52.81	6.67	0.63	105	EUR	AL
NE15	50.87	5.78	0.83	83	EUR	AL
NE31	52.26	7.01	0.5	108	EUR	AL
NE32	52.18	5.81	0.8	114	EUR	AL
NE34	52.31	7.77	0.8	91	EUR	AL
NE38	52.50	6.26	0.6	114	EUR	AL
NE40	51.97	6.73	0.4	61	EUR	AL
BJI	40.0403	116.175	0.80	110	EUR	ZG
ENH	30.2718	109.4868	0.7	37	EUR	ZG
KMI	25.1233	102.74	1.00	35	EUR	ZG
MDJ	44.6164	129.5918	1.2	118	EUR	ZG
LSA	29.7008	91.1167	1.6	126	EUR	ZG
SSE	31.0956	121.1867	1.7	121	EUR	ZG
PDA	32.379	−64.6811	1.0	338	NAM	KF
BAR	32.680	−116.672	0.9	68	NAM	LD
CAL	34.143	−118.628	1.4	86	NAM	LD
CCR	33.776	−117.106	1.2	90	NAM	LD
DGR	33.650	−117.009	1.0	85	NAM	LD
GAV	34.022	−117.512	1.3	83	NAM	LD
GSA	34.137	−118.127	1.3	75	NAM	LD
LJB	34.591	−117.848	0.8	90	NAM	LD
PLS	33.796	−117.608	0.9	79	NAM	LD
RMM	34.643	−116.624	1.1	99	NAM	LD
RPV	33.744	−118.404	1.2	92	NAM	LD
SCI	34.980	−118.547	1.5	70	NAM	LD
USC	34.019	−118.285	1.8	74	NAM	LD
VTV	34.567	−117.333	1.2	95	NAM	LD
SAO	36.765	−121.445	1.73	111	NAM	OS
BMN	40.431	−117.222	2.00	81	NAM	OS
BKS	37.877	−122.235	1.58	120	NAM	OS
MHC	37.342	−121.642	1.53	122	NAM	OS
STA	37.404	−122.174	1.54	117	NAM	OS
SVD	34.104	−117.097	1.43	88	NAM	OS
PKD	35.889	−120.420	1.60	93	NAM	OS
ZOBO	−16.3	−68.1	0.8	120	SAM	VI92
HSPV	9.87	−64.17	1.3	108	SAM	RS
MNW	9.950	−64.030	1.3	101	SAM	RS
HDC2	10.270	−84.117	0.0	000	SAM	RS
SYO	−69.0	39.6	0.7	49	ANT	KU
MAW	−67.604	62.871	0.35	6	ANT	KH
DRV	−66.665	140.010	1.2	93	ANT	KH
PMSA	−64.775	−64.048	2.15	89	ANT	KH
VNDA	−77.517	161.853	1.00	51	ANT	KH

AL: ASINA and SNIEDER, 1995; ZG: ZHENG and GAO, 1994; VI92: VINNIK *et al.*, 1992; LD: LIU *et al.*, 1995; RS: RUSSO and SILVER, 1994; OS: ÖZALEYBEY and SAVAGE, 1995; BO: BORMANN *et al.*, 1993; KU: KUBO *et al.*, 1995; KH: KUBO *et al.*, 1996.

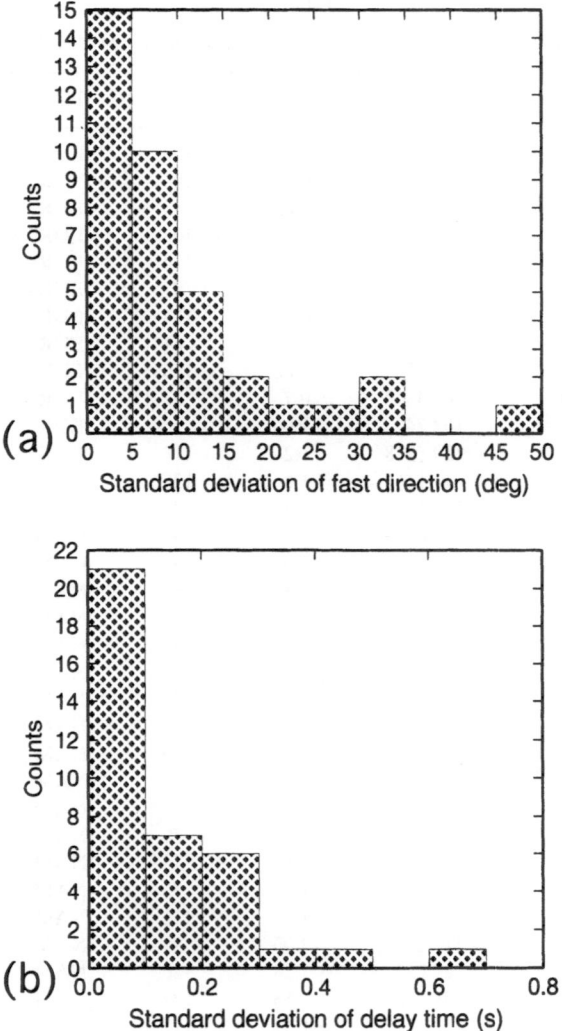

Figure 2
Histograms of standard deviations of splitting parameters at 44 stations where results of SKS splitting
are not singular: (a) fast direction (ϕ), (b) delay time (δt).

a homogeneous drag at the base of the lithosphere and unsystematic attachment of
the subduction zone in addition to no-net torque of the earth's shell (ARGUS and
GORDON, 1991). The comparison of plate velocity at stations, where SKS splitting
is analyzed is shown in Figures 7a,b,c for pairs of APM models. A typical feature
is the difference in velocity in the slow plate velocity range (0–3 cm/yr) between the
NNR-NUVEL1 model and two other models, which use the hot spot frame. In
contrast, velocities predicted by AM1–2 coincide well with velocities based on the
HS2-NUVEL1 model. This feature is important to recognize strain-rate dependent
phenomena in the asthenosphere.

Directional Deviations between APM and FPD

We examine the frequency distribution of the directional deviation angle between APM and FPD. Histograms of the whole data set show peaks of correlation between APM and FPD near 0° (Figs. 8a,b,c) for three APM models. These peaks correspond to the peak in VINNIK et al. (1992), which lead their conclusion to the dominant origin of anisotropy in the asthenosphere. However, these peaks include the effects of a large number of stations in North America, Europe and Central Eurasia. The peaks are also affected by the casual coincidence between fossil geologic trends and APMs in North America, as pointed out by SILVER and CHAN (1991) and SILVER (1996). If we exclude the data of North America as done by SILVER (1996), and adopt spatially averaged splitting parameters, we obtain a weak and slightly deviated correlation between APM and FPD as shown in Figures 9a,b,c.

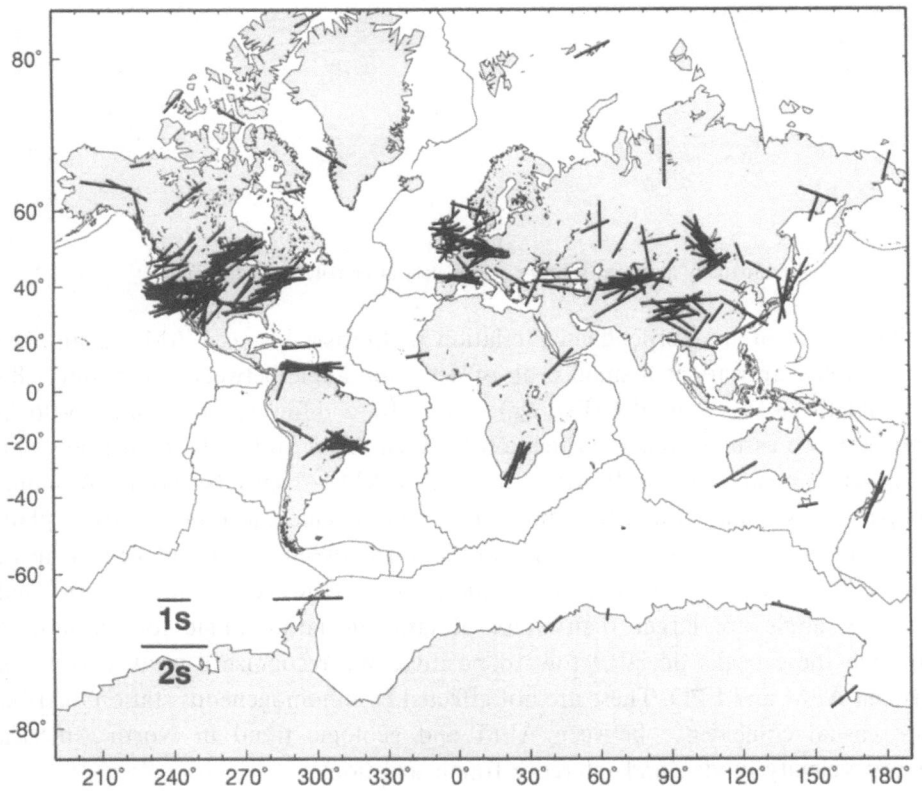

Figure 3
Compiled data set of SKS splitting at 353 stations, published before summer of 1996.

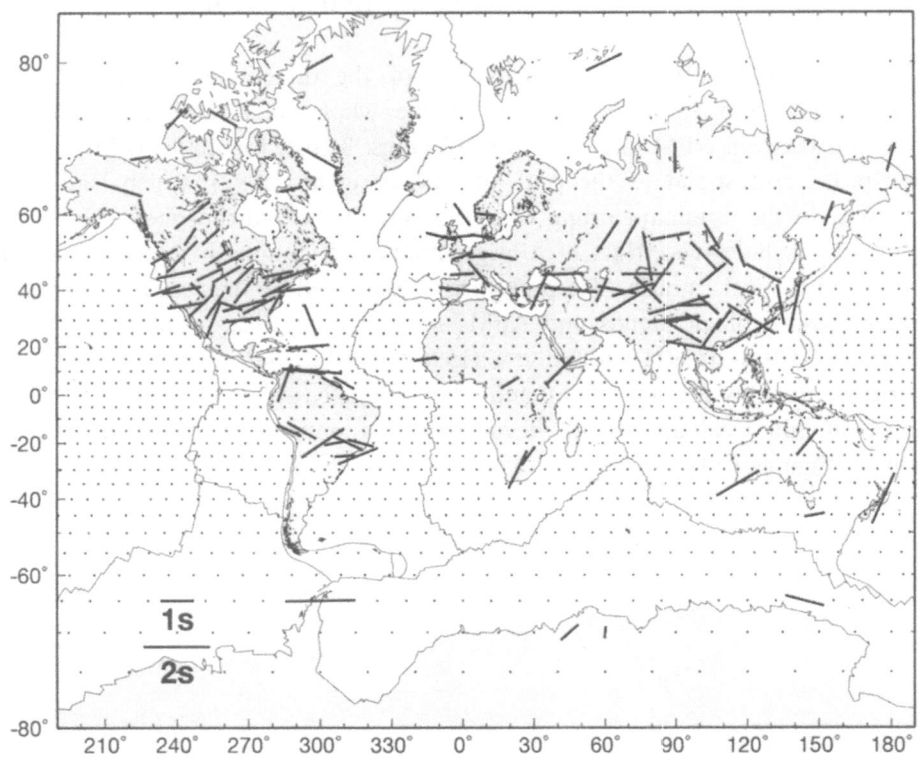

Figure 4
Spatially averaged splitting parameters at equally spaced points.

The effect of the perpendicular relation in Eurasia between AM1–2 and two other models appears as a small peak of deviation angle between −60° and −80° for only the AM1–2 model (Fig. 9a) caused by a difference of relative velocity models. If we assume simple correlation between APM and FPD, dual peaks only for AM1–2 mean that the Eurasian motion of AM1–2 may be wrong. A similar comparison was made for the oceanic lithosphere (Pacific and Indo-Australia plate) (Fig. 4 of MONTAGNER, 1994). The correlation between APM and FPD beneath continents is weaker than that of oceanic regions. However, frequencies of small deviation angle are larger than those of large deviation angle for all models, although these peaks deviated toward positive. We recognize a weak correlation between APM and FPD. These are not affected by inhomogeneous station distribution, casual coincidence between APM and geologic trend in North America, relative velocity model, and reference frame selection.

The weak correlation certainly indicates that dominant anisotropy exists in the lithosphere (see SILVER, 1996), rather than in the asthenosphere (VINNIK et al., 1992). On the other hand, we cannot reject the possibility of the existence of asthenospheric anisotropy overlapped with dominant frozen anisotropy in the lithosphere.

Plate Velocity versus Delay Time Relations of APM Correlated and Uncorrelated Anisotropy

Because of a limitation of the discussion based on directional deviation, we introduce a plot of plate velocity vs. delay time relation with the distinction of APM correlated or uncorrelated anisotropy (Fig. 10). We do not use AM1–2, because of the efficiency of the NUVEL1 model based on the F distribution (DeMetz *et al.*, 1990) and the anomalous peak which appeared in Figure 9a. We determine APM correlated and uncorrelated anisotropies according to whether the deviation angle between APM and FPD is smaller than 30° or not. Results of anisotropy in North America are included in this process because APM directions by NNR-NUVEL1 do not necessarily coincide with the trend of the Appalachian

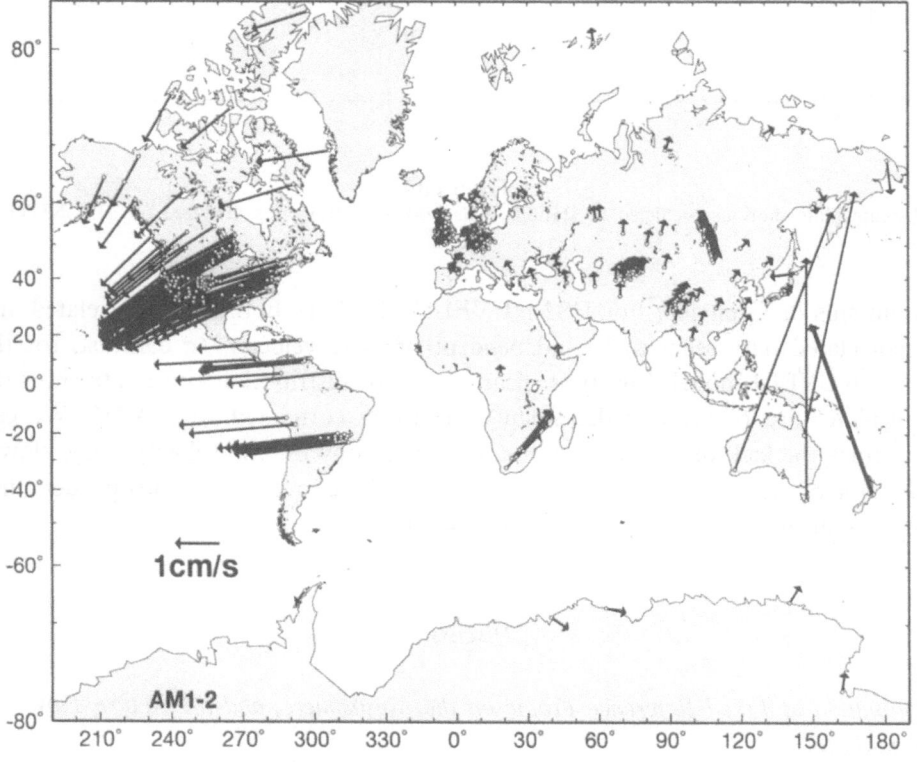

Figure 5(a)
Absolute plate motions predicted at stations, where SKS splitting was analysed for AM1–2 model.

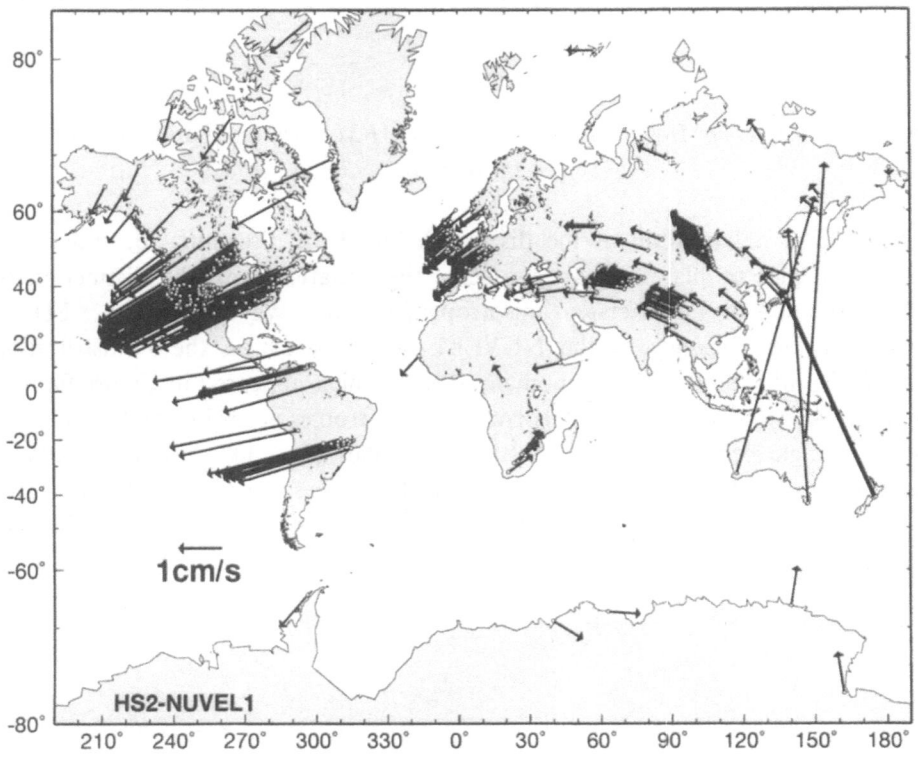

Figure 5(b)
Absolute plate motions predicted at stations, where SKS splitting was analyzed for HS2-NUVEL1 model.

mountains or Grenville. For HS2-NUVEL1 (Fig. 10a), both APM correlated and uncorrelated data seem to be independent of plate velocity. In contrast, for the NNR-NUVEL1 model (Fig. 10b), clear systematic distribution appears to show the typical feature of velocity dependent anisotropy correlated with APM. We can recognize the lack of the APM correlated anisotropy for the velocity range slower than 1.4 cm/yr. Delay time distribution of APM correlated anisotropy does not depend on plate velocity for both APM models.

Discussion

Estimation of Fixed Reference Frame on the Mesosphere, and the Return Flow Hypothesis

We discuss how the formation of anisotropy depends on the plate velocity. We need the reference frame which is fixed on the mesosphere. The selection of the

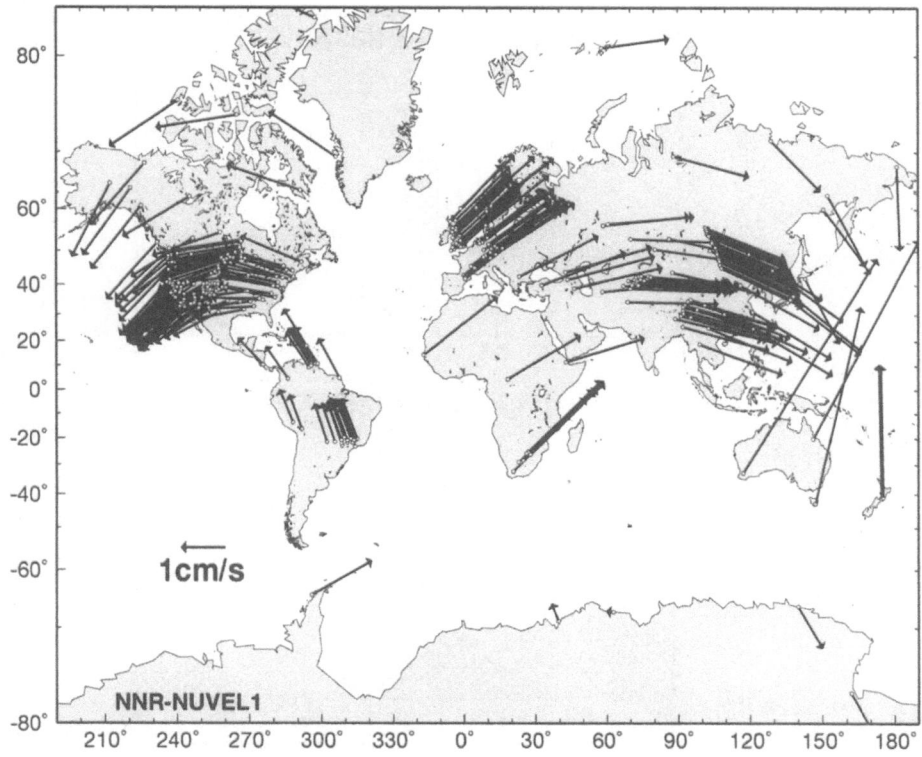

Figure 5 (c)

Absolute plate motions predicted at stations, where SKS splitting was analyzed for NNR-NUVEL1 model.

reference frame is an important factor to estimate shear deformation by present SAF beneath continents. We should admit that it is difficult to estimate the mesosphere directly beneath continent for SAF prediction, because the Pacific plate is very important to construct both the hot spot reference frame and the no-net rotation frame.

IHINGER (1995) discussed mantle flow in the upper mantle at Hawaii-Emperor seamounts. He recognized that the observed *echelon* structure of the hot spot track is due to drift by horizontal mantle flow. He proposed a mantle flow in the asthenosphere which shows an opposite direction against the lithospheric motion (Fig. 11a). The same kind of velocity relation at Hawaii is obtained from APM models with different reference frames (Fig. 11b). By correcting horizontal flow in the upper mantle, he can fix hot spot plume on the deep mantle (Fig. 11c). It is one of the estimation of reference frames fixed to the mesosphere beneath the Pacific plate. The NNR reference frame gives similar estimation to Ihinger's reference frame fixed to the mesosphere with respect to the motion of Hawaii. GORDON (1995) proposed that the difference between the hot spot frame and the NNR frame is caused by 1) horizontal drift of the hot spot in

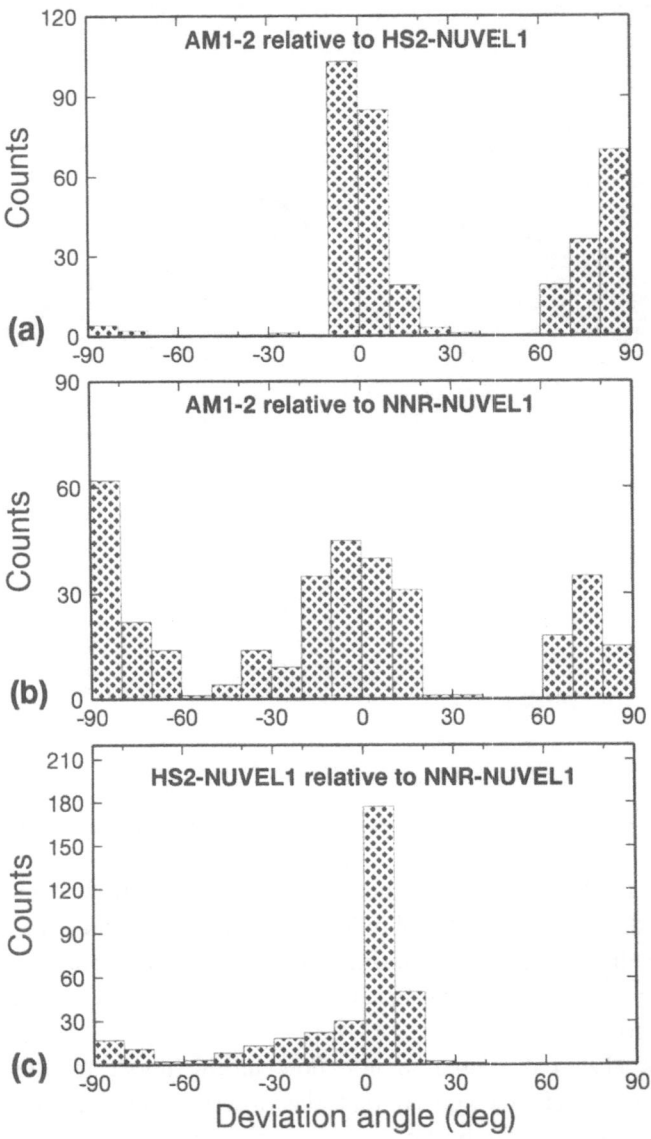

Figure 6

Comparison of the APM directions for different pairs of models at stations, where the result of SKS splitting was obtained. (a) Histogram of deviation angle between APMs measured from direction of HS2-NUVEL1 to that of AM1–2, (b) same as (a) but being measured from direction of NNR-NUVEL1 model to that of AM1–2 model, (c) same as (a) but being measured from direction of NNR-NUVEL1 to that of HS2-NUVEL1.

the upper mantle or 2) breakdown of the assumption for constructing the no-net rotation frame. Ihinger's study supports the former possibility.

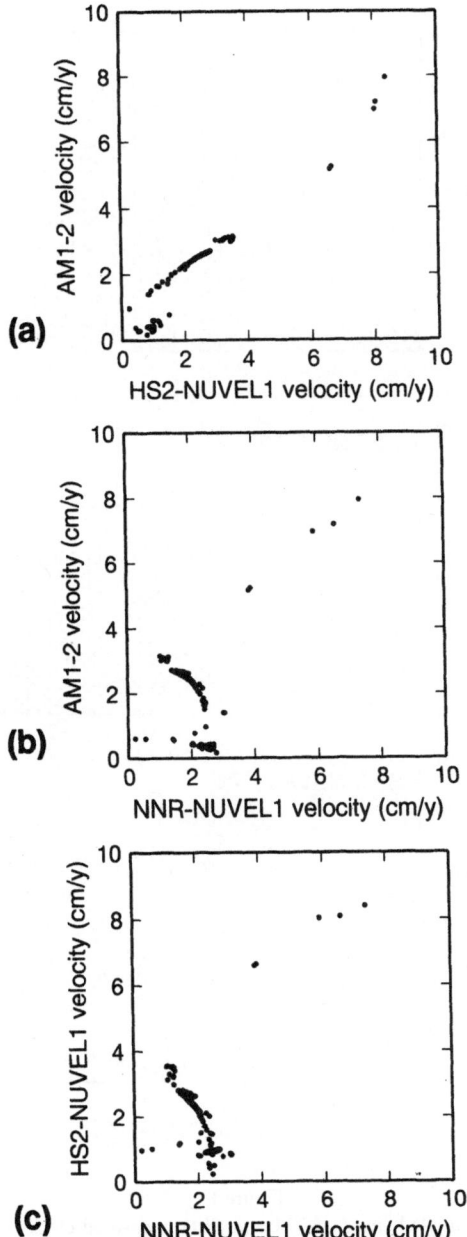

Figure 7
Comparison of magnitude of the APM velocities predicted by different models. (a) HS2-NUVEL1 vs.
AM1–2, (b) NNR-NUVEL1 vs. AM1–2, and (c) NNR-NUVEL1 vs. HS2-NUVEL1.

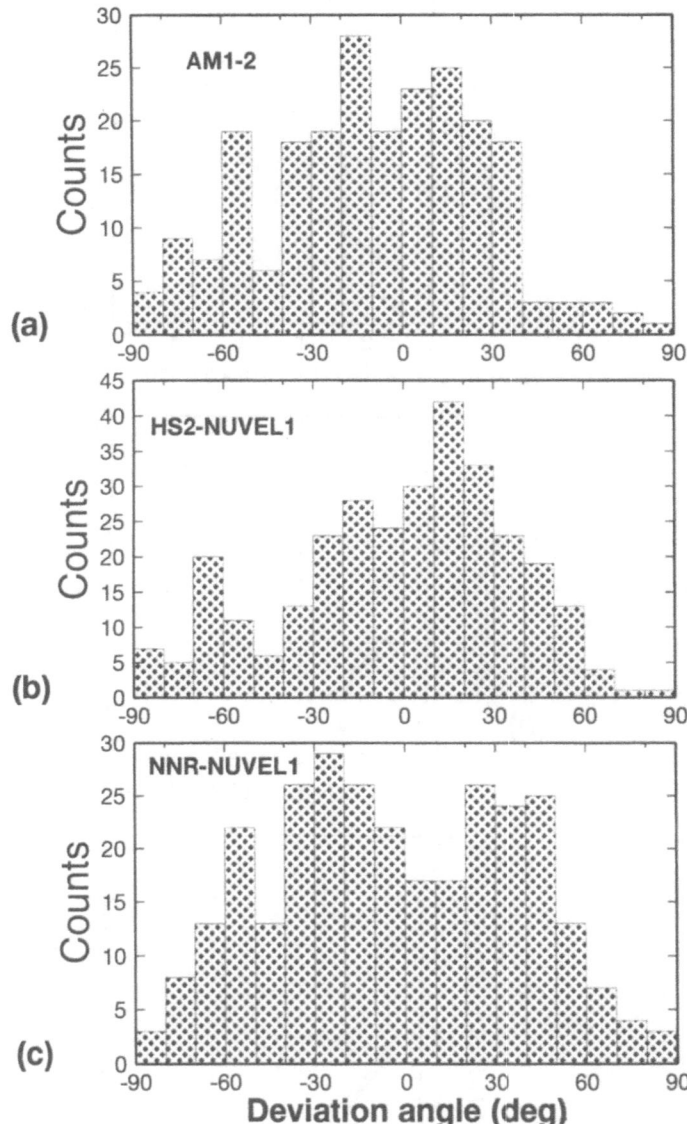

Figure 8
Histograms of the deviation angle between APM and FPD, measured clockwise from APM to FPD for
all data. (a) AM1–2 model, (b) HS2-NUVEL1, and (c) NNR-NUVEL1.

The most simple interpretation of upper mantle flow which has the opposite
direction to the lithospheric motion is return flow as shown in Figure 11c (e.g.,
ALVAREZ, 1982; ARGUS and GORDON, 1991; RICARD et al., 1991; IHINGER, 1995).

The oceanic lithosphere has monotonously increased thickness with age (TANI-MOTO and ZHANG, 1990). Systematically inclined topography of the lithosphere-asthenosphere boundary (LAB) generates a horizontal pressure gradient (Fig. 12).

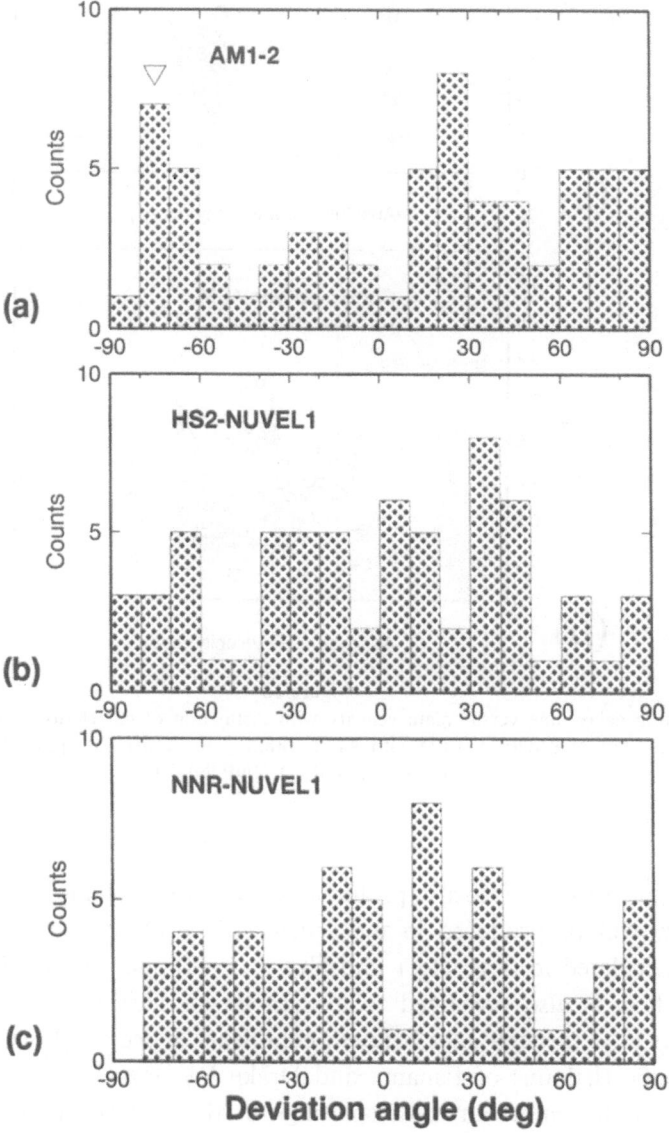

Figure 9
Histograms of the deviation angle from APM to FPD, using data of spatially averaged and excluded North American data. (a) AM1-2 model; a peak marked by reversed triangle may be the effect of Eurasian motion which is a perpendicular relation to those of other models, (b) HS2-NUVEL1 model, and (c) NNR-NUVEL1 model.

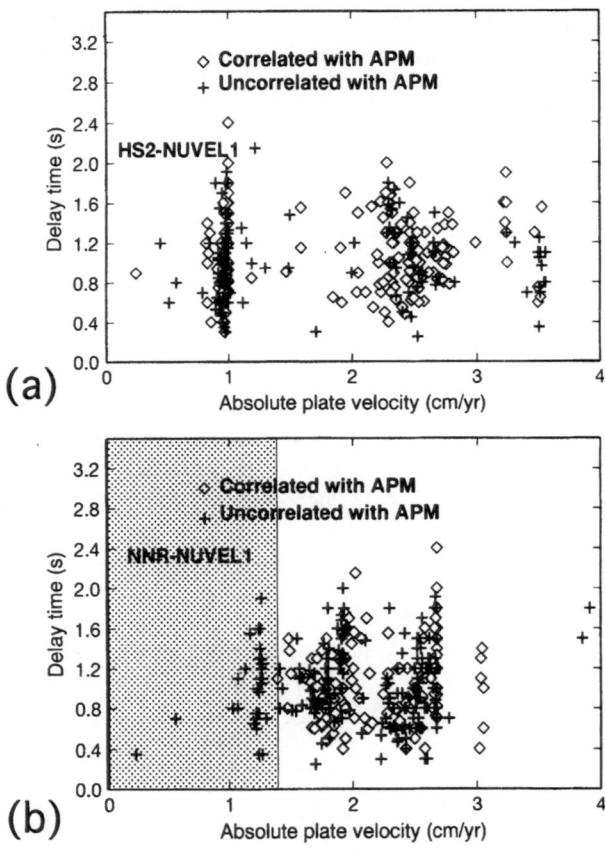

Figure 10
Plot of splitting delay time versus plate velocity with distinction of correlation with APM: diamond corresponds to correlated data of FPD with APM within ±30°, other data plotted as crosses, (a) for HS2-NUVEL1 model, and (b) NNR-NUVEL1 model.

If masses of vertical velocity profile balance within shallower upper mantle, return flow will be produced in the asthenosphere (TURCOTTE and SCHUBERT, 1982) as calculated in SCHUBERT and YUEN (1978) and CHASE (1979). The idea of return flow is also presumed to explain Pacific shrinkage (ALVAREZ, 1982). These flows pass through gaps between continents in the southeastern part of the circum-Pacific (Isthmus of Panama and Drake Passages and Australia-Antarctic Discordance). Recent shear-wave splittings in the eastern circum-Pacific region (Caribbean Sea: RUSSO et al., 1996, South-American subduction zone: RUSSO and SILVER, 1994, Antarctic Peninsula: KUBO et al., 1996) are consistent with an E-W collision between the return flow and South America, although the westward motion of South America is an alternative explanation for trench parallel anisotropy.

Because the topography of the LAB beneath continents does not show a systematic shape compared with the oceanic plate, induction of the return flow beneath continent may be weaker. The direction of the return flow turns at the trench (RUSSO and SILVER, 1994). The net rotation of the earth corresponds to the return flow limited to beneath circum-Pacific, which has a large angular velocity and area. Thus, the return flow does not affect shear deformation beneath most

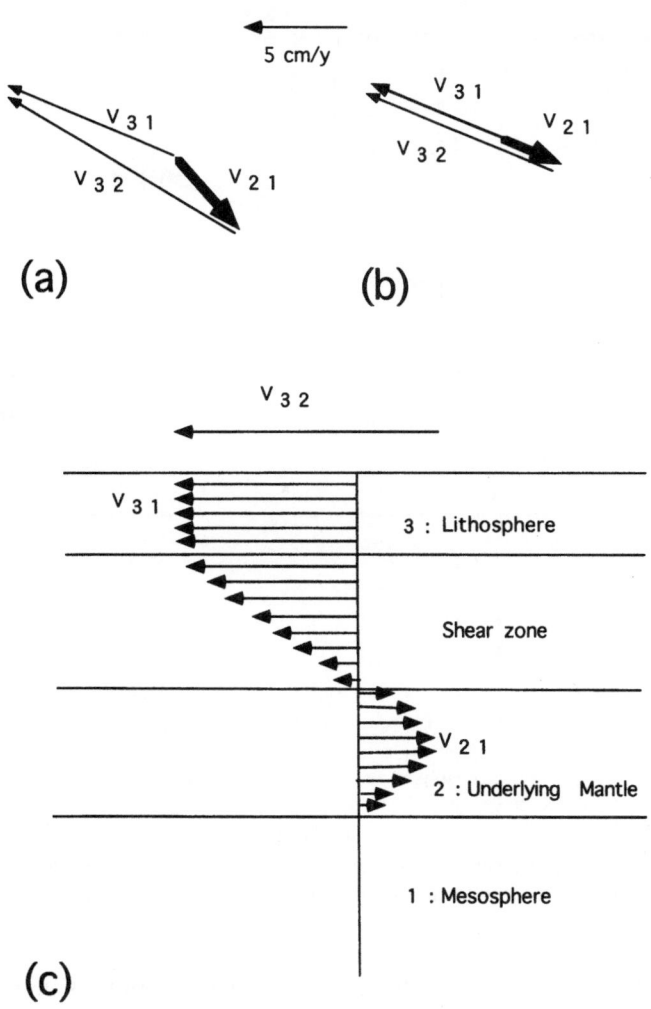

Figure 11

Velocity relations between three mantle layers (1: mesosphere, 2: asthenosphere, 3: lithosphere) at Hawaii. (a) Derived from data of Hawaiian hot spot track by IHINGER (1995), V_{32}: (12.5 cm/yr, 303°), V_{31}: (8.6 cm/yr, 292°), V_{21}: (4.3 cm/yr, 140°), (b) similar relation predicted by global APM models and NNR-NUVEL1) at Hawaii, V_{32}: (9.49 cm/yr, 302°), (HS2-NUVEL1), V_{31}: (6.95 cm/yr, 299°) (NNR-NUVEL1), V_{21}: (2.44 cm/yr, 120°) (differential rotation), (c) vertical velocity profile of lithospheric motion and return flow of underlying mantle on reference frame fixed on the mesosphere beneath the Pacific plate.

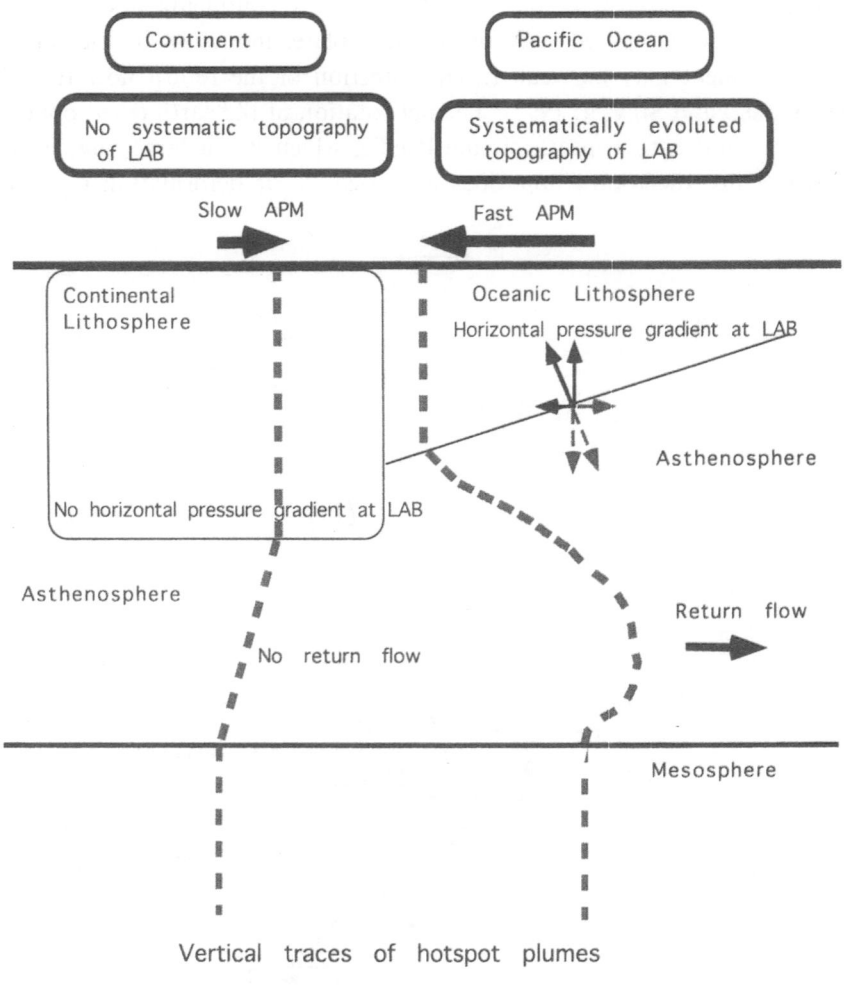

Figure 12
Schematic interaction between vertical plume and horizontal return flow beneath the Pacific plate and its effect to estimate shear deformation beneath a continent.

continents (Fig. 12). Even if the origin of net rotation is not limited in the circum-Pacific region, the preferred passage of the return flow through the channel between continents (ALVAREZ, 1982; DOGLIONI, 1994) suggests that the return flow does not strongly affect the estimation of the shear deformation at the continental basement. If the plumes of hot spots are drifted with the horizontal return flow, NNR frame becomes a better reference frame fixed to the mesosphere. Conversely, if the plume of hot spot is not affected by the horizontal mantle flow, a hot spot frame may be the best reference frame fixed to the mesosphere.

Strain Rate Dependence of Anisotropy Formation due to Present Plate Motion

To discuss the presence of asthenospheric anisotropy in addition to litho-spheric anisotropy, we attempt to detect the plate velocity dependent formation of observed seismic anisotropy. Strain rate in the asthenosphere becomes larger with increasing APM velocity and decreasing thickness of the asthenospheric shear zone. The velocities of continental plates have values between 1 and 6 cm/yr. On the other hand, if we assume that a shear zone due to the plate motion is formed above the 400 km discontinuity and the thickness of the lithosphere may vary from 100 to 300 km, then the thickness of the astheno-sphere may be between 100 and 300 km. Plate velocity relative to the meso-sphere is the fundamentally controlling parameter of strain rate due to present APM rather than the thickness of asthenosphere, although thickness variation of shear zone is not negligible.

However, the selection of the reference frame is sensitive to estimate strain rate at each of the stations. If the hot spot frame is a better reference frame, fixed to the mesosphere, no strain dependence is recognized (Fig. 10a). Thus, asthenospheric anisotropy is not required. Fossil anisotropy in the lithosphere dominates beneath continents. This conclusion is consistent with those of GA-HERTY and JORDAN (1995) and SILVER (1996). The estimation of plate velocities for hot spot fixed models is inversely correlated with that of NNR-NUVEL1 at a velocity slower than 3 cm/yr (Figs. 7b,c). If one reference frame is proper for strain-rate estimation of shear deformation, another reference frame will supply the wrong estimate. Clear lack of the APM correlated anisotropy requires forma-tion of asthenospheric anisotrpy faster than 1.4 cm/yr. The critical plate velocity 1.4 cm/yr corresponds to strain rate 0.43×10^{-14} s^{-1}, assuming the 100 km thickness of the shear zone. Plate velocity dependence of APM correlated an-isotropy appears only for NNR case (Fig. 10b). It also suggests that the NNR frame is better frame fixed to the mesosphere in addition to Ihinger's study on underlying mantle flow beneath the Hawaiian hot spot.

Critical Plate Velocity for Seismic Anisotropy Formation and Deformation Mechanism Transformation

We attempt to interpret a new possibility of asthenospheric anisotropy and its critical velocity based on the rheological property of olivine. Dislocation creep is important for the promotion of lattice preferred orientation (WENK, 1985). In contrast, pre-existing anisotropy is also diminished by a diffusion-controlled dis-location recovery process due to annealing with lapse time (KARATO et al., 1993). Thus dislocation-diffusion transition of deformation mechanisms plays an important role (KARATO, 1992; KARATO and WU, 1993) in the restriction of anisotropic zone. Two types of applciation are possible. One is the spatial (depth) variation of dislocation-diffusion transition of deformation mechanism by assuming strain rate, temperature profiles and other parameters (KARATO, 1992; KARATO and WU, 1993). Another is dislocation-diffusion transition as a function of strain rate (plate velocity) on the deformation mechanism map (Fig. 1 of KARATO, 1997). A typical deformation mechanism map is drawn in temperature-stress space with strain-rate contour. In this case, we can see the effect of plate velocity on the development of lattice preferred orientation.

As an application of the former property, several studies suggest that the depth of the dislocation-diffusion transition is a boundary between anisotropic and non-anisotropic regions, based on the relative depth relation between Lehmann discontinuity and LAB (KARATO, 1992; GAHERTY and JORDAN, 1995). If the LAB is deep as proposed by JORDAN (1975) (reaches 400 km), the dislocation-diffusion transition becomes shallower than LAB. The anisotropic zone exists only in the lithosphere as discussed by GAHERTY and JORDAN (1995). On the other hand, the thickness of the lithosphere is 200–250 km even in the cratonic area (ZHANG and TANIMOTO, 1993). For the relatively thinner lithosphere, KARATO (1992) pointed out that the dislocation creep mechanism dominates on both sides of LAB. The present-day formation of anisotropy becomes possible. The mismatch between thin and thick lithosphere may originate from problems such as isotropic-anisotropic treatment, initial model effect (PREM: DZIEWONSKI and ANDERSON, 1981) and the restrictions of anisotropy description.

The latter type of transition has not yet been applied to the formation of anisotropy, although this transition is predicted in the region of realistic strain rate due to plate motion. In our scheme, absolute depth estimations such as the depths of LAB and the diffusion-dislocation transition are not necessary to detect the direct response of the formation of anisotropy against SAF deformation. We recognized that distribution of APM correlated anisotropy requires a critical plate velocity (1.4 cm/yr) for motions on NNR frame. We deduce that the critical velocity corresponds to the dislocation-diffusion transition as a function of strain rate, assuming that plate velocity is the primarily controlling parameter for the strain rate of SAF. In the velocity range slower than 1.4 cm/yr, deformation may occur by diffusion creep. The lattice preferred orientation is not developed even if the uniform motion continues for a prolonged time. TOMMASI et al. (1996) calculate formation of asthenospheric anisotropy due to resistive shear deformation, for various plate velocities assuming the dislocation creep law. They obtained developed seismic anisotropy at all velocities. However, in a realistic situation, active flow law can change with the strain rate. If the deformation mechanism is dislocation creep, lattice preferred orientation develops with strain increasing (ZHANG and KARATO, 1995). Scattered distribution of delay times for the APM correlated data in Figure 10 may be caused by the accumulated effect of strain due to plate motion history depending on each continent.

This topic suggests that the localization of anisotropy exhibits various situations. Two extreme conclusions, which suggest the dominant origin in the lithosphere (e.g., SILVER, 1996) or in the asthenosphere (e.g., VINNIK et al., 1992; MOONEY, 1995), are not enough to describe anisotropy beneath continents. Our conclusion is important to understand the flow property of continental root. However an ambiguity remains for the selection of the reference frame. It should be concluded by further studies of relative motions between hot spots (MOLNAR and STOCK, 1987) and the origin of net rotation (STEINBERGER, 1996).

Conclusion

Global compiled data sets of SKS wave splitting demonstrate that the possibility of asthenospheric anisotropy cannot be rejected from the comparison between FPD and APM. Discussion of the strain-rate dependent formation of anisotropy due to the present SAF depends on the selection of the reference frame. If the hot spot frame is selected, no strain-rate dependent anisotropy is recognized, and observed SKS splitting mainly shows dominant lithospheric anisotropy. However, if we adopt the NNR frame, a critical plate velocity appears for APM correlated anisotropy. Asthenospheric formation of seismic anisotropy due to plate motion is required in addition to fossil anisotropy in the lithosphere. Critical plate velocity can be related to phase transformation of the deformation mechanism of upper mantle olivine.

Acknowledgments

We thank S. Pondrelli and an anonymous reviewer for their critical review and improvement of the manuscript. T. Seno critically read the manuscript. We thank A. Ibaraki for her assistance in compiling the data set.

REFERENCES

ALSINA, D., and SNIEDER, R. (1995), *Small-scale Sublithospheric Continental Mantle Deformation: Constraints from SKS Splitting Observations*, Geophys. J. Int. *123*, 431–448.

ALVAREZ, W. (1982), *Geological Evidence for the Geographical Pattern of Mantle Return Flow and the Driving Mechanism of Plate Tectonics*, J. Geophys. Res. *87*, 6697–6710.

ARGUS, D. E., and GORDON, R. G. (1991), *No-net-rotation Model of Current Plate Velocities Incorporating Plate Motion Model NUVEL-1*, Geophys. Res. Lett. *18*, 2039–2042.

BORMANN, P., BURGHARDT, P.-T., MAKEYEVA, L. I., and VINNIK, L. P. (1993), *Teleseismic Shear-wave Splitting and Deformations in Central Europe*, Phys. Earth Planet. Inter. *78*, 157–166.

BORMANN, P., GRUNTHAL, G., KIND, R., and MONTAG, H. (1996), *Upper Mantle Anisotropy beneath Central Europe from SKS Wave Splitting: Effects of Absolute Plate Motion and Lithosphere-asthenosphere Boundary Topography?* J. Geodynamics *22*, 11–32.

CHASE, C. G. (1979), *Asthenospheric Counter Flow: A Kinematic Model*, Geophys. J. Roy. Astr. Soc. *56*, 1–18.

CHRISTENSEN, N. I., and SALISBERY, M. H. (1979), *Seismic Anisotropy in the Oceanic Upper Mantle: Evidence of Bay of Island Ophiolite Complex*, J. Geophys. Res. *84*, 4601–4610.

DEMETS, C., GORDON, R. G., ARGUS, D. F., and STEIN, S. (1990), *Current Plate Motions*, Geophys. J. Int. *101*, 425–478.

DOGLIONI, C. (1994), *Foredeeps Versus Subduction Zones*, Geology *22*, 271–274.

DZIEWONSKI, A., and ANDERSON, D. L. (1981), *Preliminary Reference Earth Model*, Phys. Earth Plan. Int. *25*, 297–356.

FORSYTH, D. W. (1975), *The Early Structural Evolution and Anisotropy of the Oceanic Upper Mantle*, Geophys. J. Royal Astr. Soc. *43*, 103–162.

GAHERTY, J. B., and JORDAN, T. H. (1995), *Lehmann Discontinuity as the Base of an Anisotropic Layer Beneath Continents*, Science *268*, 1468–1471.

GORDON, R. G., *Present plate motions and plate boundaries*. In *Global Earth Physics—A Handbook of Physical Constants, 1* (ed. Ahrens, T. J.) (AGU, Washington 1995) pp. 66–87.

GRIPP, A. E., and GORDON, R. (1990), *Current Plate Velocities Relative to the Hot Spots Incorporating the NUVEL-1 Global Plate Motion Model*, Geophys. Res. Lett. *17*, 1109–1112.

HESS, H. (1964), *Seismic Anisotropy of the Uppermost Mantle under Oceans*, Nature *203*, 629–631.

IHINGER, P. D. (1995), *Mantle Flow beneath the Pacific Plate: Evidence from Seamount Segments in the Hawaiian-Emperor Chain*, Am. J. Science *295*, 1035–1057.

JAMES, D. E., and ASSUMPÇÃO, M. (1996), *Tectonic Implications of S-wave Anisotropy beneath SE Brazil*, Geophys. J. Int. *126*, 1–10.

JORDAN, T. H. (1975), *The Continental Tectosphere*, Rev. Geophys. Space Phys. *13*, 1–12.

LARSON, K. M., FREYMUELLER, J. T., and PHILIPSEN, S. (1997), *Global Plate Velocities from the Global Positioning System*, J. Geophys. Res. *102*, 9961–9982.

LIU, H., DAVIS, P. M., and GAO, S. (1995), *SKS Splitting beneath Southern California*, Geophys. Res. Lett. *22*, 767–770.

LLIBOUTRY, L. (1974), *Plate Movement Relative to Rigid Lower Mantle*, Nature *250*, 298–300.

KANESHIMA, S., and SILVER, P. G. (1994), *Anisotropic Loci in the Mantle beneath Central Peru*, Phys. Earth Planet. Inter. *88*, 257–272.

KARATO, S. (1992), *On the Lehmann Discontinuity*, Geophys. Res. Lett. *19*, 2255–2258.

KARATO, S., and WU, P. (1993), *Rheology of the Upper Mantle: A Synthesis*, Science *261*, 771–777.

KARATO, S., RUBIE, D. C., and YAN, H. (1993), *Dislocation Recovery in Olivine under Deep Upper Mantle Conditions: Implications for Creep and Diffusion*, J. Geophys. Res. *98*, 9761–9768.

KARATO, S., *Phase transformations and rheological properties of mantle minerals*. In *Earth's Deep Interior* (ed. Crossley, D. J.) (Gordon and Breach 1997) pp. 223–272.

KUBO, A., HIRAMATSU, Y., KANAO, M., ANDO, M., and TERASHIMA, T. (1995), *An Analysis of the SKS Splitting at Syowa Station in Antarctica*, Proc. NIPR Symp. Antarct. Geosci. *8*, 25–34.

KUBO, A., HIRAMATSU, Y., and KANAO, M. (1996), *Shear-wave Anisotropy by SKS Splitting in Antarctica*, 30th International Geological Congress Beijing.

MINSTER, J. B., and JORDAN, T. H. (1978), *Present-day Plate Motions*, J. Geophys. Res. *83*, 5331–5354.

MOLNAR, P., and STOCK, J. (1987), *Relative Motions of Hotspots in the Pacific, Atlantic and Indian Oceans since Late Cretaceous Time*, Nature *327*, 587–591.

MONTAGNER, J.-P. (1994), *Can Seismology Tell us Anything about Convection in the Mantle*, Rev. Geophys. *32*, 115–137.

MONTAGNER, J. P., and KENNET, B. L. N. (1996), *How to Reconcile Body-wave and Normal-mode Reference Earth Models*, Geophys. J. Int. *125*, 229–248.

MONTAGNER, J. P., and TANIMOTO, T. (1991), *Global Upper Mantle Tomography of Seismic Velocities and Anisotropies*, J. Geophys. Res. *96*, 20337–20351.

MOONEY, W. D. (1995), *Continental Roots Go with the Flow*, Nature *375*, 15.

ÖZALEYBEY, S., and SAVAGE, M. K. (1995), *Shear-wave Splitting beneath Western United States in Relation to Plate Tectonics*, J. Geophys. Res. *100*, 18135–18149.

RICARD, Y., DOGLIONI, C., and SABADINI, R. (1991), *Differential Rotation between Lithosphere and Mantle: A Consequence of Lateral Mantle Viscosity Variations*, J. Geophys. Res. *95*, 8407–8415.

RUSSO, R. M., and SILVER, P. G. (1994), *Trench-parallel Flow Beneath Nazca Plate from Seismic Anisotropy*, Science *263*, 1105–1111.

RUSSO, R. M., SILVER, P. G., FRANKE, M., AMBEH, W. B., and JAMES, D. E. (1996), *Shear-wave Splitting in Northeast Venezuela, Trinidad, and the Eastern Caribbean*, Phys. Earth Planet. Interior *95*, 251–275.

SAVAGE, M. K., and SILVER, P. G. (1993), *Mantle Deformation and Tectonics: Constraints from Seismic Anisotropy in the Western United States*, Phys. Earth Planet. Inter. *78*, 207–227.

SCHUBERT, G., and YUEN, D. A. (1978), *Mantle Circulation with Partial Shallow Return Flow: Effects of Stresses in Oceanic Plates and Topography of the Seafloor*, J. Geophys. Res. *83*, 745–758.

SILVER, P. G., and CHAN, W. W. (1988), *Implications for Continental Structure and Evolution from Anisotropy*, Nature *335*, 34–39.

SILVER, P. G., and CHAN, W. W. (1991), *Shear-wave Splitting and Subcontinental Mantle Deformation*, J. Geophys. Res. *96*, 16429–16454.

SILVER, P. G. (1996), *Seismic Anisotropy beneath the Continents: Probing the Depths of Geology*, Ann. Rev. Earth Planet. Sci. *24*, 385–432.

SOUDARIN, L., and CAZENAVE, A. (1995), *Large-scale Tectonic Plate Motions Measured with the DORIS Space Geodesy System*, Geophys. Res. Lett. *22*, 469–472.

STEINBERGER, B. M. (1996), *Motion of Hotspots and Changes of the Earth's Rotation Axis Caused by a Convecting Mantle*, Ph.D. Thesis, Harvard University.

STODDARD, P. G., and ABBOTT, D. (1996), *Influence of the Tectosphere upon Plate Motion*, J. Geophys. Res. *101*, 5425–5433.

TANIMOTO, T., and ZHANG, Y. S. (1990), *Lithospheric Thickness and Thermal Anomalies in the Upper Mantle Inferred from the Love Wave Data*, Geophys. Res. Lett. *17*, 2405–2408.

TOMMASI, A., VAUCHEZ, A., and RUSSO, R. (1996), *Seismic Anisotropy in Ocean Basins: Resistive Drag of the Sublithospheric Mantle?* Geophys. Res. Lett. *23*, 2991–2994.

TURCOTTE, D. L., and SCHUBERT, G., Chapter 6. Fluid mechanics. In *Geodynamics* (John Wiley and Sons 1982) pp. 231–237.

VINNIK, L. P., FARRA, V., and ROMANOWICZ, B. (1989), *Azimuthal Anisotropy in the Earth from Observations of SKS at GEOSCOPE and NARS Broadband Stations*, Bull. Seismol. Soc. Am. *79*, 1542–1558.

VINNIK, L. P., MAKEYEVA, L. I., MILEV, A., and USENKO, A. Y. (1992), *Global Patterns of Azimuthal Anisotropy and Deformations in the Continental Mantle*, Geophys. J. Int. *111*, 433–447.

VINNIK, L. P., GREEN, R. W. E., and NICOLAYSEN, L. O. (1995), *Recent Deformations of the Deep Continental Root beneath Southern Africa*, Nature *375*, 50–52.

WENK, H.-R., *Preferred Orientation in Metals and Rocks* (Academic Press 1985).

ZHANG, S., and KARATO, S. (1995), *Lattice Preferred Orientation of Olivine Aggregates Deformed in Simple Shear*, Nature *375*, 774–777.

ZHANG, Y.-S., and TANIMOTO, T. (1993), *High Resolution Global Upper Mantle Structure and Plate Tectonics*, J. Geophys. Res. *98*, 9793–9823.

ZHENG, S.-H., and GAO, Y. (1994), *Azimuthal Anisotropy in Lithosphere on Chinese Mainland from Observations of SKS at CDSN*, Acta Seism. Sin. *7*, 177–186.

(Received December 17, 1996, revised July 14, 1997, accepted July 17, 1997)

 To access this journal online:
http://www.birkhauser.ch

Pure appl. geophys. 151 (1998) 305–331
0033–4553/98/040305–27 $ 1.50 + 0.20/0

© Birkhäuser Verlag, Basel, 1998

❚ Pure and Applied Geophysics

Backazimuthal Variations of Splitting Parameters of Teleseismic SKS Phases Observed at the Broadband Stations in Germany

S. BRECHNER,[1,2] K. KLINGE,[1] F. KRÜGER[1,3] and T. PLENEFISCH[1]

Abstract—SKS phases observed at broadband stations in Germany show significant shear-wave splitting. We have analyzed SKS and SKKS phases for shear-wave splitting from 13 stations of the German Regional Seismic Network (GRSN), from 3 three-component stations of the Gräfenberg array (GRF) and from one Austrian station (SQTA). The data reveal strong differences in the splitting parameters (fast direction ϕ and delay time δt) from a single event at various stations as well as variations at the individual stations for events with different backazimuths. The backazimuthal variations of the splitting parameters at some stations can be explained by two-layer anisotropy models with horizontal symmetry axes. The best resolved two-layer model is the GRA1 model (upper layer: $\phi = 40°$, $\delta t = 1.15$ s; lower layer: $\phi = 115°$, $\delta t = 1.95$ s). The upper layer can be attributed to the lithosphere. Because of the magnitude of the delay time of the upper layer, the lower layer must lie within the asthenosphere. At other stations splitting parameters are consistent with an anisotropic one-layer model for the upper mantle. Stations near the Bohemian Massif show fast directions near EW. Throughout NE Germany the directions are oriented NW/SE. The reason for this direction is probably the nearby Tornquist-Teisseyre line. The observed fast axes are subparallel to this prominent Transeuropean suture zone. At stations in southern Germany near the Alps we observed ENE/WSW directions. Below some stations we also found indications of inclined anisotropic layers.

Key words: Anisotropy, shear-wave splitting, SKS waves.

Introduction

In the last decade the number of teleseismic shear-wave splitting measurements has increased exponentially worldwide. The analysis of SKS splitting (e.g., VINNIK *et al.*, 1984, 1992; KIND *et al.*, 1985; SILVER and CHAN, 1991; BARRUOL *et al.*, 1997) provides a local image of the anisotropy below the observation point with high lateral resolution. SKS and SKKS phases are highly qualified to study anisotropy in the mantle, since they are not biased by anisotropy or lateral heterogeneities at the source site. The SKS phase converts from P to S when

[1] Seismologisches Zentralobservatorium Gräfenberg der Bundesanstalt für Geowissenschaften und Rohstoffe, Krankenhausstr. 1, 91054 Erlangen, Germany.

[2] Now at: Image Science Division, Communication Technology Laboratory, Federal Institute of Technology, 8092 Zürich, Switzerland.

[3] Institut für Allgemeine und Angewandte Geophysik, Ludwig Maximilians Universität München, Theresienstr. 41, 80333 München, Germany.

penetrating the CMB from the core into the mantle. Assuming a laterally homogeneous isotropic earth, SKS should be only radially polarized after the conversion. Therefore, any SKS or SKKS energy appearing on the transverse component is an indicator for anisotropy or lateral heterogeneity under the receiver site.

Suitable epicentral distances for the analysis of SKS phases are in the range between 90° and 130° where SKS is isolated from other phases. At the lower epicentral boundary of about 85° the SKS phase overtakes S and above 90° both phases are clearly separated from each other. The upper boundary at about 130° is characterized by the interference of the SKSac (travel path only through the outer core) and the SKSdf branch (travel path also through the inner core).

Usually the SKS technique is used under the assumption of a rather simple model, the existence of only one anisotropic layer with a horizontal symmetry axis. However, triggered by the observations of backazimuthally varying splitting parameters, several authors developed new methods in the last five years, in which they account for more complex models:

On the one hand, SAVAGE and SILVER (1993) and SILVER and SAVAGE (1994) derived analytic expressions for the apparent splitting parameters in the case of two anisotropic layers. They applied their method to several stations near the San Andreas fault and could show reasonable fits to the data by two horizontal anisotropic layers. On the other hand, ALSINA and SNIEDER (1995) showed that backazimuthally varying splitting parameters observed at some Netherland stations can be explained by laterally varying anisotropy. A third group of authors (BABUŠKA *et al.*, 1984, 1993; SILENY and PLOMEROVÁ, 1996; PLOMEROVÁ *et al.*, 1996) favors the existence of an inclined symmetry axis in the lithosphere as the cause of backazimuthally varying splitting parameters. SILENY and PLOMEROVÁ (1996) and PLOMEROVÁ *et al.* (1996) derived an inversion method to retrieve the three-dimensional orientation of an inclined homogeneous anisotropic layer with hexagonal or orthorhombic symmetry axis.

Several investigations using different approaches have been performed for the Gräfenberg array (GRF) and the German Regional Seismic Network (GRSN) to explain observed SKS splitting in terms of anisotropy (see Fig. 1 for the station locations). Similar as in the approach by SILVER and SAVAGE (1994), VINNIK *et al.* (1994) explained the observed splitting parameters at GRF and GRSN by two anisotropic layers with horizontal symmetry axes. BABUŠKA *et al.* (1984, 1993) presented results based on P-travel time residuals in Central Europe, where they discussed the presence of inclined anisotropy of the subcrustal lithosphere. BORMANN *et al.* (1996) compared the apparent splitting parameters in the same area with tectonic features. They explained the observed directions of the fast axis at several stations by an asthenospheric flow which is controlled by the plate motion and the lithosphere-asthenosphere boundary topography.

In this paper we address the question of the existence and the possibility of resolving more complicated anisotropic structures such as multilayer anisotropy and/or inclined symmetry axes with the SKS technique. The data from the three-component stations of the GRF array and the three-component broadband stations of the German Regional Seismic Network (GRSN) are well suited for such an investigation due to the long recording time of the stations. This work is an extension of the work done by SAVAGE and SILVER (1993), SILVER and SAVAGE (1994) and VINNIK et al. (1994).

In the first part of this paper the theoretical background is derived. The second part gives an overview of the data, the analysis method and its resolution capability, and the quality restrictions that were applied. After the presentation of the

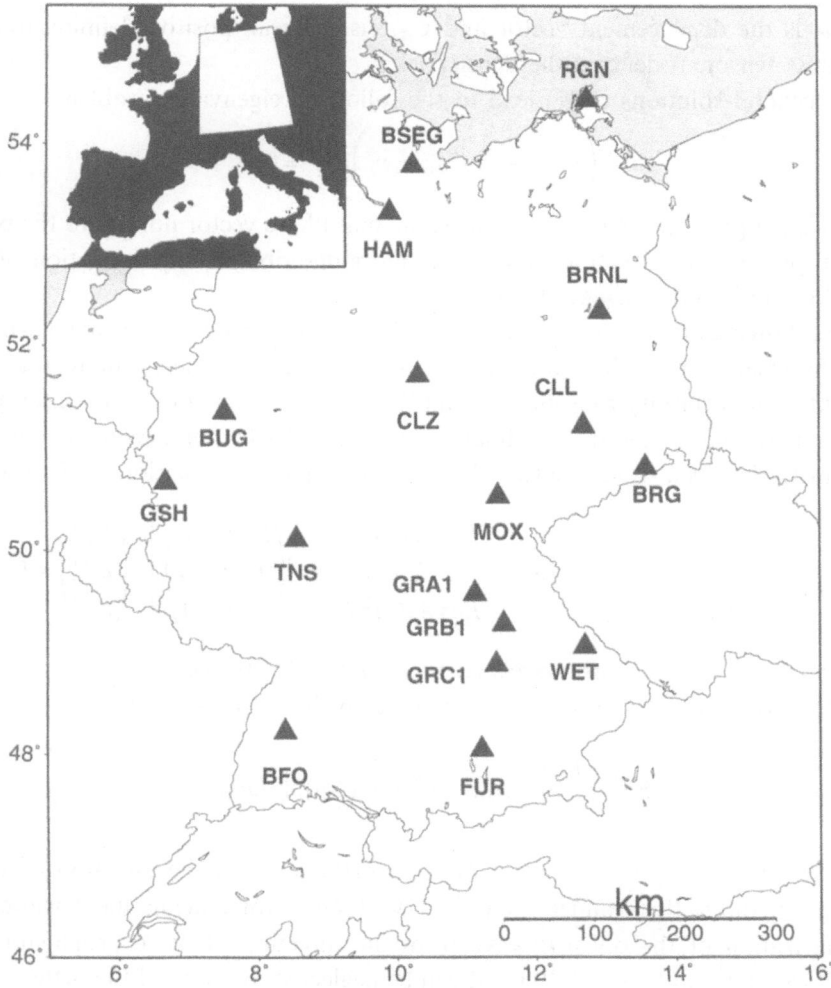

Figure 1
Distribution of the Gräfenberg array (GRF) and the German Regional Seismic Network (GRSN) three-component broadband stations in Germany.

results in the third part, the structural and tectonic implications are discussed, and it is concluded that anisotropy structure inferred from sparse data sets and the one-layer/horizontal symmetry axis hypothesis can be severely biased and should not be related directly to the earth's structure.

Theory

The governing equation of motion for general anisotropic elastic solids is

$$\rho \frac{\partial^2 u_i}{\partial t^2} = \frac{\partial}{\partial x_k} \left[C_{iklm} \frac{\partial u_m}{\partial x_l} \right] \tag{1}$$

where u_i is the displacement vector and C_{iklm} is the real, positive-definite, fourth-rank stress tensor. ρ denotes the density.

The general solutions of (1) lead to the following eigenvalue problem

$$\left(v^2 \delta_{im} - \frac{1}{\rho} C_{iklm} n_k n_l \right) E_m^{[\mu]} = 0 \tag{2}$$

where v is the phase normal velocity, n_k is the unit phase vector normal to the plane of equal phase and $E_m^{[\mu]} = 0$ are the eigenvectors describing the polarization of the wave (LEE and ALEXANDER, 1995).

The following discussion deals with the media showing anisotropic axial symmetry (transversal isotropy, hexagonal anisotropy). In this case the elastic features are invariant under arbitrary rotations around the symmetry axis. For an orientation of the x_3 axis of the Cartesian coordinate system parallel to the symmetry axis it is sufficient to discuss a wave incident in the x_1–x_3 plane and equation (2) reduces to

$$\left[\frac{1}{\rho} \begin{pmatrix} An_1^2 + Ln_3^2 & 0 & (F+L)n_1 n_3 \\ 0 & Nn_1^2 + Ln_3^2 & 0 \\ (F+L)n_1 n_3 & 0 & Ln_1^2 + Cn_3^2 \end{pmatrix} - v^2 \begin{pmatrix} 1 & 0 & 0 \\ 0 & 1 & 0 \\ 0 & 0 & 1 \end{pmatrix} \right] \cdot \begin{pmatrix} E_1^{[\mu]} \\ E_2^{[\mu]} \\ E_3^{[\mu]} \end{pmatrix} = 0 \tag{3}$$

where A, C, F, L and N are Love's coefficients (LOVE, 1927).

An eigenvector $\mathbf{E}^{[2]}$ with corresponding eigenvalue v_2^2 is obviously

$$\mathbf{E}^{[2]} = \begin{pmatrix} 0 \\ 1 \\ 0 \end{pmatrix} \quad \text{with} \quad v_2^2 = \frac{1}{\rho} (Nn_1^2 + Ln_3^2). \tag{4}$$

Therefore one eigenvector is always directed orthogonal to the direction of wave propagation and to the symmetry axis. For weak anisotropic media the deviation of the polarization of the quasi-P wave from the direction of wave propagation is small (BABUŠKA and CARA, 1991) and will be neglected. Because of the orthogonality of the eigenvectors, the second eigenvector is oriented parallel to the direction of wave propagation. The third eigenvector lies in the plane which is orthogonal to the direction of wave propagation and contains the symmetry axis.

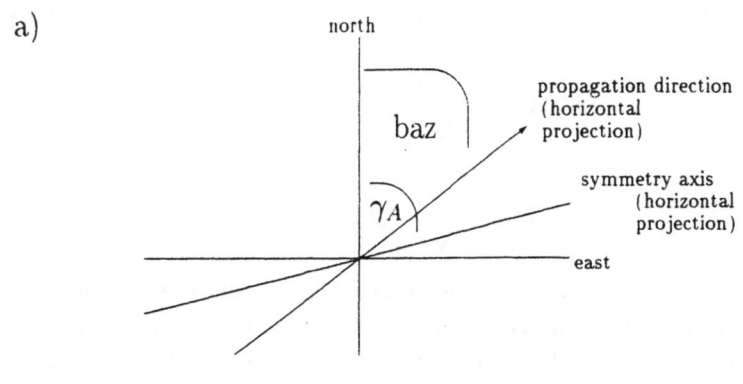

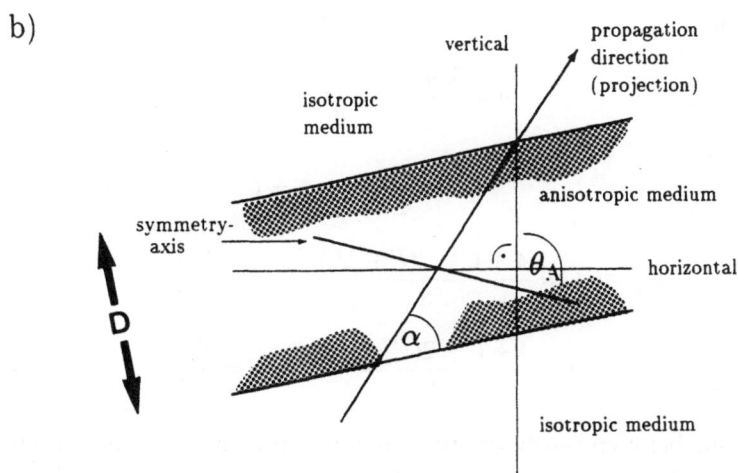

Figure 2

Definition of the angles θ_A, γ_A and α. The angles θ_L and γ_L, which characterize the vertical and horizontal orientations of the anisotropic plane-parallel layer, are defined in a manner analogous to θ_A and γ_A. a) horizontal plane, b) vertical plane containing the anisotropic symmetry axis.

In the following, differences between the transmission coefficients of the differently polarized quasi-shear waves will be neglected. For an incident wave in a single homogeneous anisotropic layer with parallel boundary surfaces and thickness, D the delay time δt between the quasi-shear waves is

$$\delta t = \frac{D \cdot (V_{qSV} - V_{qSH})}{|\sin \alpha| \cdot V_{qSV} \cdot V_{qSH}} \qquad (5)$$

where V_{qSV} and V_{qSH} denote the velocity of the quasi-vertically (qSV) and the quasi-horizontally (qSH) polarized quasi-shear waves. α is the angle between the wave propagation direction and the anisotropic layer (see Fig. 2).

Following CRAMPIN and KIRKWOOD (1981) these velocities can be expressed as

$$V_{qSH} = \sqrt{\frac{1}{\rho} \cdot (D + E \cos 4\sigma)}$$

$$V_{qSV} = \sqrt{\frac{1}{\rho} \cdot (F - G \cos 2\sigma)} \qquad (6)$$

where D, E, F, G are functions of the space variables and σ is the angle between the incident wave and the symmetry axis. σ is a function of the incidence angle, of the polarization direction of the incident wave and of the orientation of the symmetry axis. In terms of Love's coefficients (LOVE, 1927), the coefficients D, E, F, G can be expressed as

$$A = \frac{3(A + C) + 2(F + 2L)}{8}$$

$$B = \frac{A - C}{2} \qquad E = C$$

$$C = \frac{A + C - 2(F + 2L)}{8} \qquad F = \frac{N + L}{2}$$

$$D = \frac{A + C - 2(F - 2L)}{8} \qquad G = \frac{N - L}{2} .$$

Let the angle between north and the horizontal projection of the symmetry axis be θ_A. Let γ_A denote the angle between the vertical and the vertical projection of the symmetry axis. Let Φ_P denote the polarization direction and i the incidence angle of the impinging wave. $i = 0°$ corresponds to a vertically propagating wave.

Then σ can be expressed as

$$\sigma = \arccos(\sin i \cos \Phi_P \sin \theta_A \cos \gamma_A + \sin i \sin \Phi_P \sin \theta_A \sin \gamma_A + \cos i \cos \theta_A). \quad (7)$$

θ_L and γ_L are defined in a manner analogous to θ_A and γ_A to characterize the vertical and horizontal orientation of the plane-parallel anisotropic layer. Then α can be expressed as

$$\alpha = \arcsin(\cos \theta_L \cos \gamma_L \sin i \cos \Phi_P + \cos \theta_L \sin \gamma_L \sin i \sin \Phi_P - \sin \theta_L \cos i). \quad (8)$$

Substituting (8) and (7) in (6) and then (5) gives the delay time δt.

Let us exclude in the following the case of a near-vertical symmetry axis. Then equations (7) and (8) show that there exists only a weak dependence of σ and α on Φ_P for near-vertical wave-incidence (i close to 0°). In the extreme case of a perfectly vertical wave-incidence ($i = 0°$) the formula (7) and (8) are explicitly independent of Φ_P.

According to the above discussion, the polarization directions of the quasi-shear waves are fixed relative to the ray. The horizontal component of the fast polarization direction observed at the earth's surface in the case of steeply arriving SmKS waves (under the above assumption of a nonvertical symmetry axis) is therefore nearly independent of Φ_P, i.e. the backazimuthal variation of the fast polarization can be neglected.

The backazimuthal dependence of the splitting parameters for a stack of several horizontal anisotropic layers and for vertical wave propagation is derived in SILVER and SAVAGE (1994). Their method is actually more general and will be valid whenever the splitting parameters of the anisotropic layers are nearly independent of the backazimuth and incidence angle for the range available in the data (SAVAGE, 1997).

For reasons of possible future applications of the method to S waves in the following, the dependence of the splitting parameters is given as an explicit function of the polarization direction of the incoming wave. In our case of SmKS waves though it must be remembered that the incoming polarization direction actually equals the backazimuth.

In the case of two or three anisotropic layers, the observed splitting parameters are called effective or apparent. They are obtained under the assumption of one layer when in fact several layers are present. However, the effective splitting parameters ϕ_a (fast axis) and δt_a (delay time) are still meaningful. For two horizontal anisotropic layers and vertical wave propagation the apparent (suffix a) splitting parameters are

$$\tan \psi_a = \frac{a_q^2 + C_S^2}{a_q a_p + C_S C_C}$$

$$\tan \zeta_a = \frac{C_S}{a_p \sin \psi_a - a_q \cos \psi_a} \qquad (9)$$

with

$$a_p = \cos \zeta_1 \cos \zeta_2 - \sin \zeta_1 \sin \zeta_2 \cos(\psi_2 - \psi_1)$$

$$a_q = -\sin \zeta_1 \sin \zeta_2 \sin(\psi_2 - \psi_1)$$

$$C_C = \cos \zeta_1 \sin \zeta_2 \cos \psi_2 + \cos \zeta_2 \sin \zeta_1 \cos \psi_1$$

$$C_S = \cos \zeta_1 \sin \zeta_2 \sin \psi_2 + \cos \zeta_2 \sin \zeta_1 \sin \psi_1 \qquad (10)$$

and

$$\psi_{1,2} = 2\beta_{1,2}$$

$$\zeta_{1,2} = \omega \delta t_{1,2}/2$$

Table 1

Model parameters used for the theoretical calculations (see Fig. 3). i denotes the incidence angle of the SKS wave, D denotes the thickness of the anisotropic layer (4% of anisotropy is assumed), δt is the splitting time, γ_L is the backazimuth of the plane-parallel anisotropic layer with respect to north and θ_L is the inclination of the layer with respect to the upward pointing vertical axis. γ_A and θ_A denote the horizontal and vertical orientation of the fast symmetry axis analogously (see also Fig. 2). a) Model parameters for the calculations of Figure 3a (inclined one-layer models). b) Model parameters for the calculations of Figure 3b (two horizontal anisotropic layers and vertical wave propagation). c) Model parameters for the calculations for Figure 3c (two inclined anisotropic layers and symmetry axes for nonvertical wave propagation).

a. One anisotropic layer

No.	i in °	γ_L in °	θ_L in °	γ_A in °	θ_A in °	D in km
1	0	–	90	–	90	100
2	15	–	90	–	90	100
3	15	0	45	0	45	100
4	15	0	45	90	90	100
5	15	150	30	90	60	100

b. Two horizontal anisotropic layers and vertical wave propagation

Lower layer						Upper layer					
i in °	γ_L in °	θ_L in °	γ_A in °	θ_A in °	δt in sec.	i in °	γ_L in °	θ_L in °	γ_A in °	θ_A in °	δt in sec.
0	–	90	90	90	1.0	0	–	90	20	90	1.0

c. Two inclined anisotropic layers and symmetry axes for nonvertical wave propagation

Lower layer						Upper layer					
i in °	γ_L in °	θ_L in °	γ_A in °	θ_A in °	D in km	i in °	γ_L in °	θ_L in °	γ_A in °	θ_A in °	D in km
15	180	70	115	85	200	12	130	20	40	90	100

d. Used elastic parameters for the inclined models

Love's coefficients (in GPa):
$A = 225.1$, $C = 215.1$, $F = 86.0$, $L = 65.5$, $N = 70.8$
density: $\rho = 3.377$ g/cm^3

where $\beta_{1,2}$ is the angle between the incoming polarization direction and the fast polarization direction of the lower/upper layer (1,2).

The apparent splitting parameters exhibit the following characteristics (SILVER and SAVAGE, 1994):

— the effective splitting parameters show a 90° periodicity in Φ_P

— ϕ_a goes through a discontinuous jump of 90° for $a_q a_p + C_S C_C = 0$ away from this jump, ϕ_a is a monotonically increasing function of Φ_P

— δt_a reaches a maximum at a point near the discontinuous jump of ϕ_a
— the splitting parameters depend on the frequency of the incident wave.

In the case of several inclined axial-symmetric anisotropic layers, the directions of polarization within the individual layers are fixed according to the above discussion and the equations (5) and (9) can be combined to give the resulting splitting parameters. The splitting parameters no longer show a backazimuthal 90° periodicity and backazimuthal regions exist, for which ϕ_a is a monotonically decreasing function of the backazimuth.

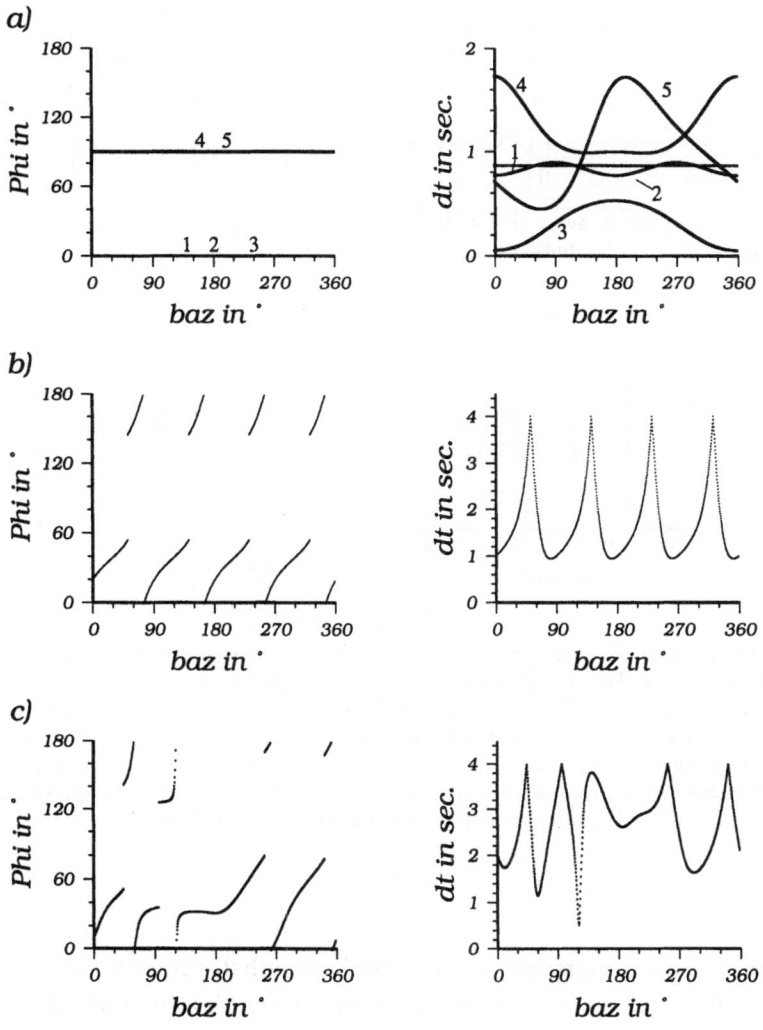

Figure 3
Backazimuthal variation of the splitting parameters (Phi = fast horizontal direction; dt = splitting time) for several models (see Table 1 for the model parameters).

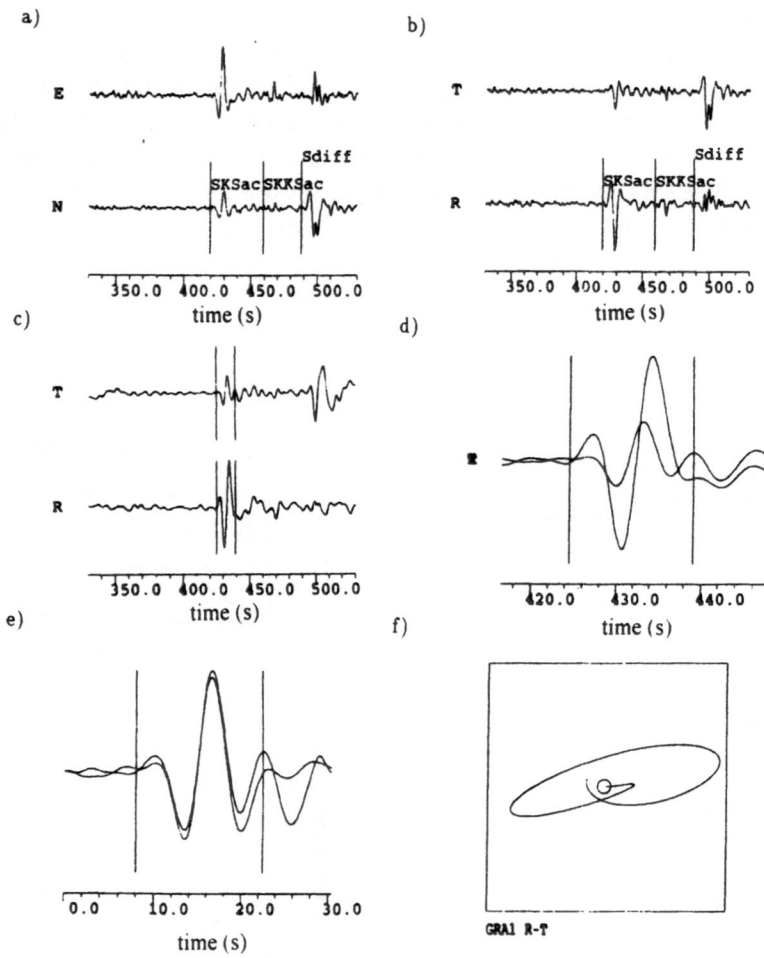

Figure 4

Representative example of a splitted SKS phase and the processing steps which have been used to derive the splitting parameters. The example refers to an earthquake from Argentina (1994, May 10) recorded at station GRA1 (epicentral distance: 102°; backazimuth: 240°). a) Original seismograms, b) rotated seismograms, c) filtered displacement proportional seismograms, d) superposition of the radial and transverse component (SKS phase zoomed), e) synthesized transverse component calculated for the splitting parameters found in the inversion ($\phi = 76°$ and $\delta t = 1.55$ s) and original transverse component, f) particle motion of the time window marked in d.

Figure 3 shows examples for the backazimuthal variation of the splitting parameters for incidence in a single anisotropic layer (3a), for vertical incidence in a system of two horizontal anisotropic layers (3b) and for nonvertical incidence in a system of two dipping anisotropic layers and symmetry axes (3c). For a more detailed discussion see BRECHNER (1996).

Data and Processing

The SKS and SKKS wave forms used in this study were recorded at the three-component stations of the GRF array, at the 14 three-component broadband stations of the GRSN and one broadband station in the Alps near Innsbruck (SQTA).

We only selected those events with clear SKS or SKKS phases respectively and with a signal on the transverse component significantly above the noise level. In the preprocessing after rotation the instrument response was removed and the seismograms were bandpass filtered between 5 s and 30 s by a Butterworth filter. For the determination of the splitting parameters δt and ϕ we used the method of VINNIK et al. (1989). The method takes advantage of the fact that the transverse component of the SKS phase behaves similar to the derivative of the radial component. Based on that relationship a filter is applied to the radial component in the frequency domain to synthesize a transverse component, which can be compared to the observed one after back transformation in the time domain. To determine the best fitting splitting parameters we used a grid search, in which the splitting parameters δt and ϕ are varied over the entire model space, and searched the global minimum of the misfit between the synthetic and the observed transverse component in a least-squares sense. Based on careful data extraction and application of different time windows of the SKS signal, we estimate that the errors of the splitting parameters are smaller than $\pm 15°$ for the fast axis and at most ± 0.5 s for the delay times. These are approximately the variances we derive for the averaged splitting parameters for small azimuth ranges with several observations.

An example of a splitted SKS phase with an energetic transverse wavetrain as well as for the individual steps of preprocessing and the application of the method described above is illustrated in Figure 4. The seismograms refer to an earthquake from Argentina (1994, May 10) recorded at station GRA1. The epicentral distance is 102° and the backazimuth is 240°. Figure 4a shows the original, 4b the rotated and 4c the filtered displacement proportional seismograms. A superposition of the radial and transverse component in a zoomed view and the particle motion of the associated time window can be seen in Figures 4d and 4f. The synthetic transverse component calculated for the splitting parameters revealed in the inversion ($\phi = 76°$ and $\delta t = 1.55$ s) and the original transverse component are plotted together in Figure 4e. A good fit is obtained.

Results

In this section the effective splitting parameters of the individual stations determined by the method of VINNIK et al. (1989) are presented. The observations are interpreted in terms of elaborated anisotropy models such as multilayer

a) GRA1

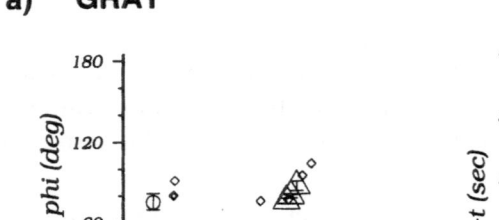

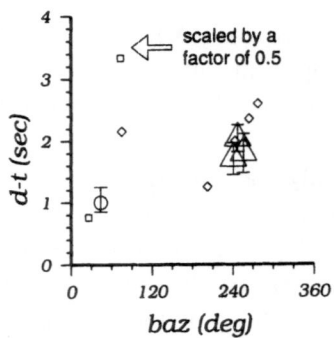

b) GRB1

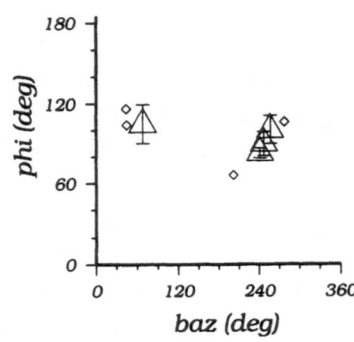

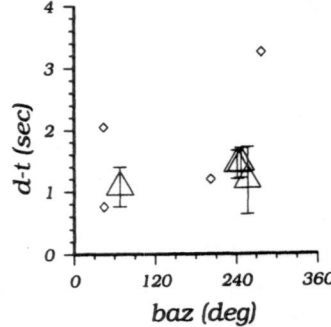

c) GRC1

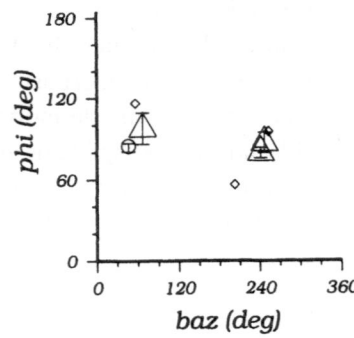

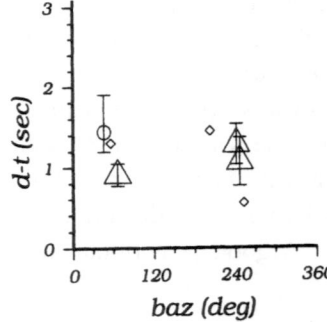

Figure 5

Effective splitting parameters ϕ and δt in dependence of the backazimuth for the three-component stations of the Gräfenberg array. The following notation has been used in the figure: Single events which are bandpass filtered from 5 to 30 s are marked by a rhombus, events which are occasionally filtered in another frequency band or which are not filtered are marked by squares (not filtered: GRA1 backazimuth $= 26°$; highpass of 30 s: GRA1 backazimuth $= 74°$). For narrow backazimuth ranges with more than one event average values and error bars denoting the standard deviation are shown (circles: 2–4 events, triangles: 5 and more events).

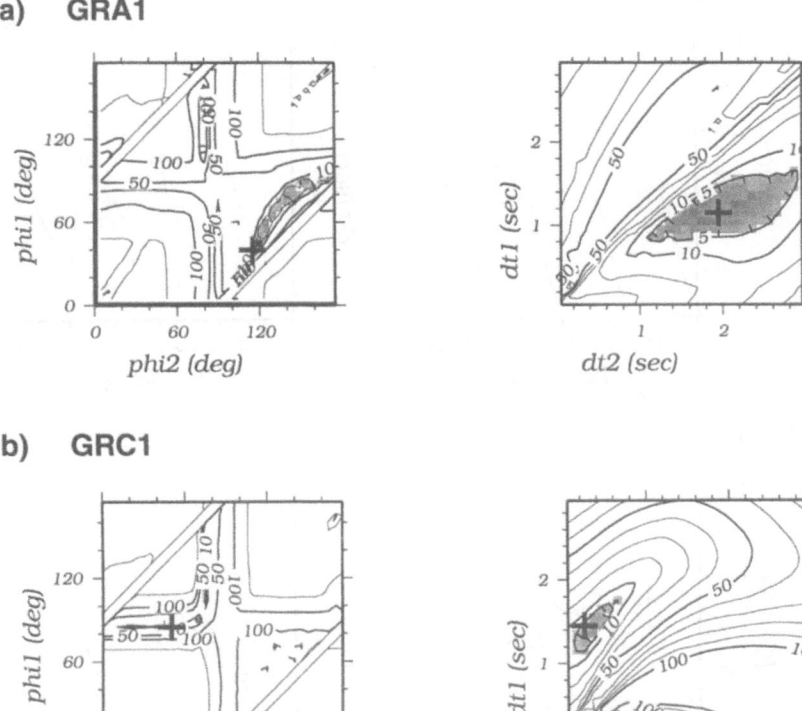

Figure 6
Results of the inversion obtained for a two-layer anisotropy model with horizontal symmetry axes: Crosses indicate the model with the smallest misfit, contour lines denote deviations of the error function from the misfit of the best model in percent. The gray-shaded areas represent regions with a deviation of at most 5% (for further explanations see text).

anisotropy or layers with inclined symmetry axis. Due to the longer operation time of the GRF stations, many more events with SKS phases of high quality than for the GRSN stations are recorded and could be used to determine the splitting parameters. The quantity of data for the GRF three-component stations allowed a detailed interpretation of the splitting parameters which is presented in the first part of this section. In the second part the results for the GRSN stations will be shown.

a) GRF Array

Between 32 and 34 events from the time period 1979–1994 with high signal-to-noise ratio for the SKS phase had been analyzed for the three-component stations of the GRF array (GRA1, GRB1, GRC1). The splitting parameters obtained for

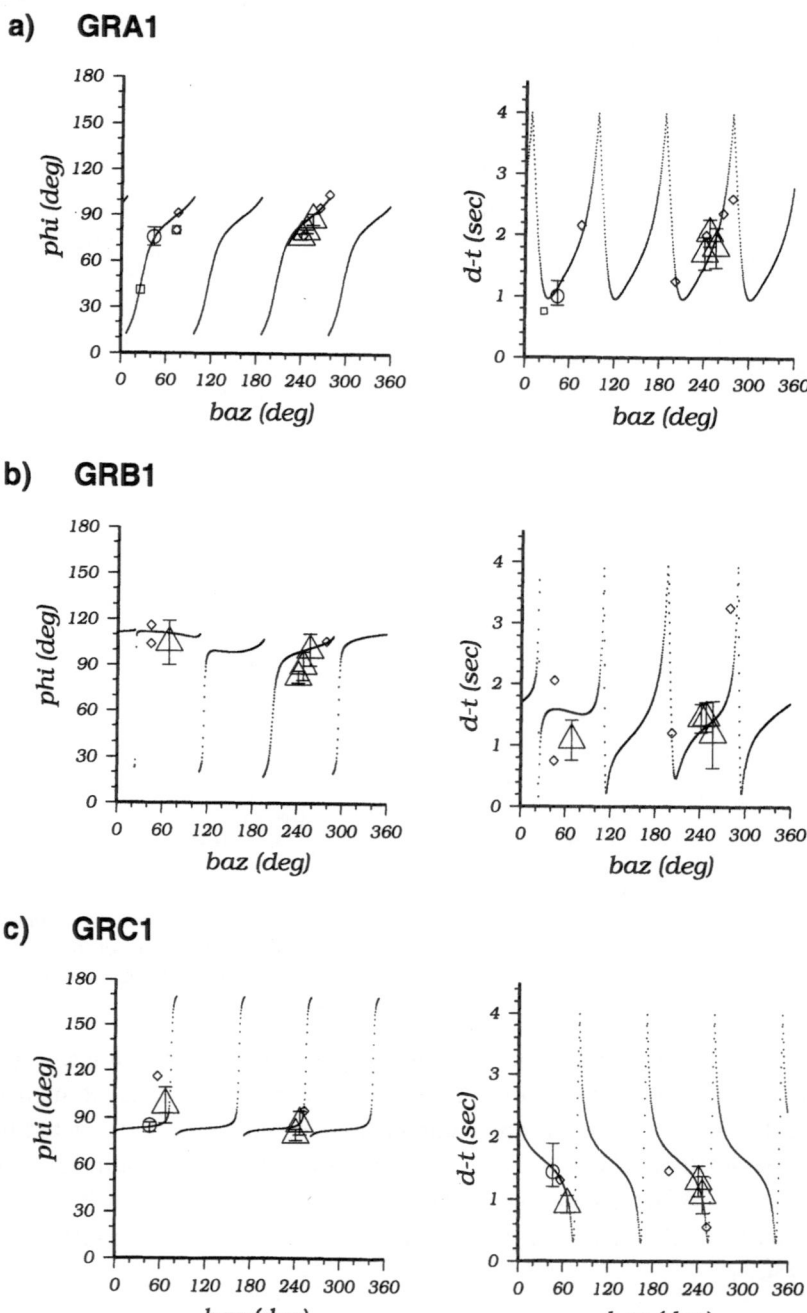

Figure 7

Curves of the effective splitting parameters ϕ and δt as a function of backazimuth and observations. The calculations of the curves for a) and c) are based on the splitting parameters which the inversion for a two-layer model with horizontal symmetry axes had revealed as the best model. b) shows a theoretical curve for a two-layer model with one horizontal symmetry axis (lower layer) and an inclined symmetry axis in the upper layer which can qualitatively represent the characteristics of the observations.

the individual events for each station are given in Figure 5 as functions of the backazimuth.

GRA1: The splitting parameters for station GRA1 (Fig. 5a) show clear backazimuthal variations of both splitting parameters. In the backazimuth range from 25° to 76° and from 239° to 278° the orientations of the fast axis and the delay times clearly rise with increasing backazimuth.

The backazimuth variation of the splitting parameters cannot be explained by one anisotropic layer with a horizontal symmetry axis, characterized by backazimuthally independent splitting parameters. The backazimuthal variation of the fast axis is also in contradiction to the constant backazimuth predicted by a one-layer model with a shallowly inclined symmetry axis. The apparent 180° periodicity is consistent with a 90° periodicity of the fast direction and therefore we favor a multilayer model with horizontal symmetry.

First we discuss the simplest case, namely a two-layer model with horizontal symmetry axes. The inversion was performed using a grid search where all four splitting parameters (ϕ_1, δt_1, ϕ_2, δt_2) were varied. Some of the splitting parameters shown in Figure 5 have been excluded in the inversions and in Figure 7 due to the following reasons: The ϕ-values for the event with a backazimuth of 202° display large error bars for all three Gräfenberg stations, but clearly resolved delay times. Therefore only the delay times of that event have been used in the inversion. Furthermore the delay time for the event of a backazimuth of 73.5° at station GRA1 has been ignored. This signal has a dominant period of about 3–5 s, significantly shorter than most other signals used (dominant period about 8–10 s). The large value of $\delta t = 6.5$ s probably indicates the direct vicinity of this backazimuth to the jumping point of the apparent delay times which may easily result in large errors.

The results of the simultaneous inversion for the delay times and the orientations of the fast axes are given in Figure 6a. The best model retrieved in the inversion shows an upper anisotropic layer with a fast axis ϕ_2 of N40°E and δt_2 of 1.15 s and a lower anisotropic layer with a fast axis ϕ_1 of N115°E and a delay time δt_1 of 1.95 s, respectively. The backazimuthal course of the splitting parameters calculated for this model is shown in Figure 7a together with the observations. A good fit to the observations is obtained. The directions of the fast axes for both layers are in good agreement with those found by VINNIK *et al.* (1994). As Figure 5a shows, the resolution is much better for the relative orientations of the symmetry axes of the lower and the upper layers than for the absolute orientations of these axes. Our estimation for the error bars is about $\pm 10°$ for the relative orientation of the axes and $\pm 30°$ for the absolute orientation.

The delay time of 1.15 s in the upper layer is probably related to anisotropy in the mantle, since a delay time exceeding 0.3 s is rarely produced by anisotropy in the crust (e.g., BARRUOL and MAINPRICE, 1993; CRAMPIN and BOOTH, 1985). Of course anisotropy in the crust contributes also to the observed effective splitting parameters. Therefore, we tested in a further step whether the inversion for a three-layer model could resolve anisotropy in the crust.

The inversion for a three-layer model gave the following parameters: N40°E and 1.7 s for the upper layer, N120°E and 2.3 s for the intermediate layer and N90°E and 0.5 s for the lowermost layer. The resulting theoretical curves for the effective splitting parameters are similar to those of the two-layer model and a significant improvement in the simulation of the observations could not be achieved. Since the delay time of 1.7 s in the upper layer cannot be attributed to the crust, there is either no anisotropy in the crust or the fast direction in the crust is the same as in the upper mantle. Since the theoretical curves for the two- and three-layer models are very similar, the two-layer model is to be preferred for the sake of simplicity.

We conclude that a one-layer model of transverse anisotropy with near-horizontal symmetry axis is clearly not compatible with the data of station GRA1. The two-layer model with horizontal symmetry axes for station GRA1 is preferred by the data.

GRB1: In Figure 5b the splitting parameters for station GRB1 are shown as functions of the backazimuth. In the backazimuth range from 200° to 260° the angle of the fast axis increases, whereas in the range from 44° to 74° no clear tendency can be observed. The delay times evidence no systematic behavior in either backazimuth range.

The increase of the fast direction between 200° and 260° clearly advocates against an anisotropic one-layer model with near-horizontal symmetry axis. Moreover, the observations are not in agreement with a horizontal multilayer model, which requires a 90° periodicity and increasing ϕ-values with backazimuth, because the ϕ-values for the backazimuth range of 40° to 70° are significantly higher than for the corresponding backazimuth range of about 240° to 260°. This decrease of ϕ with increasing backazimuth contradicts the characteristics observed for horizontal multilayer models.

The station GRB1 is located near an old continental suture zone. Therefore we have to consider the possibility of a steeply inclined fast axis. In the light of the results of ALSINA and SNIEDER (1995) it is also possible that the "strange" characteristics of the fast axis directions found for northeastern backazimuths reflect that GRB1 is located slightly south of a boundary between two lithospheric blocks manifesting different anisotropy. In favor of such an interpretation, both the apparent fast axis directions and the apparent delay times for the backazimuth range 240° to 260° show similar characteristics to those inverted for the more southern station GRC1 (see Figs. 5b and 5c).

We attempted to invert the GRB1 data for horizontal anisotropic multilayer models. Both, the two-layer and the three-layer model inversions revealed very poor fits of the observations. The strong backazimuthal variations of the fast axis direction also exclude the possibility of the existence of only one anisotropic layer with (shallowly) inclined symmetry axis. In a further step we took into account the possibility of a two-layer model with inclined axes. Starting with the two-layer model with horizontal layers and horizontal symmetry axes for station GRA1, a rotation of the upper layer in the model of GRA1 of about 50° towards the vertical and an anisotropy axis of 40° strike and 50° dip (towards the vertical) yields a considerably better fit than the theoretical curves for the models with horizontal symmetry axes. Figure 7b shows the predicted splitting parameters together with the observations. The model parameters are given in Table 1, however it has to be kept in mind that this model for station GRB1 is just one possibility among other model types (e.g., a near-vertically inclined symmetry axis or heterogeneous anisotropy), which fits the observations qualitatively better than the models with horizontal layers. Due to the great number of free parameters in the case of inclined axes and because of the lack of a complete backazimuthal coverage, a grid search inversion was not performed.

GRC1: As for the stations GRA1 and GRB1, the splitting parameters for station GRC1 exhibit backazimuthal variations (Fig. 5c). In the backazimuth range between 200° and 253° an increase of the fast direction can be detected. A similar tendency can be observed between 40° and 70° backazimuth. Only one value for backazimuth of 56° deviates slightly from this trend. However, this value (and the value at baz = 202°) belongs to a SKKS phase which can be biased due to a phase shift resulting from the bounce point at the core-mantle boundary (JAMES and ASSUMPCAO, 1996). Despite some scattering, the delay times tend to decrease with increasing backazimuth ($44° \leq$ baz $\leq 70°$, $200° \leq$ baz $\leq 253°$).

The variations in ϕ exclude a one-layer model. Both ϕ and δt are consistent with a 90° periodicity, which indicates that a horizontal multilayer model is compatible with the observations. The inversion results yield the best fit for a two-layer model (the SKKS values were neglected in the inversion). The values of the fast direction and the delay time are N85°E and 1.45 s for the upper layer and N50°E and 0.35 s for the lower layer. The result of the inversion and the theoretical curves for the splitting parameters based on this model are shown in Figures 6b and 7c. The SKKS values at backazimuths of 56° and 202° are not fitted by this model. Comparison of Figures 7a and 7c demonstrates that a model of the GRA1 type can fit all ϕ observations including this data point reasonably. Nonetheless it is impossible to find models of the GRA1 type with delay times δt decreasing in the backazimuthal range of the GRC1 data. Therefore it must be concluded that the anisotropy structure below station GRC1 can be more complex than just two horizontal anisotropic layers.

b) GRSN Stations

The first GRSN stations are in operation since 1991. The number of SKS observations for the GRSN is therefore significantly smaller than for the GRF stations. The number of events yielding reliable splitting parameters varies between

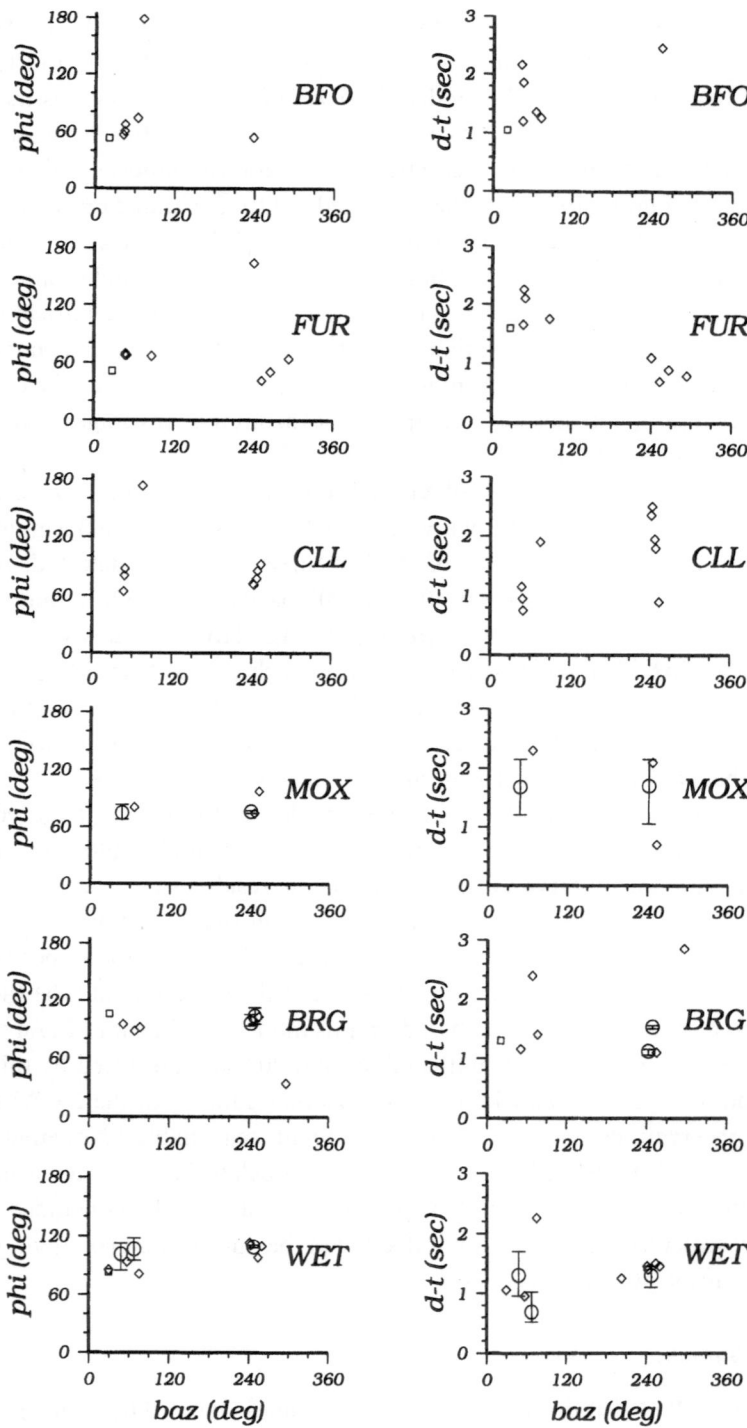

Figure 8A

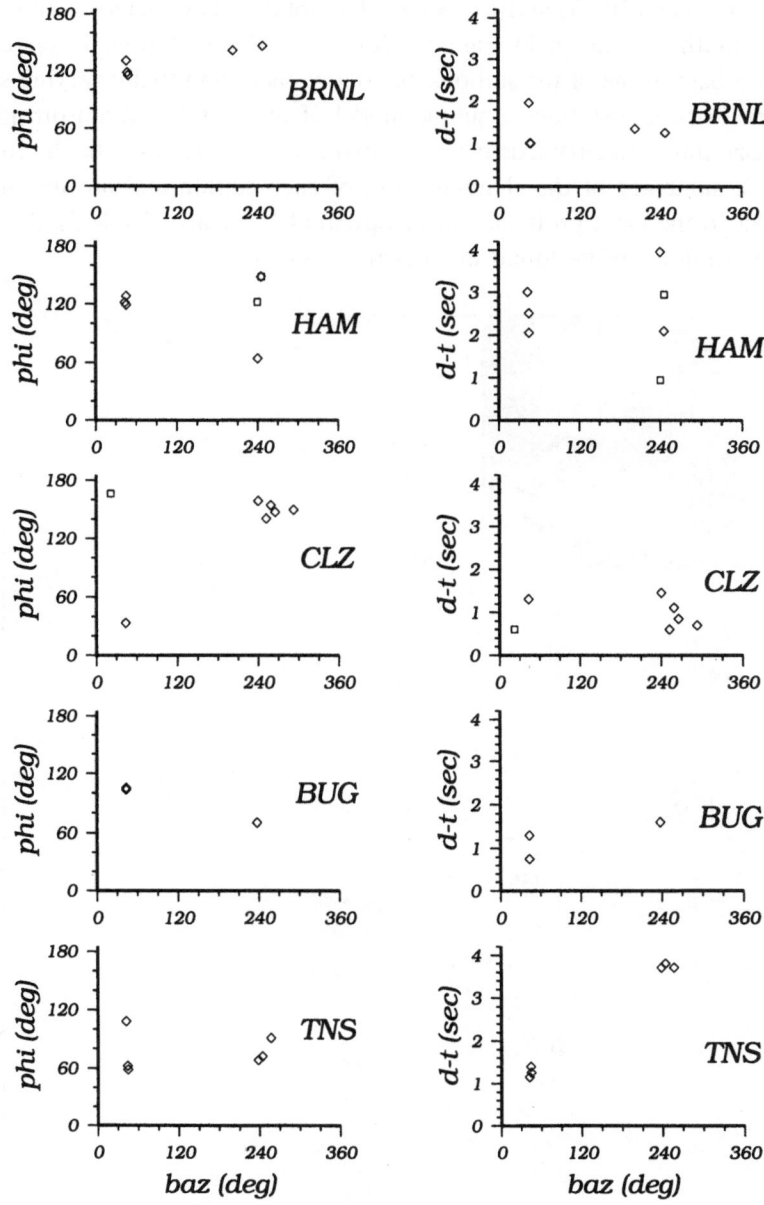

Figure 8B

Effective splitting parameters ϕ and δt in dependence on backazimuth for the stations of the German Regional Seismic Network (GRSN) (for further explanations see Fig. 5). Some $SKKS_{ac}$ events are not filtered or filtered in another frequency band. They are marked by squares (not filtered: BFO backazimuth = 27°, BRG backazimuth = 30°, WET backazimuth = 30°, CLZ backazimuth = 22°; band-pass 8 s–30 s: FUR backazimuth = 27°).

3 (stations: TNS and BUG) and 19 (WET). The splitting parameters as functions of
the backazimuth are shown in Figure 8 for each GRSN station. In spite of the
smaller data base, some of the stations also show backazimuthally varying splitting
parameters. For these stations a simple model of one horizontal anisotropic layer
with a horizontal symmetry axis cannot explain the observations. In the following
we supply a summary of the characteristics of each station and an interpretation
with respect to the most probable anisotropy model (see also Table 2). A comprehensive description can be found in BRECHNER (1996).

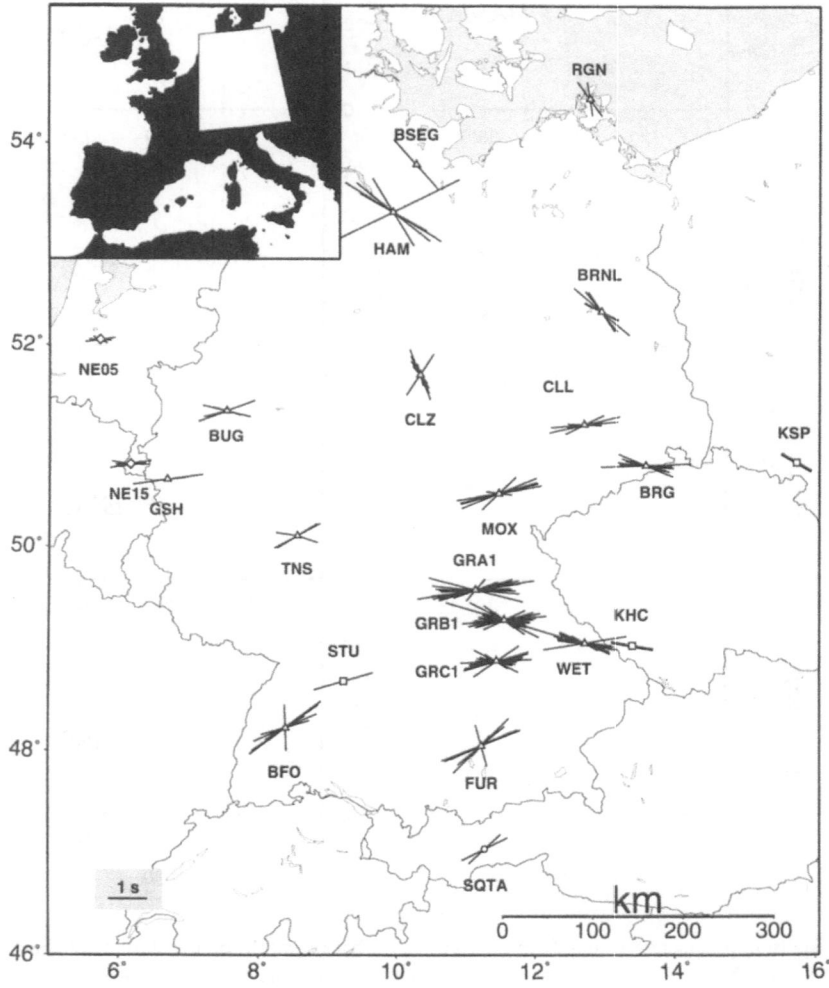

Figure 9

Observed splitting parameters for the GRF and GRSN stations. The straight lines show the directions
of polarization ϕ of the fast symmetry axes. The magnitude of the delay time δt is given by the length
of the line. In addition to the German stations the splitting parameters are also given for the Netherlands
stations (NE05, NE15) (ALSINA and SNIEDER, 1995), for the Czech stations (KHC, KSP) (MAKEYEVA
et al., 1990) and for station SQTA near Innsbruck.

BFO (Black Forest Observatory): The delay times as a function of backazimuth show a nonsystematic scattering. In the backazimuth range between 20° and 64° the ϕ-values increase, an observation advocating against a one-layer model with horizontal symmetry axis as also the ϕ-value of 178° at a backazimuth of 72°. Only one observation exists in the backazimuth range of 180° to 360°. This does not allow the determination of the periodicity or an inversion on the basis of a multilayer model. More observations are necessary for a better resolution of the anisotropy model.

FUR (Fürstenfeldbruck): The station shows a clear variation of the delay times as a function of backazimuth (1.60 s to 2.25 s for 27° < baz < 87° and 0.70 s to 1.10 s for 240° < baz < 294°). Moreover there is an increase of the ϕ-value as a function of backazimuth for backazimuths exceeding 240°. The described variations of the splitting parameters suggest a multilayer model with at least one inclined symmetry axis.

CLL (Collm): The backazimuthal variation of the ϕ-values (64° < ϕ < 92°) is in the range of the error bar, whereas the delay times show a great scatter, which could be caused by the proximity of the backazimuth to the fast axis (null direction). Within the error bars a one-layer model with near-horizontal symmetry axis is consistent with the data.

MOX (Moxa): The ϕ-values for station MOX vary between 68° and 97°. Since the scattering is of the order of the error bar assumed for the determination of a single ϕ-value, the ϕ-values can be considered as backazimuthally independent. The delay times reflect large scattering (0.70 s to 2.30 s). The consistency of the fast axis argues for a one-layer model with a horizontal symmetry axis. The mean splitting parameters are 78° for the fast axis and 1.7 s for the delay time.

BRG (Berggieshuebel): All ϕ-values are in the same range (88° ≤ ϕ ≤ 113°). The mean value for the fast direction is N99°E. With the exception of one event (baz = 68°, δt = 2.40 s), the delay times vary between 1.05 s and 1.55 s. Because of the backazimuthal constancy of the fast direction, we favor a one-layer model for BRG with a strike of about 100° for the fast axis.

WET (Wettzell): For station WET we have the largest number of records of all GRSN stations. Note that the ϕ-values (81° ≤ ϕ ≤ 118°) do not vary as a function of backazimuth, indicating the presence of only one anisotropic layer. The variation of the delay times as a function of the backazimuth may indicate the presence of a small inclination of the symmetry axis, however more data are required to confirm this statement. The mean value for the strike of the fast symmetry axis is N103°E and the mean delay time is 1.2 s.

BRNL (Berlin-Lankwitz): Due to the relatively high-noise level at station BRNL in comparison to most other GRSN stations the signal-to-noise ratio for the SKS phase is poor. Five events with relatively undisturbed SKS phases could be analyzed. The strike of the apparent fast axis varies between 120° and 140° and the delay times range from 1 s to 2 s. The observations are not in contradiction to a one-layer model.

HAM (Hamburg): HAM is the noisiest station in the data set. Five events with sufficient SKS energy on the transverse component could be selected. The investigated events display a large scatter in the delay times (2.05 s to 3.95 s) and unusually high-delay times on the average. With the exception of one value, ϕ varies between 120° and 150°. A distinction between a one-layer model and more complex models cannot be made.

CLZ (Clausthal-Zellerfeld): The ϕ-values for all but one event vary between 144° and 166° and do not show a significant dependence on backazimuth. An interpretation in terms of a one-layer model would be a reasonable interpretation.[1] However, if the deviating ϕ-value of 33° for a backazimuth of 44° cannot be attributed to heterogeneity, a multilayer model has to be considered. We performed the inversion on the basis of a two-layer model and obtained a satisfying fit of the observations with the following parameters: upper layer $\phi_1 = 165°$ and $\delta t_1 = 0.35$ s, lower layer $\phi_2 = 125°$ and $\delta t_2 = 0.75$ s. Due to the brief delay time, the upper layer can be attributed to the crust. More data are required to distinguish between both models.

BUG (Bochum University): The SKS phases observed at station BUG reveal poor signal-to-noise ratios on the transverse component. For only 3 events could reliable splitting parameters be resolved. The backazimuth of 43° gives a ϕ-value of about 105° and the backazimuth of 238° a ϕ-value of about 70°. There are too few observations to resolve this kind of anisotropy model.

TNS (Taunus): The ϕ-values ($59° \leq \phi \leq 108°$) vary more than the assumed error bar for the determination of a single ϕ-value, although they nearly belong to the same backazimuth. Since there are no observations from other backazimuths, we cannot argue for one or the other anisotropy model.

GSH (Grosshau), *BSEG* (Bad Segeberg), *RGN* (Rügen), *SQTA* (St. Querin, Tirol, Austria): Due to the short operation time of the three GRSN stations GSH, BSEG and RGN and of the temporary installed station SQTA, there are not enough resolved splitting parameters to interpret them in terms of elaborated models. The individual splitting parameters are the following: GSH (1: $\phi = 84°$, $\delta t = 0.95$ s, baz = 49°; 2: $\phi = 82°$, $\delta t = 1.90$ s, baz = 65°), BSEG ($\phi = 138°$, $\delta t = 1.80$ s, baz = 248°), RGN (1: $\phi = 171°$, $\delta t = 0.90$ s, baz = 70°; 2: $\phi = 144°$, $\delta t = 1.10$ s, baz = 251°), SQTA (1: $\phi = 64°$, $\delta t = 1.35$ s, baz = 44°; 2: $\phi = 48°$, $\delta t = 1.15$ s, baz = 70°). All observed effective splitting parameters for these stations are shown in Figure 9 together with the effective splitting parameters of all other stations. For comparison we also present the results for some Netherlands (ALSINA and SNIEDER, 1995) and Czech stations (MAKEYEVA *et al.* 1990).

Table 2

Splitting parameters for the stations of the Gräfenberg array (GRF) and the German Regional Seismic Network (GRSN) as well as a classification according to the preferred anisotropy model. For some stations (GRA1, GRC1, CLZ) an inversion for two layers could be performed. In that case the splitting parameters of the two individual layers are given in column 6–9. For two stations (BFO, FUR) a two-layer model is preferred, but no reliable result of the inversion can be given due to lack of backazimuthal coverage. Stations indicated by the term '1 layer' explicitly require a one-layer model. Stations marked by (*) are compatible with a one-layer model, but it cannot be ruled out that future observations which will increase the backazimuthal coverage will require a more complex model.

Station	No. of obs.	Preferred model	Range of $\phi(°)$	Range of $\delta t(s)$	ϕ_1 (°)	δt_1 (s)	ϕ_2 (°)	δt_2 (s)
GRA1	30	2 layers, horizontal axes	–	–	40	1.15	115	1.95
GRB1	32	complex	–	–	–	–	–	–
GRC1	31	2 layers, horizontal axes	–	–	85	1.45	50	0.35
BFO	7	2 layers	–	–	–	–	–	–
FUR	9	2 layers	–	–	–	–	–	–
CLL	6	1 layer (*)	64–92	0.75–1.95	–	–	–	–
MOX	9	1 layer (*)	68–97	0.70–2.30	–	–	–	–
BRG	10	1 layer	88–106	1.05–2.40	–	–	–	–
WET	19	1 layer	81–113	0.60–2.25	–	–	–	–
BRNL	5	1 layer (*)	115–146	1.00–1.95	–	–	–	–
HAM	5	1 layer (*)	64–148	2.05–3.95	–	–	–	–
CLZ	7	2 layers	–	–	165	0.35	125	0.75
BUG	3	no statement, too few data	70–105	0.75–1.60	–	–	–	–
TNS	3	no statement, too few data	59–108	1.15–1.40	–	–	–	–
GSH	2	no statement, too few data	82–84	0.95–1.90	–	–	–	–
BSEG	1	no statement, too few data	138	1.80	–	–	–	–
RGN	2	no statement, too few data	144–171	0.90–1.10	–	–	–	–
SQTA	2	no statement, too few data	48–64	1.15–1.35	–	–	–	–

Discussion

Our data reveal for most stations differences in the measured splitting parameters: Differences in ϕ and δt exist for measurements from a single event at various stations as well as at one station for events with different backazimuths. That means we observe lateral variations of the anisotropy and for some stations we have to consider more than one anisotropic layer and/or the existence of an inclined axis of symmetry. In the following we discuss possible models for the origin of these lateral differences and the tectonic consequences. The comparison of the observed splitting parameters within the region under investigation shows groups with different characteristics:

1. The results for CLL, MOX, BRG and WET near the Bohemian Massif are consistent with a single layer of transverse anisotropy with near-horizontal symmetry axis with the fast direction between 80° and 100° and delay times between 1 s and 2 s. Similar results in the same area were found by MAKEYEVA *et al.* (1990) and VINNIK *et al.* (1992) for the stations KHC (100°, 1.1 s) and KSP (120°, 0.9 s). For WET there are indications for a shallowly inclined layer.

2. A second group of stations is located in the middle and southern part of Germany. The upper mantle below these stations is characterized by more than one anisotropic layer. The best resolved two-layer model is the GRA1 model, for which the delay time of the upper layer can be attributed to the lithosphere. Because of the magnitude of the delay time in the upper layer the lower layer must lie within the asthenosphere. The lower layer of GRA1 has similar orientation and delay time (115°, 1.95 s) as those below BRG and WET. The relatively high delay time of the upper layer (1.15 s) reveals an anisotropic layer in the lithosphere much thicker than the crust. The model obtained here with an extended data set confirms an earlier result of VINNIK *et al.* (1994).

The fast axis direction in the lower layer below GRA1 is different than the observed Central European plate motion of about 60° observed by geodetic measurements (plate velocity between 2 and 3 cm/y; MONTAG *et al.*, 1994, 1995). This difference can be caused by the fact that the motion of the Central European plate results from the addition of several forces, i.e., the force induced by the surrounding plates and the drag force due to the asthenospheric flow. A more complicated situation exists for stations GRB1 and GRC1. For station GRB1 either one or two layers with an inclined axis or a model with lateral heterogeneous anisotropy can explain the observed variation of the apparent splitting parameters. The situation for station GRC1 is also unclear. The best horizontal two-layer model shows a very small delay time of 0.35 s in the lower layer (however, it must be mentioned that this model was not able to fit all observations). It is hard to imagine that the asthenospheric flow pattern can change on such a small lateral scale (the distance between stations GRA1 and GRC1 is about 80 km). An additional argument against a change of the anisotropy structure in the asthenosphere is the partial overlap of the Fresnel zones for SKS waves recorded at GRA1 and GRC1. Therefore we conclude that better backazimuthal coverage is needed to reveal a better and more complex model for station GRC1. This complicated situation can be attributed to the collision of two microplates, the Saxothuringian and the Moldanubian. These plates collided in the paleozoic age and their suture zone runs between the stations GRA1 and GRB1/GRC1. The investigation of this suture zone east of the GRF array is the topic of an ongoing project.

3. For the southern stations FUR, BFO and the Austrian station SQTA the effective splitting parameters are all in the same range (60°–80°, 1 s–2.5 s). Indications for a multilayer model with hints of the existence of an inclined layer appeared in the backazimuthal variation of the fast shear-wave directions as well as by the nonexisting 90° periodicity at stations FUR and BFO. The effective fast directions at these stations are suborthogonal to the general trend of the trajectories of the maximum horizontal stress (MÜLLER et al., 1992; GRÜNTHAL and STROHMEYER, 1994) and strike parallel to the Alpine mountain belt.

4. For HAM, BSEG and BRNL in Northern Germany we prefer again a one-layer model, but this result should be regarded as preliminary due to the small number of observations caused by the high-noise level in the wave-form data. The fast directions are oriented NW/SE. With the exception of a thin upper anisotropic layer—possibly related to the crust—similar splitting parameters also exist for CLZ. The reason for this direction is probably the nearby Tornquist-Teisseyre line associated with an increase in the lithosphere thickness to about 200 km (BABUSKA et al., 1987) and the continental root of the East European Platform down to 400 km, as indicated by the high shear-wave velocities in 3D-tomography models (RITZWOLLER and LAVELY, 1995). As mentioned by BORMANN et al. (1996) strong lateral variations in the depth of the lithosphere-asthenosphere boundary, especially in northeastern Germany, can influence the observed shear-wave splitting. In this region the asthenospheric mantle flow is likely to follow the trend of this line. The directions of the fast axis we inverted are in good agreement with that trend. However, to date we are not able to distinguish, if the splitting originates in the lithosphere and thereby votes for frozen anisotropy in the lithosphere or if the anisotropy indicates the present-day flow in the asthenosphere deviated at the lithospheric barrier of the East European Platform.

From our observations we conclude that it can be misleading to interpret the measured splitting parameters in terms of only one horizontal anisotropic layer. Many of the splitting parameters are influenced by more than one and/or inclined anisotropic layers and/or lateral heterogeneous anisotropy. For a better resolution of the proposed models or even more complex models (e.g., orthorhombic symmetry) much more shear-wave splitting data and better station coverage are necessary. Then it should be possible to distinguish which part of the anisotropy signal originates in the lithosphere and which part in the asthenosphere.

Acknowledgements

The authors are thankful to the GRF team for providing the high-quality digital data used in this study. P. Melichar and his collaborators provided us with the opportunity to use their station SQTA near Innsbruck. We are also indebted to L. Vinnik and G. Bock for providing programs and M. Savage, J. Plomerova and an anonymous reviewer for their helpful reviews.

REFERENCES

ALSINA, D., and SNIEDER, R. (1995), *Small-scale Sublithospheric Continental Mantle Deformation: Constraints from SKS Splitting Observations*, Geophys. J. Int. *123*, 431–448.

BABUŠKÁ, V., PLOMEROVÁ, J., and SILENY, J. (1984), *Large-scale Oriented Structures in the Subcrustal Lithosphere of Central Europe*, Ann. Geophys. *2*, 649–662.

BABUŠKA, V., PLOMEROVÁ, J., and SILENY, J., *Structural model of the subcrustal lithosphere in Central Europe*. In *The Composition, Structure and Dynamics of the Lithosphere-Asthenosphere System* (C. Froidevaux and K. Fuchs, eds.) (AGU Geophys. Series *16*, Washington DC 1987) pp. 239–251.

BABUŠKA, V., and CARA, M., *Seismic Anisotropy in the Earth* (Kluwer Academic Publishers, Dordrecht 1991).

BABUŠKA, V., PLOMEROVÁ, J., and SILENY, J. (1993), *Models of Seismic Anisotropy in the Deep Continental Lithosphere*, Phys. Earth. Plant. Int. *78*, 167–191.

BARRUOL, G., and MAINPRICE, D. (1993), *A Quantitative Evaluation of the Contribution of Crustal Rocks to the Shear-wave Splitting of Teleseismic SKS Waves*, Phys. Earth Planet. Int. *78*, 281–300.

BARRUOL, G., SILVER, P. G., and VAUCHEZ, A. (1997), *Seismic Anisotropy in the Eastern United States: Deep Structure of a Complex Continental Plate*, J. Geophys. Res. *102*, 8329–8348.

BORMANN, P., GRÜNTHAL, G., KIND, R., and MONTAG, H. (1996), *Upper Mantle Anisotropy beneath Central Europe from SKS Wave Splitting: Effects of Absolute Plate Motion and Lithosphere-Asthenosphere Boundary Topography?* J. Geodynamics *22* (1/2), 11–32.

BRECHNER, S. (1996), *Backazimuthal Variations of Splitting Parameters from Teleseismic SmKS Waves* (in German), Diploma Thesis, University of Erlangen, 133 pp.

CRAMPIN, S., and BOOTH, D. C. (1985), *Shear-wave Polarizations near the North Anatolian Fault, II, Interpretation in Terms of Crack-induced Anisotropy*, Geophys. J. R. Astr. Soc. *83*, 75–92.

CRAMPIN, S., and KIRKWOOD, S. C. (1981), *Velocity Variations in Systems of Anisotropy Symmetry*, J. Geophys. *49*, 35–42.

GRÜNTHAL, G., and STROHMEYER, D. (1992), *The Recent Crustal Stress Field in Central Europe: Trajectories and Finite-element Modelling*, J. Geophys. Res. *97*, 11805–11820.

JAMES, D. E., and ASSUMPCAO, M. (1996), *Tectonic Implications of S-wave Anisotropy beneath SE Brazil*, Geophys. J. Int. *126*, 1–10.

KIND, R., KOSAREV, G. L., MAKEYEVA, L. I., and VINNIK, L. P. (1985), *Observations of Laterally Inhomogeneous Anisotropy in the Continental Lithosphere*, Nature *318*, 358–361.

LEE, J. M., and ALEXANDER, S. S. (1995), *Seismic Anisotropy Caused by Rock Fabric*, Geophys. J. Int. *122*, 705–718.

LOVE, A. E. H., *A Treatise on the Mathematical Theory of Elasticity* (Dover Publ., New York 1927).

MAKEYEVA, L. I., PLESINGER, A., and HORALEK, J. (1990), *Backazimuthal Anisotropy beneath the Bohemian Massif from Broad-band Seismograms of SKS Waves*, Phys. Earth Planet. Int. *62*, 298–306.

MONTAG, H., REIGBER, CH., SOMMERFELD, W., and DICK, G. (1994), *Station Coordinates and Earth Rotation Parameters Based on Lageos Laser Ranging Data*, IERS Technical Notes *17*, Paris, 25–30.

MONTAG, H., REIGBER, CH., and SOMMERFELD, W. (1995), *Solution for the Terrestrial Reference Frame Based on Lageos Laser Ranging Data*, IERS Technical Notes *19*, Paris, 21–24.

MÜLLER, B., ZOBACK, M. L., FUCHS, K., MASTIN, L., GREGERSEN, S., PAVONI, N., STEPHANSSON, O., and LJUNGGREN, C. (1992), *Regional Patterns of Tectonic Stress in Europe*, J. Geophys. Res. *97*(B8), 11783–11803.

PLOMEROVÁ, J., SILENY, J., and BABUŠKA, V. (1996), *Joint Interpretation of Upper Mantle Anisotropy Based on Teleseismic P-travel Time Delays and 3-D Inversion of Shear-wave Splitting Parameters*, Phys. Earth. Plant. Int. *95*, 293–309.

RITZWOLLER, M. H., and LAVELY, E. M. (1995), *Three-dimensional Seismic Models of the Earth's Mantle*, Rev. Geophys. *33*(1), 1–66.

SAVAGE, M. K. (1997), *Seismic Anisotropy and Mantle Deformation: What have we Learned from Shear-wave Splitting Studies?* Rev. Geophys., submitted.

SAVAGE, M. K., and SILVER, P. G. (1993), *Mantle Deformation and Tectonics: Constraints from Seismic Anisotropy in the Western United States*, Phys. Earth Planet. Int. *78*, 207–227.

SILENY, J., and PLOMEROVÁ, J. (1996), *Inversion of Shear-wave Splitting Parameters to Retrieve Three-dimensional Orientation of Anisotropy in Continental Lithosphere*, Phys. Earth. Plant. Int. *95*, 277–292.

SILVER, P. G., and CHAN, W. W. (1988), *Implications for Continental Structure and Evolution from Seismic Anisotropy*, Nature *335*, 34–39.

SILVER, P. G., and CHAN, W. W. (1991), *Shear-wave Splitting and Subcontinental Mantle Deformation*, J. Geophys. Res. *96* (B10), 16429–16454.

SILVER, P. G., and SAVAGE, M. K. (1994), *The Interpretation of Shear-wave Splitting Parameters in the Presence of Two Anisotropic Layers*, Geophys. J. Int. *119*, 949–963.

VINNIK, L. P., KOSAREV, G. L., and MAKEYEVA, L. I. (1984), *Anisotropy in the Lithosphere from Observations of SKS and SKKS* (in Russian), Dokl. Acad. Nauk SSSR *278*, 1335–1339.

VINNIK, L. P., KIND, R., KOSAREV, G. L., and MAKEYEVA, L. I. (1989), *Backazimuthal Anisotropy in the Lithosphere from Observations of Long-period S Waves*, Geophys. J. Int. *99*, 549–559.

VINNIK, L. P., MAKEYEVA, L. I., MILEV, A., and USENKO, A. Y. (1992), *Global Patterns of Backazimuthal Anisotropy and Deformations in the Continental Mantle*, Geophys. J. Int. *111*, 433–447.

VINNIK, L. P., KRISHNA, V. G., KIND, R., BORMANN, P., and STAMMLER, K. (1994), *Shear-wave Splitting in the Records of the German Regional Seismic Network*, Geophys. Res. Lett. *21* (6), 457–460.

(Received November, 1996, revised July 18, 1997, accepted August 10, 1997)

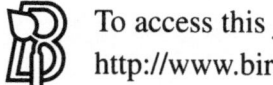

Pure appl. geophys. 151 (1998) 333–364
0033–4553/98/040333–32 $ 1.50 + 0.20/0

❚ Pure and Applied Geophysics

Anisotropic Measurements in the Rhinegraben Area and the French Massif Central: Geodynamic Implications

MICHEL GRANET,[1] ANDREAS GLAHN[2] and ULRICH ACHAUER[1]

Abstract—As part of an integrated seismic study, polarization of shear waves has been analyzed for teleseismic events recorded at a set of permanent broadband, semi-permanent long- and short-period and temporary short-period seismological stations located in two geodynamically important areas in western Europe, namely the Rhinegraben-Urach area and the French Massif Central volcanic field.

While for the semi-permanent and the permanent stations there is a good azimuthal coverage of teleseismic earthquakes which allowed us to investigate the azimuthal dependence and the spatial variation over short distances of an anisotropy direction, no even azimuthal distribution of teleseismic recordings with a clear elliptical (or linear) polarization of the *S* phases could be obtained in the case of the temporary stations.

While the mean values of the splitting parameters ϕ and δt are geographically coherent for adjacent stations, our results show a large scatter of the individual splitting parameters for the set of events used. The magnitude of the splitting time suggests that the deformation extends below the lithosphere and that the thickness of the anisotropic structure is at least 100–200 km.

For some stations located in the Rhinegraben-Urach area (ECH, RG-N, RG-S, RBG), the variations of ϕ are consistent with a two-layer anisotropic model as suggested by VINNIK *et al.* (1994) for the South German Triangle. For the stations ECH (Vosges mountains), RG-N and RG-S (Rhinegraben proper), the resulting estimates of fast direction are around N10°E–N30°E and N80°E–N100°E for the upper and lower layers, respectively. For the station RBG (Urach), the results are N60°E–N70°E and N125°E–N135°E, respectively.

In the Rhinegraben-Urach area, the estimates of the effective fast direction for a one-layer model show a rotation from a graben-related (30°) pattern to an Alpine belt-related pattern in the eastern part ($\approx$ E–W). In the French Massif Central region, the results reveal two distinct fast polarization patterns. While to the west of the Sillon Houiller, ϕ is parallel to this late-variscan transformlike fault zone and perpendicular to the variscan belt, it is to the east rather perpendicular to the Alpine belt. The results suggest a mixture of both a lithospheric and an asthenospheric component of the seismic anisotropy for the Rhinegraben-Urach as well as for the French Massif Central areas.

Key words: Seismic anisotropy, shear waves, Western Europe, mantle deformation.

[1] EOST, CNRS-UMR 7516, 5, Rue René Descartes, F-67084 Strasbourg cédex, France.
[2] Geophysical Institute, Hertzstraße, 16, D-76187 Karlsruhe, Germany (now at: Novartis Pharma Ltd., Information Technology, Bldg. 2-490.3.10, Lichtstr. 35, CH-4002 Basel, Switzerland).

1. Introduction

Data

It is generally agreed that seismic investigations play a key role in the under-standing of the mechanic/geodynamic state of the lithosphere-asthenosphere system (LAS). During the last decade, the development of new portable short-period/broadband arrays was probably one of the most significant technical improvements to assist geoscientists in studying the LAS worldwide. The study of the Urach geothermal field (Swabian Jura, SW-Germany) and as joint French/German coop-erations the study of the Upper Rhinegraben area and of the French Massif Central volcanic field, are some recent examples of the successful application of these arrays (GLAHN *et al.*, 1992, 1993; GRANET *et al.*, 1995a,b).

As small-scale high resolution teleseismic tomography studies provide significant information on the velocity perturbation of a target area, the main objective of these field experiments was to provide LAS tomographic pictures as a general outline of the structural features. However, isotropic teleseismic tomography on its own is not sufficient to properly constrain any geodynamic model deduced from imaging velocity perturbations. Another important source of information comes from the analysis of *S*-wave splitting of teleseismic data recorded by mobile short-period networks. This information on the lithospheric and asthenospheric anisotropy is complementary to tomographic images from 3-D isotropic inversion and hence allows further constraint of the models. This paper is mainly concerned with such seismic anisotropy measurements and their geodynamic interpretations. We present new and unpublished results of anisotropic data obtained from portable arrays in tectonically active regions, namely the Urach area located in the South German Triangle and the adjacent Rhinegraben area (including the Vosges and Black Forest mountains) and the French Massif Central volcanic field. Additional constraints on the interpretation come from measurements at permanent broad-band stations. In the following sections we first briefly recall the tectonic and geophysical settings of the areas under investigation, and the two methods for retrieving the splitting parameters. Next, we will present and discuss the results.

Seismic Anisotropy

In the past, analysis of shear-wave splitting for some isolated permanent stations provided only a rough idea of lateral variations and the length scales of coherent anisotropy signatures beneath continents. Teleseismic recordings of dense mobile networks in turn may be used to thoroughly investigate tectonic structures if the measured parameters of azimuthal anisotropy are consistent across the array or change across the tectonic units. However, the short deployment duration of such networks forces one to use the entire set of phases—*SKS*, *SKKS*, *S* and *ScS*—for a sufficient number and azimuthal coverage of events.

Seismic anisotropy has become a very popular tool in the last decade to investigate the structure of the subcontinental mantle. It is the best geophysical evidence for the past (inherited structures) and present dynamics (mantle flow) in the interior of the earth and is of great importance for assessing mantle deformations at depth (SILVER and CHAN, 1988; VINNIK et al., 1989; ANSEL and NATAF, 1989).

In the lithosphere-asthenosphere system, or more generally in the upper mantle, anisotropy is usually interpreted as a result of lattice preferred orientation (LPO) of its most abundant mineral, olivine, the primary upper mantle mineral, in response to finite strain. The crystallographic axes of olivine are preferentially oriented according to their shear history in a convective mantle, where preferred orientation depends critically on temperature (NICOLAS and CHRISTENSEN, 1987).

As noticed by many authors (e.g., SILVER and CHAN, 1991; VINNIK et al., 1995), anisotropy reflects deformations ranging from very recent (as in the asthenosphere due to the flow pattern) to very old (as in the "cold" lithosphere due to past and present orogenic processes). Hence, the question whether the seismic anisotropy is lithosphere- or asthenosphere-related is still strongly debated in the literature. Generally speaking, either the fast-polarization azimuth (ϕ) is parallel to the absolute plate motion (APM), this is the hypothesis of Simple Asthenospheric Flow (SAF), or it is correlated with a recent or old tectonic trend, this is the hypothesis of Vertically Coherent Deformation (VCD) (SILVER, 1996).

In this paper, we focus on a specific kind of anisotropy—polarization anisotropy—as deduced from shear-wave splitting to infer the relationship between deformations at depth and surface tectonics. A transversely isotropic medium with a horizontal axis of symmetry is assumed.

2. Tectonic and Geophysical Setting

2.1 Rhinegraben-Urach Area

Tectonic setting

The Rhinegraben area is the most prominent central segment of the European Cenozoic Rift System (ECRIS) of Oligocene age which extends from the North Sea to the French Mediterranean coast. Rifting took place in a region of pre-existing Hercynian zones of weakness. This tensile episode can be related to Alpine tectonics and the Africa-Eurasia collision. The area where the field experiment was performed, is limited to the Upper Rhinegraben extending from Basel in the South to just north of Strasbourg and limited to the west by the Vosges mountains and to the east by the Black Forest (Fig. 1). The Rhinegraben classified as a typical continental rift has primarily been formed by extensional deformation with only a very short period of active rifting (20–15 Ma). At about 20 Ma, the maximum

horizontal compressional stress changed from NNE–SSW to NNW–SSE direction
(MEIER, 1989). Today, the southern end of the Upper Rhinegraben is characterized
by a quasi-compressive, strike-slip tectonic regime, with a maximum stress-axis
oriented NW–SE well evidenced from the earthquakes focal mechanisms and *in situ*
measurements (BONJER *et al.*, 1984, 1989). The Urach geothermal field is located
within the Swabian Jura to the east of the Upper Rhinegraben area and south of

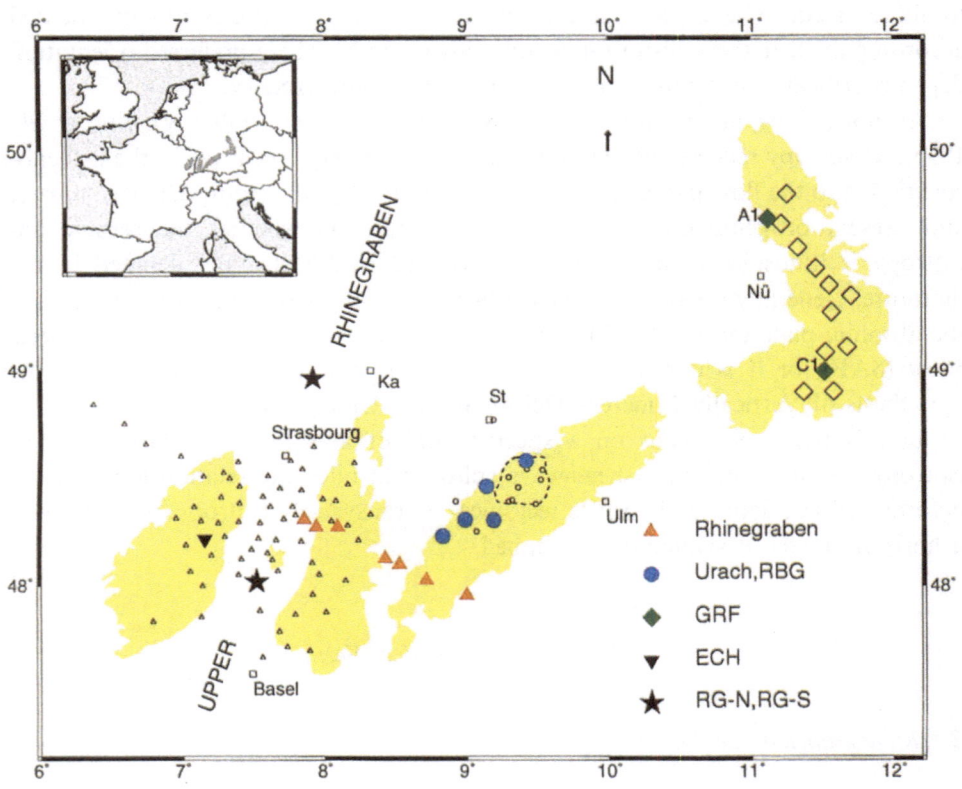

Figure 1

Map showing the location of the mobile and permanent stations which were used for the seismic studies
in the Rhinegraben and Urach tectonic areas. In the figure, superimposed on the main tectonic
structures, the stations for which anisotropic measurements have been achieved are shown as large
symbols. Inside the Rhinegraben proper, there are two mobile long-period stations (RG-N and RG-S)
identified by black stars. On the western flank, the black inverse triangle represents the ECH broadband
station. On the eastern flank, 7 red triangles indicate the location of the set of mobile stations in the
Black Forest. The blue dots correspond to the stations in the Urach region. The Urach geothermal field
proper is located inside the circle-like dashed line. To the extreme east, two green diamonds represent the
broadband stations GRF-A1 and GRF-C1. Small open triangles and open dots show the temporary
mobile arrays installed in the RG and in the Urach areas, respectively. Yellow areas show the main
tectonic regions (from west to east: Vosges mountains, Black Forest, Swabian Jura, Franconian Jura).
Ka, St, Nü indicate the location of the cities of Karlsruhe, Stuttgart and Nürnberg, respectively.

Stuttgart (Fig. 1). Next to the Urach region and to the Rhinegraben proper with its shoulders are the important tectonic features of the Molasse Basin and the Alpine foreland to the south. The Urach geothermal field located within a Miocene volcanic field has been extensively studied by geophysicists and petrologists (GLAHN et al., 1992).

Geophysical results

From a geophysical point of view, the Rhinegraben is one of the best studied regions worldwide. Readers are referred to the special issue of Tectonophysics on the Rhinegraben rift (Tectonophysics 208, 1992) and references therein. To sum up, in addition to old refraction and recent reflection seismic studies which demonstrated the thinning of the crust beneath the rift and an asymmetry of the crustal structure, a small-scale tomographic teleseismic investigation provided strong evidence for upper-mantle heterogeneities (GLAHN et al., 1993). In order to retrieve the small wavelength lithosphere-asthenosphere structure and to derive a dynamic model of the rifting process, a joint French-German field experiment was conducted from November 1988 to May 1989. An array of 25 one-component and 42 three-component seismological stations were installed in the field. Seven of the three-component stations were finally kept for the anisotropic processing (identified as BF set in Table 1). In addition, one broadband station (ECH) and two mobile long-period stations (RG-N and RG-S) provided data for this particular study.

The isotropic *P*-velocity images (anomalies are approximately 2–3% maximum, which is rather small) show a graben-related pattern to about 50 km depth and no evidence of a diapiric upwarp of the asthenosphere, contrary to a model based on the Bouguer gravity field (KAHLE and WERNER, 1980) and surface-wave analysis (PANZA et al., 1980). New surface-wave inversion models (PASSIER and SNIEDER, 1996) confirm the absence of an upwarp. The upper mantle velocity structure beneath the Moho manifests a change in the anomaly pattern at about 65–75 km depth. While above 65 km, the overall trend of the structures can be defined as parallel to the graben axis, the direction of the anomaly patterns turns into a roughly E–W direction below that depth. This might be some indication of the past extensional deformation which is expected to involve the whole lithosphere. The results demonstrated that rifting and the thermal doming stage of the Rhinegraben ceased at the end of Miocene (GLAHN and GRANET, 1992; GLAHN et al., 1993). A preliminary anisotropic study (LIOTIER et al., 1991) based on the so-called "anisotropic relative *P*-travel delay times" showed a pattern of two well-characterized areas displaying opposite behavior "fast-slow," separated by the Saxothuringian Zone-Moldanubian Region boundary.

In the Urach region, refraction and reflection seismic experiments yielded a low-velocity body in the middle crust which coincides well with the contours of isotherms, based on equal geothermic gradient. Between 1985 and 1987 a mobile array of digital short-period stations was installed in the Urach geothermal field. Only 9 stations were deployed simultaneously. Most stations were moved every 5–6

Table 1

List of the permanent and mobile seismological stations which were considered in this study with their characteristics. Station ECH (Geoscope global network) is situated in the Vosges mountains on the western flank of the Rhinegraben. Stations RG-N, RG-S are located in the Rhinegraben proper. The Black Forest set (BF set) corresponds to mobile short-period stations and is located on a profile across the Black Forest Massif on the eastern flank of the rift. Station RBG and the Urach set composed with short-period mobile stations are situated in the western part of the Swabian Jura close to the Hohenzollern graben seismic zone. Stations GRF-A1 and GRF-C1 are situated in the eastern part of the Swabian Jura. Station SSB (Geoscope global network) and the Massif Central set composed with short-period stations is located in the French Massif Central.

Station code	Station name	Type of station[1]	Period of recording	Latitude (°N)	Longitude (°E)
Rhinegraben					
ECH	Echery	permanent/BB	11.1990–	48.22	7.16
RG-N	Rhinegraben (North)	mobile/20 s	04.1991–09.1992	48.96	7.92
RG-S	Rhinegraben (South)	mobile/20 s	04.1991–09.1992	48.02	7.53
BF set	Black Forest	mobile/5 s	11.1988–05.1989	48.22[2]	8.2
XSU				*48.307*	*7.862*
XSC				*48.273*	*7.949*
XHA				*48.273*	*8.098*
XKF				*48.126*	*8.431*
XDA				*48.097*	*8.533*
XSE				*48.024*	*8.722*
XUS				*47.953*	*9.006*
Urach-Gräfenberg					
RBG	Raichsberg	semi-permanent/5 s	04.1985–07.1991	48.31	8.99
Urach set	Urach	mobile/5 s	02.1986–07.1987	48.4[3]	9.3
ALT				48.253	9.143
BEU				48.283	9.422
HAU				48.304	9.194
RSW				48.231	8.836
GRF-A1	Gräfenberg A1	permanent/BB	04.1976–	49.7	11.2
GRF-C1	Gräfenberg C1	permanent/BB	04.1976–	49.0	11.5
Massif Central					
SSB	Saint Sauveur	permanent/BB	12.1986–	45.28	4.54
MC set	Massif Central	mobile/5 s	10.1991–05.1992	45.4[4]	3.2
MC12				*45.877*	*3.282*
MC22				*45.886*	*2.432*
MC30				*45.634*	*3.278*
MC32				*45.585*	*3.519*
MC37(*)				*45.355*	*4.246*
MC61				*45.384*	*1.470*
MC66				*45.141*	*2.472*

(*) Poor quality.
[1] Period of operation: permanent, semi-permanent (> 5 years of recording), or mobile (eigenperiod: 5 s, 20 s), or BB (= broadband).
[2] Center of station set.
[3] Center of mobile network.
[4] Center of mobile network.

months in order to provide an adequate high resolution sampling of the area under investigation. However, two stations, RBG and HAU, were used as reference stations during the entire experiment. The teleseismic field experiment revealed a small-scale anomalous domain with a velocity reduction of 3% in the crust and a broad low-velocity anomaly of 4% beneath the Moho to a depth of about 50 km. While the crustal low-velocity body coincides well with a postulated old magma chamber required from xenoliths studies, the reduced P-wave velocity within the lithosphere has been explained by local enrichment of phlogopite and increased temperatures (GLAHN et al., 1992).

2.2 The Massif Central Area

Tectonic setting

The Massif Central is a part of the Hercynian belt located in the center of France. During the Oligocene a graben system as part of the European Cenozoic Rift System developed in this area (Fig. 2). This rift system is characterized by two distinct volcanic stages. The first one (65–35 Ma) occurred before the formation of the Limagnegraben and the Forezgraben. The second stage ranges from the early Miocene (20 Ma) to the late Quaternary (6000 y B.P.). Therefore, the Massif Central offers the opportunity to study Variscan lithospheric structures and their modification by rift-related igneous activity during the Tertiary.

Geophysical results

From recent refraction and wide-angle reflection seismic studies beneath the graben and the volcanic areas, the crust is thinned on the order of 2–3 km only and the P_n-wave velocity beneath the Moho discontinuity is normal ($V_{P_n} = 8.0$ km/s). In addition to past studies of Rayleigh waves dispersion and attenuation which show low S velocity and low quality factor in the 70–240 km depth range, the negative Bouguer anomaly and the high heat-flow anomaly support the presence of a body of high temperatures beneath this volcanic area (SOURIAU et al., 1980; LUCAZEAU et al., 1984; ZEYEN et al., 1997).

A teleseismic field experiment conducted in 1991–1992 within the framework of French-German cooperation allowed derivation of a small wavelength three-dimensional P-velocity model for an area of about 425 km × 300 km. An array of 79 mobile short-period digital seismic recorders was installed to complement 14 permanent stations. Only 7 three-components were finally used for the anisotropic processing (identified as MC set in Table 1). Combining the knowledge of the basement geology of the area with the timing and geochemical characteristics of Tertiary-Quaternary volcanic activity, it was demonstrated that this velocity model provides unequivocal evidence for the ascent of a thermally and chemically anomalous mantle plume beneath the Massif Central during Tertiary times (GRANET et al., 1995a,b).

MASSIF CENTRAL

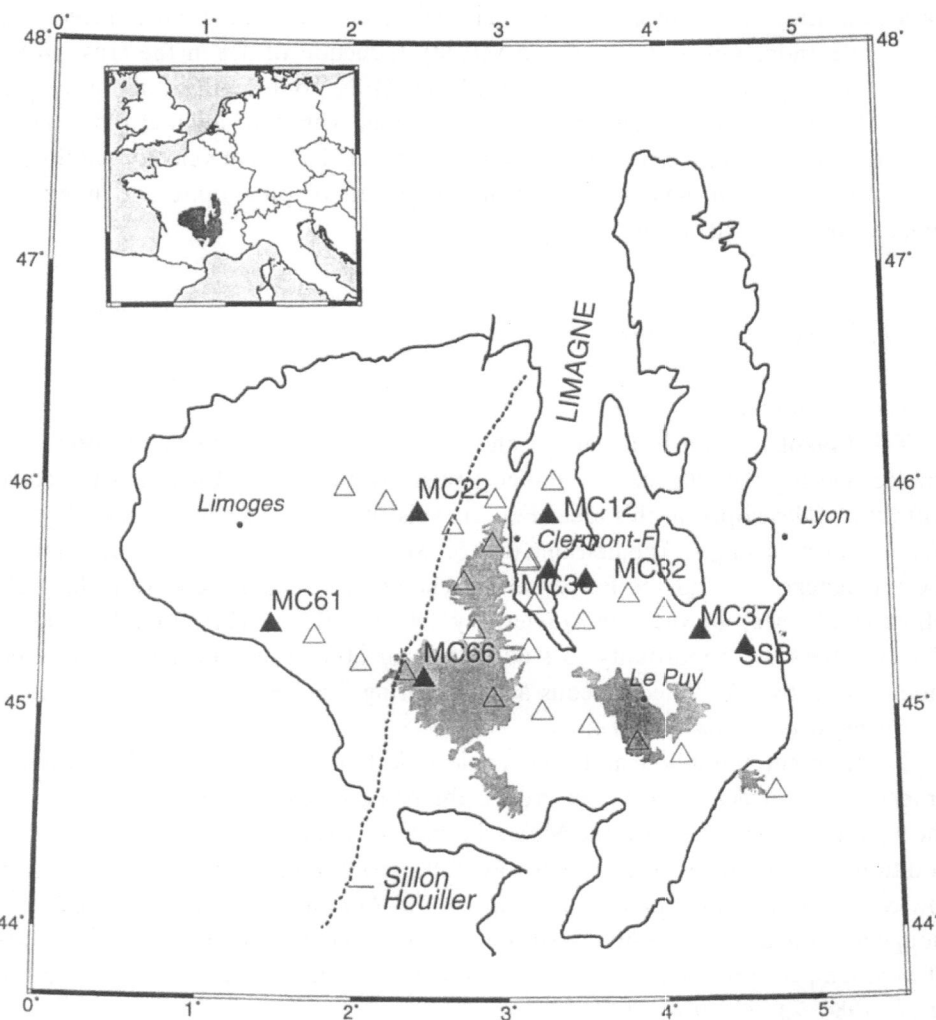

Figure 2

Map showing the location of the permanent broadband station SSB and the 3-component mobile stations installed in the Massif Central region between October 1991 and May 1992. Black triangles represent the sites where anisotropic measurements were achieved. Open triangles represent other sites. Grey areas represent the Neogene volcanic fields.

3. Methods for Retrieving the Splitting Parameters

In general, studies of the splitting of teleseismic shear waves in anisotropic media focus on the determination of the two parameters of azimuthal anisotropy, ϕ and δt. The first splitting parameter, ϕ, describes the angle between the radial direction of the incoming wave in the isotropic medium and the polarization direction of the fast shear wave. The second parameter, δt, indicates the delay time between the two quasi-shear waves.

In a first processing step, all seismograms are transformed from the original Z, N, E component system to a V, L, T system by using the theoretical backazimuth for the rotation into the horizontal plane. Thereafter, the particle motion diagram (PMD) in the horizontal plane is calculated for each event. In case the PMD displays a clear elliptical polarization which is regarded as an indication for anisotropy, either the rotation correlation method (FUKAO, 1984) or an inversion method developed by Vinnik and co-workers (VINNIK et al., 1988) are applied to retrieve the parameters of azimuthal anisotropy. Linear, or quasi-linear polarization means that no (or very weak) anisotropic effects are present or that the incident direction of the wave coincides with the fast (or slow) axis of the anisotropic medium.

Rotation Correlation Method (RCM)

In using the RCM, the radial and transverse components are rotated gradually in increments of 5° in the range between 0° and 90°. For each rotation angle the cross-correlation function is calculated. This is illustrated on Figure 3. The goal is to minimize the SKS phase on the transverse component which corresponds to a minimum of the cross-correlation function. The values of ϕ and δt which are obtained at the minimum of the cross-correlation function are finally applied to correct the radial and transverse components for the observed anisotropic effect. Under the assumption of a single transversely isotropic structure with a horizontal axis of symmetry beneath the recording station, a correction of the radial and transverse components for ϕ and δt should result in minimizing SKS on the transverse component and a linear polarization of the corrected components in the PMD.

Inversion Method

A different analysis technique was described by Vinnik and co-workers (VINNIK et al., 1988, 1989). Again, a transversely isotropic layer with a horizontal axis of symmetry beneath the station is assumed. From the observed radial component R, the transverse component T of SKS and similar phases is synthesized for numerous values of the unknown model parameter space, ϕ and δt. Assuming that the incoming shear wave is a harmonic wave, the penalty function E (ϕ, δt) is calculated on the basis of a Fourier series decomposition of the derivative of the radial component. The splitting parameters are obtained by minimizing the RMS difference between the synthetic and observed transverse component. This is illustrated on Figure 4.

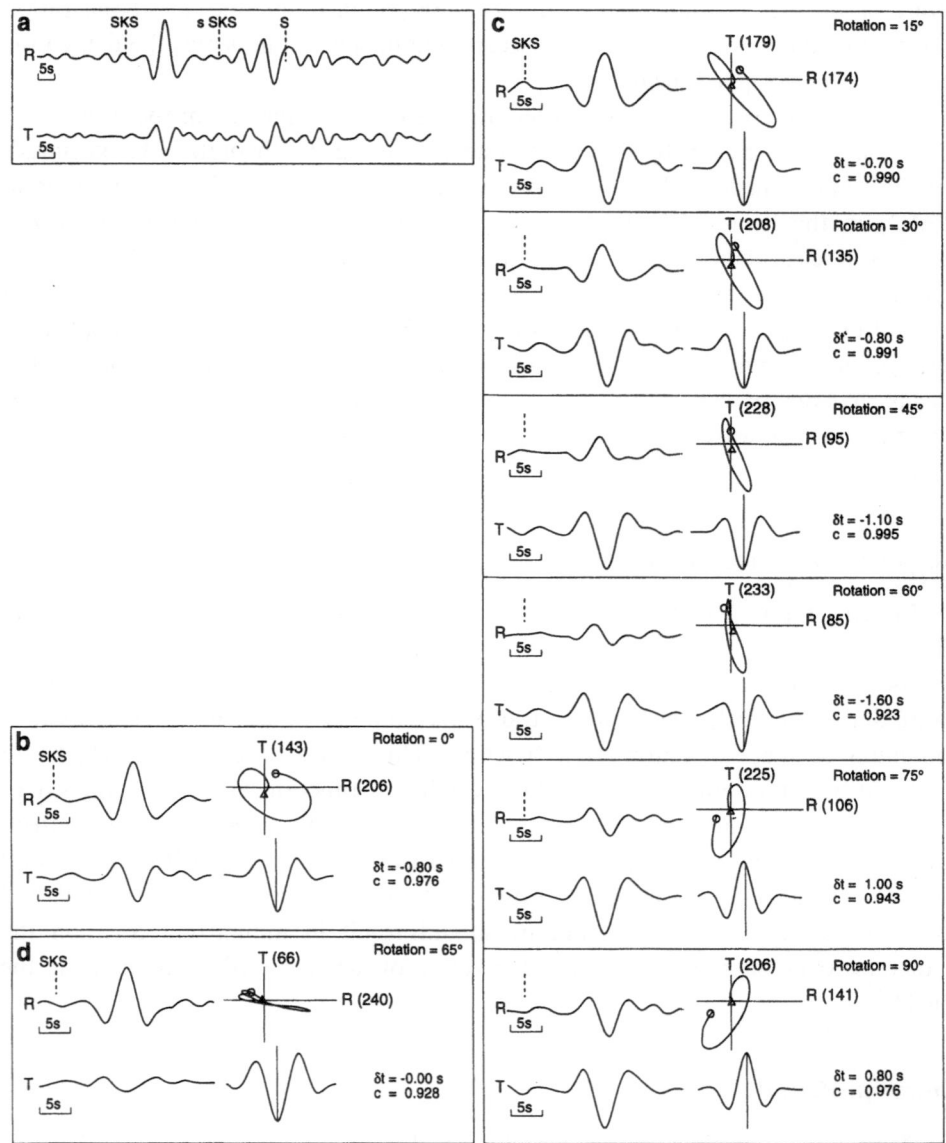

Figure 3

An illustration of the Rotation Correlation Method (RCM) by FUKAO (1984) for a Chile event recorded at Urach station RBG. Panel a: Radial (R) and Transverse (T) components with theoretical arrival times of SKS, $sSKS$ and S phases as indicated by vertical dashed bars. Traces are filtered in the frequency band (0.125 Hz, 0.5 Hz). Panel b: Particle motion diagram (PMD) of SKS phase and cross-correlation function. The amplitudes (maxima) on the R and T axes are in digits; "δt" represents the delay time in seconds and "c" is the value of the cross-correlation coefficient. Panel c: Examples of PMD and the cross-correlation function with a rotation angle varying from 0° to 90° in increments of $\Delta\varphi = 15°$. For this Chile event, the minimum of the cross-correlation function is obtained for a rotation angle of +65° (not displayed) and a delay time $\delta t = 1.3$ s. The splitting parameter ϕ is deduced from the formula (ANSEL, 1989): $\phi = \text{baz} - (+65°) + \pi/2 = 273°$ (93°). Panel d: Corrected R and T components after a time correction of $-\delta t = 1.3$ s has been applied to the slow shear wave and an inverse rotation of $-65°$.

SKS energy on T is minimized and the PMD displays a linear polarization in a radial direction.

With the Vinnik method, ϕ and δt can be retrieved for single events or for a set of events. In the latter case, a normalization is applied to prevent any bias due to the different energy content of each event. E is calculated for increments $\Delta\phi = 1°$ in the range between $0°$ and $180°$, and increments $\Delta\delta t = 0.05$ s in the range of 0 s and 2.4 s.

Neither the RCM nor Vinnik's method produce any information on the accuracy (or error bars) of the determined splitting parameters of each analyzed event (or set of events).

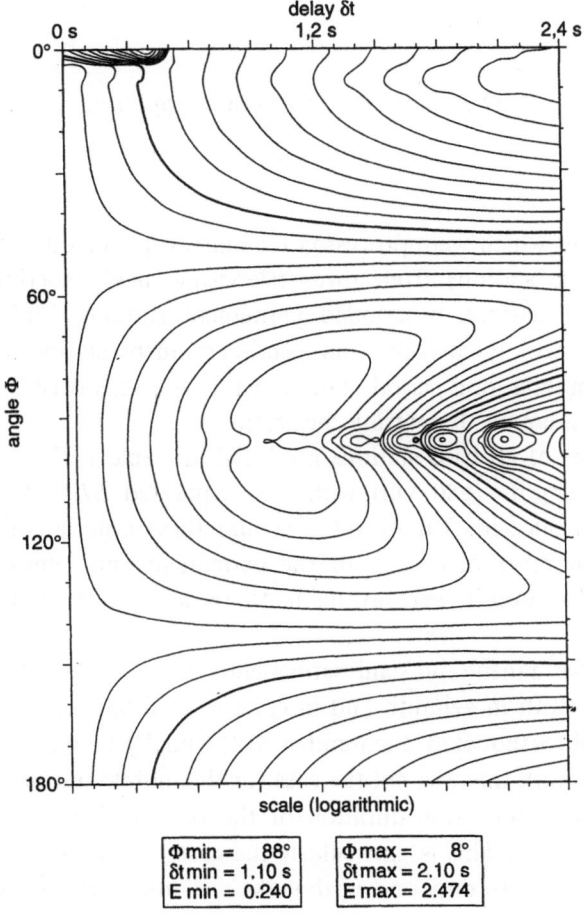

Φ min =	88°	Φ max =	8°
δt min = 1.10 s		δt max = 2.10 s	
E min = 0.240		E max = 2.474	

Figure 4

Plot of function $E(\phi, \delta t)$ in the range $0° \leq \phi \leq 180°$ and 0 s $\leq \delta t \leq 2.4$ s in order to illustrate the Inversion Method by VINNIK et al. (1988) for the Chile event recorded at RBG as analyzed in Figure 3. The minimum of E is obtained for $\phi = 88°$ and $\delta t = 1.1$ s.

In order to compare the results, a couple of events has been analyzed for both methods. A difference of $\pm 7°$ in ϕ and ± 0.3 s in δt between the two analyzing methods can be regarded as a measure for the accuracy of the parameters of azimuthal anisotropy, derived from splitting studies employing teleseismic shear waves.

The harmonic analysis method (which takes advantage of the 180° periodicity of amplitude and trave-time variations, e.g., KIND *et al.*, 1985) generates only an average value of ϕ and δt. Compared to methods in which each recording is analyzed individually by minimizing the *SKS* energy on the transverse component, the harmonic analysis does not convey a direct impression of the scatter in the anisotropic parameters. In the case of recordings well distributed in azimuths, we argue that one should apply both methods on the same dataset.

4. Observations of Shear-wave Splitting

4.1 Data

The earthquakes which were processed for anisotropic investigations have been recorded by mobile stations from three teleseismic field experiments with 5 s intermediate-period sensors, broadband permanent stations GRF-A1, GRF-C1, ECH (Geoscope), SSB (Geoscope), one semi-permanent station with 5 s sensor (RBG) and two mobile long-period stations with 20 s sensor (RG-N, RG-S). All sensors were well adapted to splitting measurements.

The coordinates of all used stations are listed in Table 1. Only large and deep events ($m_b \geq 6.0$, depths ≥ 80 km) with well separated *SKS*, *SKKS*, and *PKS* phases, have been selected. Taking into account these criteria, this resulted in a quite reasonable number of events for the permanent and semi-mobile stations, however only a few events were available to retrieve splitting parameters from mobile stations.

While for the permanent and semi-permanent stations the set of recorded events is well distributed both in azimuth and in epicentral distance (see Table 2 for the number of selected earthquakes, the number of the finally retained events and their azimuthal distribution), this was not the case for the mobile stations, due mainly to the limited recording time (the duration of the passive field experiments was 6 months as a maximum). This is particularly the case for the Black Forest and the Massif Central sets, where only one earthquake was finally considered for the data processing (Figs. 7 and 8).

As observed above, for the southern Rhinegraben area, no even azimuthal distribution of teleseismic recordings with a clear elliptical (or linear) polarization of the *SKS* phase has been obtained from the 6-month field experiment. The

Table 2

Mean values of splitting parameters (ϕ, δt) for the set of the seismological stations. For the MC set and the BF set of stations, individual estimates are indicated as these two mobile arrays provide only one event showing a clear elliptical polarization. One can see the good azimuthal distribution of the final set of data for ECH, RG-N, RG-S, RBG, the Urach set, GRF-A1 and SSB. All of these were either permanent stations or installed for a longer time on the field.

	No. of events	No. of data[1]	Azimuthal distribution	Analyzed phases	ϕ^2 (deg.)	δt^3 (s)	σ_ϕ (deg.)	$\sigma_{\delta t}$ (s)
Rhinegraben								
ECH	72	16	33–326	S, SKS, SKKS	29	1.2	46	0.37
RG-N	29	12	8–263	SKS, SKKS	34	1.2	37	0.53
RG-S	12	5	37–263	SKS	38	1.2	36	0.23
BF set	6	1	42	SKS				
XSU		1			57	1.5		
XSC		1			62	1.3		
XHA		1			72	1.4		
XKF		1			72	1.4		
XDA		1			87	1.3		
XSE		1			92	1.4		
XUS		1			92	1.2		
Urach-Gräfenberg								
RBG	64	13	4–338	S, SKS, SKKS	78	1.3	28	0.21
Urach set	26	6	13–245	S, SKS, SKKS				
ALT		5	13–245		82	1.2	37	0.19
BEU		3	13–244		73	1.3	17	0.43
HAU		6	13–245		72	1.2	27	0.26
RSW		5	13–245		65	1.2	24	0.10
GRF-A1	25	8	16–358	S, SKS	78	1.1	17	0.23
GRF-C1	12	2	20–66	S, SKS	91	1.1	9	0.0
Massif Central								
SSB	55	12	29–359	S, SKS, PKS	146	1.1	19	0.42
MC set	9	1	38	SKS				
MC12		1			77	0.4		
MC22		1			31	1.2		
MC30		1			78	0.7		
MC32		1			109	0.8		
MC37(*)		1			130	1.5		
MC61		1			30	0.8		
MC66		1			109	0.6		

(*) Poor quality.
[1] From which ϕ and δt are derived.
[2] Mean value of polarization direction of fast shear wave clockwise from North.
[3] Mean value of delay time between fast- and slow-split shear waves.

additional use of S phases with sometimes longer dominating periods resulted in large time-shifts of 2 sec and more between the two quasi-shear waves and therefore were not considered. Besides the time-shift, we observed a significant scatter in the

direction of the fast axis, which in most publications dealing with the splitting of
S phases has not been addressed in the discussion. We analyzed recordings which
were filtered in three different period bands: (2–8 s), (2–15 s), (8–15 s). A
dependence of the amplitude of the time-shift on the dominant period of the
incoming SKS (or S) phase was observed. Finally, following ANSEL (1989), all
seismograms were filtered in the range of 2 s to 8 s.

4.2 Results

Table 2 displays the splitting parameters (ϕ, δt) as obtained from the data
processing described in section 3. For single stations (ECH, RG-N, RG-S, RBG,
GRF-A1, GRF-C1 and SSB), only the mean values are reported together with
the corresponding standard errors. For the so-called BF (Black Forest) and MC
(Massif Central) sets, no information on the possible scatter is available as only
one event was processed. Details of the results will be discussed hereafter. One
can note that the number of earthquakes which were finally used for shear-
splitting measurements is always substantially smaller than the initial set of
events fitting the selection criteria. This is mainly caused by a lack of energy
however also by complex waveforms associated with the shear waves. For all
cases, the standard errors associated with the mean values of the fast-polarization
azimuth are large, thus indicating an azimuthal dependence of this fast direction
as will be shown later. Figure 5 advances a comparison of the results with those
from the literature for permanent European stations (SSB, GRF, NE05, NE15,
TNS, CLZ, MOX, CLL, BRG, KSP, KHC, WET, STU, BFO and FUR: VIN-
NIK *et al.*, 1989; FARRA *et al.*, 1991; BORMANN *et al.*, 1993; VINNIK *et al.*,
1994).

As a first result, azimuthal anisotropy in the upper mantle underneath the
southern Rhinegraben area and in the French Massif Central is observed. While
the mean values of delay time between the fast and slow shear waves range
between 1.0 s and 1.3 s in the Rhinegraben-Urach, which is consistent with the
worldwide average of splitting delay times which shows a mean distribution of
1 s, these delay times are significantly smaller in the French Massif Central
volcanic field. The fast-polarization azimuth in the Rhinegraben area/Southern
Germany ranges from about N30°E (ECH, RG-N, RG-S) to about N90°E (XSE,
XUS, GRF-C1) and from N30°E (MC61) to N146°E (SSB) in the Massif Central. Tables 4.1–4.4 detail the individual results for stations ECH, RG-N, RG-S
and RBG, for which significant azimuthal variations have been found. Due to
the lengthy time of data acquisition for this study, two methods were applied for
retrieving the splitting parameters. The correlation-rotation method was primarily
used for the stations RBG, GRF-A1, GRF-C1, the BF set and the Urach set
while VINNIK's method (1984, 1989) was applied for station ECH, RG-N, RG-S,
SSB and the MC set (Table 1).

Rhinegraben area

Individual estimates of the splitting parameters associated with the single stations ECH (broadband), RG-N, RG-S (long period) are listed in Tables 4.1, 4.2 and 4.3. For each table, the date of the earthquake, the epicentral area, the backazimuth, the type of the seismic phase and the splitting parameters are indicated.

For station ECH (Table 4.1), 16 earthquakes were processed using Vinnik's method. There is good coverage in the azimuth and mainly SKS phases were used. As the main result one notes a clear dependence of the polarization direction of the fast shear wave on the backazimuth with values ranging from $-52°$ (Peru event of 24.05.1991, baz $= 250°$) to $84°$ (Chile event of 23.06.1991, baz $= 238°$). This is

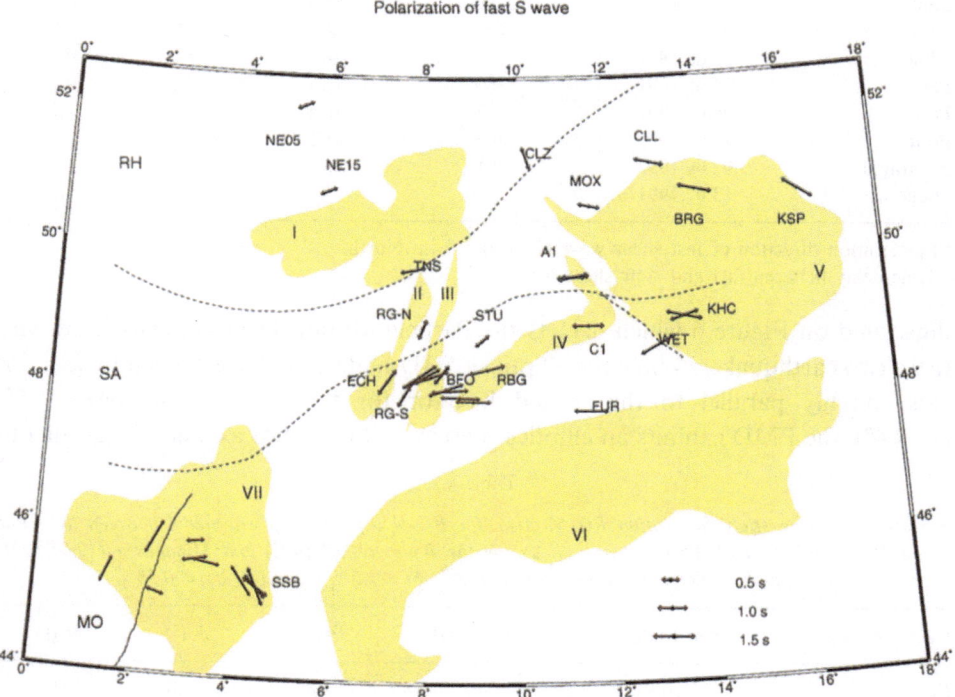

Figure 5

Splitting measurements in western Europe as obtained from this study and published results. Orientation of the arrows give their fast-polarization azimuth ϕ; line lengths are proportional to delay time δt according to the legend. Red arrows correspond to our results. Blue arrows correspond to old data from the literature: NE05, NE15, TNS, CLZ, MOX, CLL, BRG, KSP, KHC, WET, STU, BFO, FUR (VINNIK *et al.*, 1989, 1994; FARRA *et al.*, 1991; BORMANN *et al.*, 1993). We superimposed our measurements with older ones at GRF-A1 and SSB. One may note that ancient and recent data are in good agreement. Yellow areas correspond to the main tectonic units which were sampled by the data: (I) Rhenish Massif, (II) Vosges mountains, (III) Black Forest mountains, (IV) Swabian Jura, (V) Bohemian Massif, (VI) Alps, (VII) French Massif Central. Codes RH, SA, MO mean Rhenohercynian, Saxothuringian, Moldanubian Hercynian terranes, respectively.

Table 4.1

Results of splitting study for station ECH. Note the jump for the ϕ parameter observed between the Chile earthquake (23.06.1991) and the Peru earthquake (24.05.1991) within a very narrow azimuthal range. A similar jump of ϕ was found between the Mindanao (17.05.1992) and Java (02.09.1991) earthquakes. The mean values are: $\phi = 29°$, $\delta t = 1.2$ s, $\sigma_\phi = 46°$, $\sigma_{\delta t} = 0.37$ s.

Event (region)	Event (date)	Backazimuth	Phase	ϕ (°)[1]	δt (s)[2]
Vanuatu	11.10.1992	33	SKKS	60	1.0
Honshu	29.08.1992	39	SKS	72	1.0
Honshu	30.10.1992	40	SKS	60	1.8
Bonin	20.01.1992	41	SKS	72	1.0
Ryukyu	14.04.1991	50	SKS	71	1.0
Taiwan	07.05.1993	52	SKS	64	1.6
Mindanao	17.05.1992	63	SKS	52	1.3
Jawa	02.09.1992	83	SKS	8	1.3
Sumatera	02.07.1991	89	SKS	32	0.9
Sandwich	05.04.1993	197	SKS	36	1.5
Chile	23.06.1991	238	SKS	84	0.8
Peru	24.05.1991	250	SKS	− 52	0.9
Peru	06.07.1991	253	SKS	− 44	1.2
Peru	13.07.1992	263	SKS	− 32	0.7
Nicaragua	02.09.1992	281	SKS	20	1.7
Oregon	13.07.1991	326	S	− 41	1.8

[1] Polarization direction of fast shear wave clockwise from North.
[2] Time delay between fast and slow shear waves.

illustrated on Figure 6 which depicts the Particle Motion Diagram associated with these two earthquakes. While the elliptical PMD pattern is characterized by a major axis roughly parallel to the second bisector for the Chile event (baz = 238°, $\phi = 84°$), the PMD exhibits an elliptical pattern with a major axis nearly parallel to

Table 4.2

Results of splitting study for station RG-N. Also, for RG-N a jump of ϕ was found between the Chile (23.06.1991) and Peru (24.05.1991) as well as for the Bonin (20.01.1991) and Minahassa (19.05.1991) earthquakes. The mean values are: $\phi = 34°$, $\delta t = 1.2$ s, $\sigma_\phi = 37°$, $\sigma_{\delta t} = 0.53$ s.

Event (region)	Event (date)	Backazimuth	Phase	ϕ (°)	δt (s)
Fiji	09.06.1991	8	SKKS	40	0.9
Fiji	11.07.1992	13	SKKS	36	1.5
Honshu	29.05.1992	38	SKS	64	1.3
Bonin	03.05.1991	41	SKS	76	1.3
Bonin	20.01.1991	41	SKS	64	1.7
Minahassa	19.05.1991	70	SKS	−8	1.3
Minahassa	20.06.1991	70	SKS	−8	0.8
Sumatera	02.07.1991	89	SKS	76	1.2
Chile	23.06.1991	239	SKS	72	0.7
Peru	24.05.1991	250	SKS	−16	2.4
Peru	06.07.1991	254	SKS	−8	0.5
Peru	13.07.1992	263	SKS	16	0.6

Table 4.3

Results of splitting study for station RG-S. The mean values are: $\phi = 38°$, $\delta t = 1.2$ s, $\sigma_\phi = 36°$, $\sigma_{\delta t} = 0.23$ s.

Event (region)	Event (date)	Backazimuth	Phase	ϕ (°)[1]	δt (s)[2]
Honshu	11.08.1992	37	SKS	48	1.0
Honshu	30.05.1992	38	SKS	48	1.2
Minahassa	20.06.1991	70	SKS	60	1.3
Chile	23.06.1991	238	SKS	58	1.3
Peru	13.07.1992	263	SKS	− 24	1.1

[1] Polarization direction of fast shear wave clockwise from North.
[2] Time delay between fast and slow shear waves.

the first bisector for the Peru event (baz = 250°, $\phi = -52°$). However, the difference between the estimates for the same narrow azimuthal sector is generally small, thus demonstrating the accuracy of the individual measurements (e.g., Honshu-Bonin or a triplet of Peru events). From this set of results we can identify two subsets. One for which the ϕ parameter is positive with a mean value around 52 degrees—with however a notable exception (8° for the Jawa event of 02.09.1992), and the second one consisting of events showing a negative value for ϕ with a mean value of $-42°$, which differs from the other estimates by up to 90°. Such a grouping behavior is not observed for the delay times: δt values range between 0.7 s and 1.8 s. These variations must be regarded as being large. Using Vinnik's method, one does not calculate error estimates. Only subjective errors can be derived from the energy plots (minimized energy) of each single measurement.

Table 4.4

Results of splitting study for station RBG where a jump was observed between the Andreanof (18.06.1986) and Kuril (19.07.1986) earthquakes. The mean values are: $\phi = 78°$, $\delta t = 1.3$ s, $\sigma_\phi = 28°$, $\sigma_{\delta t} = 0.21$ s.

Event (region)	Event (date)	Backazimuth	Phase	ϕ (°)[1]	δt (s)[2]
Andreanof	18.06.1986	4	S	114	1.4
Kuril	19.07.1986	25	SKS	40	1.2
Ochotsk	05.07.1991	28	S	43	1.0
Eastern USSR	07.05.1987	33	S	48	0.9
Marianas	16.09.1986	41	SKS	51	1.7
Philippine	07.06.1987	60	S	100	1.2
Burma	18.05.1987	76	S	91	1.3
Nicobar	02.06.1986	88	SKS	103	1.5
Sumatera	02.07.1991	90	S	80	1.3
Argentina	01.04.1987	244	SKS	79	1.4
Chile	05.03.1987	245	SKKS	80	1.5
Chile	22.02.1988	248	SKS	93	1.3
NW Territories	23.12.1985	338	S	97	1.4

[1] Polarization direction of fast shear wave clockwise from North.
[2] Time delay between fast and slow shear waves.

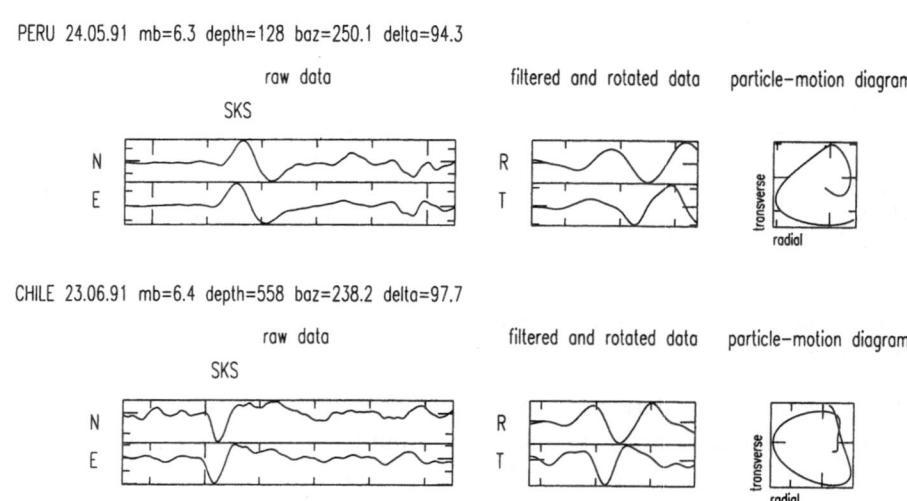

Figure 6
Seismograms and Particle Motion Diagrams (PMD) of *SKS* phases for the Peru (24.05.1991) and Chile (23.06.1991) earthquakes recorded at the single station ECH. One can note the major difference in the PMD patterns for these two earthquakes emanating from a narrow epicentral region, hence illustrating the jump in polarization direction. The length of the time windows is 8 s.

For stations RG-N (Table 4.2) and RG-S (Table 4.3), Vinnik's method was again used. Both long-period stations were installed in the field for about $1\frac{1}{2}$ years. While 12 events were finally kept for RG-N—their backazimuths ranging from 8° to 263°, only 5 events showing a clear elliptical polarization were processed for RG-S, their backazimuths ranging from 37° to 263°. As already stated for station ECH, two groups of earthquakes are evidenced for RG-N and RG-S. The first group is characterized by a positive mean ϕ value ($\sim 55°$ and $\sim 53°$ for RG-N and RG-S, respectively), while the second one is characterized by a negative value ($-10°$ and $-24°$ for RG-N and RG-S, respectively). The delay times exhibit wider variations for RG-N (of 0.5 s to 2.4 s) and small variations for RG-S (of 1.0 s to 1.3 s). As for the station ECH, one finds that the difference estimates for some events from a narrow epicentral region is small, thus proving again the accuracy of the data processing (e.g., Honshu-Bonin or a couple of Peru events).

The splitting parameters obtained for the BF set are presented in Table 2. The stations were located on the eastern shoulder of the Rhinegraben proper (see Fig. 1). From the 6 month time duration of the field experiment only one event shows a clear elliptical polarization. The mean splitting parameters are $\phi = 76°$ and $\delta t = 1.3$ s. The noticeable feature is the rotation of the fast-polarization azimuth from west to east (N57°E to N92°E).

Figure 7 shows the *SKS* records and the corresponding particle motion diagrams (PMD) for the Honshu, 27 April 1989, earthquake. The polarization of the

SKS phase is clearly elliptical for stations XSU, XKF and XSE. A nearly linear PMD is obtained for stations XHA and XDA which deviates by about 45° from the radial direction. While the elliptical form of the PMD means that the *T* component is shifted with respect to *R* by approximately a quarter period, a linear polarization might be explained by either lateral heterogeneity even the deviation from the radial polarization is large or by weak anisotropy.

Urach and Gräfenberg regions

Individual estimates of the splitting parameters are listed only for station RBG (Table 4.4). We present the splitting parameters for 13 events, providing a good azimuthal coverage from the initial set of 64 events. Again, as observed for the

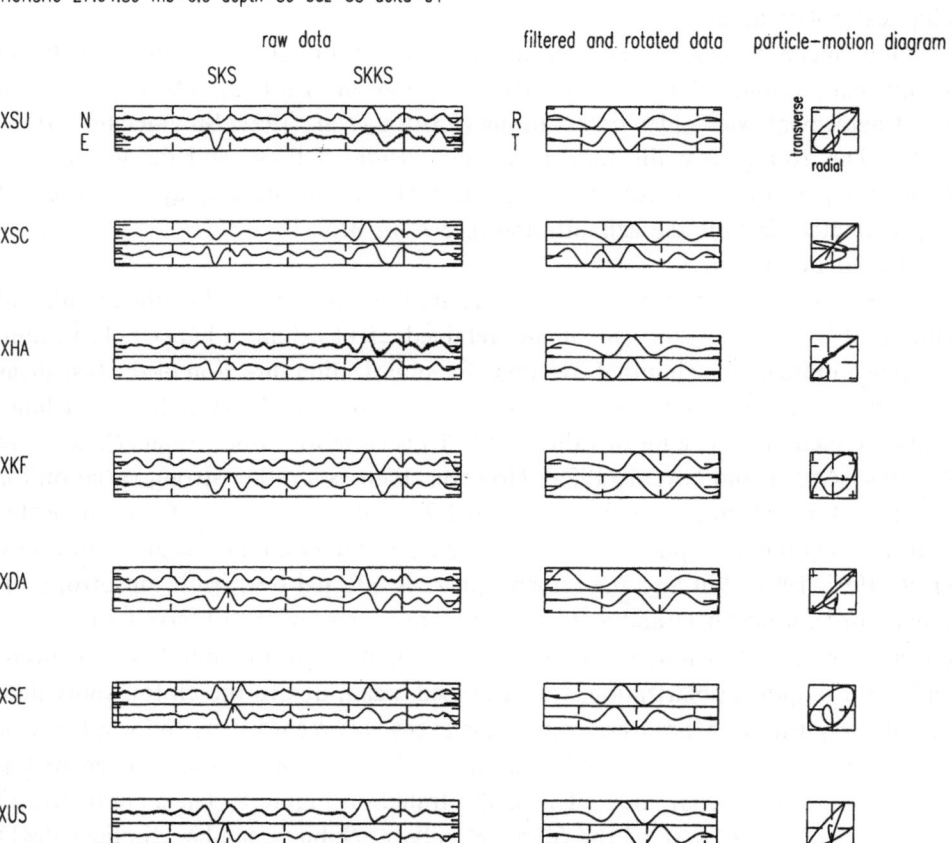

HONSHU 27.04.89 mb=6.0 depth=89 baz=38 delta=91

Figure 7

Seismograms of *SKS* phases for the Honshu (27.04.1989) earthquake recorded at the BF set. The left panels show the original raw data (north and east components). From top to bottom the records are displayed from the extreme west (XSU) to the extreme east (XUS). The central panels exhibit the filtered (0.125 Hz, 0.5 Hz) and rotated data: *R* and *T* are the radial and transverse components, respectively. The right panels show the Particle Motion Diagrams (PMD) given with respect to the orientation for the *T* and *R* axes, as indicated. The length of the time windows is 8 s.

Rhinegraben area, there is a dependence of the polarization direction of the fast shear wave on the backazimuth. Two subsets of data are evidenced. One, with a mean value of ϕ of around N90°E, the second with a mean value of ϕ of around N45°E. The second group of events corresponds to an azimutal range of 25°–41°.

The splitting parameters for stations in the Urach region (stations ALT, BEU, HAU and RSW) are shown in Table 2. Only a few events were finally considered. There is again a strong dependence of the polarization azimuth of the fast shear wave on the backazimuth. The resulting estimates (obtained by the combined processing of all records) are similar. When processing the data, we noted a tendency that stations away from the low-velocity anomaly observed in the crust and upper mantle (as seen in the tomographic images, see GLAHN *et al.*, 1992) show similar elliptical polarization, while stations on top of the anomaly show no elliptical polarization.

Only mean values of the splitting parameters obtained for the broadband permanent stations GRF-A1 and GRF-C1 are given (Table 2). The results are in good agreement with older observations (FARRA *et al.*, 1991; SILVER and CHAN, 1991). The average fast direction is mainly orientated E–W and the mean delay time is equal to 1.1 s for both stations. The estimates display a scatter of approximately 20° for the azimuth and 0.25 s for δt.

Massif Central area

Mean values of the splitting parameters are listed in Table 2. For the broadband station SSB, from the 20 events finally retained, eight exhibit a linear polarization. Despite a quite good azimuthal coverage, the polarization direction of the fast shear wave does not show a large scatter in contrast to that observed for the Rhinegraben-Urach area. The mean value (N146°E) is close to former results (VINNIK *et al.*, 1994; ANSEL and NATAF, 1989). However, there is a rather strong variation for the delay time which ranges from 0.5 s to 1.8 s. The mean value of 1.1 s indicates that not only the lithosphere, with a thickness around 70 km in this area (SOBOLEV *et al.*, 1996, 1997), but also the asthenosphere is affected by seismic anisotropy. In other words, lithosphere and asthenosphere are sources of the observed anisotropy effects. For the set of mobile stations in the Massif Central region, only one event with clear elliptical polarization was processed, again due to the all too short time of data acquisition. Moreover, even if this event was recorded by the whole set of the 3-components stations, only 6 stations exhibited a polarization pattern with a particle motion far from linear. This is illustrated on Figure 8. The stations MC22 and MC61, located west of the Sillon Houiller, exhibit a similar elliptical PMD pattern with a major axis roughly parallel to the second bisector. On the other hand, stations MC12, MC30, MC32 and MC66, located east of the Sillon Houiller, also show a self-consistent PMD, which, however, is quite different than the one observed for western stations. Hence, the splitting results yield a distinct difference between two regions on each side of the Sillon Houiller fault, which appears as a main boundary between a stable lithosphere to the west and a more heterogeneous

BONIN ISLAND 20.01.92 mb=5.9 depth=536 baz=38 delta=96

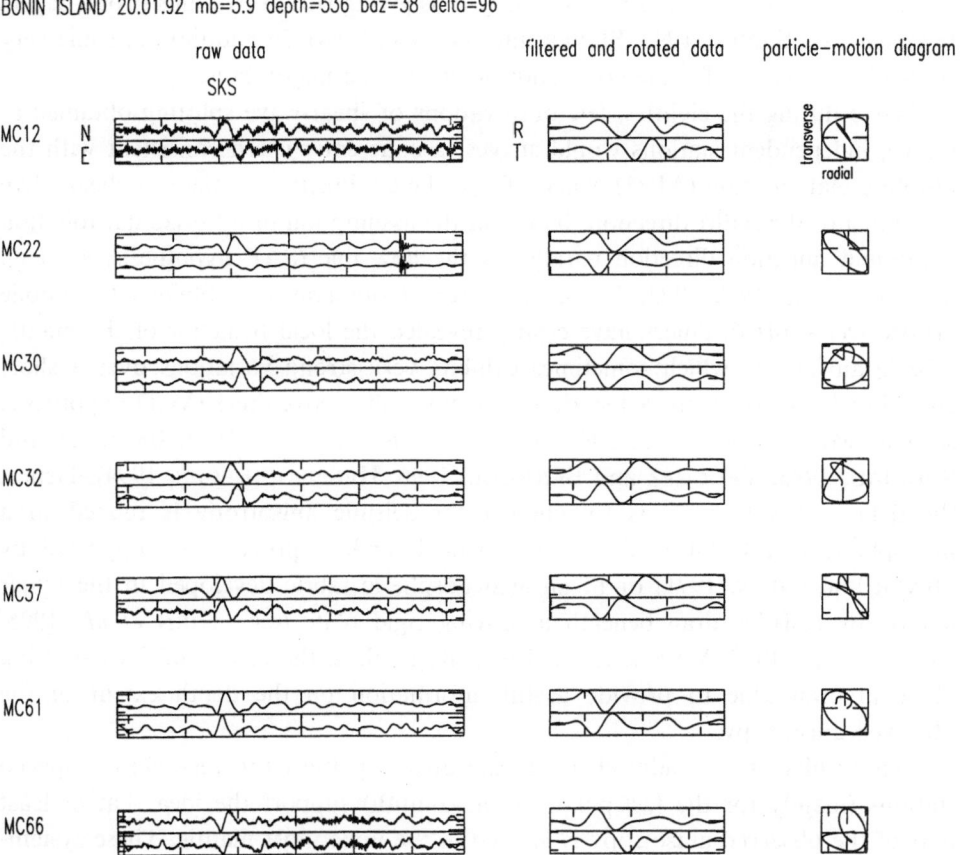

Figure 8

Same as Figure 7, but for the Bonin (20.01.1992) earthquake recorded at the MC set. The length of the time windows is 8 s.

reworked lithosphere to the east. The group of stations east of the *SH* yields a fast direction between 77° and 130°, while the other group, west of the *SH*, shows a different pattern with a fast direction aligned with the strike of the fault ($\sim$ N30°E). Contrastingly, the delay times do not manifest such a clear tendency.

5. Discussion and Tectonic Implications

SILVER (1996) presented an exhaustive review of the plausible causes of seismic anisotropy observed in subcrustal areas. In summary, there are as main hypotheses the so-called vertically coherent deformation (VCD: crust and subcontinental mantle deform coherently in orogenies) and the simple asthenospheric flow (SAF: subcontinental mantle anisotropy is dominated by asthenospheric flow). Both VCD and SAF can also operate simultaneously thus giving rise to two anisotropic layers (SAVAGE and SILVER, 1993; SILVER and SAVAGE, 1994). In addition, in some very particular areas, small-scale convection might be the major cause.

While during the eighties, few observations of shear-wave splitting obtained at scarce, independent stations would suggest a coherent pattern consistent with the absolute plate motion (APM), where the predicted direction of the fast shear wave is parallel to the APM direction, based on the assumption of a horizontal foliation plane and lineation direction parallel to the flow line (SAF hypothesis: see e.g., VINNIK *et al.*, 1989, 1992, 1995), many recent observations obtained by mobile arrays over short distances have clearly revealed the local behavior of the mantle seismic anisotropy which sometimes exhibits very strong variations over a short spatial scale at closely spaced stations in tectonically active areas (VCD hypothesis: see e.g., MCNAMARA *et al.*, 1994; SILVER and KANESHIMA, 1993; BARRUOL and SOURIAU, 1995; VAUCHEZ and BARRUOL, 1996; HELFFRICH *et al.*, 1994). Hence, the debate is still open as to whether the seismic anisotropy is related to a lithospheric mantle fabric due to the plate boundary process (as supported by SILVER and CHAN, 1991) or to an asthenospheric fabric developed in the upper mantle by resistive drag beneath a moving rigid plate (e.g., HIRN *et al.*, 1995; VINNIK *et al.*, 1992; VINNIK *et al.*, 1995). Regardless, the results of SKS splitting alone are not able to provide useful information on the depth extent of the observed anisotropy.

Our results of a rapidly changing anisotropy pattern for very closely spaced stations (mainly for the fast-polarization azimuth) support the idea that at least part of the observed anisotropy is located in the uppermost mantle. These systematic variations on a local scale are not consistent with a "pure" SAF hypothesis.

Our observations are plotted on Figure 5. We note that the directions of ϕ apparently follow coherently the large-scale geologic trends of western Europe and thus reflect geological variations which are parallel to the fabric. Apart from the most active tectonic areas (volcanic Neogene region in the French Massif Central and the Rhinegraben proper), the stations display a systematic variation from roughly a NNE direction (N30°E) in the west (stations MC22 and MC61) to nearly an east-west direction in the east (GRF-A1 and GRF-C1). This rotation of approximately 60° over 800 km closely follows the general geologic trends of the Paleozoic Hercynian orogenic belt in this zone, suggesting a close correlation between the orientation of large-scale surface deformation and the orientation of ϕ. VINNIK *et al.* (1994) argued that the values of ϕ are locally perpendicular to a smoothed model of the direction of maximum horizontal stress (GRÜNTHAL and STROMEYER, 1992) and that the observed anisotropy may reflect the asthenospheric flow induced by the Alpine Orogeny. At GRF (A1 and C1), the fast-polarization azimuth is practically parallel to the predominant Hercynian deformation, suggesting that it might be the primary cause of the anisotropic fabric.

In the Rhinegraben area, the broadband stations ECH and BFO, located on the western (Vosges mountains) and on the eastern flanks (Black Forest mountains), respectively, clearly exhibit a graben related pattern with a fast-polarization azimuth aligned with the graben strike, thus evidencing a coherence of ϕ with young tectonic structures. Assuming that pure shear deformation is the most important factor controlling the deformation pattern in the lithosphere during rifting and collision processes, the observed ϕ direction suggests a relation to the period of rifting which is the most recent, significant episode of deformation in the Rhinegraben region. Thus, the fast-polarization azimuth trend in the seismic anisotropy pattern for the rift area is, at least partially, controlled by a contribution from recent tectonics. In other words, as suggested by VAUCHEZ and NICOLAS (1991), one might suspect that the lithosphere beneath stations RG-N, RG-S and ECH (also BFO, although it was not analyzed by us) has a fabric owing to the deformation associated with transcurrent lithosphere shear zones. Another explanation would be to postulate an ancient "rift-parallel" mantle flow which accounts locally for the lithospheric fabric. Note also that the rupture of the continental lithosphere takes place along pre-existing Hercynian zones of weakness.

In the French Massif Central region, the rapid change in splitting properties across the Sillon Houiller fault again suggests that a significant part of the anisotropy derives from the subcrustal lithosphere. Moreover, while the fast-polarization azimuth west of the Sillon Houiller correlates well with the strike of this late-variscan fault (BURG et al., 1990), again suggesting that deformations near major strike-slip faults is the prevailing mechanism to generate anisotropy with a fast axis parallel to the strike of the fault, this fast-polarization azimuth shows east of the fault a roughly N100°E trend which correlates remarkably well with the low velocity pattern at upper mantle levels obtained from a teleseismic tomography study (GRANET et al., 1995a,b), probably associated with the asthenosphere if one considers the thin lithosphere as suggested in recent papers (SOBOLEV et al., 1996, 1997). Hence, the reorientation of the mantle fabric due to the rising plume, responsible for the rough E–W orientation of ϕ, seems to be a plausible cause.

BABUŠKA et al. (1993) calculated diagrams of the azimuth-incidence-angle-dependent terms of relative P residuals for some European permanent seismological stations. In the Massif Central, 4 stations were analyzed. The diagrams show a more or less consistent bipolar pattern of spatial variations of P residuals with a NNW dipping fast direction. This was interpreted as caused by large dipping anisotropic structures in the subcrustal lithosphere. However, in order to better constrain the model of seismic anisotropy in this area, we must study both P- and S-wave anisotropy at closely spaced seismological stations (PLOMEROVÁ et al., 1996).

According to SILVER and CHAN (1991), the values of the averaged delay times, δt_{avr}, i.e., the difference between the arrival time of the fast and slow shear waves, provide an estimate of the thickness of the anisotropic structure. Following SILVER

(1996), in the case of anisotropy and a coherent strain pattern in the LAS, one second of delay time corresponds to an apparent layer thickness of nearly 115 km, assuming a dimensionless intrinsic anisotropy of 0.04, based on the assumption that the upper mantle has about 60% olivine content (MAINPRICE and SILVER, 1993). Considering only averaged delay times (BF set and MC set are not concerned), our measurements indicated in general values of δt_{avr} of a little more than 1 s (Table 2), with the largest values of 1.3 s observed at the stations RBG and BEU both located above the Urach geothermal field. Large variations of δt_{avr} with the ray azimuth and incidence angle would suggest either that the axes of anisotropy are not both in a horizontal plane, or that several structures with a different anisotropy are superimposed. We do not observe large variations of δt_{avr} in contrary to the azimuthal dependence and the spatial variations of the fast-polarization pattern. These rather small variations of δt_{avr} across western Europe suggest that the thickness of the anisotropic structure was not significantly affected by active tectonics. Hence, in the studied areas, the values of δt_{avr} are associated with mantle deformation to a depth of some 120 km for both the stable and the active areas, allowing for a crustal contribution of about 0.2 s.

However, as pointed out, there is a clear dependence of the fast-polarization azimuth on the backazimuth for most of the stations with abundant observations (ECH, RG-N, RG-S, RBG). For these stations, all characterized by a complicated splitting pattern (Tables 4.1 (ECH), 4.2 (RG-N), 4.3 (RG-S), 4.4 (RBG)), there is strong evidence for the operation of multiple processes. Among these different processes either a model made of two distinct anisotropic layers (SAVAGE and SILVER, 1993; SILVER and SAVAGE, 1994) or a single anisotropic layer with dipping axes (BABUŠKA et al., 1993; PLOMEROVÁ et al., 1996) may explain our observations. Assuming a single homogeneous layer of hexagonally symmetric material with a horizontal symmetry axis, the splitting parameters are nearly independent of arrival angle and initial polarization. Since our observations evidence a dependence on these variables, one must think of a more complicate process. In the case of a dependence on initial polarization with a $\pi/2$ periodicity, the presence of two layers can be expected.

Figure 9a presents the results of a two-layer modelling for station RBG. From our splitting measurements and following a method initially proposed by SAVAGE and SILVER (1993), we modelled an anisotropic structure consisting of an upper layer characterized by a fast axis oriented N65°E and a delay of 0.8 s, and a lower layer characterized by a fast axis oriented N130°E and a delay of 1.0 s. For comparison, the model of FARRA et al. (1991) which was proposed for station GRF-A1, has been plotted as well on Figure 9a (dashed lines; see legend for details). For the upper layer, our modelling illustrates a correlation of the fast-polarization azimuth (N65°E) with the trajectories of the present-day crustal stress field (recent anisotropy) and/or with the strike of the Hercynian orogenic belt in Central Europe. The estimated layer of thickness ($\delta t = 0.8$ s corresponds to nearly

90 km) can easily be reconciled with the seismic and thermal states of the lithosphere, as the size of δt for this upper layer is consistent with a plausible lithosphere thickness (80–100 km) and with a lithosphere-asthenosphere transition as suggested by the tomographic study (GLAHN et al., 1993). For the lower layer, while the fast-polarization azimuth (N130°E) correlates well with the direction of asthenospheric flow according to the absolute plate motion model by MINSTER and JORDAN (1978), no such correlation is observed when compared with the model by GRIPP and GORDON (1990). The amplitudes of δt are consistent with the thickness of the asthenosphere as obtained by surface wave studies (YANOVSKAYA et al., 1990).

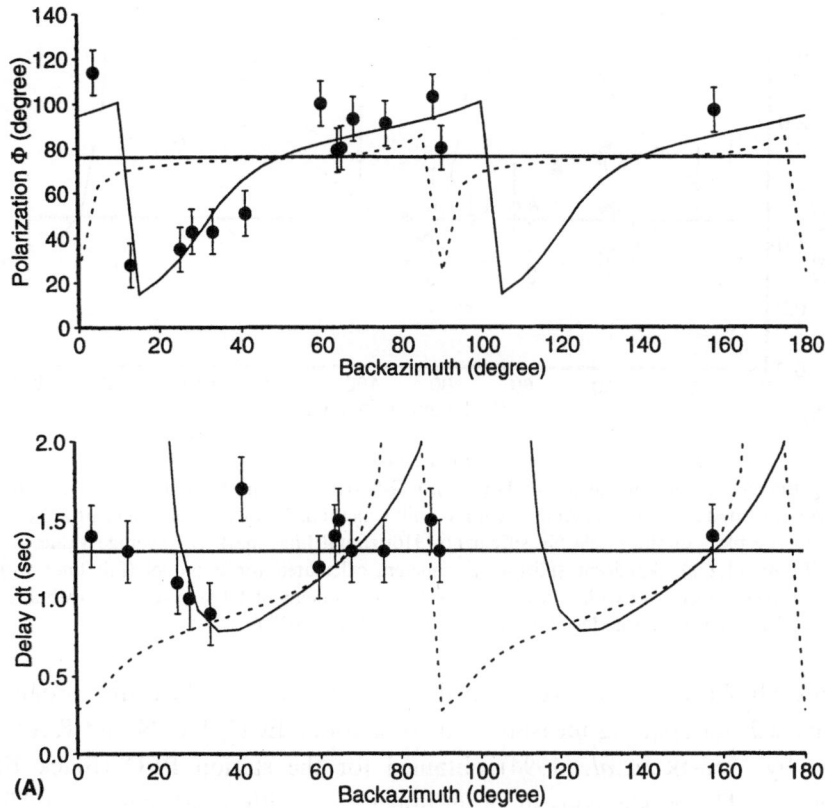

Figure 9a

Splitting parameters (ϕ, δt) of Urach station RBG as a function of backazimuth (modulo 180°), determined with the RCM, are indicated by black dots. The error bars given here in the figure ($\pm 10°$ for ϕ and ± 0.2 s for δt) are taken from a statement by ANSEL (1989), who estimated errors from tests with synthetic data. Horizontal lines mark the average value of $\phi = $ N78°E and $\delta t = 1.3$ s. Apparent splitting parameters calculated for models with two anisotropic layers: model 1 (full line) with a fast axis of N65°E and a delay of 0.8 s in the upper layer, and a fast axis of N130°E and a delay of 1.0 s in the lower layer. Model 2 (dashed line) with a fast axis of N25°E and a delay of 0.28 s in the upper layer, and a fast axis of N90°E and a delay of 1.0 s in the lower layer (model of FARRA et al., 1991, for GRF).

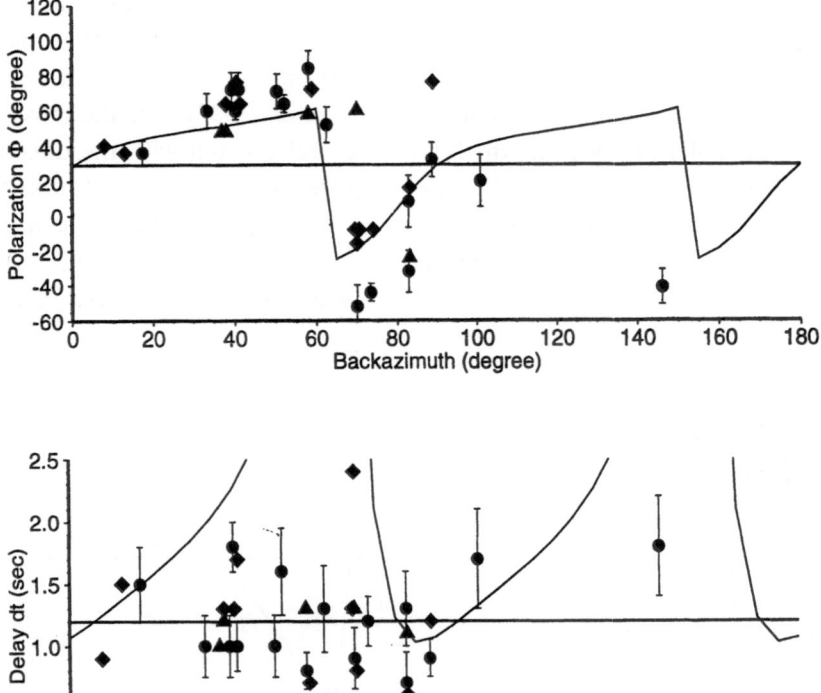

Figure 9b

Splitting parameters (ϕ, δt) of stations ECH (dots), RG-N (triangle) and RG-S (diamond) as a function of backazimuth (modulo 180°) determined with Vinnik's method. Error bars of ϕ and δt are estimated from the energy plots of the single measurements. Horizontal lines mark the average values of ECH ($\phi = \text{N29°E}$, $\delta t = 1.2$ s). Apparent splitting parameters calculated for a model (full line) with two anisotropic layers: upper layer with a fast axis of N10°E and a delay of 1.1 s, lower layer with a fast axis of N82°E and a delay 1.9 s (model of VINNIK *et al.*, 1994, for BFO).

Figure 9b shows a two-layer modelling for the Rhinegraben area proper. We superimposed our splitting measurements of stations ECH, RG-N and RG-S with a model by VINNIK *et al.* (1994) obtained for the station BFO (Black Forest Observatory). This model consists of an upper layer with a fast axis of N10°E and a delay of 1.1 s and a lower layer with a fast axis of N82°E (roughly E–W) and a delay of 1.9 s. However, while our fast-polarization azimuths fit quite well this two-layer modelling (Fig. 9b; top panel), there is a considerable misfit for the delay times (bottom panel). In order to check the consistency of our splitting measurements, we performed a two-layer modelling separately using the data of the station ECH and RG-S. While the same results as those of VINNIK *et al.* (1994) were retrieved for ECH, the model for RG-S consisted of an upper layer with a fast axis of N20°E and a delay time of 0.8 s, and a lower layer with a fast axis of N100°E and a delay time of 1.5 s. This small discrepancy is probably due to insufficient data for station RG-S ($n = 5$) compared to station ECH ($n = 16$). Hence, it is likely from these modellings that a one-layer model cannot explain the anisotropy pattern observed in the Rhinegraben-Urach area due to its dependence on backazimuth and the observed $\pi/2$ periodicity. In addition, these variations of ϕ and δt with the backazimuth might suggest that there are not only several structures with a different azimuthal direction of the fast axis (Figs. 9a,b), but also that the axes of anisotropy are not always in the horizontal plane (e.g., PLOMEROVÁ et al., 1996). For instance, in western and central Europe, one can imagine that different Hercynian terranes may have an inclined anisotropy with the "a" axis parallel to possible (relict) subduction zones of Hercynian or pre-Hercynian age.

Figure 9c illustrates the splitting measurements as a function of backazimuth for station SSB. From individual estimates of the splitting parameters, no $\pi/2$ periodic-

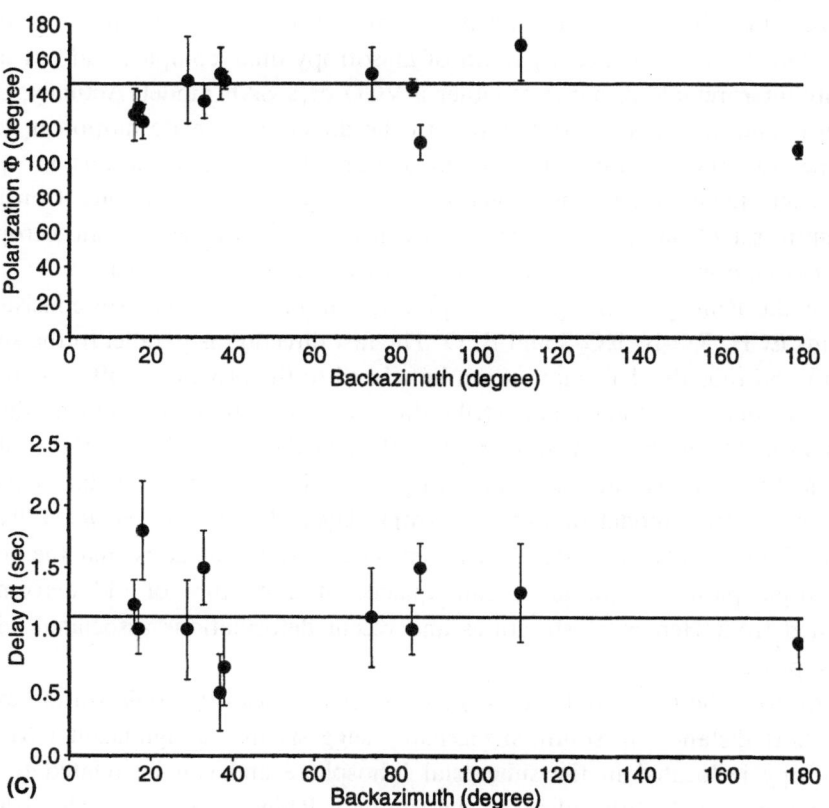

Figure 9c

Splitting parameters (ϕ, δt) of Geoscope station SSB (dots) as a function of backazimuth (modulo 180°) determined with Vinnik's method. Error bars of ϕ and δt are estimated from the energy plots of the single measurements. Horizontal lines mark the average values of $\phi = $ N146°E and $\delta t = 1.1$ s.

ity is observed and a two-layer modelling is rather speculative. Our preferred interpretation of the results for station SSB is that the observed pattern is related to volcanism. The mean fast direction (146° clockwise from North) is similar to the one obtained by VINNIK *et al.* (1989) who suggested a shallow origin for the source of anisotropy, in the lithosphere of the Massif Central.

Since the splitting parameters represent an integrated effect along the entire ray path, it is always difficult to distinguish between lithospheric and asthenospheric contributions. We believe that more work should be accomplished in the direction of multi-layer anisotropic theory.

6. Final Remarks

Studying seismic anisotropy provides constraints on the evolution of the structures of the lithosphere-asthenosphere system developed during past and recent deformation events. We studied two domains of the Hercynian belt in western Europe which have been affected by recent tectonics of Oligocene and Neogene times. Our results evidence a pattern of anisotropy quite complex and not unequivocally interpretable in terms of either a VCD or a SAF model. Actually, there is no clear and systematic correlation with the direction of plate motion in western Europe. On the contrary, at a regional scale there are variations in splitting parameters from west to east which can reflect geological variations. Considering the entire set of data, we observe a local behavior of the seismic anisotropy and clear correlation with recent movements in tectonically active areas.

For the Rhinegraben region the splitting parameters (ϕ, δt) are consistent for stations ECH, RG-N, RG-S and BFO. The fast direction is parallel to the strike of the rift and thus the deformation can be related to the process of rift formation. In fact, deformations along major strike-slip faults may be considered as the main mechanism to generate an anisotropy with a fast axis parallel to the faults. To explain the strong variations of the fast-polarization azimuth, our data have been compared with a model of two anisotropic layers by VINNIK *et al.* (1994). The result of this two-layer modelling as seen on Figure 9b suggests that the complex anisotropic pattern is rather a consequence of a mixture of old deformations inherited from Hercynian structures and recent deformations associated with the rifting process.

However, the rapid (but smooth) change of the mean fast-polarization azimuth over short distances in southern Germany suggests that a significant part of the anisotropy is located in the subcrustal lithosphere and can be related to recent deformations of the lithosphere in the southern Rhinegraben area. The azimuthal variations of splitting parameters (ϕ, δt) could in the case of station RBG, in the Urach region, be explained by a model of two anisotropic layers with fast-polarization azimuths (upper layer: $\phi \approx$ N60°E–N70°E; lower layer: $\phi \approx$ N125°E–N135°E).

The upper layer shows a correlation of the fast-polarization azimuth with the trajectories of the present-day crustal stress field (recent anisotropy) and/or with the strike of the Hercynian mountain belt in western and central Europe. However, while the fast direction in the lower layer correlates with the direction of asthenospheric flow according to the model of absolute plate motion by MINSTER and JORDAN (1978), it is not consistent with the model by GRIPP and GORDON (1990). The question remains in debate.

In the French Massif Central area, two domains are evidenced. One to the west of the Sillon Houiller where the deformation along this strike-slip transformlike fault is likely the main mechanism, the other to the east, most probably related to the volcanic flow.

Clearly, there are two main sources of seismic anisotropy in the upper mantle, and both contributions must be considered. One source is located within the lithosphere and, in that case, the anisotropy must be mainly interpreted as "frozen-in" in the uppermost mantle, the other source is located in the asthenosphere, directly linked with the mantle flow. Generally speaking, it is reasonable to expect that the whole lithosphere-asthenosphere system is responsible for the observed anisotropy. Inside the Hercynian domain the results signify that a coherent deformation developed in the lithospheric mantle, suggesting that the fabric was preserved and may have a major influence on the tectonic behavior of the subcrustal lithosphere. On the other hand, in areas where the lithosphere was reactivated by recent tectonics, the seismic anisotropy signature may be correlated with the last tectonic episode.

Acknowledgments

We thank everyone who contributed to the collection of seismological data during the field experiments and those involved in seismological observations without whom such studies would not have been possible. We are also grateful to Yves Liotier who provided the events recorded at stations RG-N and RG-S (grant INSU-CNRS). The contribution of Monique Blanck for figure preparation was appreciated. The paper benefited from critical reviews by two anonymous reviewers. This work was supported by the French Lithoscope Program (CNRS-INSU) and the SFB-108 project (Germany).

REFERENCES

ANSEL, V. (1989), *Mantle Anisotropy as Deduced from Shear-wave Splitting*, Ph.D. Thesis, Université de Paris-Sud, Centre d'Orsay, 302 pp. (in French).

ANSEL, V., and NATAF, H. C. (1989), *Anisotropy beneath 9 Stations of the Geoscope Broadband Network as Deduced from Shear-wave Splitting*, Geophys. Res. Lett. *16*, 409–412.

BABUŠKA, V., PLOMEROVÁ, J., and ŠILENY, J. (1993), *Models of Seismic Anisotropy in Deep Continental Lithosphere*, Phys. Earth Planet. Int. *78*, 167–191.

BARRUOL, G., and SOURIAU, A. (1995), *Anisotropy beneath the Pyrenees Range from Teleseismic Shear-wave Splitting: Results from a Test Experiment*, Geophys. Res. Lett. *22*, 493–496.

BONJER, K. P., GELBKE, C., GILG, G., ROULAND, D., MAYER-ROSA, D., and MASSINON, B. (1984), *Seismicity and Dynamics of the Upper Rhinegraben*, J. Geophys. *55*, 1–12.

BONJER, K. P., FABER, S., and APOPEI, I. (1989), *Seismizität als Zugang zu räumlichen und zeitlichen Anomalien der Spannungen in der Lithosphäre*, SFB108, Spannung und Spannungsumwandlung in der Lithosphäre, Berichtsband (A) 1987–1989, Univ. Karlsruhe, pp. 66–131 (in German).

BORMANN, P., BURGHARDT, P. T., MAKEYEVA, L. I., and VINNIK, L. P. (1993), *Teleseismic Shear-wave Splitting and Deformations in Central Europe*, Phys. Earth Planet. Int. *78*, 157–166.

BURG, J. P., BRUN, J. P., and VAN DEN DRIESSCHE, J. (1990), *Le Sillon Houiller du Massif Central français: Faille de transfert pendant l'amincissement crustal de la chaîne varisque?* C. R. Acad. Sci. Paris 311, série II, 147–152.

FARRA, V., VINNIK, L. P., ROMANOWICZ, B., KOSAREV, G. L., and KIND, R. (1991), *Inversion of Teleseismic S Particle Motion for Azimuthal Anisotropy in the Upper Mantle: A Feasibility Study*, Geophys. J. Int. *106*, 421–431.

FUKAO, Y. (1984), *Evidence for Core-reflected Shear Waves for Anisotropy in the Earth's Mantle*, Nature 309, 695–698.

GLAHN, A., SACHS, P. M., and ACHAUER, U. (1992), *A Teleseismic and Petrological Study of the Crust and Upper Mantle beneath the Geothermal Anomaly Urach/SW-Germany*, Phys. Earth Planet. Int. *69*, 176–206.

GLAHN, A., and GRANET, M. (1992), *3-D Structure of the Lithosphere beneath the Southern Rhine Graben Area*, Tectonophysics *208*, 149–158.

GLAHN, A., GRANET, M., and the Rhine Graben Teleseismic Group (1993), *Southern Rhine Graben: Small-wavelength Tomographic Study and Implications for the Dynamic Evolution of the Graben*, Geophys. J. Int. *113*, 399–418.

GRANET, M., STOLL, G., DOREL, J., ACHAUER, U., POUPINET, G., and FUCHS, K. (1995a), *Massif Central (France): New Constraints on the Geodynamical Evolution from Teleseismic Tomography*, Geophys. J. Int. *121*, 33–48.

GRANET, M., WILSON, M., and ACHAUER, U. (1995b), *Imaging a Mantle Plume beneath the Massif Central (France)*, Earth and Planet. Sc. Lett. *136*, 281–296.

GRIPP, A. E., and GORDON, R. G. (1990), *Current Plate Velocities Relative to the Hotspots Incorporating the NUVEL-1 Global Plate Motion Model*, Geophys. Res. Lett. *17* (8), 1109–1112.

GRÜNTHAL, G., and STROMEYER, D. (1992), *The Recent Crustal Stress Field in Central Europe: Trajectories and Finite Element Modelling*, J. Geophys. Res. *97*, 11805–11820.

HELFFRICH, G., SILVER, P. G., and GIVEN, H. (1994), *Shear-wave Splitting Variation over Short Spatial Scales on Continents*, Geophys. J. Int. *119*, 561–573.

HIRN, A., JIANG, M., SAPIN, M., DIAZ, J., NERCESSIAN, A., LU, Q. T., LÉPINE, J-C., SHI, D. N., SACHPAZI, M., PANDEY, M. R., MA, K., and GALLART, J. (1995), *Seismic Anisotropy as an Indicator of Mantle Flow beneath the Himalayas and Tibet*, Nature 375, 571–574.

KAHLE, H. G., and WERNER, D. (1980), *A Geophysical Study of the Rhinegraben-II. Gravity Anomalies and Geothermal Implications*, Geophys. J. R. Astron. Soc. *62*, 631–647.

KIND, R., KOSAREV, G. L., MAKEYEVA, L. I., and VINNIK, L. P. (1985), *Observations of Laterally Inhomogeneous Anisotropy in the Continental Lithosphere*, Nature 318, 358–361.

LIOTIER, Y., GRANET, M., and GLAHN, A. *Investigation of azimuthal anisotropy in the southern Rhine Graben area from P-residuals and SKS phases observation.* In Supplement to EOS, October 29 (Am. Geophys. Un., Washington, D.C. 1991) p. 306.

LUCAZEAU, F., VASSEUR, G., and BAYER, R. (1984), *Interpretation of Heat Flow Data in the French Massif Central*, Tectonophysics *103*, 99–119.

MAINPRICE, D., and SILVER, P. G. (1993), *Interpretation of SKS Waves Using Samples from the Subcontinental Lithosphere*, Phys. Earth Planet. Int. *78*, 257–280.

MEIR, L. (1989), *Ein Modell für die Tiefenstruktur und Kinematik im Bereich des nördlichen Rheingrabens*, Ph.D. Thesis, Univ. Karlsruhe (in German).

MINSTER, J-B., and JORDAN, T. H. (1978), *Present-day Plate Motions*, J. Geophys. Res. *83*, 5331–5352.

MCNAMARA, D. E., OWENS, T. J., SILVER, P. G., and WU, F. T. (1994), *Shear-wave Anisotropy beneath the Tibetan Plateau*, J. Geophys. Res. *99*, 13655–13665.

NICOLAS, A., and CHRISTENSEN, N. I., *Formation of anisotropy in upper mantle peridotites—A review*. In *Composition, Structure and Dynamics of the Lithosphere-asthenosphere System* (FUCHS, K., and FROIDEVAUX, C., eds.) (Am. Geophys. Un., Washington, D.C. 1987) pp. 111–123.

PANZA, G. F., MUELLER, S., and CALCAGNILE, G. (1980), *The Gross Features of the Lithosphere-asthenosphere System in Europe from Seismic Surface Waves and Body Waves*, Pure appl. geophys. *118*, 1209–1213.

PASSIER, M. L., and SNIEDER, R. K. (1996), *Correlation between Shear-wave Upper Mantle Structure and Tectonic Surface Expressions. Application to Central and Southern Germany*, J. Geophys. Res. *101*, 25293–25304.

PLOMEROVÁ, J., ŠILENY, J., and BABUŠKA, V. (1996), *Joint Interpretation of Upper Mantle Anisotropy Based on Teleseismic P-travel Time Delays and Inversion of Shear-wave Splitting Parameters*, Phys. Earth Planet. Int. *95*, 293–309.

SAVAGE, M. K., and SILVER, P. G. (1993), *Mantle Deformation and Tectonics: Constraints from Seismic Anisotropy in the Western United States*, Phys. of the Earth and Planet. Int. *78*, 207–227.

SILVER, P. G., and CHAN, W. W. (1988), *Implications for Continental Structure and Evolution from Seismic Anisotropy*, Nature *335*, 34–39.

SILVER, P. G., and CHAN, W. W. (1991), *Shear-wave Splitting and Subcontinental Mantle Deformation*, J. Geophys. Res. *96*, 16429–16454.

SILVER, P. G., and KANESHIMA, S. (1993), *Constraints on Mantle Anisotropy beneath the Precambrian North America from a Transportable Teleseismic Experiment*, Geophys. Res. Lett. *20*, 1127–1130.

SILVER, P. G., and SAVAGE, M. K. (1994), *The Interpretation of Shear-wave Splitting Parameters in the Presence of Two Anisotropic Layers*, Geophys. J. Int. *119*, 949–963.

SILVER, P. G. (1996), *Seismic Anisotropy beneath the Continents: Probing the Depths of Geology*, Ann. Rev. Earth Planet. Sci. *24*, 385–432.

SOBOLEV, S. V., ZEYEN, H., STOLL, G., WERLING, F., ALTHERR, R., and FUCHS, K. (1996), *Upper Mantle Temperatures from Teleseismic Tomography of French Massif Central Including Effects of Composition, Mineral Reactions, Anaharmonicity, Anelasticity and Partial Melt*, Earth Planet. Sci. Lett. *139*, 147–163.

SOBOLEV, S. V., ZEYEN, H., GRANET, M., ACHAUER, U., BAUER, C., WERLING, F., ALTHERR, R., and FUCHS, K. (1997), *Upper Mantle Temperatures and Lithosphere-asthenosphere System beneath the French Massif Central Constrained by Seismic, Gravity, Petrologic and Thermal Observations*, Tectonophysics *275*, 143–164.

SOURIAU, A., CORREIG, A. M., and SOURIAU, M. (1980), *Attenuation of Rayleigh Waves across the Volcanic Area of the Massif Central, France*, Phys. Earth Planet. Int. *23*, 62–71.

VAUCHEZ, A., and NICOLAS, A. (1991), *Mountain Building: Strike-parallel Displacement and Mantle Anisotropy*, Tectonophysics *185*, 183–201.

VAUCHEZ A., and BARRUOL, G. (1996), *Shear-wave Splitting in the Appalachians and the Pyrenees: Importance of the Inherited Tectonic Fabric of the Lithosphere*, Phys. Earth Planet. Int. *95*, 127–138.

VINNIK, L. P., KOSAREV, G. L., and MAKEYEVA, L. I. (1984), *Anisotropy of the Lithosphere from the Observations of SKS and SKKS*, Proc. Acad. Sci. USSR *278*, 1335–1339 (in Russian).

VINNIK, L. P., KOSAREV, G. L., and MAKEYEVA, L. I. (1988), *Azimuthal anisotropy of the lithosphere from observations of long period converted waves*. In *Structure and Dynamics of the Lithosphere According to Seismic Data* (Nercesov, I. L., ed.) (Nauka, Moscow 1988) 221 pp. (in Russian).

VINNIK, L. P., FARRA, V., and ROMANOWICZ, B. (1989), *Azimuthal Anisotropy in the Earth from Observations of SKS at Geoscope and NARS Broadband Stations*, Bull. Seismol. Soc. Am. *79*, 1542–1558.

VINNIK, L. P., MAKEYEVA, L. I., MILEV, A., and USENKO, Y. (1992), *Global Patterns of Azimuthal Anisotropy and Deformations in the Continental Mantle*, Geophys. J. Int. *111*, 433–447.

VINNIK, L. P., KRISHNA, V. G., KIND, R., BORMANN, P., and STAMMLER, K. (1994), *Shear-wave Splitting in the Records of the German Regional Seismic Network*, Geophys. Res. Lett. *21*, 457–460.

VINNIK, L. P., GREEN, R. W. E., and NICOLAYSEN, L. O., (1995), *Recent Deformations of the Deep Continental Roots in Southern Africa*, Nature *375*, 50–52.

YANOVSKAYA, T. B., PANZA, G. F., DITMAR, P. D., SUHADOLC, P., and MUELLER, S. (1990), *Structural Heterogeneity and Anisotropy Based on 2-D Phase Velocity Patterns of Rayleigh Waves in Western Europe*, Rend. Fis. Acc. Lincei *9*, 127–135.

ZEYEN, H., NOVAK, O., LANDES, M., HIRN, A., FUCHS, K., and PRODEHL, A. (1997), *Refraction Seismic Investigation of the Northern French Massif Central (France)*, Tectonophysics *275*, 99–117.

(Received May 7, 1997, revised/accepted September 15, 1997)

To access this journal online:
http://www.birkhauser.ch

Pure appl. geophys. 151 (1998) 365–394
0033–4553/98/040365–30 $ 1.50 + 0.20/0

Pure and Applied Geophysics

Seismic Anisotropy and Velocity Variations in the Mantle beneath the Saxothuringicum-Moldanubicum Contact in Central Europe

J. Plomerová,[1] V. Babuška,[2] J. Šílený[1] and J. Horálek[1]

Abstract—We report on results of a passive seismic experiment undertaken to study the 3-D velocity structure and anisotropy of the upper mantle around the contact zone of the Saxothuringicum and Moldanubicum in the western margin of the Bohemian Massif in central Europe. Spatial variations of *P*-wave velocities and lateral variations of the particle motion of split shear waves over the region monitor changes of structure and anisotropy within the deep lithosphere and the asthenosphere. A joint interpretation of *P*-residual spheres and shear-wave splitting results in an anisotropic model of the lithosphere with high velocities plunging divergently from the contact of both tectonic units. Lateral variations of the mean residuals are related to a southward thickening of the lithosphere beneath the Moldanubicum.

Key words: Upper mantle anisotropy, inversion of shear-wave splitting parameters, *P*-residual spheres, joint interpretation of body wave observations.

1. Introduction

The western part of the Bohemian Massif is characterized by the juxtaposition of two major tectonic units, the Moldanubicum and the Saxothuringicum (see Figs. 1 and 5). The knowledge of the mutual relation of these units is one of the keys for understanding the orogenic structure and evolution of the Variscides of central Europe. This was also one of the basic scientific targets of the KTB Programme (Kontinentales Tiefbohrungprogramm, e.g., EMMERMANN *et al.*, 1993) which covered, besides the deep drilling, a broad spectrum of scientific studies from all disciplines of earth sciences. The KTB Programme contributed greatly to the understanding of the crustal structure and processes underway in the upper and middle crusts.

The most important problems of the continental geology concern the relationship between surface structure and geodynamic processes at depth. For their solution we must understand how the entire lithosphere is structured and how it

---

[1] Geophysical Institute, Czech Academy of Sciences, 14131 Praha 4, Czech Republic; jpl@ig.cas.cz
[2] UNESCO, Division of Earth Sciences, 1, rue Miollis, 75732 Paris, France; v.babuska@unesco.org

interacts with the underlying convecting mantle. As one of the future goals we must be able to define the 3-D configurations of individual parts of the continental plates and how these have changed with time. This is a great challenge for geophysics and especially for seismology.

Variscan orogeny in Europe was set by the Cambro-Ordovician breakup of the northern margin of Gondwana into several microcontinental fragments separated by evolved rifts or narrow oceans (FRANKE, 1989). The Variscides of central Europe may represent a collision zone characterized by two systems of paleosubductions divergent relative to the suture between the Moldanubicum and the Saxothuringicum (BABUŠKA et al., 1993). The surface trace of this prominent, deep tectonic boundary follows the WSW-ENE trend of Variscan units in central Europe and according to previous teleseismic P-residual studies (e.g., BABUŠKA and PLOMEROVÁ, 1992) separates prevailingly the north-northwestward orientation of dipping anisotropic structures in the subcrustal lithosphere from the southward orientations.

It has been recognized by a number of seismological studies during the last decade that effects of seismic anisotropy are of the same order of importance as velocity heterogeneities in the deep continental lithosphere (e.g., BABUŠKA et al., 1984; ANDERSON, 1989). Anisotropy of physical properties is inherent to rock-forming minerals and their systematic preferred orientation is reflected in the large-scale anisotropy of physical parameters. In the case of the subcrustal lithosphere and sublithospheric upper mantle, the recorded anisotropy results mainly from the strong preferred orientation of the olivine developed in ultramafic upper-mantle rocks. The seismic anisotropy thus provides information regarding the fabric and the deformation of the upper mantle.

Important information about P-velocity anisotropy within the lithosphere can be obtained from teleseismic P-residual spheres which reveal that part of the relative travel-time residuals which depends on azimuth and incidence angle (BABUŠKA et al., 1984, 1988). However, the spatial dependence of P residuals is not commonly attributed to anisotropic propagation. Only azimuthal variations of the residuals with π-periodicity have been associated with upper mantle anisotropy (e.g., DZIEWONSKI and ANDERSON, 1983). Doubts about anisotropy derived from P residuals disappear if shear-wave splitting is detected, which is generally accepted as direct proof of anisotropy. Two directions of orthogonal polarizations and the delay time between both arrivals put constraints upon the mantle anisotropy. For a comprehensive discussion of observations of anisotropy in continental regions the reader is referred to BABUŠKA and CARA (1991) and SILVER (1996).

Various geophysical studies have been aimed at studying the deep structure of the lithosphere and the asthenosphere beneath the western part of the Bohemian Massif. These include the wide-angle reflection seismics (e.g., GEBRANDE et al., 1989), crustal studies beneath the GRF-array (KRÜGER and WEBER, 1992; KRÜGER, 1994), upper mantle tomographic studies (FABER et al., 1986; PLOM-

EROVÁ and BABUŠKA, 1988), as well as the *SKS* splitting measurements (KIND *et al.*, 1985; SILVER and CHAN, 1988; VINNIK *et al.*, 1994; BORMANN *et al.*, 1996). In this paper we have analyzed both lateral and spatial variations of relative residuals of teleseismic *P* waves and variations of the splitting parameters of teleseismic shear waves. The analysis aimed at constructing a self-consistent 3-D anisotropic model of the upper mantle beneath the broader contact of the Moldanubicum (M) and Saxothuringicum (S) in central Europe, i.e., at an anisotropic model which is compatible with both the observed teleseismic *P* delay times and the shear-wave splitting.

2. Location of Stations and Data

To study in detail lateral variations of seismic velocities within the lower lithosphere and especially lateral variations of seismic anisotropy, a passive seismic experiment was conducted during eight months of 1992. Four digital seismic

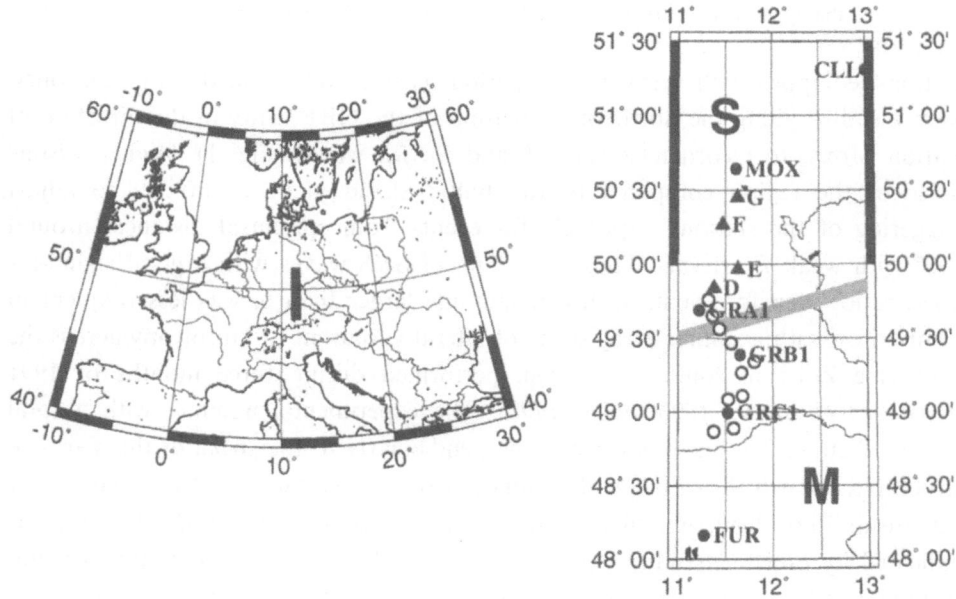

Figure 1

Location of stations (right) along with an insert showing the studied region (left) within central Europe. Four stations D, E, F and G marked by triangles are temporary digital 3-c stations installed during the field experiment in 1992. Circles denote the broadband stations of the German Regional Network of Stations—full circles stand for 3-c stations, empty circles for 1-c vertical stations. A surface trace of the Saxothuringicum (S) and Moldanubicum (M) contact is marked schematically by a grey band.

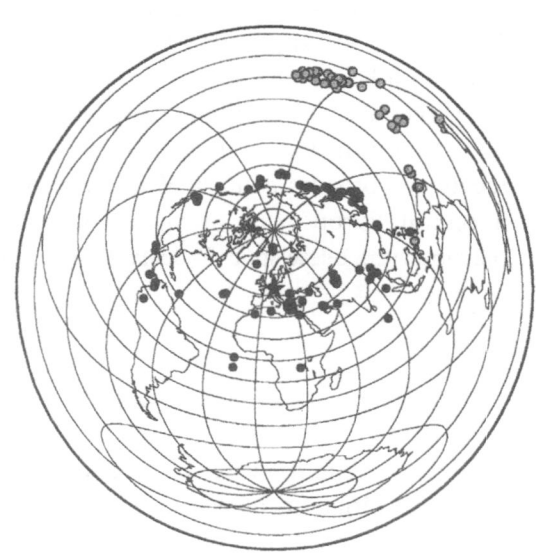

Figure 2
Distribution of 162 teleseismic events with *P* (black circles) and *PKP* (grey circles) used in the study of
the upper mantle structure around the S/M contact (star) in central Europe.

stations, equipped with vertical short-period sensors and 3-c medium-period ones,
were installed along the northern elongation of the GRF array in the direction of
station Moxa, to record teleseismic *P* and *S/SKS* waves (Fig. 1). Higher seismic
noise of the region compared to the Baltic Shield made it difficult to adjust
triggering of the stations, especially for events from epicentral distances around
100° with weak *P* arrivals. Thus, recording of *SKS* waves with a good signal-to-
noise ratio, even for events with a magnitude larger than 6, was less frequent in
comparison with a similar array study of lateral variations of anisotropy across the
Protogine Zone in southern Sweden, performed during three months of 1991
(PLOMEROVÁ et al., 1993, 1996, 1998). The experimental antenna with a total
length of about 250 km was oriented perpendicularly to the strike of the Variscan
structures, for which a divergent inclination of the seismic anisotropy within the
subcrustal lithosphere, as well as a significant increase in the mean *P* velocities, was
indicated by previous studies (e.g., BABUŠKA et al., 1984, 1995; PLOMEROVÁ and
BABUŠKA, 1988; BABUŠKA and PLOMEROVÁ, 1989). Stations MOX, G and F are
located in the Saxothuringicum, four stations of subarray GRF-A (GRA1–4) and
stations D and E very close to the surface trace of the S/M transition, and nine
stations of subarrays GRF-B (GRB1–5) and GRF-C (GRC1–4) and station FUR
in the Moldanubicum.

Approximately 160 teleseismic events were recorded during the experiment by
four mobile stations and they could be used in our study. These data, comprising

teleseismic P and PKP arrivals (Fig. 2) and shear waveforms (SKS, $SKKS$, S, ScS), complement data recorded at permanent observatories in the region: FUR, GRF-array, MOX and partly CLL. Altogether, the experimental array was formed by twenty stations of which 16 belong to the German Regional Seismic Network (GRSN). Stations FUR, GRA1, GRB1, GRC1 and MOX are 3-component digital broadband (BB) stations; other stations of GRF-subarrays are 1-component vertical BB stations. Four experimental stations were equipped with short-period vertical seismometers SM3 ($T_S = 1.7$ s) and modified medium-period 3-c seismometer S5S ($T_S = 10.5$ s), and these stations used a Lennartz PCM 5800 data acquisition system. The system has a flat velocity response in an interval 0.2–10.5 s, dynamic range 126 dB, resolution 10 bits and sampling rate 20 ms.

3. Method

To retrieve a general 3-D orientation of anisotropic structures and lateral variations of seismic velocities within the subcrustal lithosphere, we applied a method of joint interpretation of upper mantle anisotropy based on an analysis of teleismic P and PKP travel-time residuals, and inversion of shear-wave splitting parameters (ŠÍLENÝ and PLOMEROVÁ, 1996; PLOMEROVÁ et al., 1996).

The analysis of P residuals computes at each station a static term and directional terms of relative residuals. In this procedure, complicating effects of source regions and deep mantle paths are minimized by a normalization and grouping of the events. Delays due to different crustal structure-thickness, average velocity, and the same for sediments, are corrected. The static term reflects a mean velocity within the upper mantle beneath a station, and its lateral variations can be related to lateral variations of the lithosphere thickness (BABUŠKA et al., 1987). Dependencies of relative residuals on azimuths and incidence angles provide information pertaining to spatial variations of relative P velocities in the subcrustal lithosphere beneath the stations, and we relate them to the anisotropic structure of the subcrustal lithosphere. For more details of the method we refer to e.g. BABUŠKA et al. (1984, 1988).

To show the spatial P-velocity variations we construct for each station a residual diagram, in which directional terms of the relative P-wave arrival residuals are plotted on a stereographic projection of the lower hemisphere below each station ("residual spheres"). As a directional mean residual (BABUŠKA et al., 1988) forms a zero level in each diagram, the residual patterns of different stations, which mark relatively high- and low-velocity directions in the subcrustal lithosphere, can be easily compared. To express clearly the differences in velocities beneath the stations for various directions of propagation and to have a simple base to quantify the diagrams, we smoothed the residual spheres by a two-dimensional linear filter with the azimuth and incidence angle as parameters. In most cases we do not observe the full anisotropy of the medium, because only a part of the hemisphere is surveyed by teleseismic P waves, whose range of azimuths is limited by the world's seismicity, and whose incidence angles at the M discontinuity vary approximately between 20° and 50°. As vertical and "shallow" incidences are missing in the diagrams, the magnitude of the anisotropy can be greater than that which is observed, if real velocity extremes are oriented outside the space surveyed.

Fast and slow quasi-shear waves, propagating along the same path, are generated in anisotropic media, and their detection is considered as proof of the anisotropy (ANDO, 1984). SKS waves, propagating through the outer core as a P wave, are not affected by an anisotropic structure in the source region and thus their splitting reflects the anisotropy in the mantle on receiver sides. To prove the upper mantle anisotropy, the azimuth of polarization of the fast SKS waves and the time delay δt between two split SKS waves are measured (e.g., SILVER and CHAN, 1988; VINNIK *et al.*, 1989). Both characteristics depend on the orientation of the anisotropic structure and on the magnitude of anisotropy and its symmetry. In comparison with the spatial distribution of teleseismic P waves, directions of propagation of SKS are quite strongly limited due to their subvertical propagation through the upper mantle. If we leave the assumption of a hexagonal approximation of the upper mantle anisotropy with horizontal "fast" symmetry axis and accept the idea of a general orientation of anisotropy or its presence in more than one layer, then both splitting characteristics depend on the direction of propagation of shear waves, including the angle of propagation through the anisotropic layer. SAVAGE and SILVER (1993) showed that S phases also can be used in these studies, if they exhibit "an internal consistency with the SKS." This broadens a fan of shear-wave "incidences" and the mantle beneath a receiver is probed more comprehensively. A broadening of a fan of rays surveying the upper mantle is inevitable for the detection of a 3-D orientation and magnitude of seismic anisotropy, as in several regions we have indications that the orientation of the fabric varies relative to the horizontal (FURLONG and OWENS, 1997).

Here we present alternative 3-D anisotropic models with plunging symmetry axes based on the inversion of the shear-wave splitting parameters. To analyze shear-wave particle motion we applied a method by SILVER and CHAN (1991) generalized for 3-D, by using not only horizontal components of the shear waveforms, but also vertical components of the records. We use the fast SKS and/or S polarization vectors ($\mathbf{u}_S$), defined by spherical angles ϕ_S and Θ_S (measured upward from the positive z axis oriented downward), and the time differences δt between two split shear waves. To fit variations of these parameters with backazimuths and angle of propagation shear waves within the upper mantle, we apply a method of inversion of the splitting parameters to retrieve the 3-D orientation of an inclined anisotropic medium (ŠÍLENÝ and PLOMEROVÁ, 1996). Starting models consist of one-layer homogeneous hexagonal or orthorhombic media with plunging symmetry axes. On the basis of papers which review the anisotropy of olivine ultramafites (CHRISTENSEN, 1984; NICOLAS and CHRISTENSEN, 1987) we selected a typical periodotite aggregate with preferred orientation of olivine crystals. The aggregate composed of 17% of olivine oriented in orthorhombic symmetry, 51.3% of olivine oriented in hexagonal symmetry and 31.7% of isotropic mixture exhibits the P velocities of 8.73, 7.98 and 8.10 km/s along the a', b' and c' axes, respectively. The elastic constants of the hexagonal model (b', (a', c')) were computed by rotating the orthorhombic aggregate with the b' axis fixed and the a' and c' axes regularly moving around the b'. The elastic constants averaged by this rotation describe a hexagonal anisotropy with P velocity 7.9 km/s along the symmetry axis (b') and 8.425 km/s in the plane (a', c') perpendicular to the b' (BABUŠKA et al., 1993). The quantity to be minimized in the inversion is the weighted residual sum R

$$R = \sum_{i=1}^{N} \frac{|{}^{i}\mathbf{u}_{S}^{\mathrm{obs}} - {}^{i}\mathbf{u}_{S}|^{2}}{(\sigma_{i}^{\mathrm{ang}})^{2}} + \frac{|\delta t_{i}^{\mathrm{obs}} - \delta t_{i}|^{2}}{(\delta_{i}^{t})^{2}} \tag{1}$$

where $\mathbf{u}_{S}$ and δt are the orientation of the fast shear-wave polarization and the time delay, respectively, σ are standard error estimates provided by the procedure which determines the splitting parameters. We also tested the stability of the solutions by exploring all solutions, for which the residual sum remains below a selected percentage of the minimum R_{min} found from (1); here we took 20% level.

The main advantage of the new 3-D anisotropic models is that they provide synthetic P-residual spheres which are similar to the observed ones (BABUŠKA et al., 1993; PLOMEROVÁ et al., 1996, 1997), which is not the case if anisotropic one- or two-layer models with horizontal symmetry axes are used (e.g., MAKEYEVA et al., 1990; SAVAGE and SILVER, 1993; OZALAYBEY and SAVAGE, 1994; BORMANN et al., 1993; VINNIK et al., 1994; SILVER, 1996).

4. P-velocity Variations

4.1 Lateral Variations of the Static Term of P Residuals

We use the static term of P residuals to model the "seismic" lithosphere as a high-velocity layer overlying the low-velocity asthenosphere (BABUŠKA et al., 1987). Lateral variations of mean velocities of teleseismic P waves propagating steeply through the upper mantle indicate an increase of the lithospheric thickness to the south-southeast from the contact zone of the Moldanubicum and Saxothuringicum (BABUŠKA et al., 1987). PLOMEROVÁ and BABUŠKA (1988) discussed effects due both to a southward increase of the mantle velocity and a thickening of the lithosphere. They showed that an unrealistic increase of the average velocity in the subcrustal lithosphere (e.g., from 8.2 km/s to 9.1 km/s for 80-km-thick lithosphere) over a horizontal distance of about 100 km would be required to explain the 0.6 s difference in the mean relative residuals. They assigned the change to the deepening of the lithosphere-asthenosphere transition by approximately 50 km beneath the GRF-array. The magnetotelluric measurements along two crossing profiles, one parallel to the GRF-array and the other perpendicular to it, confirmed in gross features both the depth of the transition and its topography (PRAUS *et al.*, 1990).

The normalization procedures used for computing relative residuals deserve special attention in each investigated region and each data set. As the normalization influences the relative residuals, both the mean values (static terms) and especially the resulting pattern of the residual spheres, various reference levels of the normalization were tested. The residuals were normalized relative to an average residual computed from all stations, which recorded an event with residuals less than ± 2 s relative to the average residual in one geographic segment of grouped hypocenters. Ten stations were considered as minimum per event. Moreover, pairs FUR and MOX or FUR and CLL, as representations of structures beneath the southern and northern parts of the antenna with the opposite orientation of anisotropy in the lithosphere (BABUŠKA *et al.*, 1984), were also used to test the stability and the effects of the normalization on the results.

The new experimental data set confirmed a significant decrease of the relative mean residuals, from north to south, which represents an increase of apparent P velocities. A value of -0.55 s between the northernmost A3 and southernmost C2 stations of the GRF-array (Fig. 3) is identical with that found in considerably older data set -0.6 ± 0.1 s (PLOMEROVÁ and BABUŠKA, 1988) and, as mentioned above, interpreted as a steep deepening of the lithosphere-asthenosphere boundary to the south beneath the GRF array. KRÜGER and WEBER (1992) suggested an alternative explanation for the southward decrease of the average relative residuals. They developed crustal models with a low-velocity sedimentary wedge with a lower boundary dipping to NNE at about 0.8°. Using extremely low velocities of P waves in the sediments (2.2–2.4 km/s), the authors claim that a pattern of the array mislocation vectors and about half of the amplitude of the effects observed, including the difference in the mean residuals, can be reproduced by such models derived from geological evidence, borehole measurements and refraction surveys. The models register relative travel-time residuals of 0.2–0.3 s between the subarray central stations A1 and C1, including the very extreme model. This travel-time difference is less than half of the observed difference in the average residuals. Variations in relative P residuals are too large to be explained only by heterogeneities in the crust. Taking a 0.2 s reduction in the mean residual difference due to the sediments, if the realistic crustal model by KRÜGER and WEBER (1992) is considered, then -0.35 s remains to be explained. Such relative mean residual difference results in a lithosphere thickening of about 30 km (BABUŠKA *et al.*, 1987), which deepens beneath the Moldanubicum and the Alpine foredeep to the south. The thickening corresponds to the general relief of the lithosphere-asthenosphere transition (L/A) in central Europe (Fig. 4). North of the S/M contact the L/A relief is flat in the NW direction and deepens to the northeast. The mean residual of station CLL is included in Figure 2, although the station is not located directly in the northern continuation of the antenna. The mean residual at CLL reflects the lithosphere thickening to the northeast in the Saxothuringicum and to the East-European Platform.

We also computed the relative residuals from the experimental data set, directly considering the crustal model beneath the GRF-array developed by KRÜGER and WEBER (1992), (Fig. 3). The residual difference (A3–C2) reduces to −0.30 s, which represents a decrease of 45% compared to the difference, if no corrections for the sedimentary wedge in the crust are introduced. We also tested a crustal model assuming the sedimentary wedge continues beneath the experimental stations D and E, situated close to the GRA1 station. However for such model, the residuals are too low at these stations. They differ by −0.2 s from nearby stations: A3 to the south and F to the north. The difference between stations A3 and C2 also remains

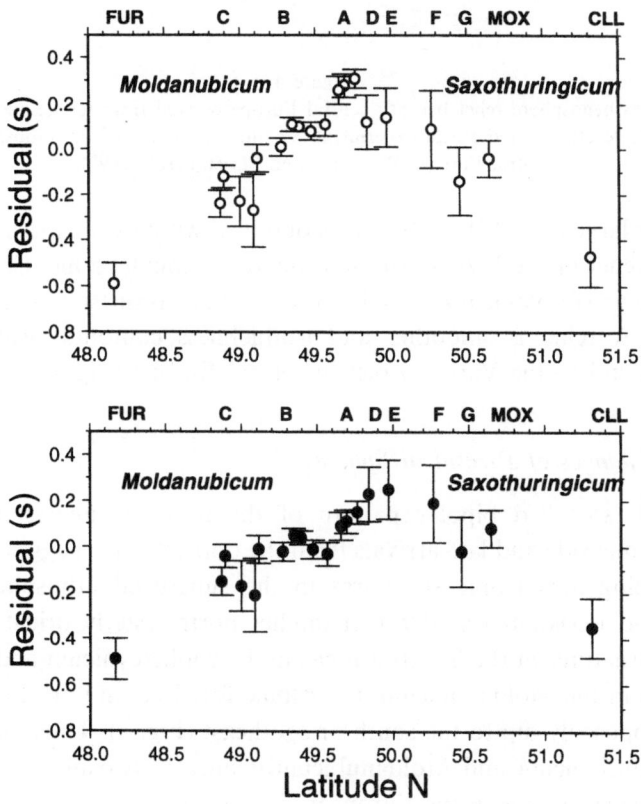

Figure 3
Lateral changes of static terms of the relative residuals along the experimental antenna in projection on the 11.5 E longitude. Corrections for sediments according to KRÜGER and WEBER (1992), for more details see the text, were applied in computations presented in the lower part of the figure (full circles).

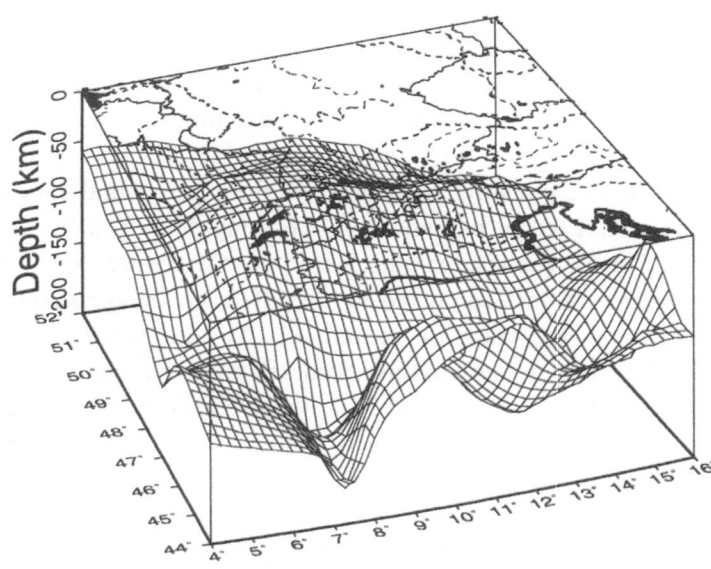

Figure 4
The lithosphere-asthenosphere relief beneath central Europe derived from teleseismic *P* residuals after incorporating the results from the field experiment '92 into the model of the lithosphere thickness in central Europe (BABUŠKA and PLOMEROVÁ, 1992).

in this case as large as −0.30 s. Residuals obtained within the experiment were used to refine the relief of the L/A transition around the contact zone. Compared to the previous models (PLOMEROVÁ and BABUŠKA, 1988; BABUŠKA and PLOMEROVÁ, 1992), the L/A relief is smoother and isothickness contours clearly follow the WSW-ENE trend of the Variscan belt in central Europe (Fig. 4).

4.2 *Lateral Changes of P-residual Spheres*

A general, so-called bipolar pattern of the residual spheres, showing early arrivals from one side and late arrivals from the opposite one, suggests the existence of large dipping anisotropic structures in the subcrustal lithosphere of Europe (BABUŠKA and PLOMEROVÁ, 1992). It implies northwesterly oriented dips of the anisotropic structures in the Saxothuringicum, Rhenohercynicum and in the Massif Central while in the Moldanubicum, the Alpine foredeep and most of the Alps the structures supposedly dip to the south or southeast (Fig. 5). Thus the contact zone of the Saxothuringicum and Moldanubicum delineates two major orientations of the anisotropic structures in the European lithosphere. A surface trace of the S/M contact is located beneath the northern subarray A of the GRF, striking in the NE-SW direction. The suture supposedly dips to the south (MENGEL, 1992).

Figure 6 exhibits smoothed residual spheres with absolute residuals of recorded events. Along all the antenna, the waves arriving from the south and southwest are faster than those arriving from other directions. However, waves recorded at stations in the southern part, station FUR and subarray C, are much faster. The importance of the selection of a proper normalization, especially for a study of anisotropy, has often been discussed (e.g., BABUŠKA et al., 1984; BABUŠKA and PLOMEROVÁ, 1992). If the normalization is efficient, i.e., effects of mislocations, errors in origin time of events and effects of a structure of focal zone and the deep mantle are minimized, then the resulting pattern of the residual spheres is stable and independent of data sets (e.g., PLOMEROVÁ, 1997). After processing the relative residuals, we can examine variations of the azimuth-incidence angle dependent terms of the relative residuals.

Figure 7 presents smoothed residual spheres showing spatial variations of the directional terms of P residuals recorded during the experiment. Although the data included in these spheres are very limited, the pattern of the spheres differs at southern stations from that at the northern stations, similarly to what was found in previous studies, based on large European data sets (e.g., BABUŠKA and PLOMEROVÁ, 1992) or in a regional study (FABER et al., 1986). In general, stations in the

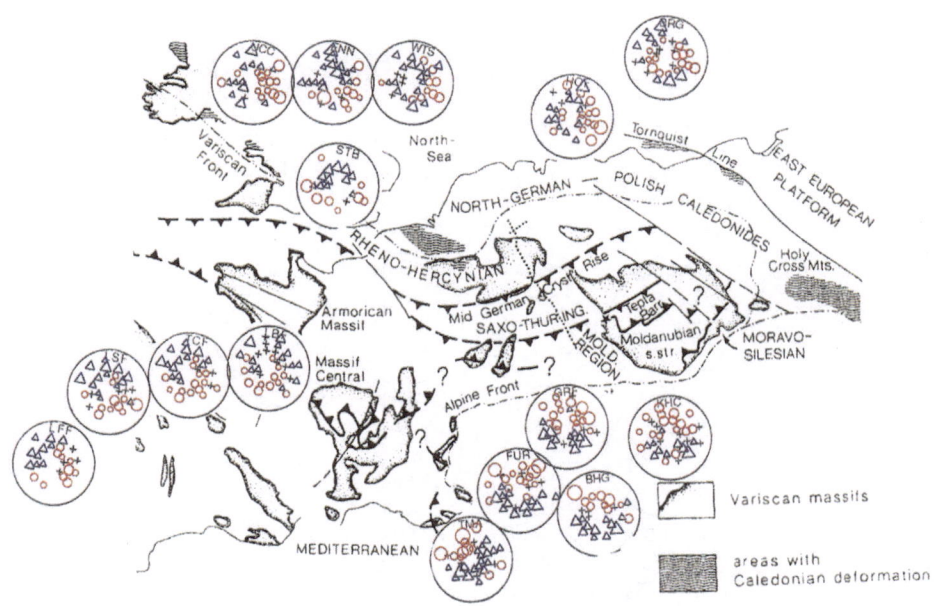

Figure 5

Main structural elements in the Variscan Belt of central Europe (after FRANKE, 1989) and residual spheres showing a typical pattern of smoothed directional terms of the relative residuals. Blue triangles represent early arrivals (the high-velocity directions within the lithosphere) and red circles the late arrivals. The size of the signs is proportional to the residual value. Black plus signs mark residuals within ⟨−0.1 s, 0.1 s⟩ interval. The maximum incidence angle is 50° at the M discontinuity at a reference depth of 33 km, which corresponds to an epicentral distance of about 20°. The perimeter marks the maximum angle of 50°.

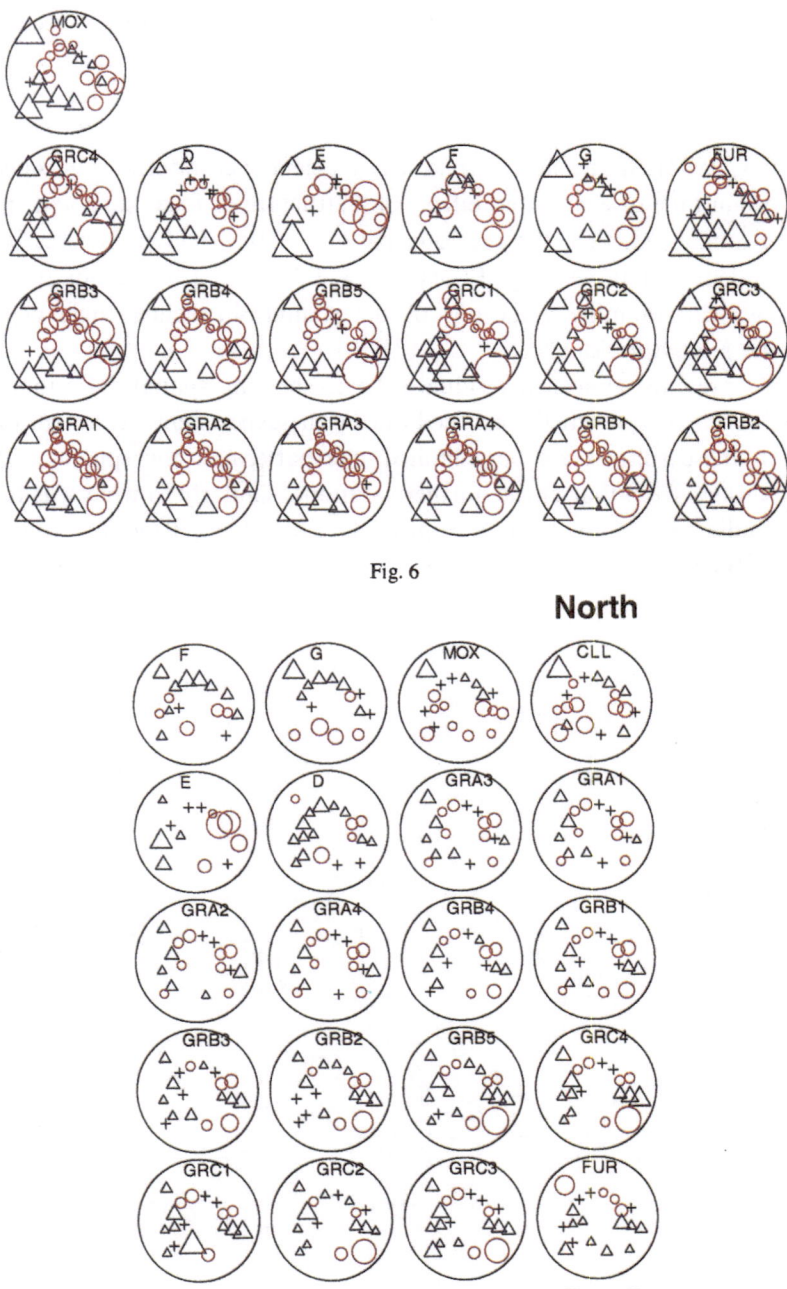

Figure 6

Residual spheres with absolute *P* residuals of 116 single events recorded during the field experiment, computed with the use of the J-B model. Corrections for crustal sediments according to KRÜGER and WEBER (1992) were applied. The residual spheres were smoothed by a two-dimensional linear filter with parameters of 30° in azimuths and 10° in incidence angles, and with increments of 30° and 5°, respectively. Only residuals with absolute values less or equal to 3.0 s were used. The spheres are cut at incidences of 60°. For more details also see the caption of Figure 5.

Moldanubicum, namely FUR and subarrays C and partly B and A, are character-ized by relatively early arrivals for waves emanating from southern or southwestern azimuths. The residual pattern changes dramatically between the experimental stations E and F. While in stations D and E the early arrivals prevail for westward propagations, at stations F and G the early arrivals are observed mainly from the north. The stations in the Saxothuringicum (MOX and CLL) keep this pattern. In comparison with sudden velocity and anisotropy changes related to the relatively narrow and steep Protogine zone in Sweden (PLOMEROVÁ et al., 1993, 1998), the change from the "southern" to the "northern" pattern takes place over a larger region, which suggests an assumption that the contact between the Saxo-thuringicum and Moldanubicum also dips in the subcrustal lithosphere, similarly to that in the crust (MENGEL, 1992). However, a substantial larger data set would be necessary to reveal a detailed geometry of the contact through the entire lithosphere.

Although "zero" levels differ in various types of normalizations, the general feature of the spheres remains unchanged. Small size disturbances in the diagrams can be attributed to very local heterogeneities or other effects that could not be eliminated via normalization and processing of the data. At first sight, the inclina-tion of the L/A transition derived from the static terms could partly provide an explanation of the observed residual pattern (Figs. 6 and 7). Nonetheless the effect of the inclined boundary between the high-velocity lithosphere and the low-velocity asthenosphere is mostly compensated by subtracting a directional mean at each station. Moreover, the explanation without considering the anisotropy could be only used on a very local scale and hardly could be implemented on the large European scale (Figs. 4 and 5), as this would require nearly regular alternation of the low- and high-velocity heterogeneities within the upper mantle, particularly within the litho-sphere.

5. Shear-wave Splitting

5.1 Analysis and Inversion of Published Models and Shear-wave Splitting Parameters

Previous investigations of the upper mantle anisotropy beneath central Europe and namely the GRA1 station (KIND et al., 1985; SILVER and CHAN, 1988;

Figure 7
Residual spheres with directional terms of the relative residuals grouped into 27 azimuth-epicentral distance segments of 30° and 10°, respectively. The residual terms were smoothed by a two-dimensional linear filter with parameters of 30° in azimuths and 10° in incidence angles, and with increments of 30° in azimuth and 10° in incidence angles. Only residuals with absolute values less or equal to 1.5 s were used. The residual spheres are shown for stations in tetrads from the north to south. D, E, F, and G (upper left corner) are the field stations installed for the experiment. For more details see the captions of Figures 5 and 6.

BORMANN *et al.*, 1996), based on particle motion analysis of the *SKS* waves, determined the eastward azimuth of 80° as a direction of polarization of the split fast *S* wave and associated it with the high-velocity direction in the upper mantle. This azimuth is close to the strike of the Variscan structures and oblique to the APM (BORMANN *et al.*, 1996). BORMANN *et al.* (1996) assign the majority of the observed anisotropy to the sublithospheric part of the mantle, and lateral changes in the apparent fast shear-wave polarization azimuths within all of western Europe relate to the effects of absolute plate motion and the topography of the lithosphere-asthenosphere boundary (BABUŠKA and PLOMEROVÁ, 1992).

Independence of the splitting parameters on the backazimuth or polarization of the shear waves, is the main requirement for interpreting the upper mantle anisotropy by a single layer with hexagonal symmetry and with the "fast" symmetry axis fixed in the horizontal plane. Moreover, accumulation of shear-wave splitting data proved in many regions (for a review see SILVER, 1996) that the splitting is not independent of backazimuth. SAVAGE and SILVER (1993) developed a more advanced analysis of the shear-wave splitting, resulting in two-layer anisotropic structures within the upper mantle, with "fast" horizontal symmetry axes oriented in different azimuths.

To explain variations of the fast *S* polarization azimuth and split time of the *SKS* waves with backazimuth, VINNIK *et al.* (1994) proposed a two-layer anisotropic model beneath several stations of the German Regional Seismic Network, including the GRF-array, namely beneath the GRA1 station. Two anisotropic layers are described by thicknesses and the fast directions of 83 km and 35° for the upper layer and 300 km and 105° for the lower one. Hexagonal symmetries are assumed to have the horizontal symmetry axes oriented in the azimuth of the high velocity directions. We computed the synthetic *P*-residual sphere for such a model (Fig. 8) and compared it with the observed one (Figs. 5 and ,7). It is evident that there is a large discrepancy between patterns of the synthetic and observed spheres. While the observed spheres exhibit a bipolar pattern (early arrivals from one side and late arrivals from the opposite one), with the high velocities from the south, the synthetics have completely different symmetry, expecting the high velocity directions only for nearer teleseisms from the west and east. This demonstrates a clear incompatibility between the shear-wave model and the *P*-residual observation.

VINNIK *et al.* (1994) tested their method with a two-layer hexagonal model also for a 45° dip of the "fast" symmetry axis in the upper layer and compared it with the basic model with a horizontal orientation of both axes. There is only a slight difference between estimated variations of the fast *SKS* polarization azimuth and split time for both models as presented by the authors. Therefore, contrary to the authors, we think that the model with the horizontal symmetry axes cannot be preferred to explain variations of splitting data for the GRA1.

GRA1

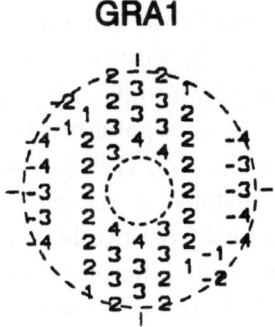

Figure 8

Synthetic *P*-residual sphere (residuals in tenths of a second) computed for a two-layer anisotropic model with hexagonal symmetry at the GRA1 station (VINNIK *et al.*, 1994) with horizontal "fast" symmetry axis. The outer and inner circles correspond to the 50° and 10° of incidence angles, respectively.

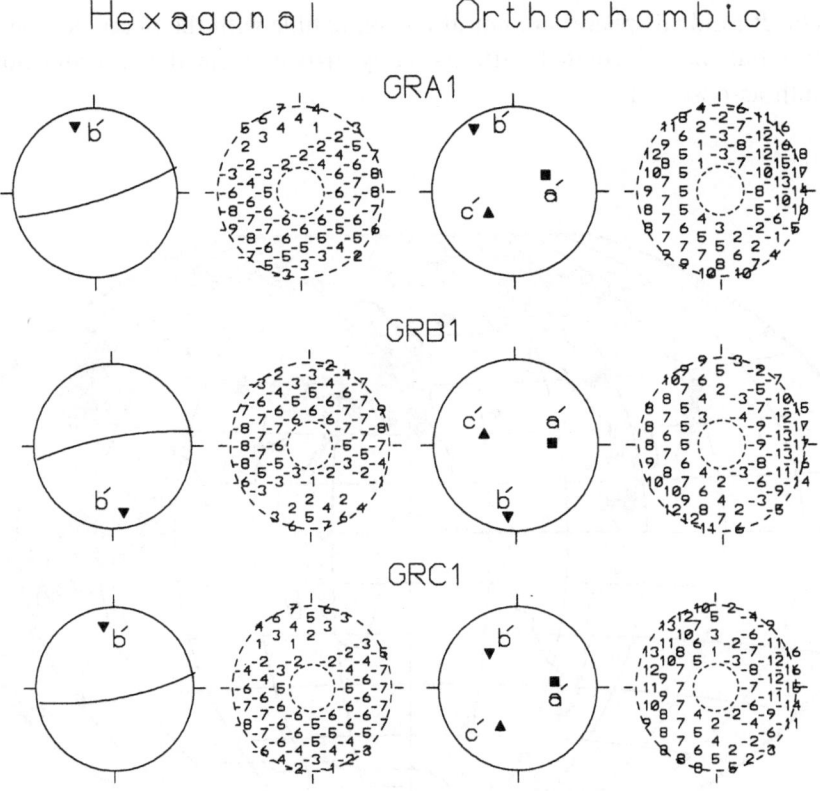

Figure 9

Orientations of hexagonal and orthorhombic models resulting from the inversions of splitting parameters evaluated by VINNIK *et al.* (1994) for the 3-c stations of the GRF-array in the projection of the lower hemisphere. The triangles mark symmetry axes and the curve the dipping high-velocity plane (*a′*, *c′*) in the case of the hexagonal model. Spheres with figures represent corresponding synthetic *P* residuals (in tenths of a second) for incidence angles between 20° and 50°.

Furthermore, we performed the inversion of shear-wave splitting parameters for GRA1, GRB1 and GRC1, evaluated from horizontal components of shear waves as given in a manuscript of a paper by VINNIK *et al.* (1994) (personal communication). Figure 9 shows the resulting anisotropic inferences along with corresponding synthetic *P*-residual spheres. Using data published by SAVAGE and SILVER (1993) for the western U.S., PLOMEROVÁ *et al.* (1996) demonstrated that the inversion method was also applied successfully in the case when the fast polarization azimuths and δt are only determined from two horizontal components. In the case of the hexagonal approximation of the lithospheric anisotropy, the inversions with the GRF-array data result in almost vertically plunging fast velocity planes (a', c') striking at about 80°, i.e., with nearly horizontal orientation of the "slow" symmetry axis b' pointing to the south or north. The corresponding *P*-residual synthetics tend to have the bipolar pattern, but not similar to the observed ones. Results for the inversions with the orthorhombic model are more stable and show the a'-axis dipping eastward at about 40° beneath all three stations. However, the synthetic *P*-residual spheres are not at all compatible with the observed ones. They exhibit about the $\pi/2$ rotated pattern—early arrivals from the east and not from the south as observed.

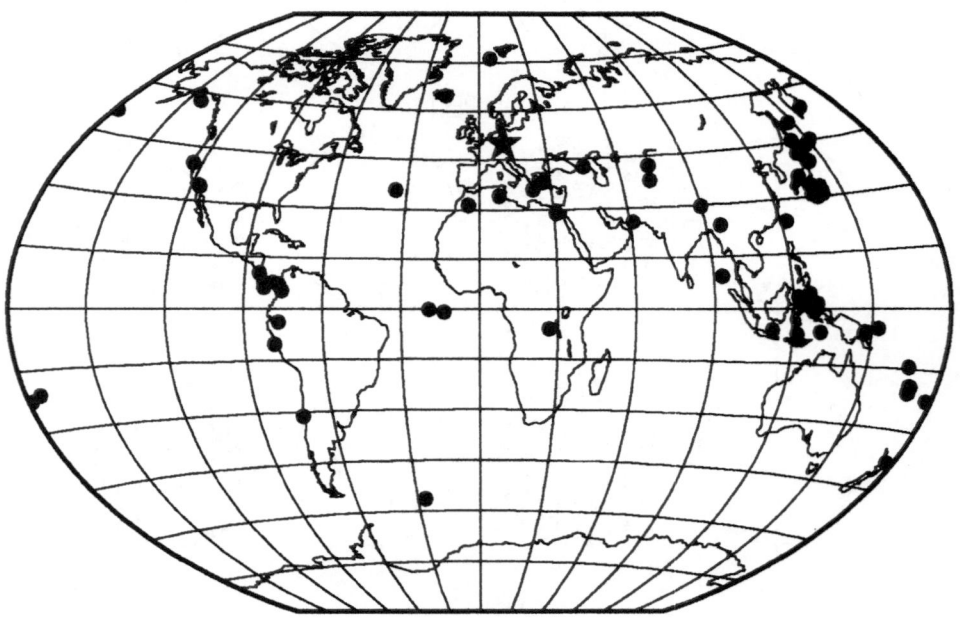

Figure 10
Distribution of events whose shear-wave splitting parameters were analyzed and used in the study of the upper mantle structure around the S/M contact (star) in central Europe.

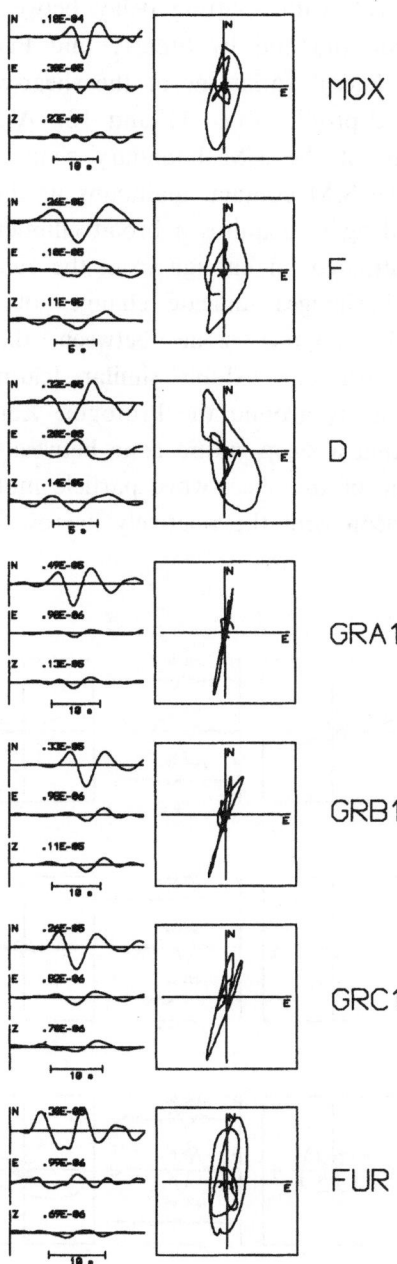

Figure 11
Example of distinct lateral variations of the polarization pattern of the S wave from an epicentral distance of 80° with a backazimuth of 4°.

5.2 Analysis and Inversion of Shear-wave Data from the Field Experiment

During the 1992 experiment 203 three-component shear waveforms of 89 teleseismic events (Fig. 10), with splitting delay between 0.6–4.0 s, served as an input of the inversion (method by ŠíLENÝ and PLOMEROVÁ, 1996). First we present examples of lateral variations of the shear-wave particle motion at several stations along the profile (Figs. 11 and 12). A broad elliptical particle motion at stations north of the S/M boundary changes to the linear one at station A1 just above the S/M contact, maintains its linearity to the south (at stations B1 and C1) and again acquires a broad ellipticity at the southernmost station FUR, which is situated far enough from the contact zone in the Moldanubicum. These lateral changes indicate changes of the upper mantle anisotropy related to the contact zone between the Moldanubicum and Saxothuringicum. The variations exhibit similar features to those observed across the experimental array around the Protogine Zone in southern Sweden, most probably representing a steep suture zone between two continental plates. The characteristic change of the shear-wave particle motion takes place over a larger region in comparison with the relatively narrow Protogine zone in Swe-

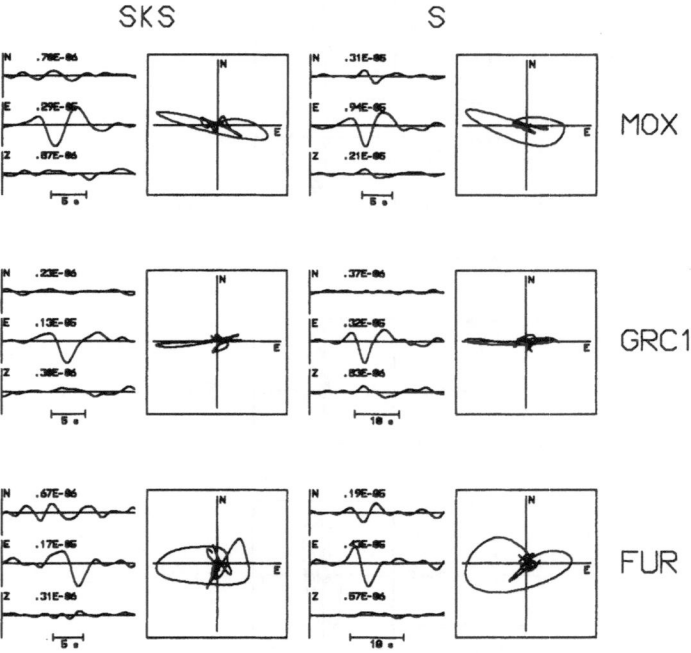

Figure 12

Example of similarity of lateral changes in the shear-wave particle motion between *S* and *SKS* waves of a single event at an epicentral distance of 92° with a backazimuth of 266°.

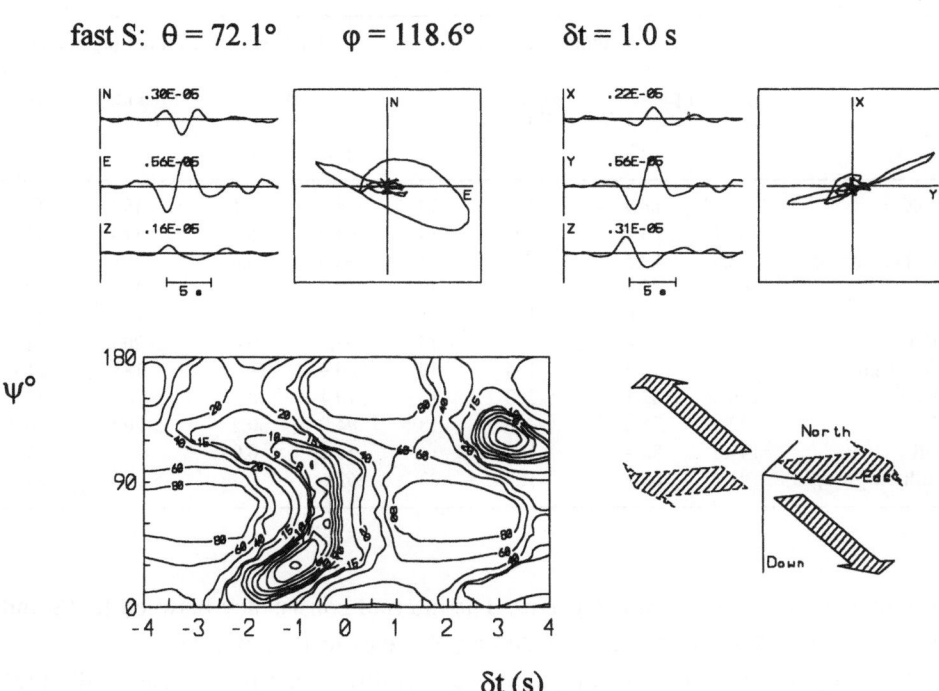

N. Peru July 3, 1992   station E

fast S: θ = 72.1°   φ = 118.6°   δt = 1.0 s

ψ°

δt (s)

Figure 13

Evaluation of shear-wave splitting parameters Ψ and δt by eliminating the effects of anisotropy. Upper left, recorded waveforms and particle motion displaying an elliptical polarization; upper right, the shear-wave motion in the (x = fast wave, y = slow wave) plane perpendicular to the shear-wave ray (z, perpendicular to the (x, y) plane) and with eliminated time delay δt between the fast and slow waves resulting in the quasilinear particle motion. Lower left, relative isolines of the quantity minimized during the process eliminating the splitting: for an S wave it is the least eigenvalue of the cross-correlation matrix of the S-wave motion components in the plane perpendicular to the S-wave ray; for an SKS wave it is the energy of the wave in the transverse component. The horizontal axis of the isoline plot gives time delay δt; the vertical axis gives the angle Ψ specifying the orientation of the fast and slow shear-wave polarization in the (x, y) plane by two spherical angles φ and Θ. Lower right, axonometry display of the orientation of polarization of the fast (arrows with solid outline) and slow (arrows with dashed outline) shear waves in the geographical Cartesian coordinate system.

den (PLOMEROVÁ *et al.*, 1993), which again supports the idea of a gradually dipping contact of the S/M through the entire lithosphere.

Figure 12 demonstrates that both *SKS* and *S* exhibit distinct similarity in variations of the particle motion. This similarity confirms that effects due to the laterally varying anisotropic structure of the lithosphere below the region are decisive in the shear-wave splitting. Slight differences in the particle motion can be attributed to differences in the direction of propagation within the anisotropic

Table 1

Anisotropic structure of the subcrustal lithosphere beneath the Saxothuringicum-Moldanubicum contact (results of the inversion of the shear-wave splitting parameters)

| | Hexagonal model ($k_p = 5\%$) | | | | Orthorhombic model ($k_p = 9\%$) | | | |
| | (a', c') plane | | thickness (km) | N | a' axis | | thickness (km) | N |
	φ_d°	α_d°			φ_d°	α_d°		
MOX	171.1	86.4	190	14	264.3	4.1	197	17
F	161.1	84.7	>200	14	104.3	6.2	170	12
Northern group	353.9	82.5	152	21	98.3	3.6	161	21
E	165.0	82.5	160	19	250.2	7.7	147	19
D	176.7	81.6	196	15	77.3	32.2	171	15
GRA1	209.6	86.4	>200	14	293.2	9.4	>200	14
Central group	194.1	78.3	194	19	260.9	11.7	185	21
GRB1	156.5	83.1	>200	12	270.4	8.2	>200	10
GRC	22.0	75.6	>200	18	94.9	30.2	>200	20
FUR	149.0	82.4	164	14	58.1	6.0	173	15
Southern group	191.0	85.1	187	22	263.6	8.3	> 200	24

structures with inclined symmetry axes. The internal consistency between the S and SKS particle motions justify also including S waves in the analysis.

An example of evaluating the shear-wave splitting parameters using all three components of the seismograms is shown in Figure 13. As described in Section 3, we invert the variations of splitting parameters to infer the anisotropic structure of the subcrustal lithosphere with the symmetry axes arbitrarily oriented in 3-D. Table 1 supplies results of the inversions of the splitting parameters, both for the single stations and for 3 groups created from northern, central and southern stations along the profile of the station. A general feature of the resulting orientations of the anisotropic structures is a very steep dip of the high-velocity plane (a', c') of the hexagonal model with the strike changing from ENE to ESE (α_d stands for the dip measured from the horizontal, ϕ_d is the dip-azimuth of the (a', c') plane, N is number of events). All solutions less then $1.2 \times R_{\min}$ (1), shown in Figure 14, are close to the $R_{\min}$, thus demonstrating the stability of the solution which is lower only for the results of the "central" group. In comparison with the hexagonal model, the stability of the solutions of the orthorhombic model is in general lower, namely as to the orientation of the b' and c' axes (ŠÍLENÝ and PLOMEROVÁ, 1996). The strike of the a'-axis is to the east on average, with a well-defined dip of 30° only beneath the C1 and D stations. Beneath other stations the a'-axis is tilted less than 10° from the horizontal. The results obtained from the shear-wave data recorded during the field measurements and analyzed in 3-D, are close to the results obtained from data if only horizontal components of shear waves were analyzed (see Fig. 9).

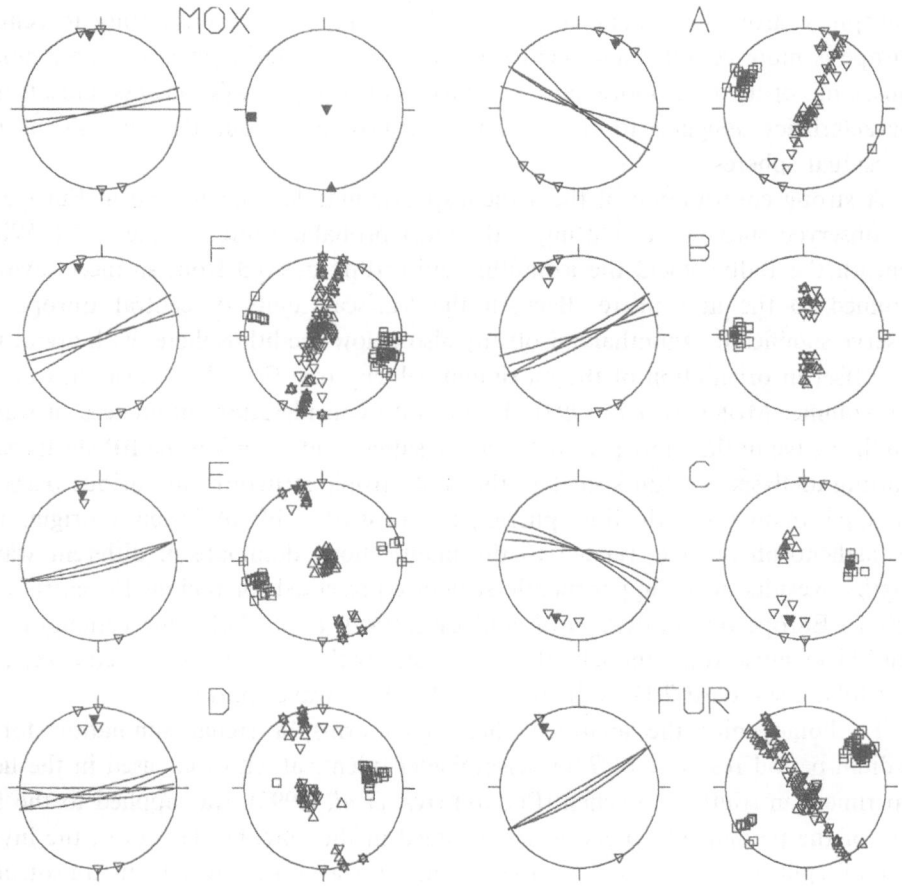

Figure 14

Orientations of hexagonal (left) and orthorhombic (right) models resulting from the inversions of splitting parameters of shear waves recorded during the experiment in projection of the lower hemisphere (full symbols and thick curves). Open symbols and thin curves represent all acceptable solutions less than $1.2 \times R_{min}$. Square stands for the a' axis, inverted triangle for the b' axis and triangle for the c' axis.

6. Self-consistent 3-D Anisotropic Models of the S/M Contact—Discussion

It is evident that models with a subvertically dipping high-velocity plane (a', c') in the case of the hexagonal approximation, and to the east oriented dipping a'-axis in the case of the orthorhombic approximation, are not compatible with inferences of anisotropic structure with inclined symmetry axes within the lithosphere, as derived from P-residual spheres. Thus, contrary to a similar study in southern Sweden (PLOMEROVÁ et al., 1993, 1996, 1998), the inversion of splitting parameters, assuming one anisotropic layer with either hexagonal or orthorhombic symmetry, did not result in a self-consistent 3-D anisotropic model of the subcrustal lithosphere around the contact zone of the S/M. The tectonic setting in central Europe is more complex and, evidently, we must search for an anisotropic model consisting of two or more layers, whose upper layer will possess anisotropic characteristics assigned to the subcrustal lithosphere from the analysis of the *P*-residual spheres.

A strong contribution of the asthenospheric mantle beneath central Europe to the observed shear-wave splitting is the most probable source of the effect. While beneath the Baltic shield the azimuthal anisotropy derived from surface waves is confined to the lithosphere. Beneath the Variscan units of central Europe we observe significant azimuthal anisotropy also below the lithosphere, with about the $\pi/2$ offset in orientation of the maximum velocity (see Fig. 3b in BABUŠKA *et al.*, this volume; MONTAGNER, 1994). Taking into consideration differences in wavelengths active in the short-period *P*-wave residual studies and in the BB shear-wave splitting analysis, we can speculate that anisotropic structures in various parts of the upper mantle, i.e., the lithospheric part, most probably of frozen-in origin, and the asthenospheric one, related to the mantle flow, dominate in different wavelengths. Results of the upper mantle structure presented for regions in central and western Europe by GRANET *et al.* and BRECHNER *et al.* (both this volume) favor multi-layer anisotropic models, though mostly with horizontal symmetry axes, as they follow standard 2-D evaluations of the shear-wave splitting.

To homogenize the analyzed shear waveforms, a signal simulation for a medium-period response of 7.6 s seismometer, identical with that used in the field experiment in southern Sweden (PLOMEROVÁ *et al.*, 1993), was applied to the BB data of the permanent observatories included in the antenna. However, the inversion of splitting parameters of the simulated waveforms results in anisotropic inferences close to those found in the original broadband data. It seems that a contribution to the observed BB shear-wave splitting deriving from the anisotropic sublithospheric mantle is very significant in central Europe (BORMANN *et al.*, 1993) and masks the lithospheric contribution in the frequency range of teleseismic shear waves.

Although the total amount of inverted splitting inputs should be sufficient (ŠÍLENÝ and PLOMEROVÁ, 1996), the distribution of distant events recorded in central Europe as to the back azimuth is not ideal (see Fig. 10). The majority of waves, especially *SKS*, arrive from directions along the strike of the Variscan structures. Therefore, the resulting anisotropic models are less sensitive to retrieving the inclination of the structures. The distribution of events in combination with other effects, such as the significant contribution from the sublithospheric mantle, do not allow us, in this particular case, to reliably retrieve by the inversion, a tilt of the symmetry axes of the anisotropic structures within the subcrustal lithosphere.

There is no doubt that two-layer anisotropic models of the upper mantle structure in central Europe describe its lithospheric and sublithospheric parts

(BABUŠKA *et al.*, this volume) better than one-layer models. BABUŠKA and PLOM-
EROVÁ (1995) estimated the magnitude of anisotropy of the subcrustal lithosphere
both from P residuals and from published shear-wave splitting delay times.
Although estimated shear-wave anisotropies k_S are within the range of anisotropies

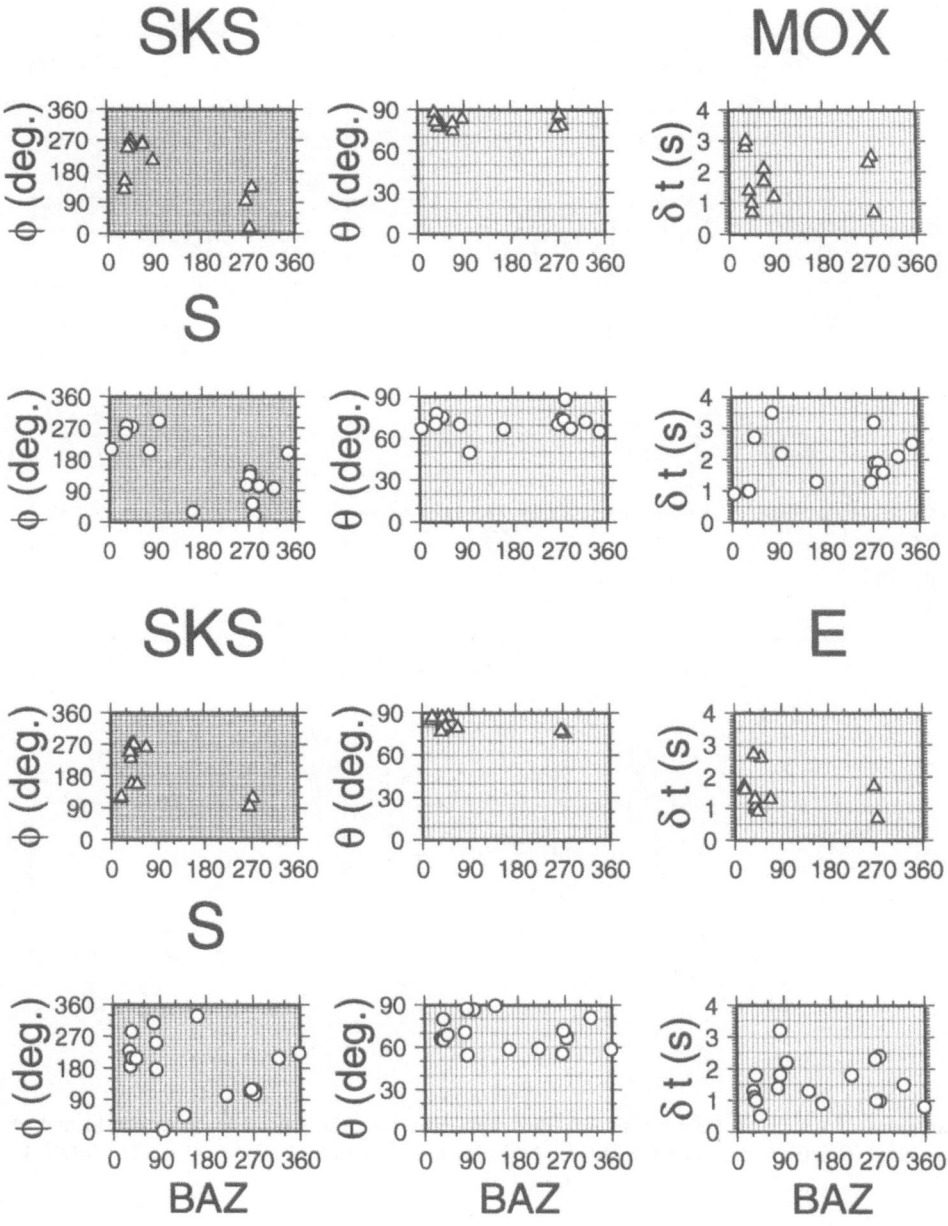

Figure 15

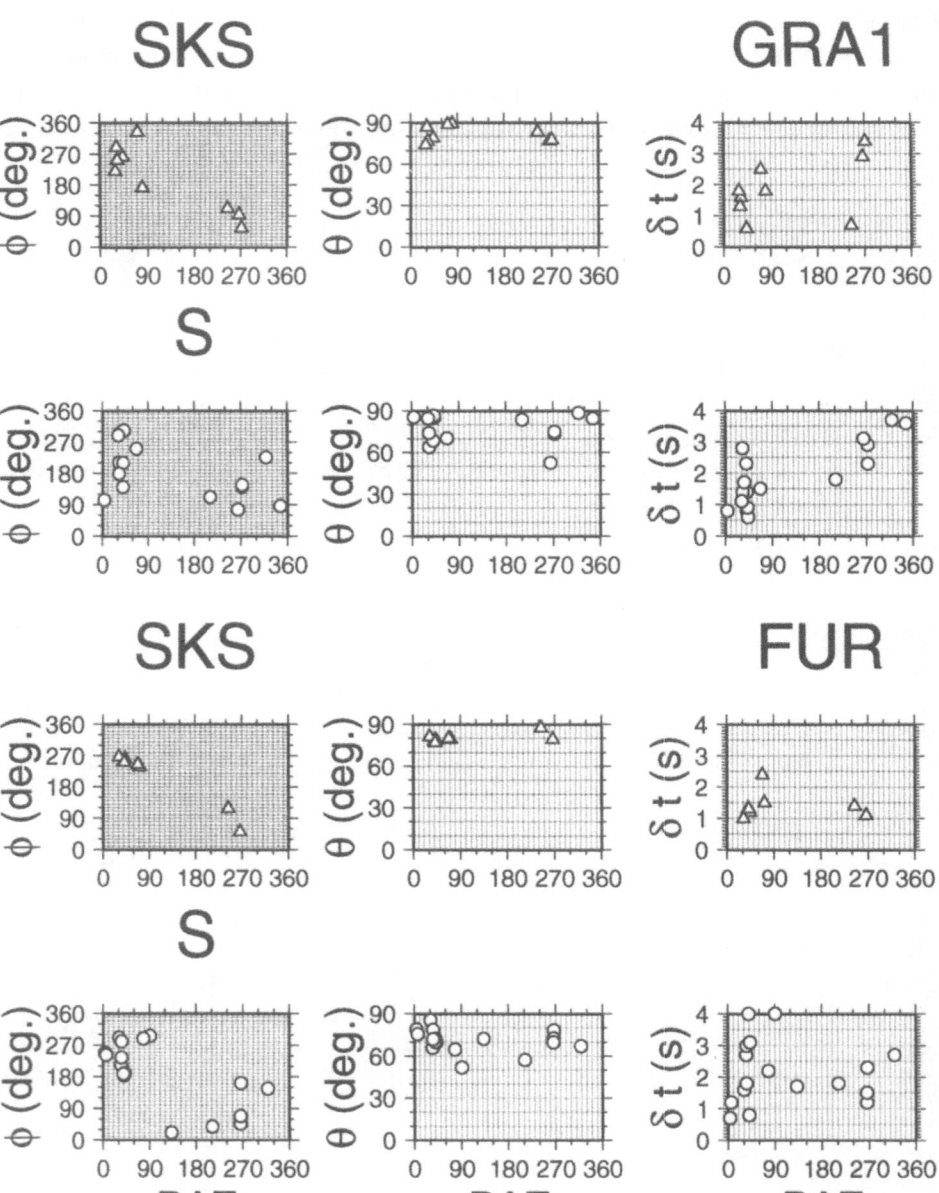

Figure 15
Variations of the fast shear-wave velocity directions given by spherical angles ϕ (azimuth) and Θ (measured from the vertical axis upward), and splitting time δt with backazimuth at stations MOX, E, GRA1 and FUR.

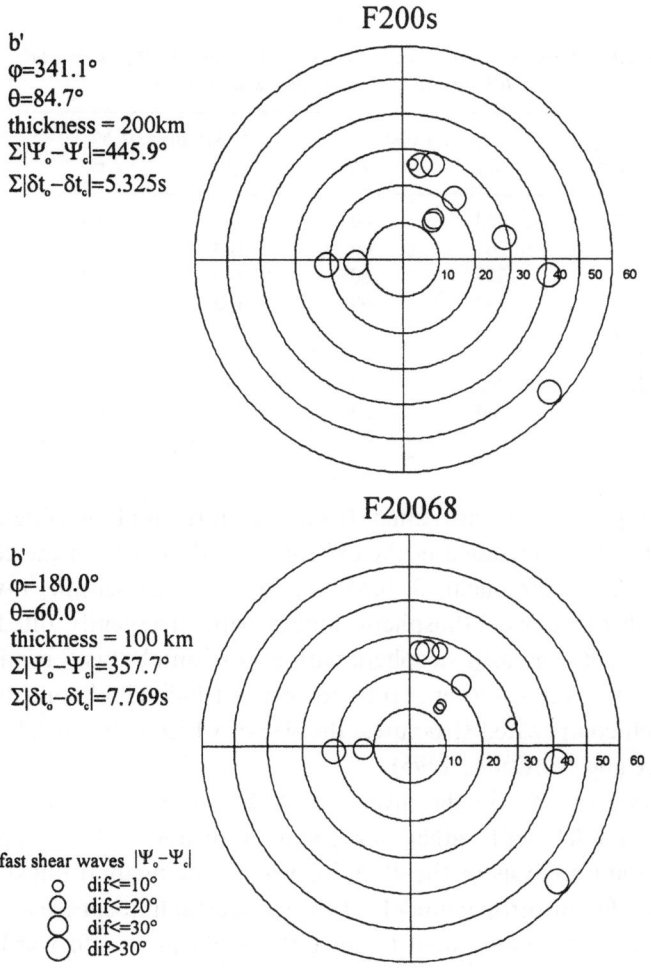

Figure 16

Example of deviations between observations and synthetics for models beneath station F retrieved by the inversions of the shear-wave splitting parameters (a) and by the forward task (b). $\Sigma|\Psi_o - \Psi_c|$ expresses sum of differences between observed and synthetic fast shear-wave directions Ψ and $\Sigma|\delta t_o - \delta t_c|$ are differences in δt. Orientation of the hexagonal models with "slow" symmetry axis b' is given by the spherical angles ϕ and Θ.

calculated by MAINPRICE et al. (1993) for ultramafic rocks of lherzolite and harzburgite compositions, the observed anisotropy must be split into lithospheric and sublithospheric contributions of the upper mantle.

A test of a possible dependence of the fast shear-wave polarization direction, given by spherical angles ϕ and Θ, and of δt on the backazimuth of arriving waves (BAZ, Fig. 15), showed no simple variations with the $\pi/2$ periodicity which would correspond to the horizontal orientation of the "fast" symmetry axes in the two layers (SAVAGE and SILVER, 1993; VINNIK et al., 1996). This also indicates that the symmetry axes of hexagonal or orthorhombic anisotropy related to the lithospheric

Table 2

Results of forward modelling of seismic anisotropy in the subcrustal lithosphere with hexagonal symmetry

Station	(a', c') plane		Symm. axis b'	
	ϕ_d°	α_d°	ϕ°	Θ°
MOX	340	60	160	60
F	360	60	180	60
E	320	60	140	60
D	340	60	160	60
GRA1	180	30	360	30
GRB1	200	60	20	60
GRC1	200	60	20	60
FUR	160	60	340	60

part of the upper mantle are tilted from the horizontal in different azimuths beneath all 20 stations included in the experiment. Moreover, in the central part of the antenna, above the inclined S/M contact, the teleseismic rays propagate consecutively through two lithospheric blocks with divergently tilted symmetries, underlaid by anisotropic asthenosphere with a horizontal "fast" symmetry axis. It is difficult to detect the splitting and/or to sort out individual contributions to the splitting in such complicated structures, the size of which is below adequate Fresnel zones (ALSINA and SNIEDER, 1995).

Currently, we can use in the inversion of the shear-wave splitting parameters only one-layer models with either hexagonal or orthorhombic symmetry. As we illustrated by joint analysis of the *P* and *S* waves, the method failed to determine self-consistent 3-D anisotropic models of the subcrustal lithosphere around the S/M contact. Therefore, we endeavored to solve the problem as a forward task. In this case, starting one-layer models were derived from the results of the inversion of splitting parameters (Table 1) and from the *P*-residual spheres which assumed a thickness of the subcrustal lithosphere of 100 km (BABUŠKA and PLOMEROVÁ, 1992). We selected a subset of the best determined shear-wave splitting parameters at each station and searched for such orientations of the olivine aggregate with hexagonal symmetry (with the "slow" symmetry axis *b'*; BABUŠKA *et al.*, 1993) which meet the observed variations of the fast shear-wave directions and δt. Table 2 presents anisotropic models of subcrustal lithosphere with plunging symmetry axes which are compatible with both *P* and *S* observations. The plunge of the high-velocity plane (a', c') of the hexagonal model is given by a dip-azimuth ϕ_d and a dip of the plane from the horizontal α_d. While differences between the observed and synthetic fast shear-wave directions (Fig. 16b) are reduced by 20% in comparison with those for models resulting from the inversion (Fig. 16a), differences between the splitting times remain higher. This is most probably a consequence of the presence of the asthenospheric contribution (BABUŠKA *et al.*, this volume), which is not taken into account in the forward task, but which affects the minimization process of the inversion and is reflected also in the required thickness of the anisotropic layer (Table 1).

The models presented in Table 2 are characterized by divergently dipping high velocities relative to the S/M contact. In our interpretations, the dipping structures may represent two systems of paleosubductions divergent relative to the suture between the Moldanubicum and the Saxothuringicum. MATTE (1991), in his analysis of the crustal evolution of the Variscan belt, remarks also about the double lithospheric subduction system operating from 380 to 300 Ma, which was inherited from previous oceanic subductions. The general fan-like aspect of the belts resulted from the obduction of the oceanic crust in opposite directions onto the passive margins on both sides of the orogen.

7. Conclusions

The processing of teleseismic data recorded during the field experiment results in:

1) a detailed mapping of the lithosphere-asthenosphere relief around the S/M contact which manifests a southward lithospheric thickening beneath the Moldanubicum;

2) an analysis of systematic spatial variations of the smoothed directional terms of relative P residuals confirmed the existence of large dipping anisotropic structures in the subcrustal lithosphere. The high velocities dip almost perpendicularly to the WSW-ENE strike of the surface trace of the S/M contact zone. They plunge to the north-northwest in the region north of the suture and prevailingly to the south beneath stations south of it;

3) distinct lateral changes of the shear-wave splitting parameters monitor the complex anisotropic structure of the upper mantle beneath the contact.

Joint analysis and interpretation of the shear-wave splitting parameters and P-residual spheres result in a 3-D self-consistent anisotropic model of the subcrustal lithosphere around the S/M contact with divergently plunging high-velocity directions and anisotropic sublithospheric mantle with the horizontal high velocities oriented E-W, on average. The Variscides of central Europe may thus represent a collision zone characterized by two systems of paleosubductions divergent relative to the suture between the Moldanubicum and the Saxothuringicum.

Acknowledgements

Special thanks are extended to D. Seidl, K. Klinge and K. Stammler (Zentralobservatorium Gräfenberg) for their kind assistance in finding suitable localities for the temporary seismic stations and their cooperation during the evaluation of records of permanent stations of the German Regional Seismic Network. The Geophysical Institute of Slovak Acad. Sci. in Bratislava supplied us with two field stations for the experiment. We are also grateful to P. Jedlička and F. Hampl for installation and maintenance of the stations. Picking the *P*-arrival times for some stations by J. Pospíšil and operating programs for the shear-wave splitting analysis by K. Špaček, as well as the assistance of A. Slancové and D. Kouba in forward-task computations and finalization of graphical outputs are greatly appreciated. We thank K. Furlong and an anonymous reviewer for their comments and suggestions which significantly improved the manuscript. The research was partly supported by Grants Nos. 31225, A312115 and A3012604 of the Czech Academy of Sciences.

REFERENCES

ALSINA, D., and SNIEDER, R. (1995), *Small-scale Sublithospheric Continental Mantle Deformations: Constraints from SKS Splitting Observations*, Geophys. J. Int. *123*, 431–448.

ANDERSON, D. L., *Theory of the Earth* (Blackwell Sci. Publ. 1989) 336 pp.

ANDO, M. (1984), *ScS Polarization Anisotropy around Pacific Ocean*, J. Phys. Earth *32*, 179–195.

BABUŠKA, V., PLOMEROVÁ, J., and ŠÍLENÝ, J. (1984), *Large-scale Oriented Structures in the Subcrustal Lithosphere of Central Europe*, Ann. Geophys. *2*, 649–662.

BABUŠKA, V., PLOMEROVÁ, J., and ŠÍLENÝ, J., *Structural model of the subcrustal lithosphere in Central Europe*. In *The Composition, Structure and Dynamics of the Lithosphere-asthenosphere System* (C. Froidevaux, and K. Fuchs, eds.), AGU Geophys. Series, vol. 16 (Washington, DC 1987) pp. 239–251.

BABUŠKA, V., PLOMEROVÁ, J., and PAJDUŠÁK, P., *Lithosphere-asthenosphere in central Europe: Models derived from P residuals*. In *Proceedings of the Fourth Workshop on the European Geotraverse (EGT) Project: The Upper Mantle* (G. Nolet and B. Dost, eds.) (European Science Foundation, Strasbourg 1988) pp. 37–48.

BABUŠKA, V., and PLOMEROVÁ, J., *Seismic anisotropy of the subcrustal lithosphere in Europe: Another clue to recognition of accreted terrances?* In *Deep Structure and Past Kinematics of Accreted Terranes* (J. W. Hillhouse ed.) Geophys. Monograph 50, IUGG, vol. 5 (Washington, DC 1989) pp. 209–217.

BABUŠKA, V., and CARA, M., *Seismic Anisotropy in the Earth* (Kluwer, Dordrecht 1991) 217 pp.

BABUŠKA, V., and PLOMEROVÁ, J. (1992), *The Lithosphere in Central Europe—Seismological and Petrological Aspects*, Tectonophys. *207*, 141–163.

BABUŠKA, V., PLOMEROVÁ, J., and ŠÍLENÝ, J. (1993), *Models of Seismic Anisotropy in Deep Continental Lithosphere*, Phys. Earth Planet. Inter. *78*, 167–191.

BABUŠKA, V., and PLOMEROVÁ, J. (1995), *Regional Fabric of the Deep Lithosphere in Central Europe from Seismic Anisotropy*, Proc. of the EUROPROBE Symposium TESZ, Liblice 1993, Studia geof. et geod. *39*, 219–226.

BABUŠKA V., MONTAGNER, J.-P., PLOMEROVÁ, J., and GIRARDIN, N. (1998), *Age Dependent Large-scale Fabric of the Mantle Lithosphere as Derived from Surface-wave Velocity Anisotropy*, Pure appl. geophys. *151*, 257–280.

BORMANN, P., BURGHARDT, P.-T., MAKEYEVA, L. I., and VINNIK, L. P. (1993), *Teleseismic Shear-wave Splitting and Deformations in Central Europe*, Phys. Earth Planet. Inter. *78*, 157–166.

BORMANN, P., GRÜNTHAL, G., KIND, R., and MONTAG, H. (1996), *Upper Mantle Anisotropy beneath Central Europe from SKS Wave Splitting: Effects of Absolute Plate Motion and Lithosphere-asthenosphere Boundary Topography?* J. Geodyn. *22*, 11–32.

BRECHNER, S., KLINGE, K., KRÜGER, F., and PLENEFISCH, T. (1998), *Backazimuthal Variations of Splitting Parameters of Teleseismic SKS-phases Observed at the Broadband Stations in Germany*, Pure appl. geophys. *151*, 305–331.

CHRISTENSEN, N. I. (1984), *The Magnitude, Symmetry and Origin of Upper Mantle Anisotropy Based on Fabric Analyses of Ultramafic Tectonites*, Geophys. J. R. Astr. Soc. *76*, 89–111.

DZIEWONSKI, A. M., and ANDERSON, D. L. (1983), *Travel Times and Station Corrections for P Waves at Teleseismic Distances*, J. Geophys. Res. *88*, 3295–3314.

EMMERMANN, R., LAUTERJUNG, J., and UMSONST, T. (1993), *Contributions to the 6th Annual KTB-Colloquium*, Geoscientific Results. KTB Report 93–2, Hannover.

FABER, S., PLOMEROVÁ, J., and BABUŠKA, V. (1986), *Deep-seated Lateral Velocity Variations beneath the GRF Array Inferred from Mislocation Patterns and P Residuals*, J. Geophys. *60*, 139–148.

FRANKE, W. (1989), *Variscan Plate Tectonics in Central Europe—Current Ideas and Open Questions*, Tectonophys. *169*, 221–228.

FURLONG, K. P., and OWENS, T. J. (1997), *3-D Mantle Fabric beneath Tibet*, Annales Geophysicae, Supp. to vol. 15, *C61* (abstract).

GEBRANDE, H., BOPP, M., NEURIEDER, P., and SCHMIDT, T., *Crustal structure in the surroundings of the KTB drill site as derived from refraction and wide-angle seismic observations*. In *German Continental Deep Drilling Program* (KTB) (Emmermann, R. and Wohlenbert, J., eds.) (Springer Verlag, Berlin 1989) pp. 151–176.

GRANET, M., GLAHN, A., and ACHAUER, U. (1998), *Anisotropic Measurements in the Rhinegraben Area and the French Massif Central: Geodynamic Implications*, Pure appl. geophys. *151*, 333–364.

KIND, R., KOSAREV, G. L., MAKEEVA, L. I., and VINNIK, L. P. (1985), *Observations of Laterally Inhomogeneous Anisotropy in the Continental Lithosphere*, Nature *318*, 358–361.

KRÜGER, F., and WEBER, M. (1992), *The Effect of Low-velocity Sediments on the Mislocation Vectors of the GRF Array*, Geophys. J. Int. *108*, 387–393.

KRÜGER, F. (1994), *Sediment Structure at GRF from Polarizations Analysis of P Waves of Nuclear Explosion*, Bull. Seismol. Soc. Am. *84*, 149–170.

MAINPRICE, D., VAUCHEZ, A., and MONTAGNER, J.-P., Eds. (1993), *Dynamics of the Subcontinental Mantle: From Seismic Anisotropy to Mountain Building*, Phys. Earth Planet. Inter. *78* (Special Issue), 157–354.

MAKEYEVA, L. I., PLEŠINGER, A., and HORÁLEK, J. (1990), *Azimuthal Anisotropy beneath the Bohemian Massif from Broad-band Seismograms of SKS Waves*, Phys. Earth Planet. Inter. *62*, 298–306.

MATTE, P. (1986), *Tectonics and Plate Tectonics Models for the Variscan Belt of Europe*, Tectonophys. *126*, 329–374.

MENGEL, K., *Evidence from xenoliths for the composition of the lithosphere*. In *A Continent Revealed* (D. Blundell, R. Freeman, and S. Mueller, eds.), (Cambridge University, Cambridge 1992) pp. 91–102.

MONTAGNER, J.-P. (1994), *Can Seismology Tell us Anything about Convection in the Mantle?* Rev. Geophys. *32*, 115–137.

NICOLAS, A., and CHRISTENSEN, N. I., *Formation of anisotropy in upper mantle peridotites—A review*. In *The Composition, Structure and Dynamics of the Lithosphere-asthenosphere System* (C. Froidevaux, and K. Fuchs, eds.), AGU Geophys. Series, vol. 16 (Washington, DC 1987) pp. 111–123.

OZALAYBEY, S., and SAVAGE, M. K. (1994), *Double-layer Anisotropy Resolved from S Phases*, Geophys. J. Int. *117*, 653–664.

PLOMEROVÁ, J., and BABUŠKA, V. (1988), *Lithosphere Thickness in the Contact Zone of the Moldanubicum and Saxothuringicum in Central Europe*, Phys. Earth Planet Inter. *51*, 159–165.

PLOMEROVÁ, J., ŠÍLENÝ, J., BABUŠKA, V., HORÁLEK, J., PAJDUŠÁK, P., POUPINET, G., GRANET, M., KULHÁNEK, O., and ARVIDSSON, R. (1993), *A Small Array Study of S-wave Anisotropy of the Deep Lithosphere across the Protogine Zone in Southern Sweden*, Proceedings of the XXIII General Assembly of the ESC, Prague 1992, pp. 307–312.

PLOMEROVÁ, J., ŠÍLENÝ, J., and BABUŠKA, V. (1996), *Joint Interpretation of Upper Mantle Anisotropy Based on Teleseismic P Travel-time Delays and 3-D Inversion of Shear-wave Splitting Parameters*, Phys. Earth. Planet. Inter. *95*, 293–309.

PLOMEROVÁ, J., ŠÍLENÝ, J., ARVIDSSON, R., BABUŠKA, V., GRANET, M., KULHÁNEK, O., and POUPINET, G. (1998), *An Array Study of Lithospheric Structure across the Protogine Zone, Southern Scandinavia—Signs of a Paleocontinental Collision*, in preparation.

PLOMEROVÁ, J. (1997), *Seismic Anisotropy in Tomographic Studies of the Upper Mantle beneath Southern Europe*, Annali di Geofisica *XL*, 111–121.

PRAUS, O., PĚČOVÁ, J., PETR, V., BABUŠKA, V., and PLOMEROVÁ, J. (1990), *Magnetotelluric and Seismological Determination of the Lithosphere-asthenosphere Transition in Central Europe*, Phys. Earth Planet. Inter. *60*, 212–228.

SAVAGE, M. K., and SILVER, P. G. (1993), *Mantle Deformation and Tectonics: Constraints from Seismic Anisotropy in Western United States*, Phys. Earth Planet. Inter. *78*, 207–227.

SILVER, P., and CHAN, W. W. (1988), *Implications for Continental Structure and Evolution from Seismic Anisotropy*, Nature *335*, 34–39.

SILVER, P., and CHAN, W. W. (1991), *Shear-wave Splitting and Subcontinental Mantle Deformation*, J. Geophys. Res. *98*, 16429–16454.

SILVER, P. G. (1996), *Seismic Anisotropy beneath the Continents: Probing the Depths of Geology*, Ann. Rev. Earth Planet. Sci. *24*, 385–432.

ŠÍLENÝ, J., and PLOMEROVÁ, J. (1996), *3-D Inversion of Shear-wave Splitting Parameters to Retrieve Anisotropy of Continental Lithosphere*, Phys. Earth. Plant. Int. *95*, 277–292.

VINNIK, L. P., KIND, R., KOSAREV, G. L., and MAKEYEVA, L. I. (1989), *Azimuthal Anisotropy in the Lithosphere from Observations of Long-period S Waves*, Geophys. J. Int. *99*, 549–559.

VINNIK, L. P., KRISHNA, V. G., KIND, R., BORMANN, P., and STAMMLER, K. (1994), *Shear-wave Splitting in the Records of the German Regional Seismic Network*, Geophys. Res. Lett. *21*, 457–460.

(Received May 9, 1997, revised October 8, 1997, accepted October 27, 1997)

To access this journal online:
http://www.birkhauser.ch

Pure appl. geophys. 151 (1998) 395–405
0033–4553/98/040395–11 $ 1.50 + 0.20/0

Pure and Applied Geophysics

Anisotropy beneath the Iberian Peninsula: The Contribution of the ILIHA-NARS Broad-band Experiment

J. Díaz,[1] J. Gallart,[1] A. Hirn[2] and H. Paulssen[3]

Abstract—The presence of anisotropy beneath the Iberian Peninsula and its main distinctive features can be established through the analysis of teleseismic shear-wave splitting observed in the ILIHA-NARS experiment. In this experiment, an homogeneous data set is provided by a network of 14 broad-band stations deployed over the entire peninsula for about one year. Even if technical problems led to an amount of data smaller than expected, significant variations in the inferred fast velocity direction are observed for stations located in different Iberian domains. The stations in Central and East Iberia show a fast velocity direction oriented roughly E–W, coincident with previous results in Toledo. A clearly different NE–SW direction is observed in the Ossa-Morena zone, supporting the image from a previous regional experiment. The observed delay times lie between 0.5 and 1 s. Although large-scale mechanisms, such as the absolute plate motion of Eurasia, can be invoked to explain the origin of anisotropic features in many sites, the regional variations observed in some domains imply that differentiated origins of the anisotropy have to be considered, probably related to the particular tectonics in the area. An interesting example of this fact is provided by the stations in the Betic chain; the fast velocity direction inferred for a station located in the limit of the External Betics (South Iberian domain), oriented N80°E, is clearly different from the N15–35°E direction observed in the Internal Betics (Alboran crustal domain), the origin of which has to be related to the Alpine building of the chain.

Key words: Anisotropy, shear-wave splitting, Iberian Peninsula, regional variations.

Introduction

The Iberian Peninsula provides a suitable framework to explore the presence of lithospheric anisotropy and to try to discern the mechanism responsible for the lattice preferred orientation (LPO) of the mantle minerals. A large area of the central and western Iberian Peninsula is covered by the Variscan Iberian Massif, a large, old and geologically stable block of continental lithosphere (DALLMEYER and MARTÍNEZ GARCÍA, 1990). On the other hand, the margins of Iberia have suffered a number of more recent tectonic events: late Triassic/Late Jurassic rifting

[1] Departament de Geofísica, Institut de Ciències de la Terra, CSIC, c/Lluís Solé Sabarís s/n, 08028 Barcelona, Spain.
[2] Laboratoire de Sismologie, Institut de Physique du Globe, 4, Pl. Jussieu, 75252 Paris Cedex 05, France.
[3] Institute voor Aardwetenschappen, P.O. Box 80021, 3508 TA Utrecht, The Netherlands.

to the west, Eocene collision with Europe to the north, Neogene rifting of the Valencia Trough to the east and the Miocene Betic orogeny and Neogene Alboran Sea extension to the south (VEGAS and BANDA, 1982) (Fig. 1). Evidence of the presence of anisotropy in the subcrustal levels of the Iberian lithosphere has been reported in a number of studies that will be reviewed in the next section. However, these anisotropy observations are geographically too scarce to allow for a comparative discussion of the anisotropic parameters.

Conceived as a part of the Iberian Lithosphere Heterogeneity and Anisotropy (ILIHA) project, in the ILIHA-NARS experiment fourteen broadband stations of the portable Network of Autonomously Recording Stations (NARS) were deployed for a period of one year (PAULSSEN, 1990) to record body and surface waves of teleseismic and local earthquakes (Fig. 1). The objective of this experiment was to provide a comprehensive set of seismological data which would allow the establishment of the main structural features of the Iberian lithosphere and upper mantle. In our study we have analysed the splitting of teleseismic SKS and SKKS waves recorded by the ILIHA-NARS array in order to investigate the presence of anisotropy beneath the Iberian Peninsula, as well as the existence of possible regional variations in the anisotropic parameters.

Previous Observations of Anisotropy in the Iberian Peninsula

Prior to the present study, information regarding the presence of anisotropy beneath the Iberian Peninsula was restricted to three specific regions: SW Iberia, Toledo (central Iberia) and the area around the ECORS profile in the eastern Pyrenees.

By analysing the azimuthal variations of the P-wave velocities from the ILIHA deep seismic sounding profiles, DÍAZ et al. (1993) inferred a fast velocity direction oriented roughly N10°E beneath the Ossa-Morena zone (SW Iberia). This anisotropy was restricted to well-defined depth levels between 60 and 90 km. Splitting observations on teleseismic shear waves recorded by a portable network installed in SW Iberia, in the same area explored by the ILIHA DSS profiles, suggested a NE–SW to E–W fast velocity direction (DÍAZ et al., 1996). The possible origin of this anisotropy has been discussed in ABALOS and DÍAZ (1995). Using data from the permanent station located in Toledo (central Iberia), SILVER and CHAN (1988) and VINNIK et al. (1989) inferred, also from the analysis of teleseismic shear-wave splitting, a fast velocity direction oriented close to N90°E. Finally, BARRUOL and SOURIAU (1995) reported a fast velocity direction beneath the eastern Pyrenees, oriented N100°E from a temporary network around the ECORS profile. At a larger spatial scale, MAUPIN and CARA (1992), using the surface waves recorded by the ILIHA-NARS network, have shown that anisotropy should be present beneath depths of 100 km, although the data could not constrain a fast velocity direction. These different results are summarised and included in Figure 1.

The ILIHA-NARS Data Set

The deployment of the ILIHA-NARS array was designed to cover the most significant tectonic units of the Iberian Peninsula (Fig. 1). Unfortunately, technical problems related to event detection, malfunctioning of some horizontal components and high natural noise level have reduced the number of useful recordings for anisotropy studies, especially on the stations located in NW Iberia. In our study we have only selected events with epicentral distances larger than 85° in which well-defined SKS or SKKS arrivals can be identified. Most of these arrivals provide evidence of anisotropy, allowing the determination of the fast velocity direction and the time delays induced by the anisotropy.

As in most of the published SKS studies, we assume an anisotropic model with hexagonal symmetry and horizontal symmetry axis. Under this hypothesis, a single measurement should be enough to fix the properties of the anisotropic media. Other realistic hypotheses (orthorhombic symmetry, hexagonal symmetry with inclined axis of symmetry) cannot be discussed due to the lack of good azimuthal coverage. The anisotropic parameters have been retrieved using classical cross correlation (ANDO, 1984) and/or minimum amplitude difference techniques (DÍAZ, 1994). The original signal is projected in different coordinate system orientations (e.g., every 5°) and the fitting of the two horizontal wave forms is analysed in each step. The coordinate system for which the two components are closer, is identified as corresponding to the fast and slow velocity directions. Only the records with good fitting between the fast and slow components and an acceptable minimisation of the energy in the transverse component have been retained. The resolution of the method can be estimated in $\pm 10°$ for the fast velocity direction, ϕ, and ± 0.05 s for the time delay, ∂t. Table 1 shows the records retained as well as the anisotropic parameters obtained, which are also reported in Figure 1 (white arrows). Clear arrivals that do not exhibit evidence of anisotropy are interpreted as reaching the station with a backazimuth coincident with the fast or slow propagation directions. These directions are reported as thin black arrows in Figure 1. An example of the data analysis technique is shown in Figure 2.

Anisotropy Results in Different Tectonic Domains of Iberia

The anisotropic parameters inferred from station 17, located in the Central Variscan Massif of Iberia (Fig. 1), are consistent with the previous results presented by SILVER and CHAN (1988) and VINNIK et al. (1989). The orientation of the fast velocity direction (N80°E to N105°E) has been related by SILVER and CHAN (1988) to the E–W strike of the Variscan orogeny over the Toledo area. This interpretation implies that the anisotropy should be "frozen" in the lithosphere, the lattice preferred orientation being induced by the last significant orogenic episode in the

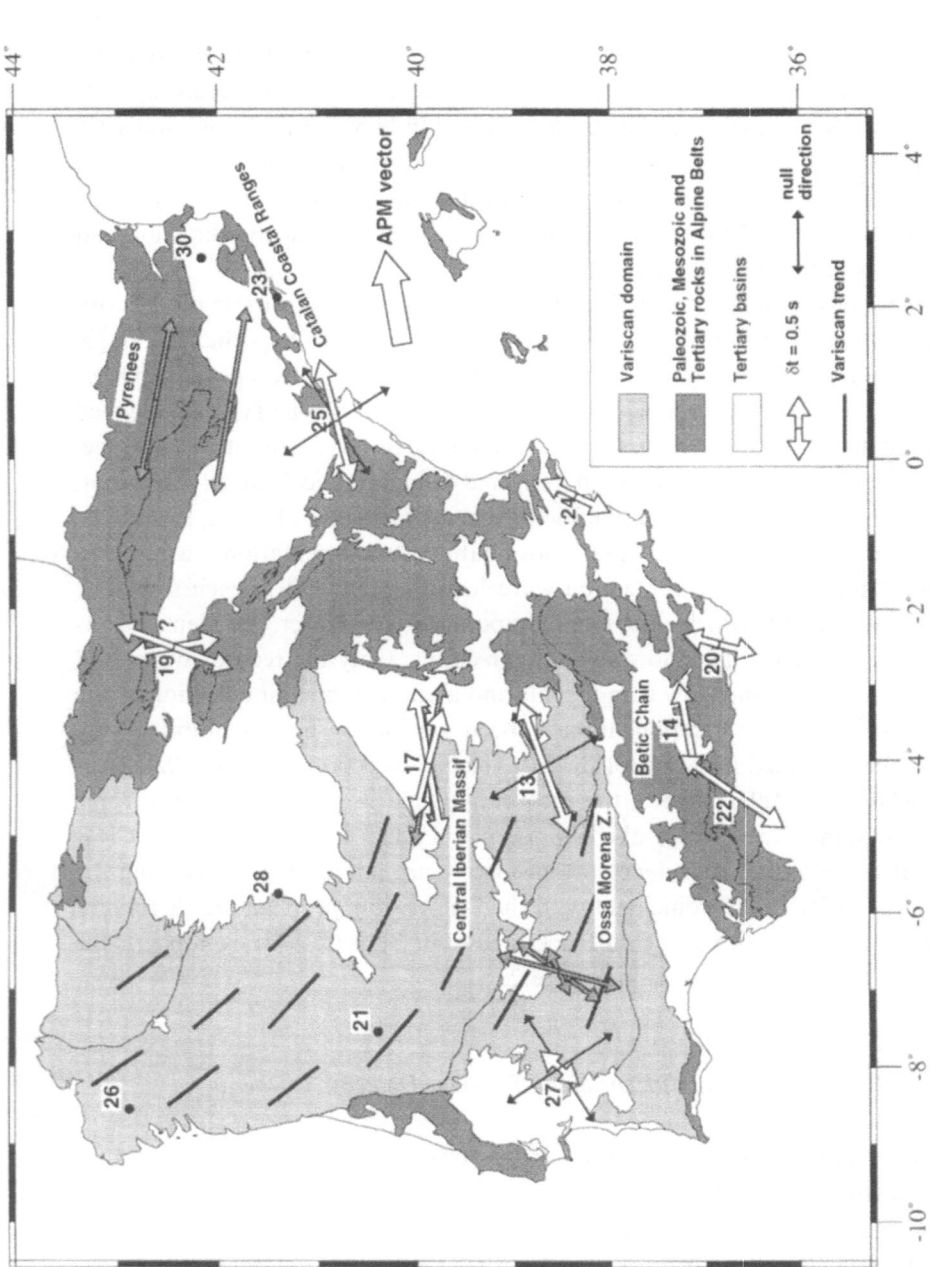

Figure 1

Anisotropic results for the Iberian Peninsula inferred from the present analysis of ILIHA-NARS data. White arrows show the fast velocity direction obtained and are proportional to the time delay. Black thin arrows show the "null measurements". Numbers identify the stations. Question mark for station 19 stands for low quality data. Previous results reported are included (grey arrows, for references, see text). Main tectonic units and Variscan trends are also sketched, as well as the absolute plate motion (APM) vector.

Table 1

Events retained in this study and anisotropic results obtained (fast velocity ϕ and splitting time ∂t)

Station	Event (yr/day)	Dist (°)	Focal Depth (km)	Phase	Baz. (°)	ϕ (°)	∂t (s)
13	88/223	155	141	SKKS	20	70	1.1
13	89/41	120	43	SKKS	60	null	–
14	88/223	155	141	SKKS	23	80	0.8
14	88/251	104	490	SKS	35	80	0.5
17	88/227	91	39	SKS	235	105	0.9
17	88/251	100	490	SKS	35	80	1.2
19	88/210	87	295	SKS	235	−10	0.8
19	88/251	99	490	SKS	35	20	1.0
20	89/41	120	43	SKKS	62	15	0.6
20	88/210	84	295	SKS	235	15	0.5
22	88/223	157	141	SKKS	21	35	1.0
24	89/41	116	43	SKKS	63	25	0.6
25	89/41	114	43	SKKS	63	null	–
25	88/223	151	141	SKKS	28	75	1.0
25	88/227	95	39	SKS	238	null	–
27	89/41	121	43	SKKS	57	null	–
27	88/223	156	141	SKKS	11	40	0.4

area. However, the fast velocity direction (FVD) inferred for station 13, located also in the Central Variscan Massif 100 km south of station 17, shows a difference of about 40° with the Variscan trend in that area.

Station 27 is also located in the Variscan domain, but in a different structural unit, the Ossa-Morena zone, which has been interpreted as an imbricated margin of the Iberian autochthon (QUESADA, 1991). For this station, the fast velocity direction is oriented N40°E, significantly different from that of the previously reported stations. The delay times are also different; 0.4 s for station 27 versus 1.0 s for stations 17 and 13. This result at station 27 is consistent with that presented by DíAZ et al. (1996) from a teleseismic network covering an area located in the same tectonic unit, about 80 km east of that station. The fast velocity direction cannot be related to the Variscan main trend in the area, which is oriented roughly NW–SE (Fig. 1). ABALOS and DíAZ (1995) have argued that the Variscan orogeny may not have affected the whole Ossa-Morena Zone, while there is evidence showing that the Cadomian one affected the entire lithosphere. The Cadomian stretching direction is oriented N30°E (ABALOS and DíAZ, 1995), in proximity to the fast velocity direction inferred from the present data.

Station 25, located in the Catalan Coastal Ranges, constrains a fast velocity direction oriented N75°E (Fig. 2). This direction is very similar to the results inferred from the stations on the Central Variscan Massif, although the tectonic evolution is significantly different. The Catalan Coastal Ranges, oriented mainly NE–SW, are compressive structures created during the Paleogene in the eastern margin of Iberia, as the result of the NNE–SSW orientation of the convergence between the Eurasian and the African plates. Since late Miocene, the compressive direction changed to

NNW–SSE. Therefore, the fast velocity direction inferred for this station (N75°E) could be interpreted as being induced by this Alpine compressive episode. Later on, the Catalan Coastal Ranges have also been affected, since the Oligocene-Miocene transition, by extensional processes that produced a number of Neogene normal faults, oriented ENE–WSW (ROCA and GUIMERÀ, 1992). However, on a lithospheric scale the Neogene extension has mainly affected the Valencia Trough area, having limited expression inland (ZEYEN and FERNÁNDEZ, 1994). In any case, the inferred fast velocity direction does not reflect this episode, since for extensional processes ϕ is expected to be aligned with the extension direction (e.g., SILVER, 1996).

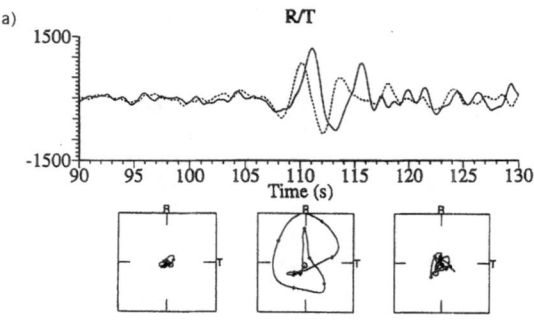

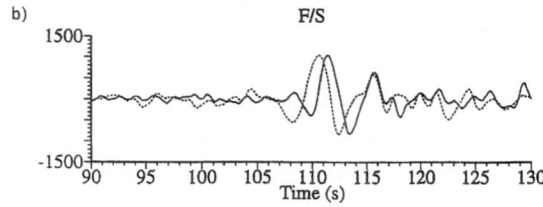

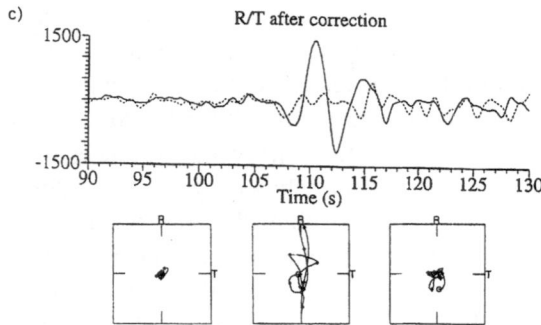

Figure 2

Data analysis of the SKKS phase from the event 88/223 for station 25. a) Radial (solid line) and transverse (dashed line) components and its particle motion diagram in the horizontal plane (5 s per square). b) Projection to the fast and slow directions derived from our analysis. c) Signal after correction of the anisotropic effect, back in the radial/transverse projection. The particle motion diagram now shows a good linearity in contrast with the original data.

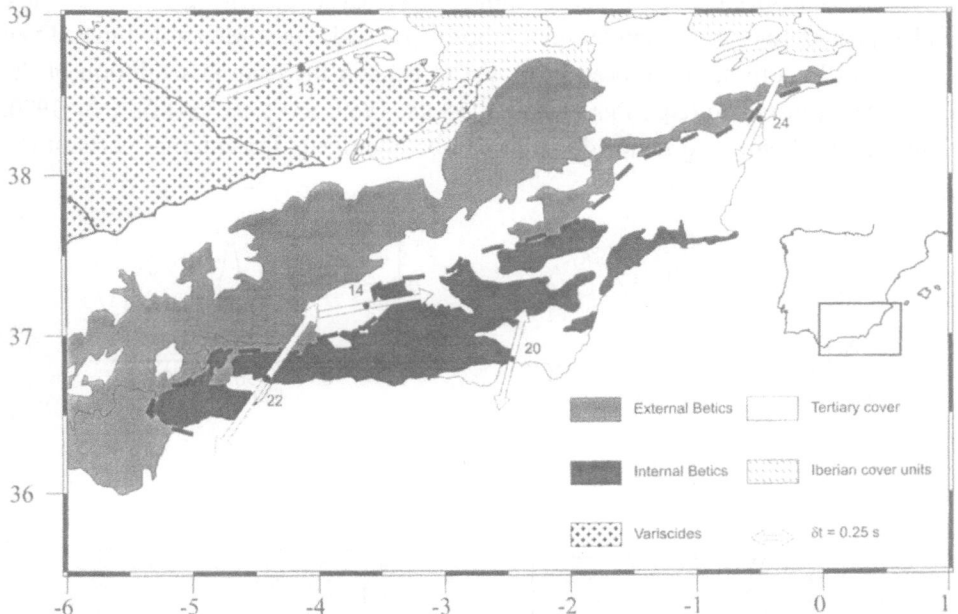

Figure 3

Anisotropic results inferred in the Betics-Alboran domain. Dashed line shows the limit between the Alboran and South-Iberian domains.

Station 19, located south of the western edge of the Pyrenees, displays a roughly N–S fast velocity direction, although more data would be needed to confirm this result, as the data quality is lower than in the remaining stations. This direction clearly differs from most other results and may be related to the particular Alpine tectonic history of this area, affected both by the formation of the E–W Pyrenean chain and the NW–SE Celtiberian chain.

The ILIHA-NARS experiment also provides for the first time, remarkable anisotropy results concerning the Betics-Alboran domain. The Granada station (14) in the External Betics, near the limit of the Internal units, shows a fast velocity direction oriented roughly E–W, similar to the results obtained in the Central Variscan Massif. In contrast, the stations in the Internal Betics (20, 22, 24) express a fast velocity direction clearly different, oriented NNE–SSW (Fig. 3). In Figure 4 this difference is illustrated by comparing, for stations 14 and 22 recording the same event, the fitting of the projections to the fast and slow directions (upper panels) and to the fast and slow directions inferred for the other station (lower panels). The figure demonstrates that the solution for the two stations is clearly different. Therefore, even if this remarkable feature should be confirmed by an enlarged data set, contrasted anisotropic properties seem to characterise these two main Betic domains.

The westernmost end of the Alpine orogenic belt is marked by the Betic and Rif chain and the Gibraltar Arc surrounding the Alboran basin (García-Dueñas *et al.*,

1992). It is associated with the boundary between two large plates that have undergone complex geodynamics on a lithospheric scale since Mesozoic times. Even if different hypotheses have been proposed to explain the tectonic evolution of the Betics-Alboran region (e.g., PLATT and VISSERS, 1989; SANZ DE GALDEANO, 1990; SEBER *et al.*, 1996), two major domains can be identified. The internal parts of the

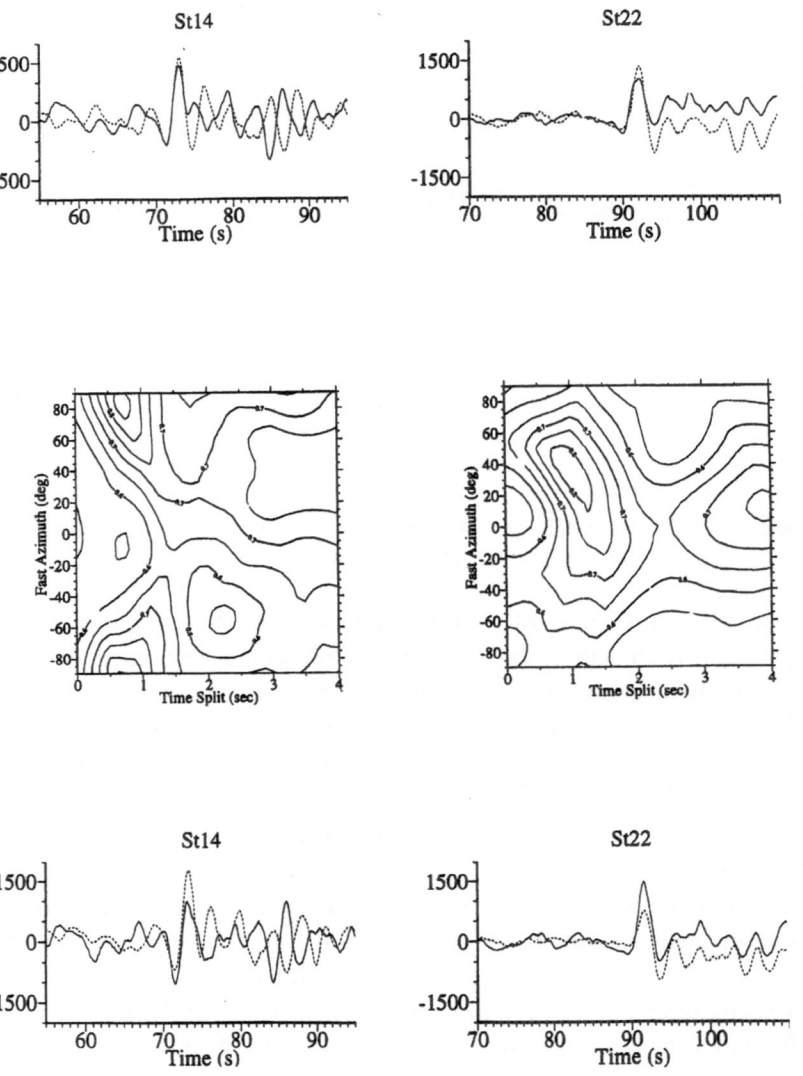

Figure 4

Data analysis of event 88/223 for station 14 (left panels), located in the External Betics (south Iberian domain) compared to station 22 (right panels), located in the Internal Betics (Alboran domain). The upper panels show the projection to the fast and slow directions obtained for each station (see Table 1) once the anisotropic effect has been corrected. The middle panels show the (ϕ, ∂t) dependency of the correlation coefficient. The lower panels show, for each station, the projection to the fast and slow directions obtained for the other station, and illustrate that such a test results in a clearly degraded fitting.

Gibraltar Arc (the so-called Alboran crustal domain, which includes the Internal Betics and Alboran Sea basin) are formed by paleogeographical domains not belonging to the Iberian Peninsula. The northern external part (i.e., the south Iberian domain, which includes the External Betics) is the Mesozoic margin of Iberia (e.g., SANZ DE GALDEANO, 1990). The Alboran crustal domain is character-ised at present by higher seismic activity and strong changes in lithospheric structure (CARBONELL and GARCÍA-DUEÑAS, 1997) with respect to the south Iberian domain, more similar to the stable Iberia. In this area, tomographic studies have also shown the presence of an anomalous lithospheric body, interpreted as a piece of subducted or delaminated lithosphere (BLANCO and SPAKMAN, 1993; SEBER *et al.*, 1996) resulting from the orogenic collapse.

These significant differences between the Alboran and the south Iberian domains are also observed in our anisotropy results. The fast velocity directions in the External Betics are directed along the Mesozoic margin of SE Iberia and are similar to those obtained more internally to this margin, in the Central Massif. The fast velocity directions change clearly for stations located in the Internal Betics (Fig. 3). Therefore, the anisotropy pattern supports the structural differences between these two tectonic domains.

Discussion and Conclusions

The analysis of the splitted teleseismic shear waves recorded in the ILIHA-NARS experiment provides for the first time anisotropy constraints beneath different tectonic domains of the Iberian Peninsula from a homogeneous seismic network. Although the present data set is still rather limited, providing only a few measurements for each location and, therefore, the results cannot be considered as conclusive, they evidence the existence of significant variations in the anisotropy on a regional scale.

A classical approach is to relate the origin of the anisotropy with the last main tectonic event in the area. According to this hypothesis, the major trends of the Variscan orogeny could explain the inferred fast velocity direction for some locations (station 17 at Toledo, station 25 in the Catalan Coastal Ranges, Pyrenees). In some of these cases, more recent tectonic episodes can also be invoked as responsible for the anisotropy pattern: Alpine orogeny for the Pyrenees, Paleogene compression in the Catalan Coastal Ranges. In this latter case it must also be assumed that the Neogene extension had no major influence, on a lithospheric scale, inland on the eastern Iberian margin, as the anisotropic properties of the lithosphere have not been modified accordingly. On the other hand, this hypothesis fails to explain the results at some Variscan locations (station 27 in the Ossa-Morena zone, station 13 south of Toledo) (Fig. 1).

Another classical hypothesis submitted to explain the origin of the LPO of the mantle minerals, assumes that it is due to the progressive simple shear induced by the motion of the continental plate over a stationary mantle or, conversely, of the drag of the asthenospheric flow on the base of the lithosphere (e.g., VINNIK *et al.*, 1989). Under this hypothesis, the fast velocity direction is expected to be aligned with the absolute plate motion (APM) and should remain stable along large regions. The APM direction is poorly known for Eurasia (MINSTER and JORDAN, 1978; GRIPP and GORDON, 1990) and therefore this hypothesis is difficult to discuss. The most recent calculations of the APM in the Eurasian plate (Morgan, personal communication), show an APM motion oriented roughly N75°E in the Iberian Peninsula (Fig. 1). This is consistent with the general ENE–WSW trend for the fast velocity direction in large regions of Iberia (stations 17 and 13 in the Central Massif, station 27 in the Catalan Coastal Ranges) but it cannot explain the results at some specific domains (Betic Chain, Ossa-Morena zone). ALSINA and SNIEDER (1995) have proposed that present-day flow in the mantle, at smaller scale than APM, can also contribute to the observed anisotropic pattern and could explain regional differences.

Some of the stations where multiple measurements are available (stations 17, 19) show some dispersion in the fast velocity direction, which cannot be related to the resolution of the method. This feature has also been reported in other regional scale anisotropic experiments (e.g., BARRUOL and SOURIAU, 1995), and is sometimes hidden by statistics. In our opinion, this fact has to be interpreted as reflecting the presence of a more complex anisotropic structure, probably including dipping axis of anisotropy and/or a symmetry system other than hexagonal, which cannot be resolved with our present data.

The observed delay times are in most cases close to the usual value of 1 s. However, some stations present smaller values and in other cases there is a significant variation in the delay time measurements at single stations (see Table 1). This fact cannot be explained by the classical hypotheses of the origin of anisotropy and may be attributed to the existence of more complex anisotropic patterns.

The regional variations in the anisotropic parameters documented by the ILIHA-NARS data set imply that differentiated origins of the anisotropy have to be considered in some areas, in relation to their particular lithospheric geodynamics. A clear example of this is the Betic Chain, where a strong contrast appears in the anisotropic results between the south Iberian domain and the Alboran crustal domain. In the former the results are similar to those obtained for sites on the stable Iberia, whereas in the latter the different anisotropic properties must be related to the complex Alpine geodynamics.

REFERENCES

ABALOS, B., and DÍAZ, J. (1995), *Correlation between Seismic Anisotropy and Major Geological Structures in SW Iberia: A Case Study on Continental Lithosphere Deformation*, Tectonics *14*, 1021–1040.

ALSINA, D., and SNIEDER, R. (1995), *Small-scale Sublithospheric Continental Mantle Deformation: Constraints from SKS Splitting Observations*, Geophys. J. Int. *123*, 431–448.

ANDO, M. (1984), *ScS Polarisation Anisotropy around the Pacific Ocean*, J. Phys. Earth *32*, 179–195.

BARRUOL, G., and SOURIAU, A. (1995), *Anisotropy beneath the Pyrenees Range from Teleseismic Shear-wave Splitting: Results from a Test Experiment*, Geophys. Res. Lett. *22*, 493–496.

BLANCO, M. J., and SPAKMAN, W. (1993), *The P-wave Velocity Structure of the Mantle below the Iberian Peninsula: Evidence of Subducted Lithosphere below Southern Spain*, Tectonophys. *221*, 13–34.

CARBONELL, R., and GARCÍA-DUEÑAS, V. (1997), *The Crust Beneath the Betics Mountain Chain: An ESCI View*, Earth Planet. Sci. Lett. (submitted).

DALLMEYER, R. D., and MARTÍNEZ GARCÍA, E. (eds.), *Pre-Mesozoic Geology of Iberia* (Springer-Verlag, Berlin, Heidelberg, New York, 1990), 416 pp.

DÍAZ, J., *Anisotropie sismique et deformation dans le manteau superieur: approche conjointe par sismique refraction et ondes S telesismiques en Iberia. Methodes d'analyse d'ondes S et leur resultat au Tibet.* (Thése de Doctorat de l'Universite Paris VII. Paris, 1994).

DÍAZ, J., HIRN, A., GALLART, J., and SENOS, L. (1993), *Evidence for Azimuthal Anisotropy in Southwest Iberia from Deep Seismic Sounding Data*, Phys. Earth Planet. Int. *78*, 193–206.

DÍAZ, J., HIRN, A., GALLART, J., and ABALOS, B. (1996), *Upper-mantle Anisotropy in SW Iberia from Long-range Seismic Profiles and Teleseismic Shear-wave Data*, Phys. Earth Planet. Int. *95*, 153–166.

GARCÍA-DUEÑAS, V., BALANYÀ, J. C., and MARTINEZ-MARTINEZ, J. (1992), *Miocene Extensional Detachments in the Outcropping Basements of the Northern Alboran Basin (Betics) and their Tectonic Implications*, Geo-Mar. Lett. *12*, 88–95.

GRIPP, A. E., and GORDON, R. G. (1990), *Current Plate Velocities Relative to Hotspots Incorporating the Nuvel-1 Global Plate Motion Model*, Geophys. Res. Lett. *17*, 1109–12.

MAUPIN, V., and CARA, M. (1992), *Love-Rayleigh Wave Incompatibility and Possible Deep Upper Mantle Anisotropy in the Iberian Peninsula*, Pure appl. geophys. *138*, 429–444.

MINSTER, J. B., and JORDAN, T. H. (1978), *Present-day Plate Motions*, J. Geophys. Res. *83*, 5331–5354.

PAULSSEN, H. (1990), *The Iberian Peninsula and the ILIHA Project*, Terra Nova *2*, 429–435.

PLATT, J. P., and VISSERS, R. L. M. (1989), *Extensional Collapse of Thickened Continental Lithosphere: A Working Hypothesis for the Alboran Sea and Gibraltar Arc*, Geology *17*, 540–543.

QUESADA, C. (1991), *Geological Constraints on the Paleozoic Tectonic Evolution of Tectonostratigraphic Terranes in the Iberian Massif*, Tectonophys. *185*, 225–245.

ROCA, E., and GUIMERÀ, J. (1992), *The Neogen Structure of the Eastern Iberian Margin: Structural Constraints on the Crustal Evolution of the Valencia Though*, Tectonophys. *203*, 203–218.

SANZ DE GALDEANO, C. (1990), *Geologic Evolution of the Betic Cordilleras in the Western Mediterranean*, Miocene to Present, Tectonophys. *172*, 107–119.

SEBER, D., BARANZANGI, M., IBENBRAHIM, A., and DEMNATI, A. (1996), *Geophysical Evidence for Lithospheric Delamination beneath the Alboran Sea and Rif-Betic Mountains*, Nature *379*, 785–790.

SILVER, P. G. (1996), *Seismic Anisotropy Beneath the Continents: Probing the Depths of Geology*, Ann. Rev. Earth Planet. Sci. *24*, 385–432.

SILVER, P. G., and CHAN, W. W. (1988), *Implications for Continental Structure and Evolution from Seismic Anisotropy*, Nature *355*, 34–39.

VEGAS, R., and BANDA, E. (1982), *Tectonic Framework and Alpine Evolution of the Iberian Peninsula*, Earth Evol. Sci. *2(4)*, 320–343.

VINNIK, L. P., FARRA, V., and ROMANOWICZ, B. (1989), *Azimuthal Anisotropy in the Earth from Observations of SKS at Geoscope and NARS Broadband Stations*, B.S.S.A. *79*, 1542–1558.

ZEYEN, H. J., and FERNÁNDEZ, M. (1994), *Integrated Lithospheric Modelling Combining Thermal, Gravity and Isostasy Analysis: Application to the NE Spanish Geotransect*, J. Geophys. Res. *99*, 18089–18102.

(Received October 29, 1996, revised May 13, 1997, accepted May 20, 1997)

 To access this journal online:
http://www.birkhauser.ch

Pure appl. geophys. 151 (1998) 407–431
0033–4553/98/040407–25 $ 1.50 + 0.20/0

Pure and Applied Geophysics

Variation of Shear-wave Residuals and Splitting Parameters from Array Observations in Southern Tibet

ALFRED HIRN,[1] JORDI DÍAZ,[2] MARTINE SAPIN[1] and JEAN-LOUIS VEINANTE[1]

Abstract—A tight array of seismographs spanning a 500 km traverse of southern Tibet resolved anisotropy from *SKS* with a spatial variation of its direction and an increase northward of the splitting delay, as well as of its first arrival residual. Both waves split by velocity anisotropy are slow relative to *P* and their waveform analysis may be interpreted to suggest attenuation anisotropy. The array here provides examples of residuals and splitting of other *S* waves which do not tightly conform to the anisotropy assumed in the simplest model of olivine of transverse isotropy with horizontal symmetry axis. *S* waves are also split, with parameters which vary along the array, and hence are relevant to near-receiver structure like those of *SKS*. Their splitting delay, for non-vertical incidence and polarization, appears larger than that for *SKS*. Residuals of *S*-wave first arrivals and splitting delays increase less northwards for *S* than *SKS*. Anisotropy with a slow vertical axis may account for these observations. Its origin may be related to horizontal shear or flow in low-velocity layers.

Key words: Shear waves, anisotropy, Tibet.

1. Introduction

We use observations by a portable array of twenty-three 3-component, 5-second seismographs deployed with a station spacing of 30 km along a line traversing southern Tibet (Fig. 1; HIRN *et al.*, 1995a) to illustrate peculiarities of records of shear waves at teleseismic distance, including anisotropic splitting. Other splitting data in Tibet have been obtained before by MCNAMARA *et al.* (1994) at stations with about 120 km spacing, on a traverse shifted northwards with respect to ours and an overlap over the Lhasa block and southern Quang Tang. NI and SANDVOL (1995) also analyzed splitting at some sites in the southern Lhasa block and south of the Indus-Tsangpo suture. Over the northern part of the study of MCNAMARA *et al.* (1994), in northern Qang Tang and across the Kun Lun, HERQUEL *et al.* (1995) and GUILBERT *et al.* (1996) reported anisotropy over a line of 40 km-tight spaced stations.

---

[1] Laboratoire de Sismologie Expérimentale, Département de Sismologie UA 195 CNRS, Institut de Physique du Globe de Paris, 4 place Jussieu, F-75252 Paris.
[2] Institut de Ciencies de la Terra "Jaume Almera", CSIC, c/Lluis Solé Sabaris s/n, E-08028 Barcelona.

Although reported results are in general consistent among studies where they overlap, as for the central Lhasa block in the three first mentioned surveys, the spatial extent or density of each survey has led to different emphases in their interpretations.

—At the scale of the broadest survey, McNAMARA *et al.* (1994) consider that the fast polarization parallels surface geologic features and argue that anisotropy indicates that mantle deformation is coherent with the current surface deformation.

—Our denser array would allow sampling of the variation across major features of surface geology in the Lhasa block and its boundaries (HIRN *et al.*, 1995a, b). The direction of anisotropy is found not to align with, nor to change across, the Karakorum-Jiali fault strike-slip zone in the Lhasa block nor with the Bang Gong Nu Jiang suture at its northern boundary. The Indus-Tsangpo suture between the Lhasa block and the Tethyan Himalayas to the south appears to separate regions of contrasting directions, none of which being that of surface geology. Such a spatial

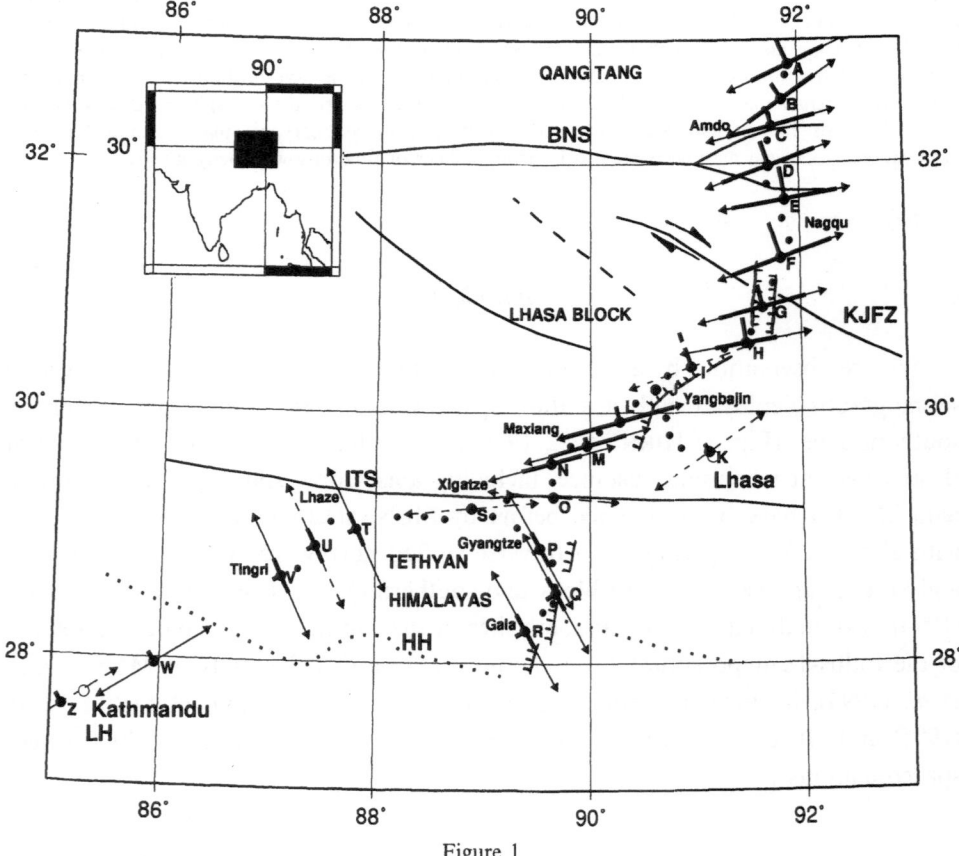

Figure 1

Sketch location map of southern Tibet. Stations of HIRN *et al.* (1995b) also used here, with *SKS* results: splitting directions (arrows), arrival time residuals with respect to *P* (length of segment along arrow) and splitting delays (length of segment orthogonal to arrow).

pattern of anisotropy does not support different models of mantle deformation assumed elsewhere. We documented a joint spatial variation of the splitting delay and of the relative shear to compressional wave arrival residuals northward of the Indus-Tsangpo suture, or into the zone of major recent extension of the Yangbajain graben and Maxiang Tertiary volcanism. Accordingly, we suggested that in this region anisotropy is induced by crystal reorientation under differential shear on horizontal planes or orientation of fluid-filled defects by ductile flow.

—In modeling results of *SKS* anisotropy published for a number of regions, PLOMEROVÁ et al. (1995) propose that in general the fast direction has a significant dip. In Tibet they consider the range of the results obtained at some of the stations of MCNAMARA et al. (1994) as corresponding to a dependence of splitting delays and fast orientations on backazimuth, which they attribute to dipping axis anisotropy.

—Following proposals of BABUŠKA et al. (1993), GUILBERT et al. (1996) consider the azimuthal distribution of *P* residuals in northern Tibet together with the *SKS* fast direction. They report a fast direction for *P* orthogonal to the Kun Lun fault. On the other hand, they report the fast *SKS* polarization to be aligned with the fault. Their final model is of anisotropy in the lithosphere with the fast velocity directed orthogonal to the Kun Lun fault, and dipping outwards of the fault zone on either side.

—LAVÉ et al. (1996) elaborate on the original *SKS* anisotropy results of MCNAMARA et al. (1994), HIRN et al. (1995) and GUILBERT et al. (1996), discussing them with respect to the model of present-day motion of Tibetan crust relative to stable Asia as inferred from the slip-rates on major active faults. They note that the strong *SKS* anisotropy aligns with the major strike-slip faults, as in the case of the Kun Lun fault of GUILBERT et al. (1996), but contend instead that anisotropy is caused by vertical-plane strike-slip motion with the magnitude of anisotropy indicating that these faults extend into the lithospheric mantle. They regard the weaker *SKS* anisotropy away from major faults to be caused by shear on horizontal planes imposed on the asthenosphere by the movement of rigid lithospheric blocks extruded along the major strike-slip faults quantified by AVOUAC and TAPPONNIER (1993).

—DAVIS et al. (1997), taking a broader view to also include *SKS* splitting results in Tien Shan (MAKEYEVA et al., 1992), Mongolia and Southern Siberia (GAO et al., 1994), conclude that the splitting observations agree better with those predicted by calculations of internal lithospheric deformation (HOUSEMAN and ENGLAND, 1996). As their work was made available to us after submission of the present paper, we comment in more detail, in a note added during revision of the present paper, after presenting the previous independent conclusions of our original manuscript.

The seismic data analyzed in the present paper allows us to consider the splitting of other shear waves besides *SKS*, as well as to document a dependence of transit time residuals on angle of incidence. The description of the velocity anisotropy can hence be expanded out of the horizontal plane to which it was confined by the sole consideration of *SKS* splitting. The velocity projected along the vertical appears significantly slower. An explanation could be sought in terms of dipping axes anisotropy which have components along the vertical, but such a type of model can hardly be constrained further with the limited coverage of the data. Another line of interpretation, the simplest yet, is that the vertical is a principal axis of anisotropy, which then appears as the slowest. Hence these data bring additional support to the interpretation of the cause of anisotropy as being shear in horizontal planes or flow in lower velocity material at rather shallow levels in the crust and uppermost mantle, weak layers located higher than the usual asthenosphere elsewhere. This suggestion was made before, only on the different basis of the spatial variation of *SKS* splitting delay with the *SKS/PKP* residual (HIRN *et al.*, 1995b). The data by themselves have several implications with respect to the other, different views expressed recently regarding the regional deformation and geodynamics by LAVÉ *et al.* (1996) and DAVIS *et al.* (1997) from interpreting the same first-order parameters of the horizontal direction of fast *SKS* anisotropy in the assumption of hexagonal symmetry with axis in the horizontal plane.

2. *S and SKS Anisotropy Variation*

2.1. *Array Measurements of Spatial Variations of S: Larger Splitting Delays at Nonvertical Incidence*

Developments in the interpretation of teleseismic shear-wave splitting over the last decade have been based on observations of *SKS* (VINNIK *et al.* 1994; KIND *et al.*, 1985; SILVER and CHAN, 1988). These waves arrive with pure *SV* polarization because of their *P* to *S* conversion after traversing the core, and with a very steep incidence. Hence their vibration is nearly confined to the horizontal plane and their splitting samples anisotropy in that plane. For such a case, a single observation uniquely defines azimuthal anisotropy with a horizontal fast axis of symmetry, a model which has been commonly taken to represent the *in situ* mantle anisotropy. In order to test the uniqueness of this type of model, other angles of incidence and polarization orientations must be sampled. *S* waves can provide them, however their use at single stations is not straightforward since they may sample anisotropy anywhere along the whole raypath. Instead, if *S* from a same earthquake can be analyzed across an array, and a spatial variation in the splitting is detected, it allows one to attribute the anisotropy to the part of the path nearest the receivers. Teleseismic *S* which can be used are however limited to those with angles of incidence in the mantle small enough for the observation at the surface to be kept in the shear-wave window where particle motion at the free surface is linearly related to the incident wave particle motion.

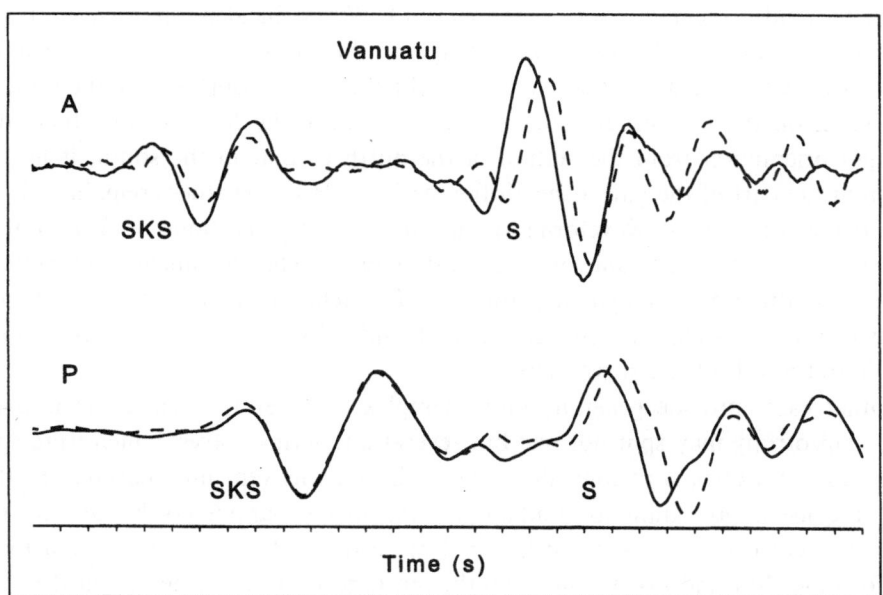

Figure 2
Examples of seismograms containing *SKS* and *S* from Vanuatu event (backazimuth 109°, distance 87°). Bold trace is projection on the fast direction of *SKS* anisotropy, 70°E for the top figure, northern station *A* in the southern Qang Tang, and 20°W for the bottom figure, southern station *P*, south of the *ITS*. Dotted traces are the orthogonal projections, on the slow direction of *SKS*, shifted forward by the respective *SKS* splitting delays, 0.6 s in the north on top and 0.2 s in the south at bottom. Note that the second wave, *S*, has also same waveforms on the two components, indicating splitting on same axes as *SKS*, but which remain shifted in time, indicating larger magnitude of splitting delay.

Along the traverse from the southern Qang Tang to the Himalayas (HIRN *et al.*, 1995a,b), *SKS* was found with the following three characteristics: i) at all sites splitting is detected, ii) anisotropy is oriented differently on either side of the *ITS*, and iii) the splitting delay is larger north of the Lhasa block and Qang Tang. In the following examples of several earthquakes, *S* splitting similarly changes orientation across the *ITS*, like *SKS*, which suggests that it represents anisotropy under the receivers. The orientations of *S* splitting do not significantly deviate from those of *SKS*, but the magnitudes of the splitting delays are larger for *S*. Since *S* is not at vertical incidence and *SV* polarized, these observations suggest that velocity anisotropy may have a different component out of the horizontal plane.

A first example, in Figure 2 displays representative seismograms north and south of the Indus-Tangpo suture, which contain the *SKS* and *S* of Vanuatu events. For each record the horizontal particle motion has been projected on the fast and slow orthogonal directions of *SKS* splitting (fast direction is along 70°E in the north and 20°W in the south), and the slow component shifted early by the corresponding splitting delays (0.6 s and 0.2 s respectively in the north and south), so as to render the two split components of *SKS* coincident in waveform and arrival time. For *S*, the two components in this projection also have similar waveforms, which indicates that the principal axes for *S* anisotropy are the same in projection on the horizontal plane as for *SKS*. Since the *S* anisotropy then also changes orientation from the station in the north to that in the south, it has its origin in the part of the path beneath the receivers. However, there remains a large time delay between the waveforms of the orthogonal projections of *S* after they have been shifted by the value of *SKS* splitting delay. This documents a magnitude of the *S* splitting delay larger than for *SKS*. It reaches a similar value of 0.9 s for the two stations displayed, with repect to 0.6 and 0.2 s for *SKS*, respectively in the North and South of the survey line.

Other cases provide generally more complex *S* waveforms, since the near-receiver anisotropy may split not one but several interfering waves, which were split in previous anisotropic regions along the path or underwent multipathing. In this case, the usual procedure for finding the anisotropy parameters by seeking the maximum correlation between orthogonal projections of the waveform as a function of time shift and projection azimuth, can hardly succeed when applied to the whole complex waveform. In the case the seismograph responds to short periods, and the splitting delay is a significant proportion of the signal period as in studies of upper crustal anisotropy from local earthquakes, anisotropy parameters can be derived similarly from simple analysis of the particle motion diagrams (CRAMPIN *et al.*, 1980). An example for a Fiji event is shown as Figure 3. At the onset the particle motion of *S* is not elliptical, since a linear motion can be clearly seen in the first part of its particle motion diagram. Here its polarization is far from the direction predicted from the focal mechanism which is almost along the backazimuth, but is instead oriented along 70°E, and hence with the same fast axis as *SKS* anisotropy at this station. The beginning of the following orthogonal motion of the corresponding slow split wave is indicated, but gets rapidly complicated by later waves originating from previous splitting or multipathing. For this case also, consistently from the same earthquake over the array, *SKS* and *S* are shown again to have similar fast direction but *S* has a larger delay, as in the previous case.

The splitting delay for *S* is larger than for *SKS*, by up to three times at the southern station. Although *S* has a larger angle of incidence than *SKS*, its path through an anisotropic layer is longer only in proportion to the inverse cosines of incidence. This is not sufficient to account for the observed difference which is not expected for the usual model considered in *SKS* studies of transverse isotropy with a horizontal symmetry axis. Keeping this same type of hexagonal symmetry of anisotropy but allowing it to depart from horizontal, one interpretation of the dependence on incidence is to relate it to a dip of the plane containing the principal axes. A strong dependence of *S* splitting on backazimuth should accordingly be observed. Unfortunately, the observations available do not span the whole range of

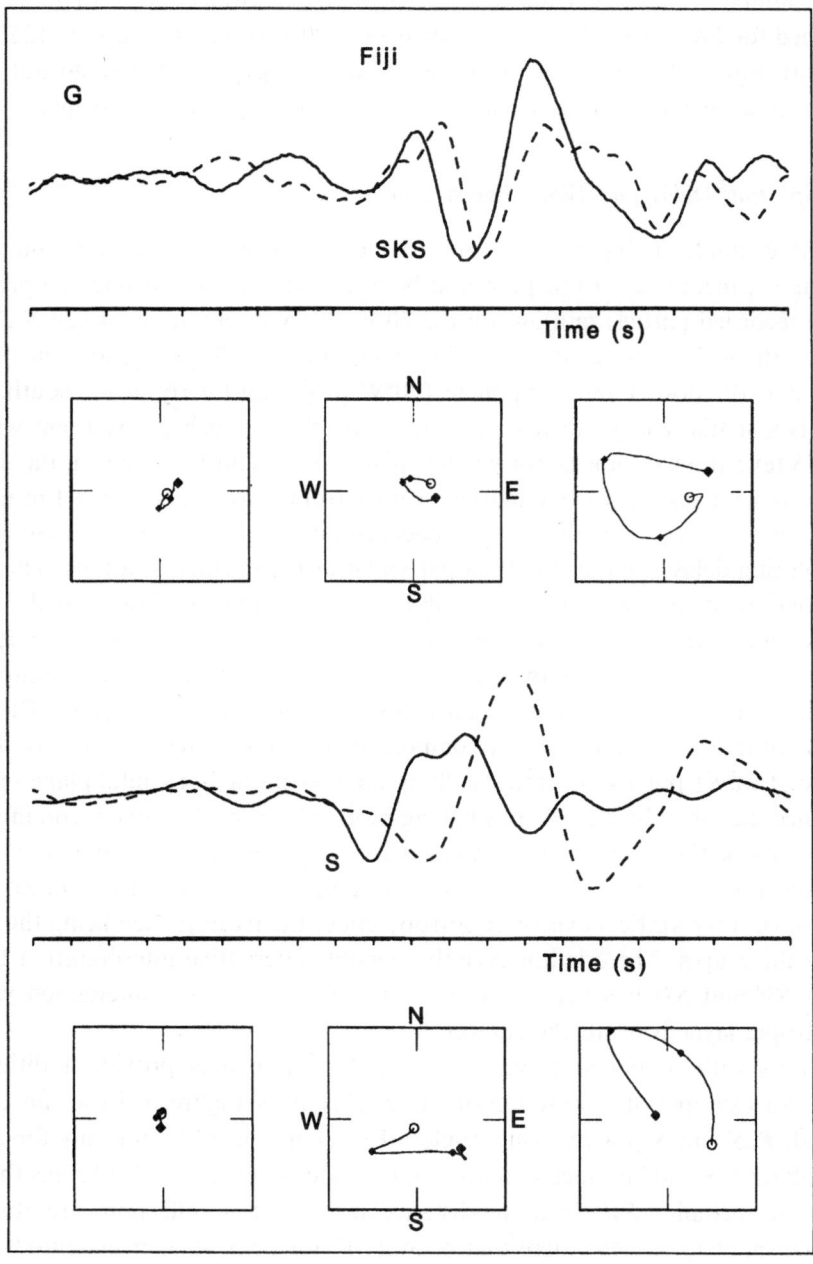

Figure 3

Examples of *SKS*, and of *S* for Fiji event (backazimuth 109°, distance 102°) at station *G* with more complex waveform than the previous example, projected on 70°E. Particle motion diagrams in N-S, E-W, showing initial linear polarization of *S* on 70°E.

backazimuths, nonetheless the *S* splitting is systematically larger than the values obtained for *SKS*, from directions as diverse as 20°E from Alaska, 105–125°E from Vanuatu-Fiji, 140°E from Sulawesi and 80°W for Egypt, that they do not suggest an oriented phenomenon as would result from dipping axis anisotropy.

2.2. Splitting Delay and Wave Polarization

The example in Figure 4 is for a Minahassa event with a focal mechanism causing the incoming *S* to be polarized between *SH* and *SV*. Orthogonal projection of the recorded particle motions on the 70°E–20°W orientations of fast *SKS* north and south of *ITS* shows the solid-line component (70°E projection) early in the north and the dotted-line component (20°W projection) early in the south. Hence there is a spatial change of fast anisotropy direction, which is consistent with that of *SKS* (except at station *Q*, for whch explanations might be sought in the different positions with respect to boundaries of anisotropic structures of the Fresnel zone for the waves arriving with different backazimuths). Here again, the magnitude of the splitting delay is larger for *S*, in particular at the southern stations. The sample seismogram of station *E* (Fig. 5) exhibits a conspicuous amplitude on the vertical component, convenient for recognizing the arrival of an *SV*-polarized part of the wave. This arrival is late with respect to the onset on the horizontal components, meaning that the slowest axis of anisotropy is along the vertical axis. The earlier waves without detectable vertical components are those which are resolved into a couple of waves polarized orthogonally to another in the horizontal plane in Figure 4. Hence they may be regarded as having been split from the same incoming phase, which may be the *SH* part of the wave. This complexity may be accounted for with an orthorhombic symmetry instead of the hexagonal model usually assumed in *SKS* studies, with the vertical axis of anisotropy under the receiver then being the slowest of the three axes. There is however the possible alternative interpretation that the lag of *SV* and *SH* has been acquired along the path before interaction with the anisotropic layer beneath the station.

Waves with particular geometry and polarization may provide additional insight. An example of core-reflections is displayed as Figure 6. From an event in Talaud, *PcS* and *ScS* arrive with angles of incidence as *SKS*, but only for *PcS* are the polarizations alike, since in reflection the mode conversion also brings this wave strictly to incoming *SV*. Since the distance at which core reflections are observable is much smaller, here 40°, than that of over 80° for *SKS*, smaller magnitude events, and with less attenuation during propagation provide shorter period signals which have been currently used in a former development of anisotropy studies (e.g., ANDO, 1984) before broadband instruments fostered the *SKS* studies. The large focal depth determines lengths of the down- and up-going parts of the path that are sufficiently different to prevent *ScP* from interfering with *PcS*. For *ScS*, the incidence is quite similar, with the major difference being that the wave keeps the

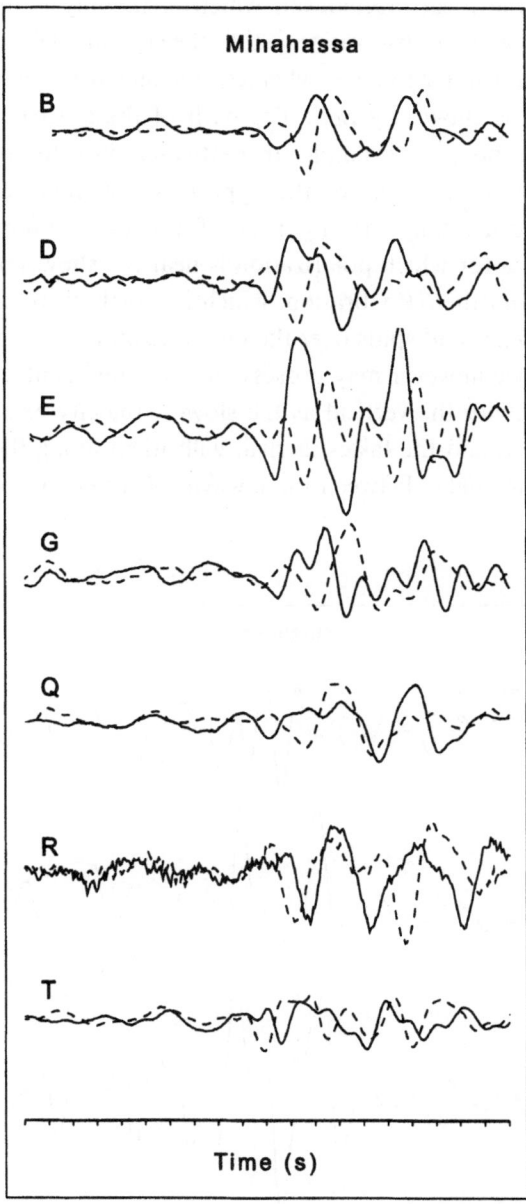

Figure 4
Example of *S* from the Minahassa event (backazimuth 132°, distance 42°). Traces of several stations are plotted in the projections on 70°E, bold and 20°W, dotted. Note that first arriving split wave changes from the bold direction in the north to the dotted one in the south, bottom traces.

polarization due to the focal mechanism, which commonly is different from *SV* polarization. The subsequent observation is that the splitting delay of *PcS* is slightly smaller than a second, similar to *SKS*, whereas at a similar incidence a significantly larger splitting delay is shown by *ScS*. The path of these waves is very different from that of *S* in the previous example. In particular since the path is very steep from the core to the surface, one of the hypotheses of interpretation, involving anisotropy in the horizontally grazing part of the ray, cannot apply. Among examples of *ScS*, those for which polarization is near *SV* show small splitting. *PcS* and *SKS* are built from the *SV* vibration, which is mostly in the horizontal plane, due to the steep incidence and splits over the two horizontal axes of anisotropy. *ScS* with the same incidence however may possess an *SH* component. In an orthorhombic symmetry system with the vertical as the slowest velocity axis, the *SV* component will be at the last, and the fastest arrival will orient along the horizontal axes of anisotropy. The time delay between these waves of different polarization will be

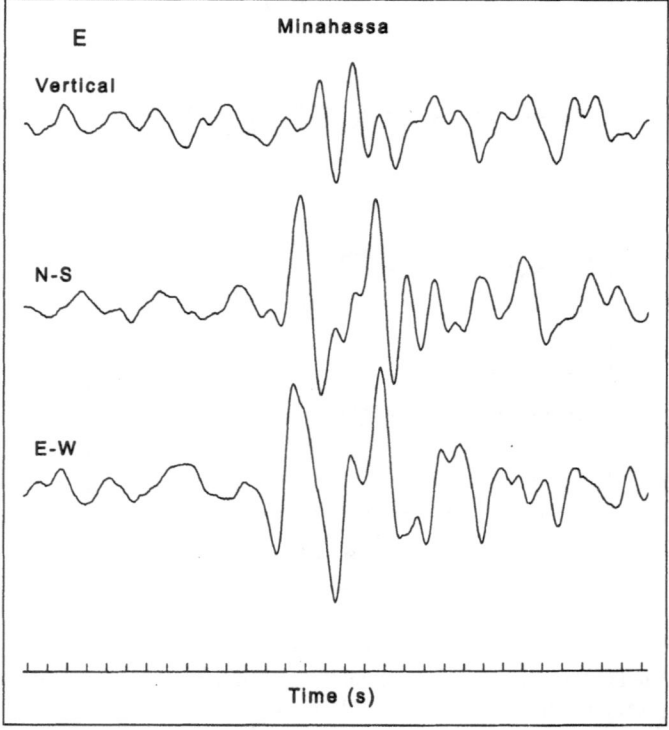

Figure 5

Three-component seismogram of *S* of Minahassa at station *E*, illustrating a strong amplitude on the vertical component indicating an *SV* polarized wave. This wave arrives later than two split waves without a vertical component, the first nearly on E-W (more precisely, the WSW-ENE direction of *SKS*), the second nearly N-S.

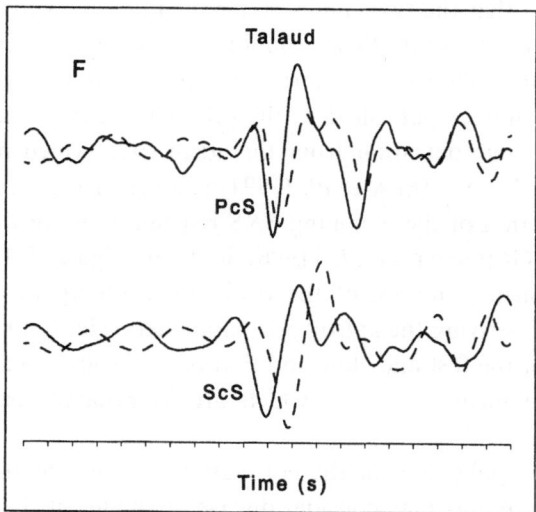

Figure 6

Projection on 70°E azimuth, bold, and orthogonal, dotted, of seismogram at station *F* from Talaud event (backazimuth 120°, distance 42°). The two waves are core-reflections with similarly steep incidence, but *PcS* on top has theoretically pure *SV* incoming polarization due to *P*- to *S*-mode conversion on reflection, whereas *ScS* at bottom is polarized between *SV* and *SH* due to the focal mechanism. Note that for each wave the similarity of the two components shows that the orientation of anisotropy is significantly the same as for *SKS*, but that their time shift, hence splitting delay, is larger for *ScS*, at bottom.

more important than if the two horizontal axes of anisotropy with the fast and intermediate velocity values alone governed the wave propagation.

3. Anisotropy Parameters from Different Criteria or Neighboring Events

3.1. *The Consistency of Anisotropy Parameters Retrieved with Different Criteria*

Variations among fast directions and splitting delays of *SKS* from different events at the same station (McNamara *et al.*, 1994) have been attributed by Plomerova *et al.* (1995), to a dependence on backazimuth, which they interpret in favor of a model of anisotropy with dipping axes. Usual processing of our data in some cases also leads to diverse splitting results at the same station, but before interpreting them, we may return to the seismograms in order to analyze them with different criteria and discuss what may induce such results. Most methods of retrieving anisotropy parameters from seismograms use a procedure of searching for the maximum correlation between waveforms on orthogonal projections as a function of the time shift and the azimuth of the projection (Silver and Chan,

1991). This assumes that the two split waves have the same waveform. Anisotropy parameters derived by fitting them should equivalently, when applied to the observed seismograms, allow one to retrieve the polarization of the incoming wave, radial for *SKS*. Examples published in the different studies, presumably the best fitting ones, show some deviation from this ideal case, which may contain additional information. MCNAMARA *et al.* (1994) recover in their Figure 3f a reasonably linear polarization of the incoming *SKS* but fail by more than 10° to recover the backazimuth, GUILBERT *et al.* (1996) in their Figure 5 leave a significant ellipticity of the particle motion of the retrieved incoming wave unexplained by their processing for deriving the anisotropy parameters. Hence the two criteria, that fitting waveforms of the fast and slow and that of retrieving the linearity and radial polarization of the incoming wave, which are in principle equivalent, are not simultaneously met.

In the practical application of the correlation criterion of the two split waveforms, different viewpoints may consider diverse attributes of the waveforms. Their maximum amplitude for instance may be considered more important to be matched, rather than the whole waveform, since this may be more corrupted by noise, as it includes the parts of the time series with small amplitude. Values differing by 10° for the direction and 20% for the splitting delay can be obtained depending on which parameter is used to describe the similarity of two orthogonal projections: their overall shapes, their maximum amplitudes, or their apparent periods (DIAZ, 1994; HIRN *et al.*, 1995a). Using these data with high signal-to-noise ratio and time resolution, and a bandwidth extended to high frequency which also allows description of the details of the waveforms better than in the broadband data previously used for anisotropy studies, we suggested that this may not indicate scatter but rather that the signals do not meet the assumptions made for processing them. The possibility of anisotropy in attenuation appears to allow the analysis of the diverse signal attributes to be rendered internally consistent. Such a phenomenon may then also be considered in the interpretation of the discrepancy of results obtained from different criteria for a same signal, and on the other hand in investigating reasons for the scatter of results at a same station for different signals and the same criterion.

3.2. The Range of Anisotropy Parameters Derived from Neighboring Events

Only few, discrete parts of the worldwide seismicity distribution are able to provide *SKS* to a particular station. Backazimuths to earthquakes from one of those regions giving *SKS* for a particular station may differ slightly. When anisotropy parameters derived for them are found to differ, a dipping axis of anisotropy is an attractive interpretation, since such a geometry allows for sharp changes in polarization and delays for small changes of the ray direction. Other differences between earthquakes of the same region, like their magnitude and the dominant period of the signal reaching the station, may be tested as alternative reasons for a difference in retrieved anisotropy parameters by examining the consistency of results obtained with the two criteria of measuring the satisfactory retrieval of anisotropy parameters.

Among several Vanuatu events recorded at station A in the Qangtang just south of the Tanggula Pass, the correlation criterion applied to fitting the full waveform as a function of projection azimuth and delay retrieves a value of 45°E for the azimuth of the fast direction for the example in Figure 7b of event 2, instead of 65°E in Figure 7a for event 1, representative of most data. The backazimuths of the two events here are the same to within one degree, which excludes in practice that

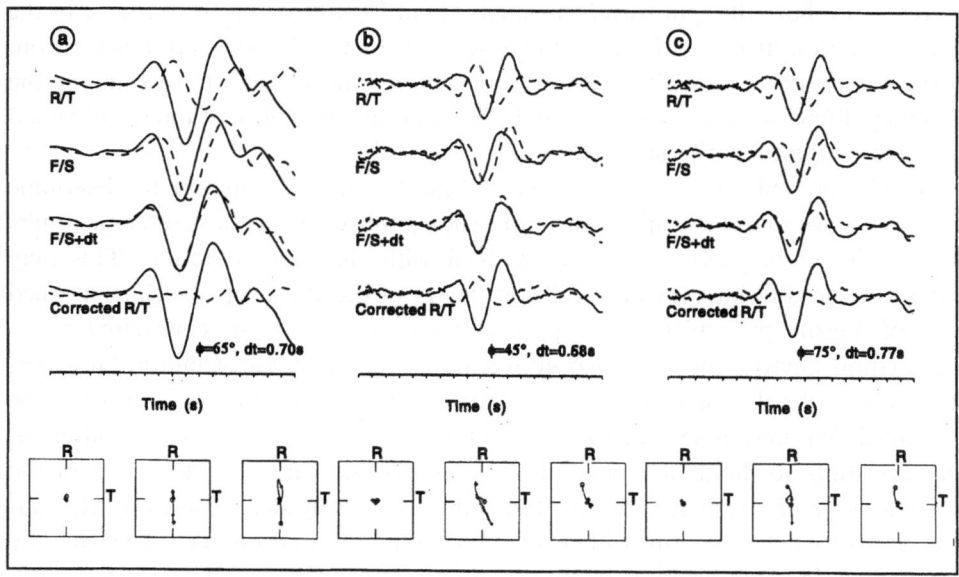

Figure 7

Station A in southern Qang Tang. *SKS* waves from Vanuatu events, backazimuth 109°, distance 87°. R/T are waveforms projected on radial, bold and transverse, dotted orthogonal components. F/S projections on fast (azimuth phi noted on each frame) and slow orthogonal direction. $F/S + dt$, the same, with the slow wave shifted early by the best-fit splitting delay. Corrected R/T is the projection of the previous, i.e., incoming waveform corrected for anisotropy with these parameters, on the radial and transverse components, the second, dotted should have zero amplitude. Bottom are frames of 3 s duration of the particle motion diagrams in the horizontal plane of this recovered incoming waveform. a) Results for event 1. Note good fit of F/S waveforms, giving a nearly zero value of the corrected T and corresponding recovered particle motion along the radial direction. b) Results for event 2, with the same distance and backazimuth. Best correlation of waveforms F/S is obtained for significantly different anisotropy parameters. Corresponding corrected T is however significantly different from zero and recovered particle motion is not oriented along radial direction. c) Same event 2 as in b). Here the anisotropy parameters have been derived in order to best retrieve radial polarization of incoming wave, minimizing the amplitude on the dotted transverse in the corrected R/T and orienting the particle motion diagram. The anisotropy parameters obtained are nearer those for event a), however the fit of waveforms F/S is significantly worse, with the slow, dotted having smaller amplitude.

this dramatic change be interpreted as being due to a particular inclination of the principal axes. More commonly these results may be taken as scatter and incorporated in an estimation of mean value and standard deviation. A check of the conformity to the criterion of restoration of linearity and radial azimuth of the incoming wave is satisfactory for event 1 in Figure 7a, but unsatisfactory for event 2. The restored particle motion diagram of the incoming wave, with the parameters obtained for the best fit waveform correlation displayed in Figure 7b does not retrieve the backazimuth. Taking in Figure 7c, this alternative criterion of backazimuth retrieval on that same seismogram of event 2 leads in fact to derive anisotropy parameters which no longer differ significantly from those derived for event 1. However it is obvious in Figure 7c that these parameters then do not at all allow fulfillment of the criterion of similarity of waveforms of the fast and slow waves. Since both the split waveform similarity in Figure 7b and the retrieval of the particle motion diagram of the incoming wave in Figure 7c are good, but far from perfect, it would be difficult to choose, from the technical point of view, among the two very different parameter sets only by comparing the quality of the fits obtained for each of the two criteria.

In the present case, taking the backazimuth retrieval criterion to determine anisotropy parameters would seem more successful to avoid the scatter in results from different neighboring events obtained with the other criterion. This may reduce the need to introduce a specific geometry of the anisotropy, such as inclined axes of anisotropy, which in the particular case could not be considered as an explanation anyway since the two events are in the same backazimuth. However, the cause of the difference in signals, if not in retrieved parameters, still has to be explained. We may note that the pulse widths from the two events are different, corresponding to different magnitudes and source mechanisms. Then, one of the possible ways of interpreting these observations is to consider anisotropic attenuation. For the slow wave, this will decrease the amplitude and broaden the signal, by preferential loss of the high frequency part in the case of a broadband width incoming signal, or mainly decrease the amplitude for signals of more peaked spectra. Since event 2 has a shorter pulse width than event 1, attenuation would decrease the amplitude of its slower split wave, with respect to the fast one, more than for event 1. Consistently, in Figure 7c, the projection of the signal of event 2 on the axes defined from backazimuth retrieval which are also those found for event 1, the slow wave has a significantly weaker amplitude than the fast one.

4. *S-wave First Arrival Residuals, Spatial Variation*

Residuals of arrival times have been used at different levels in the discussion of anisotropy which, due to experimental limitations, have mainly been from observations of *P* waves. SILVER and CHAN (1992) noted for a group of stations in the eastern U.S. an inverse correlation between mean P residual and SKS delay. They interpreted it in support of their preferred model of transverse isotropy, with a horizontal axis, in which anisotropy is due to frozen-in deformation of the lithosphere. The interpretation of the splitting data in Tibet by MCNAMARA *et al.* (1994) assumes the same anisotropy model, although no residual variations between the stations is given in support. NI *et al.* (1995) do not report a change of travel-time residuals of compressional waves for their stations across the *ITS* but suggest that a 1 s delay northward may exist for shear arrivals. They attribute it to variation in structure, but this would be specifically in Poisson's ratio, hence not inconsistent with the large and sharp variation of this parameter resolved with a tight array of stations (HIRN *et al.*, 1995a,b). For positions of receivers increasing in latitude towards the north into the Lhasa and Qang Tang blocks, we documented an increase of the relative arrival residual of shear with respect to compressional wave arrivals, as well as of the magnitude of the SKS splitting delay. We proposed a model of shallow partial melt contribution to anisotropy (HIRN *et al.*, 1995a,b, 1996) in order to account for this correlation which is unexpected in the anisotropy cases generally considered, of frozen-in lithospheric lattice preferred orientation (SILVER and CHAN, 1991) or of deeper asthenospheric flow (VINNIK *et al.*, 1992). In such weak layers, literally the asthenosphere, but which here may have a crustal component, temperature would promote crystal reorientation under differential horizontal shear across them. Partial melt could even allow anisotropy in the form of fluid-filled defects oriented in a ductile flow.

BABUSKA *et al.* (1984), from permanent seismological observatories and many events, describe P residual variations with incidence and azimuth, which they attribute to anisotropy with a dipping high velocity axis. They advocate (BABUSKA *et al.*, 1993) that mantle anisotropy could not be reduced to hexagonal symmetry with horizontal axis, as considered in SKS studies, and provide means for a joint interpretation of the two types of observations (PLOMEROVÁ *et al.*, 1996). They made a case for inclined axis anisotropy, seen as the tilting of slabs of olivine aggregates (originally oriented in the formation of the oceanic lithosphere) by later subduction in successive accretion of oceanic terranes to form the continents.

On the array across southern Tibet, relative residuals for both shear and compressional waves may be measured accurately for teleseismic events which triggered and produced clear comparable waveforms at several stations. For 13 earthquakes in diverse azimuths, with an average of 12 shear arrival readings each, that is half the stations, the residual of the shear waves increases consistently northwards with respect to that of the compressional wave arrival. For waves with steep angles of incidence, SKS and core-reflected phases, the variation along the line of their residuals relative to P, appears in general larger than is the case for direct teleseismic S. Such dependence of residual on steepness of incidence can be illustrated with better accuracy when both types of shear waves are observed for the same event, such as Vanuatu in Figure 8, Talaud in Figure 9 and from a Kuriles event

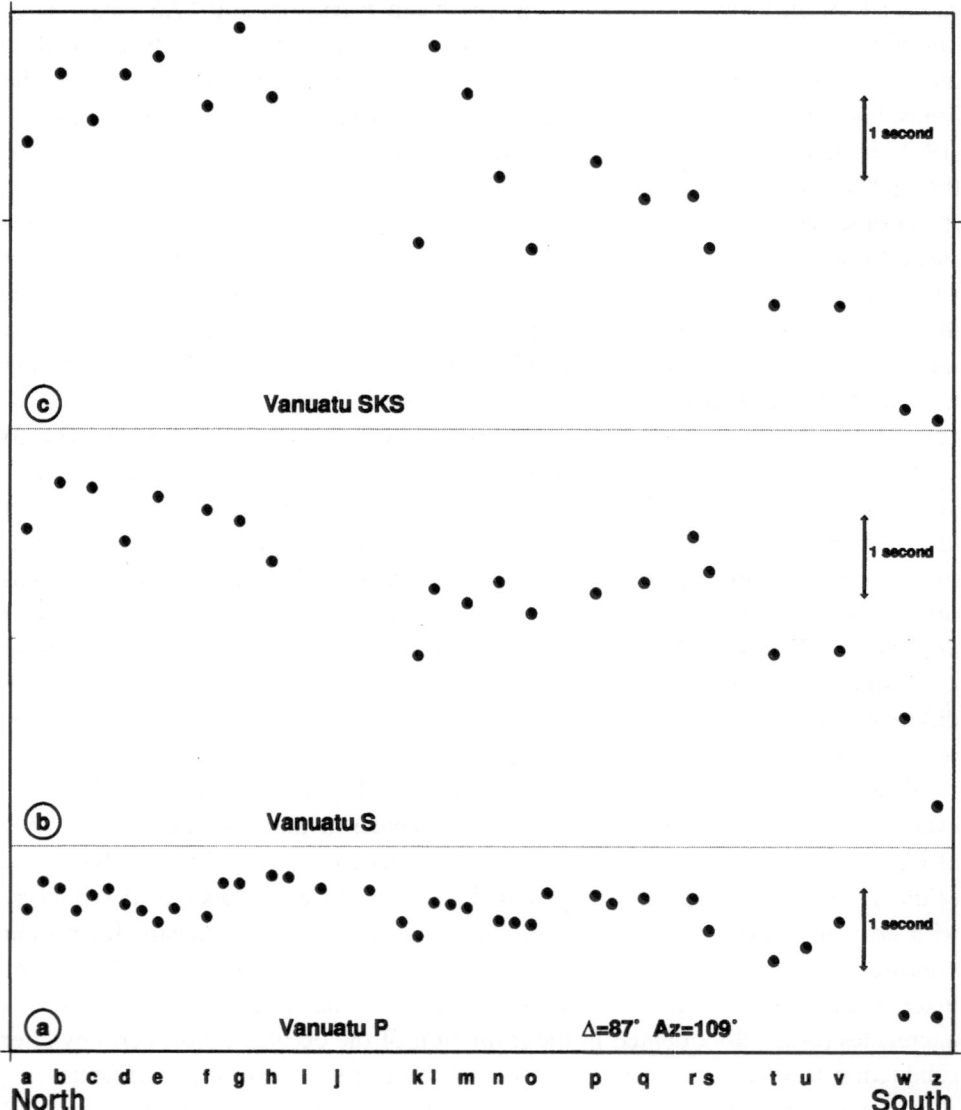

Figure 8

Vanuatu event. Relative residuals along the array of diverse waves. Note that although both shear waves have residuals increasing more than for *P* northwards, left, this effect is larger for the steep incidence *SKS* than for *S*.

(backazimuth 43°, distance 67° in Fig. 6 of HIRN *et al.*, 1996). Along the line, a relative change of the velocity within a layer, or of the depth of an interface between two layers of constrasting velocities, induces a larger effect on differential residuals, the longer the raypath, and therefore the less steep its incidence. The larger range of variation observed along the array for the wave with steeper incidence is hence in contradiction with the expectation for the effect of such structural heterogeneity in an isotropic model. However, because of their different angles of incidence, the two waves do not sample the same part of the upper mantle and an explanation by a particular 3-D heterogeneity cannot be excluded because of the poor coverage in azimuths. Since anisotropy is attested by shear-wave splitting, other interpretations are suggested in which the fastest propagation velocity should not be vertical to agree with the observations. It is possible that models of anisotropy with hexagonal symmetry dipping axes may be found to account for the observations, although the data in the diverse azimuths may be difficult to satisfy. At the present stage, we would prefer a more general model of anisotropy with orthorhombic symmetry with a slow vertical axis. In such a model the angle of incidence for which the fast velocity is maximum, and the contrast between fastest and slowest velocities greatest, is shifted away from the vertical, hence *SKS* is more delayed and less split than *S*, as observed.

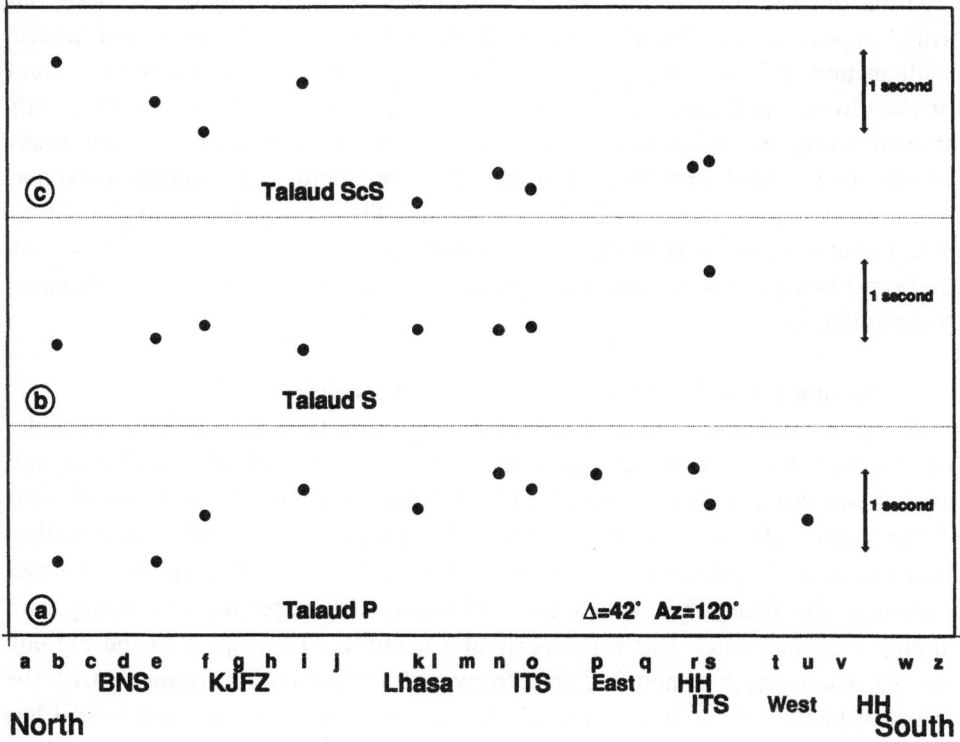

Figure 9

Talaud event. Relative residuals along the array of diverse waves. Note that the shear-wave residuals increase northwards more than *P*, and this effect is larger for the steep incidence core-reflected phases than for *S*.

5. Conclusions and Discussion

5.1. Summary of Observations

The shear-wave residuals and splitting delays do not have a coverage in azimuth and incidence which allows a unique determination of the velocity anisotropy. Their relative values and the observation of their spatial variation across an array provide hints of characteristics of this anisotropy which can be summarized with respect to predictions of the model commonly assumed in interpretations of single station measurements.

—Spatial variation of the splitting delay with respect to the shear-wave first arrival residual

Teleseismic *P*-arrival residuals over a group of stations in the eastern U.S. were reported by SILVER and CHAN (1991) to have an inverse correlation with *SKS* splitting delays, suggesting the magnitude of those to be proportional to the thickness of the high velocity mantle lithosphere. Across southern Tibet, spatial variation of residuals, here of shear waves, and of *SKS* splitting delays along our array, appear to be directly correlated instead: northwards both first arrival residuals and splitting delays increase (HIRN *et al.*, 1995a, b). In order to account for the direct correlation, the anisotropic layer of variable thickness or differently strained along the observation line, would need to be of material with lower velocity than normal, or rather with higher Poisson's ratio, which may be viewed as due to partial melt or at least high temperature. This interpretation implied shear on horizontal planes, with three axes of anisotropy: two in the foliation plane with the fastest being in the transport direction, and one along the vertical, orthogonal to the foliation.

—Splitting delay dependence on the angle of incidence

Array measurements across Tibet allow us to identify *S*-wave anisotropy under the receivers. They provide examples of shear-wave observations at incidence and polarization different from those of *SKS*, which do not agree with the predictions of the commonly assumed model for *SKS* splitting. *SKS*, with quasi-vertical incidence and *SV* polarization, appears with later first arrival residuals but lesser splitting delays than other shear waves. Among models suggested, anisotropy with dipping axis may cause this for a particular azimuth with respect to the dipping axis. Alternatively, in a model of anisotropy with orthorhombic symmetry with the slowest velocity axis along the vertical, the incidence for which the first arrival has earliest residuals and largest splitting delays shifts away from the vertical, which may account for the observation that *SKS* is less split than *S*. Furthermore, examples would suggest that along the steep part of the ray extending to the receiver, the *SH* component of the wave which is vibrating in the horizontal plane may split along the same axes in that plane as the vertically incident *SKS*. A later arrival component of *S* or *ScS* would come from its *SV* part, also being sensitive to a slow velocity of a vertical axis of the anisotropy, and may be split itself depending on azimuth.

—Spatial variation of relative splitting delays of *SKS* and *S*

Another observation is that the *S* splitting delay varies less in space along the array than that of *SKS*, remaining large even towards the south and through the Indus-Tsangpo suture. In the orthorhombic model this suggests that on the two sides of this boundary, the slowest velocity is along the vertical and contrasts strongly with the velocity in the horizontal plane. Instead, the contrast between the two horizontal velocities and their orientation changes strongly across the Indus-Tsangpo suture. North of it where the fast direction orients WSW-ENE and the contrast in the two horizontal velocities is large, if anisotropy is related to shear deformation parallel to the horizontal plane, the strain ellipse would be very elongated, or if anisotropy would relate to ductile flow, this flow should be large, in strong contrast to south of the boundary.

—Relative spatial variation of first residuals for *SKS* with respect to other angles of incidence or polarization

The relative residuals of shear waves increase northwards with respect to those of *P*. The residuals of arrivals of waves with very steep incidence, *SKS* and core-reflected have a larger northward increase than those of *S* waves of less steep incidence. There may be trivial reasons for the differences which relate to the 3-D heterogeneity, since because of their different incidences the two waves do not sample identical paths in the upper mantle. There is no evidence relevant to such isotropic velocity contrast in space. Alternatively, these observations may be accounted for in the medium with orthorhombic symmetry already considered from other lines of evidence. With a slow vertical axis, the incidence for which the fast velocity is maximum shifts away from the vertical. At variance with the hexagonal model, vertical incidence corresponds to a minimum of the fast velocity and when the magnitude of anisotropy increases as here northwards, the first arrival residual of *SKS* increases more than the one for *S*.

5.2. Elements for a Model of Anisotropy

Seismic body-wave anisotropy has been attributed to diverse causes in specific contexts which result in similar, relatively simple, hexagonal or orthorhombic symmetries. One cause is micro-scale orientation of structures such as periodic thin layers for instance as formed by deposition in sedimentary basins. Preferred orientation of other objects may follow the principal compressive stress or strain direction; the case of dry or fluid-filled cracks which results in extensive dilatancy anisotropy (CRAMPIN, 1978), in environments comprising fluid, water, vapor and possibly partial melt. Another source is the intrinsic anisotropy of mantle forming minerals, specifically olivine. It may be by itself a cause of macro-scale anisotropy if minerals are orderly oriented upon crystallization, or if later strain imposes structure and order on them in the form of lattice-preferred orientation.

In crack-controlled anisotropy the fastest and most split shear wave has a propagation direction along the axis of fast compressional wave propagation in the crack plane. In contrast, in an olivine crystal or ordered aggregate, the shear wave is slowest and least split in propagation along the (a) axis of fastest compressional velocity. The fastest and most split shear wave has a propagation direction between the (a) and (b) axes of fastest and intermediate compressional velocity. In deformed rocks containing olivine, the character of the anisotropy, orthorhombic or hexagonal symmetry and the directions of its principal directions depend on the mechanism of deformation. Pure shear, uniaxial compression or extension may both result in transverse isotropy, although with either a slow or fast axis of symmetry. Simple shear will result in the fast axis being in the direction of mineral lineation. Shearing deformation is in fact commonly revealed by mantle xenoliths, however they exhibit fabrics in which the respective positions of the intermediate and of the slow axes are different, depending on temperature and strain rates (NICOLAS and CHRISTENSEN, 1987). They may be randomly distributed in the plane orthogonal to the fast axis, reducing anisotropy to hexagonal. Orthorhombic anisotropy of different orientation is observed, depending on temperature: under low temperatures the slow axis is in the shear-induced foliation like the fast axis, whereas under high temperature this slow axis is orthogonal to the foliation plane.

In the case of Tibet most of the above-mentioned causes for anisotropy may be discussed, and furthermore the anisotropic material may have dipping axes. *SKS* splitting interpretations have commonly considered only a simple model of transverse isotropy with horizontal axis of symmetry. Even then, processes as different as vertically coherent deformation of the lithosphere (SILVER and CHAN, 1991) and absolute plate motion over the asthenosphere (VINNIK et al., 1992) have been interpreted from results worldwide in that same assumption on the symmetry and attitude of the anisotropy. MAINPRICE and SILVER (1993) measured and interpreted specific samples of upper mantle resulting from deformation. In their model of anisotropy, the fastest and most split shear wave is the one propagating along the intermediate axis, the direction orthogonal to the lineation in the plane of foliation. For deformed mantle olivine two particular cases of orientation of LPO may be related to vertical and horizontal plane shear, respectively. Internally coherent deformation of the lithosphere is considered as a likely general cause of continental mantle anisotropy by SILVER and CHAN (1991) as well as for Tibet by McNAMARA et al. (1994), but also the anisotropy close to major faults in Tibet of LAVÉ et al. (1996) may be viewed as lattice-preferred orientation, LPO, with a vertical foliation plane. According to such a model, the vertical incidence shear wave would then be the fastest and have the largest splitting delay, which is not supported by the additional data presented here on S and their comparison with SKS. LPO with a horizontal foliation plane is instead associated with the horizontal shear or ductile flow model which we suggested to account for the correlation of SKS residuals with splitting delays revealed along our array (HIRN et al., 1995b). It is also taken over in the model of block motions over the asthenosphere of LAVÉ et al. (1996) to account for splitting far from block boundaries. The examples of observations presented in previous sections of the spatial variation along the array of the first arrival residuals and splitting delays of S, when compared to those of SKS, consistently allow resolution between the two models, also in favor of the one of LPO related to horizontal plane shear.

5.3. Implications on the Evolution of Tibet

For the Lhasa and southern Qang Tang region we inferred (HIRN et al., 1995a, b) that fast SKS directions were aligned nearer to the direction of average crustal transport inferred by AVOUAC and TAPPONNIER (1993), than to that of their major faults: Karakoram-Jiali Fault Zone or the Banggong-Nujiang suture, or that of crust carried by small-scale mantle flow (Molnar, 1993), or that of coherent mantle and crust deformation (SILVER and CHAN, 1988; MCNAMARA et al., 1994). Hence we suggested that anisotropy was related to horizontal shear between layers, or ductile flow in low velocity layers. LAVÉ et al. (1996) more precisely assessed from an analysis of slip on the major faults seen at the surface and a resulting model of a few small plates for central Asia, the directions of crustal transport with respect to fixed Eurasia in the region encompassing the anisotropy surveys of MCNAMARA et al. (1994) and GUILBERT et al. (1996). They contended that there is a regional correlation of lithospheric transport, implicitly the same as that derived from surface observations, with fast directions of anisotropy, since the distribution of their difference in azimuth is narrowly peaked. This difference is, however, 30° in a possible range of 180°. For instance, where our station line crosses the 110°E oriented KJFZ, their direction of crustal transport is 40°E, but the fast axis of anisotropy is 30° clockwise from it, at 70°E (although splitting only defines the orientation, not the direction).

In their interpretation of anisotropy as due to shear on horizontal planes imposed by the extruding lithospheric blocks on the passive underlying astheno-sphere, LAVÉ et al. (1996) then consider in addition to the strain that they could analyze from surface features, the absolute plate motion, APM of Eurasia. They infer that for tectonic transport to be rotated by 30° in order to align with fast anisotropy, it would require that the large Eurasian plate must move by as much as 2 cm/yr southward over the deep mantle. However the APM of both GRIPP and GORDON (1990) and MINSTER and JORDAN (1978), although different, have a northward component instead, which does not support such a model.

Without evidence to resolve better whether the APM of Eurasia should be large or small, this difference of 30° may be viewed differently. The motion of the lithospheric blocks proposed by LAVÉ et al. (1996) as inducing the fast axis of anisotropy by imposing horizontal shear on the passive mantle, does this specifically along their direction of motion. However, the transport direction they define is that of the upper crust. The specific characters of anisotropy we documented lead us to regard it as due to flow in weak ductile layers. Although the coupling of deeper motion with that of the brittle cover occurs through these ductile layers, the direction of transport need not be the same at all levels. In the case of flow of material in ductile weak layers, it is the absolute direction of this flow which will give the direction of anisotropy, rather than the differential motion of this flow with respect to their top or base which may move too. The N70°E anisotropy would then mark the direction of the flow, carrying its top in an overall N40°E direction, the difference in motion being presumably due to the particular lateral boundary conditions at the surface for the brittle outer shell.

A consequence of this interpretation of seismic anisotropy could then be that material of the ductile layers flows with a more easterly direction than that inferred from brittle tectonics of the upper crust. With respect to overall Eurasia, more material than that corresponding to the extruding brittle upper part would be transported to the northeast from underneath the Tibetan plateau, and may contribute to the expansion of it in this direction. This suggestion is clearly dependent on a number of parameters of both the motions inferred from tectonics and geodynamics, but also of the seismic anisotropy which is still far from being sufficiently documented and constrained. The complexity of anisotropy in space implied by the examples presented needs more complete data coverage than available in order to be described satisfactorily. The relevance of the information contained in details of the seismic data for the structure and deformation at depth, which is central to evolutionary models of Tibet and of the continents, warrants more complete data acquisition efforts including array measurements of teleseismic P and S residuals and splitting.

Note Added in Revision

The availability during revision of the present paper, of the manuscript submitted by Davis, England and Houseman to Journal of Geophysical Research leads to an additional discussion with respect to our analysis and conclusions.

The seismic data analyzed in the present paper allow us to document the significance of a component of velocity anisotropy out of the horizontal plane, by considering the splitting of other shear waves than *SKS*, as well as from documenting a dependence of transit time residuals on angle of incidence. This has implications with respect to the different views expressed recently regarding the regional deformation and geodynamics by interpreting anisotropy in the horizontal plane, solely from the *SKS* splitting (LAVÉ et al., 1993; DAVIS et al., 1997).

We concur with DAVIS *et al.* (1997) who, referring to our conclusion from the sharp spatial change in anisotropy direction across the Indus-Tsangpo suture and the High Himalayas (HIRN *et al.*, 1995b) and using their data in Mongolia (GAO *et al.*, 1994), constrain the anisotropic layer to be located above 200 km depth, in correspondence with a Fresnel zone width of 120 km for 0.1 Hz frequency waves. The higher frequency content in the additional observations presented here, including those of *S*, would even tend to reduce this depth for southern Tibet.

These observations support our previous interpretation that among causes of anisotropy dominant in southern Tibet-Himalayas, we may exclude those which predict a regional continuity in direction, like that considering absolute plate motion, for which furthermore none of the diverse proposed directions for Eurasia fits any of the two anisotropy directions measured in southern Tibet. The amount or velocity of displacement with respect to stable Eurasia, or shear of the lithospheric blocks over the asthenosphere has also been considered as a cause for anisotropy for instance by LAVÉ *et al.* (1996) north of the Indus-Tsangpo suture. This velocity is shown to increase from northern Tibet into India (AVOUAC and TAPPONNIER, 1993; England and Molnar, cited in DAVIS *et al.*, 1997). DAVIS *et al.* (1997) hence contend that it cannot be the cause for anisotropy since the observed magnitude of anisotropy does not vary in that sense. This contention is in agreement with conclusions from our data set, where south of the suture and across the Himalayas the expected larger values of the splitting magnitudes are not observed. Even if we excluded the region south of the Indus-Tsangpo suture from the tentative interpretation, as LAVÉ *et al.* (1996) do, we estimated that the difference between the horizontal direction of the fast *SKS* anisotropy north of the Indus-Tsangpo and the direction of upper crustal flow is meaningful, so that an alternative to their interpretation is needed, a conclusion to which those of DAVIS *et al.* (1997) concur.

Searching for a single mechanism to account for anisotropy over all of Central Asia, DAVIS *et al.* (1997) base their analysis on the only generally available parameters of splitting from *SKS*: orientation of the principal axes assumed to lie in the horizontal plane. They find the internal lithospheric deformation due to indentation by India to be the most appropriate cause of anisotropy to explain the spatial variation of a fast *SKS* direction. An implication of such a model, which the available *SKS* data they use cannot test, however, is that anisotropy would then have a third principal axis, in fact the fastest one, along the vertical. In the present paper we reported evidence from the comparison of *SKS* and *S* splitting delays, and also from their corresponding first arrival residuals, and their relative sense of variation, which we regarded as suggesting that there is a third principal axis of anisotropy along the vertical. However it appears to correspond to the slowest velocity. Hence the interpretation of anisotropy we developed, admittedly for the restricted part of our Tibetan survey north of the Indus-Tsangpo suture, is obviously different from vertically-coherent deformation of the lithosphere. Instead we have correspondingly made the suggestion that horizontal plane flow is here the mode of deformation inducing anisotropy in weak layers (in this sense asthenosphere) at shallow levels which elsewhere correspond to the lithospheric mantle and crust. This is not a world- or continent-wide explanation of anisotropy, but its local dominance would already have a geodynamical meaning. Its parameters would imply a difference in the transport of the upper crust and deeper domain, resulting from or leading to differential thickening.

Acknowledgements

We acknowledge the INSU-CNRS (Paris) and MGMR (Beijing) for having supported the data acquisition. We thank J. Plomerová for stimulation to write up results presented, P. Molnar for having provided a stimulating though kind review, signalling the paper of P. Davis, whom we thank for a copy in advance of publication.

REFERENCES

ANDO, M. (1984), *ScS Polarization Anisotropy around the Pacific Ocean*, J. Phys. Earth. *32*, 179–195.

AVOUAC, J. P., and TAPPONNIER, P. (1993), *Kinematic Model of Active Deformation in Central Asia*, Geophys. Res. Lett. *20*, 895–898.

BABUSKA, V., PLOMEROVA, J., and SILENY, J. (1984), *Large-scale Oriented Structures in the Subcrustal Lithosphere of Central Europe*, Ann. Geophys. *2*, 649–662.

BABUSKA, V., PLOMEROVA, J., and SILENY, J. (1993), *Models of Seismic Anisotropy in the Deep Continental Lithosphere*, Phys. Earth Planet. Int. *78*, 167–191.

CRAMPIN, S. (1978), *Seismic Wave Propagation through a Cracked Solid: Polarization as a Possible Dilatancy Diagnostic*, Geophys. J. R. Astr. Soc. *53*, 467–496.

CRAMPIN, S., EVANS, R., UCER, B., DOYLE, M., DAVIS, J. P., YEGORKINA, G. V., and MILLER, A. (1980), *Observation of Dilatancy-induced Polarization Anomalies and Earthquake Prediction*, Nature *286*, 874–877.

DAVIS, P., ENGLAND, P., and HOUSEMAN, G. (1997), *Comparison of Shear-wave Splitting and Finite Strain from the India-Asia Collision Zone*, J. Geophys. Res., in press.

DIAZ, J. (1994), *Anisotropie sismique et déformation dans le manteau supérieur. Approche conjointe par sismique réfraction et ondes S télésismiques en Ibérie. Méthodes d'analyse d'ondes et leur résultat au Tibet*. Thèse de Doctorat de l'Université Paris VII, pp. 173.

ENGLAND, P., and HOUSEMAN, G. (1989), *Extension during Continental Convergence, with Application to the Tibetan Plateau*, J. Geophys. Res. *94*, 17561–17579.

GUILBERT, J., POUPINET, G., and JIANG, M. (1996), *A Study of Azimuthal P residuals and Shear-wave Splitting across the Kunlun Range (Northern Tibetan Plateau)*, Phys. Earth and Planet. Int. *95*, 167–174.

HERQUEL, G., WITTLINGER, G., and GUILBERT, J. (1995), *Anisotropy and Crustal Thickness of Northern Tibet. New Constraints for Tectonic Modelling*, Geophys. Res. Lett. *22*, 1925–1928.

HIRN, A., NERCESSIAN, A., SAPIN, M., LÉPINE, J. C., SACHPAZI, M., FERRAZZINI, V., JIENG, M., LU, Q. T., SHI, D. N., MA, K., PANDEY, M. R., DIAZ, J., and GALLART, J. (1995a), *Lesser Himalayas to Qang Tang: A 500 km Teleseismic Deployment to Test Geodynamic Models*, J. Nepal Geol. Soc. *11*, 39–52.

HIRN, A., JIANG, M., SAPIN, M., DIAZ, J., NERCESSIAN, A., LU, Q. T., LÉPINE, J. C., SHI, D. N., SACHPAZI, M., PANDEY, M. R., MA, K., and GALLART, J. (1995b), *Seismic Anisotropy as an Indicator of mantle flow beneath the Himalayas and Tibet*, Nature *375*, 571–574.

HIRN, A., NERCESSIAN, A., SAPIN, M., LÉPINE, J. C., DIAZ, J., and JIANG MEI (1996), *Increase in Melt Fraction along a South-north Traverse below the Tibetan Plateau: Evidence from Seismology*, Tectonophysics *273*, 17–30.

KIND, R., KOSAREV, G. L., MAKEYEVA, L. I., and VINNIK, L. P. (1984), *Observations of Laterally Inhomogeneous Anisotropy in the Continental Lithosphere*, Nature *318*, 58–361.

LAVÉ, J., AVOUAC, J. P., LACASSIN, R., TAPPONNIER, P., and MONTAGNER, J. P. (1996), *Seismic Anisotropy beneath Tibet: Evidence for Eastward Extrusion of the Tibetan Lithosphere*, Earth Planet. Sci. Lett. *140*, 83–96.

MAINPRICE, D., and SILVER, P. G. (1993), *Interpretation of SKS Waves Using Samples from the Subcontinental Lithosphere*, Phys. Earth and Planet. Int. *78*, 257–280.

MCNAMARA, D. E., OWENS, T. J., SILVER, P. G., and WU, F. T. (1994), *Shear-wave Anisotropy beneath the Tibetan Plateau*, J. Geophys. Res. *99*, 13655–13665.

MOLNAR, P. (1992), *Crust in Mantle Overdrive*, Nature *358*, 105–106.

NI, J., and SANDVOL, E. (1994), *Seismic Azimuthal Anisotropy across the Tsangpo Suture from INDEPTH II. Broadband Data (Abstract). EOS*, Trans. Am. Geophys. Un., 75–44, 628.

NI, J., KIND, R., NABELEK, J., and INDEPTH II BROADBAND TEAM (1995), *New Seismological Results from the INDEPTH II Broadband Experiments, Implications for Evolutionary Models (Abstract). EOS*, Trans. Am. Geophys. Un., 76–46, 392.

NICOLAS, A., and CHRISTENSEN, N. I. (1987), *Formation of anisotropy in upper mantle peridotites—A review. In Composition, Structure and Dynamics of the Lithosphere—Asthenosphere System* (Fuchs, K. and Froidevaux, C., eds), AGU Mon. 16, 11–123.

PLOMEROVÁ, J., SILENY, J., and BABUŠKA, V. (1995), *Seismic Anisotropy of the Continental Lithosphere*, IUGG XXIst Assembly, Boulder, Co., Abstract vol., p. A167.

PLOMEROVÁ, J., SILENY, J., and BABÜSKA, V. (1996), *Joint Interpretation of Upper-mantle Anisotropy Based on Teleseismic P Travel-time Delays and Inversion of Shear-wave Splitting Parameters*, Phys. Earth Planet. Int. *95*, 293–309.

SILVER, P. G., and CHAN, W. W. (1988), *Implications for Continental Structure and Evolution from Seismic Anisotropy*, Nature *335*, 34–39.

SILVER, P., and CHAN, W. W. (1991), *Shear-wave Splitting and Subcontinental Mantle Deformation*, J. Geophys. Res. *96*, 16429–16454.

VINNIK, L. P., MAKEYEVA, L. I., MILEV, A., and USENKO, A. Yu. (1984), *Global Patterns of Azimuthal Anisotropy and Deformations in the Continental Mantle*, Geophys. J. Int. *111*, 433–447.

VINNIK, L. P., KOSAREV, G. L., and MAKEYEVA, L. I. (1984), *Anizotropia litosfery po nyabludeniam voln SKS, SKKS*, Dokl. Akad. Nauk, USSR, *278*, 1335–1339.

(Received November 1, 1996, revised July 1, 1997, accepted July 5, 1997)

Pure appl. geophys. 151 (1998) 433–442
0033–4553/98/040433–10 $ 1.50 + 0.20/0

⌐Pure and Applied Geophysics

Upper Mantle Anisotropy in Victoria Land (Antarctica)

S. PONDRELLI[1] and R. AZZARA[1]

Abstract—During the Austral Summers 1993–94 and 1994–95, we operated two temporary seismic arrays in the Victoria Land region, Antarctica. The first was located around the Mt. Melbourne volcano, near Terra Nova Bay, and the second southward, along the ACRUP1 Geotraverse. The aim of these experiments was to provide additional constraints on the Moho geometry and on lithosphere-asthenosphere structure in this zone. For this reason, a number of techniques were applied to recorded teleseismic data. In this paper we describe the results of the analysis on SKS- and S-wave splitting, which show the presence of seismic anisotropy. Computed fast polarization directions range, on average, between 131° and 166°, and delay times are on the order of 1 s. We presume that anisotropy is mainly located in the upper mantle, although we cannot detect, at present, if a contribution from the crust does exist. The fast polarization direction we determine is quite parallel to the opening of the Ross Sea, an active rift system, but also to the Absolute Plate Motion direction. Therefore, we assume that the identified anisotropy may be induced by the extension, activated by plate motion and is still related to it.

Key words: Seismic anisotropy, upper mantle, Antarctica.

Introduction

Antarctica has a strategic position and history for the global geodynamics reconstruction. At the same time, Antarctica is a difficult region to be directly studied. Geophysics, and in particular seismology, have provided considerable information on the structure of this continent (COOPER *et al.*, 1989; STERN and ten BRINK, 1989; BEHRENDT *et al.*, 1991; ROULT and ROULAND, 1994).

During the period 1993–1995, two experiments of passive seismology were developed in Victoria Land, to study the structure of the lithosphere-asthenosphere system. Well established techniques (i.e., receiver function or travel-time residuals) have been applied to recorded data, looking for the depth of the Moho, velocity anomalies and the presence of anisotropic bodies beneath the area between the Ross Sea and the Eastern Antarctica, beyond the Transantarctic Mountains (Fig. 1; CIMINI *et al.*, 1995; DI BONA *et al.*, 1997; PONDRELLI *et al.*, 1997).

Studying seismic anisotropy, which is mainly due to strain-induced orientation of highly anisotropic crystals, is a powerful tool to recognize mantle and crustal structures directly related to plate motion or to more local stress fields (SILVER,

---

[1] Istituto Nazionale di Geofisica, Via di Vigna Murata 605, 00143 Rome, Italy.

1996 and references therein). Anisotropy causes shear-wave splitting: a shear-wave incident to an anisotropic layer splits into two polarized orthogonal waves with different velocities. The determination of polarization direction and time delay between the two waves allows characterization of the anisotropy. On the hypothesis that anisotropic minerals, such as olivine and pyroxene, are oriented in the directions of mantle lineations (MAINPRICE et al., 1993), the fast polarization direction of a deep anisotropy (lower crust, upper mantle) is usually considered to be perpendicular to the collisional direction in orogenic belts (SILVER and CHAN, 1988) and parallel to the extensional direction in regions of crustal rifting. In fact, the correspondence between these directions studied worldwide is rather irregular (VINNIK et al., 1992). The anisotropy is also present within the crust, where it is generally inferred by aligned cracks in the shallower part (CRAMPIN and BOOTH, 1985) and mainly by rock fabrics at greater depth, directly related to deformation (BARRUOL and KERN, 1993).

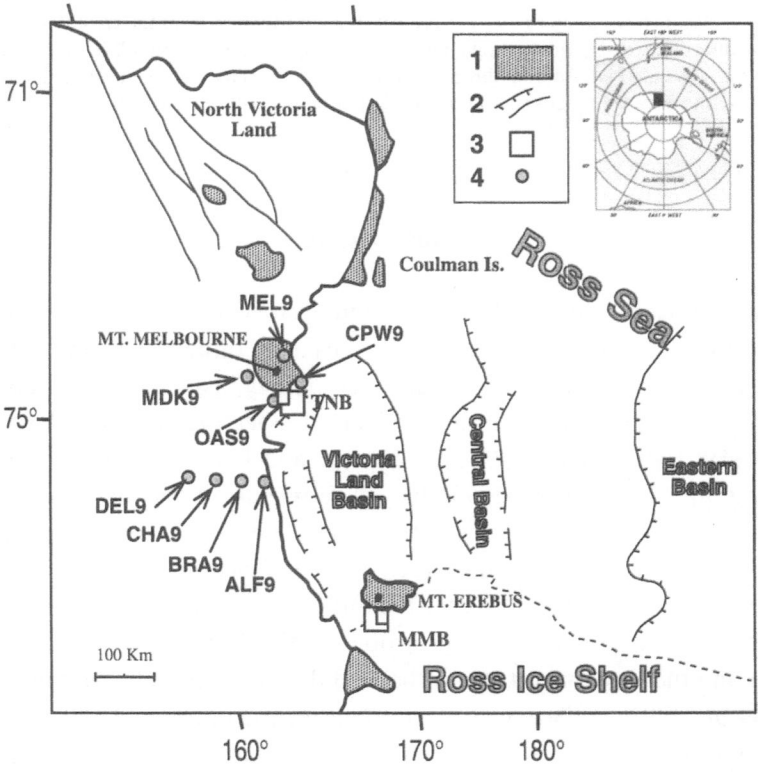

Figure 1

Sketch map of Victoria Land. Symbol Caption: 1—McMurdo Volcanics (Mesozoic-Quaternary); 2—Tectonic structures; 3—Bases (TNB is Terra Nova Bay Base, MMB is McMurdo Base); 4—Seismographic stations.

Shear-wave splitting data for Antarctica are decidedly scarce, due to the low number of seismographic stations present there. Maps of upper mantle velocity and anisotropy anomalies were published by ROULT and ROULAND (1994) from a surface wave dispersion analysis. They identified anomalies that were associated with known structural features, such as the low-velocity zone probably connected with the Ross Sea rift. Beneath the craton (Eastern Antarctica) they also recognized a weak anisotropy that became stronger in the Ross Sea-Transantarctic Mountains area. Shear-wave splitting analyzed by KUBO et al. (1996) pertaining to data recorded by permanent stations located in Antarctica, detected a frozen anisotropy in the lithosphere, due to ancient tectonic episodes. In this paper we present the results of shear-wave splitting of S and SKS phases recorded at eight sites occupied with our 4 stations in the Victoria Land. Experiments and data are first described. Thereafter, evidence of the presence of a mantle anisotropy beneath this region is shown, and at the end of the discussion we conclude that this anisotropy is related to the Ross Sea rift system.

Experiments and Data

In these passive seismological experiments, 4 stations have been used. Each of them includes a three-components Lennartz sensor (3D-5s) with eigenfrequency of 0.2 Hz, a RefTek 72A07 digitizer with a dynamic range of 24 bits, a hard disk of 560 Mbytes, a GPS receiver and a 70 Watt solar panel connected to a 12V–100Ah battery. The stations operated in a continuous mode with a sampling rate of 50 sps. Their installation required particular care, to protect them from very low temperature and strong wind (CIMINI et al., 1995). These temporary stations were deployed in two different areas (Table 1 and Fig. 1): Terra Nova Bay and the Mt. Melbourne volcano area during both Austral summers (November–December 1993; November–December 1994); in the Southern Victoria Land, along the ACRUP1 Geotraverse (DELLA VEDOVA et al., 1997) only 20 days in January 1994. The data detection was substantial, despite the short recording time (4 months in the first area and less than one month in the other) and the background noise shows, for all sites, a low level (CIMINI et al., 1995). Applying different techniques to recorded teleseisms, some interesting results have been obtained. The Moho depth, computed with the receiver function technique, ranges between 17 and 26 km in the area of Mt. Melbourne and between 25 and 43 km along the ACRUP1 Geotraverse, with a clear thickening of the crust moving inland. Moreover, evidence of the presence of seismic velocity anomalies has been found (DI BONA et al., 1997; PONDRELLI et al., 1997).

To detect anisotropy, we used teleseismic SKS and S phase with an epicentral distance between 70° and 110° (Table 1). SKS phases sample the anisotropy beneath the stations, while S phases could also record a source-side anisotropy. We

Table 1

Events used in this study (NEIC locations and magnitude). The used phase and the array that recorded each event are in the last two columns

Date (Mo/Dy/Yr)	Time (hh:mm:ss)	Lat. (deg.)	Long. (deg.)	Depth (km)	Magn.	Range (deg.)	Backaz. (degr.)	Phase	Array
11/05/93	22:37:20	−3.20	148.24	16.	5.6 mb	72.	343.	S	TNB
11/10/93	00:03:25	−4.68	151.91	113.	5.6 mb	70.	350.	S	TNB
11/11/93	10:13:59	−4.57	153.08	80.	5.7 mb	70	348.	S	TNB
11/14/93	01:59:19	−22.66	−68.62	108.	5.8 mb	114.	340.	SKS	TNB
11/22/93	03:00:55	5.87	126.32	33.	5.8 mb	83.	322.	S	TNB
11/27/93	06:11:23	38.61	141.19	108.	6.0 mb	114.	340.	SKS	TNB
11/28/93	10:50:27	−5.61	110.25	565.	5.6 mb	75.	303.	S	TNB
11/29/93	20:28:43	10.25	126.51	33.	5.6 mb	88.	322.	SKS	TNB
12/09/93	04:32:22	0.48	125.98	33.	6.3 mb	78.	320.	S	TNB
12/09/93	11:38:31	0.43	125.89	33.	6.1 mb	78.	320.	S	TNB
12/10/93	08:59:37	20.85	121.28	19.	5.8 mb	99.	319.	SKS	TNB
12/12/93	17:03:21	0.31	125.96	33.	5.8 mb	78.	320.	S	TNB
01/10/94	15:53:50	−13.31	−69.39	589.	6.4 mb	86.	128.	SKS	ACRUP
01/19/94	01:53:36	−3.21	136.02	33.	6.0 mb	73.	330.	S	ACRUP
01/21/94	02:24:31	1.01	127.73	33.	7.2 Ms	78.	322.	S	ACRUP
01/21/94	18:00:18	−4.82	103.74	89.	6.0 mb	78.	297.	S	ACRUP
11/14/94	19:15:31	13.53	121.09	33.	6.1 mb	92.	318.	SKS	TNB
11/15/94	20:18:11	−5.61	110.20	559.	6.2 mb	75.	303.	S	TNB
11/20/94	16:59:06	−2.0	135.93	24.	5.7 mb	74.	330.	S	TNB
12/07/94	03:37:55	−23.46	−66.74	243.	5.6 mb	77.	132.	S	TNB
12/10/94	03:39:31	−23.58	−70.52	37.	5.7 mb	76.	129.	S	TNB
12/12/94	07:41:55	−17.50	−69.65	151.	5.8 mb	83.	129.	S	TNB
12/15/94	23:56:11	−3.25	139.79	110.	5.5 mb	73.	334.	SKS	TNB

used SKS for events with an epicentral distance greater than 85°. However, to reach a significant number of data and a better azimuthal coverage, additionally S phases have been used. Any source-side anisotropy correction was too difficult. Therefore, to attribute an S-phase splitting to an anisotropy located beneath the station, we required the consistency of results obtained by SKS and S analyses for events originating from different directions. As a consequence, a good azimuthal coverage, such as in our case (except for the SW quarter), is necessary. The analyzed data have a period (about 10 s), and thus a wavelength in the upper mantle (about 40–45 km), which allows consideration of the effect of the crust (from 17 to about 40 km thick) as marginal.

All data, filtered between 0.02 and 0.2 Hz, have been analyzed by three different methods: first, by the visual inspection of particle motion, to recognize a time window over which the splitting analysis is performed, then by the decomposition of the covariance matrix and finally, utilizing the cross correlation method. This process is well explained in ZHANG and SCHWARTZ (1994), and MARGHERITI et al. (1996) described its application to teleseismic data. The results have been weighted

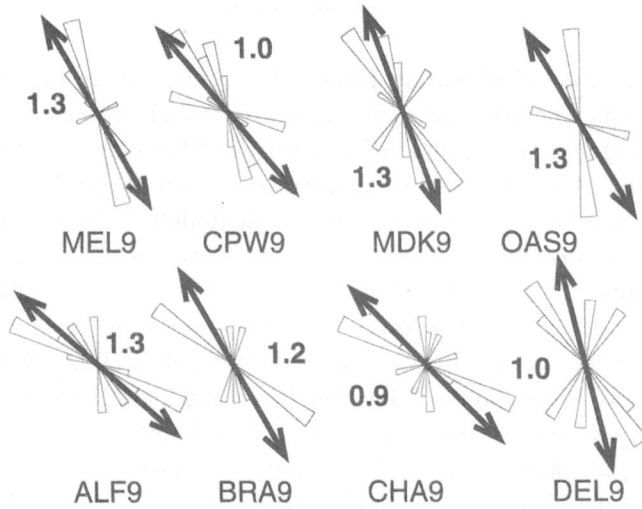

Figure 2

Rose diagrams and average values (thick arrow) of fast polarization direction for all stations. In each rose diagram, the longer the slice, the more frequent the direction. The time delay in seconds is reported in bold.

on the basis of the used phase (weight 1.0 to SKS phases, weight 0.6 to S phases with a fast polarization direction within $\pm 30°$ from that of SKS and weight 0.3 to S phases with other fast polarization directions), but even on the basis of the cross-correlation coefficient and of the quality of linearization after the rotation in the fast direction. For each station a frequency distribution is shown and the average fast direction and delay time are evaluated (Fig. 2 and Table 2).

Table 2

Coordinates, average fast polarization direction and time delay, associated with the variances, for each studied site

	Latitude (degrees)	Longitude (degrees)	Fast Direction	Delay Time (sec)
OAS9	−74.692	164.100	150° ± 27	1.3 ± 0.7
CPW9	−74.628	165.495	139° ± 28	1.0 ± 0.5
MEL9	−74.315	165.019	151° ± 24	1.3 ± 0.4
MDK9	−74.395	163.971	160° ± 29	1.3 ± 0.5
ALF9	−75.899	162.579	131° ± 21	1.3 ± 0.7
BRA9	−75.826	160.500	150° ± 25	1.2 ± 0.4
CHA9	−75.749	158.304	134° ± 35	0.9 ± 0.3
DEL9	−75.697	157.075	166° ± 28	1.0 ± 03

The Anisotropy

From the analysis of teleseismic phases, the presence of seismic anisotropy in the Victoria Land is demonstrated: data recorded by each station reveal S and SKS splitting. In Figure 3 an example of an analyzed SKS phase is shown: this event, occurred in the Philippine region, recorded by station OAS9, has a $m_b = 6.1$.

Fast direction frequency distribution for all stations is shown in Figure 2. The results for sites close to Terra Nova Bay are shown in the upper part. The fast polarization direction ranges between 140° and 180° (NNW-SSE) and the delay time between 1.0 and 1.3 s. In the lower part of Figure 2 the results for sites along ACRUP1 Geotraverse are shown: the fast direction polarization ranges between 110° and 150° (WNW-ESE) and the delay time between 0.9 and 1.3 s. If we consider the average fast polarization direction (Fig. 2 and Table 2), the results are more homogeneous, all ranging between 130° and 160°. These data show a dispersion, visible from rose diagrams and variances in Table 2. It is probably due to a contribution of source-side anisotropy from S phases that even so should be smoothed by the lower weight given to these phases. On the other hand, it also could be caused by the presence of different anisotropic layers beneath the area (SAVAGE and SILVER, 1993) or to an inclination of a single anisotropic layer symmetry axis (PLOMEROVÁ et al., 1996). Unfortunately, the available data at present, do not allow us to better constrain these hypotheses. We presume that anisotropy is located in the upper mantle because there the different ray paths share a common segment. The time delays have values corresponding to those traditionally associated with mantle anisotropy, and they vary from site to site independently from the crustal thickness. Moreover, the 1.2 s average delay time corresponds to an anisotropic layer of 100–150 km thick on average, both assuming a peridotite composition with about 70% of olivine (NICOLAS and CHRISTENSEN, 1987; SAVAGE et al., 1990) or a simple average anisotropy of 4–6% (MAINPRICE and SILVER, 1993). All these observations are in good agreement with a primarily mantle anisotropy. We could also assume that the dispersion direction could be due to still unknown anisotropy located within the crust, that is strongly complex in this area, even if the characteristics of the data used discourage this hypothesis. Only a study of P-to-S teleseismic phases will probably provide more information regarding the presence of crustal anisotropy, in that local and regional seismicity is, unfortunately, too low (in number and magnitude of events) to be useful.

Although the dispersion is large, the mean fast polarization direction which characterizes this area is mainly NW-SE (Fig. 4). This direction may be related to different geodynamic features. The Absolute Plate Motion (APM) for the Antarctic plate shows a NNW-SSE direction in this area (DE METS et al., 1990; GRIPP and GORDON, 1990), quite close to the fast directions we find, however other regional characteristics must be considered. Actually, the Ross Sea extension, started in the Mesozoic (BEHRENDT et al., 1991) and still working, occurs along a

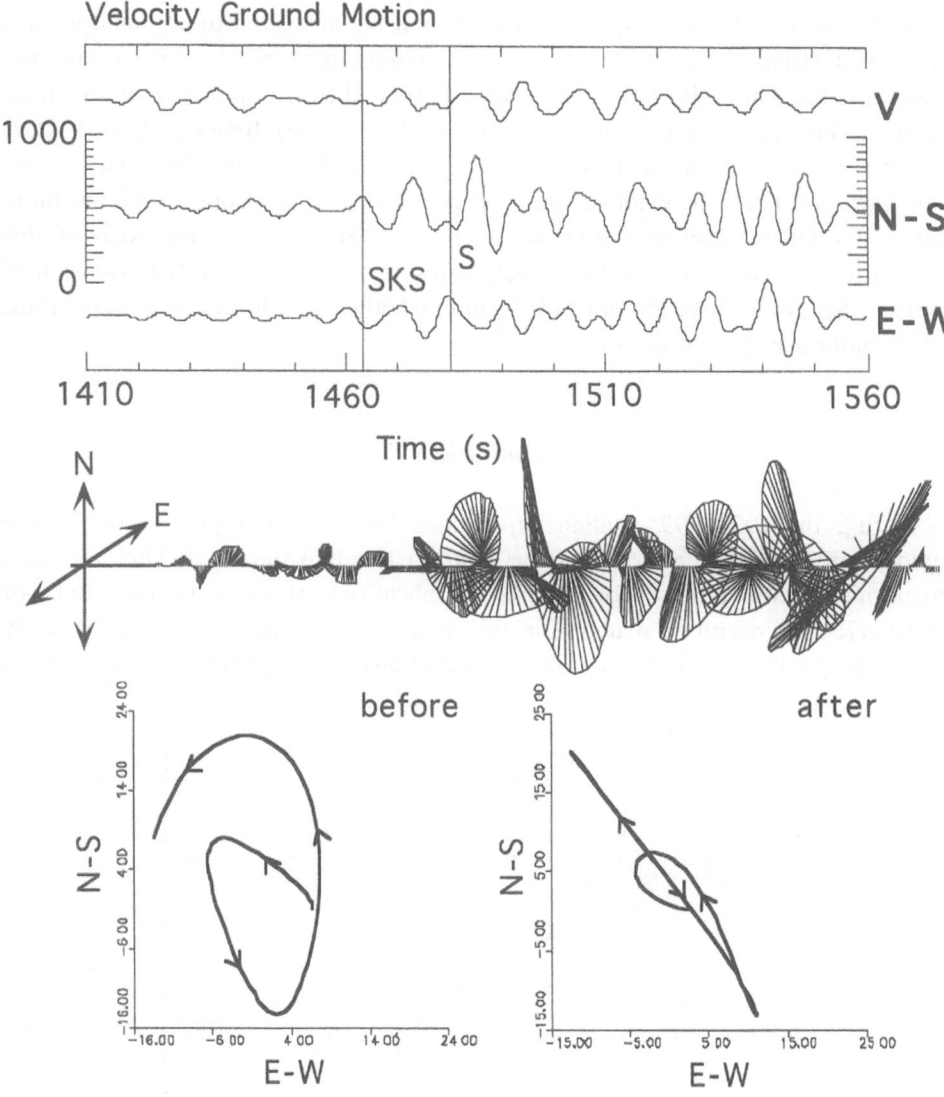

Figure 3
Example of an analyzed event, from the Philippine region, which occurred on 14/11/94 at 19:21 ($m_b = 6.1$). Other parameters in Table 1. Top: velocity ground motion components (the time is in seconds after the origin time). Center: horizontal polarization. Bottom: particle motion during 15 seconds selected in the upper graph, before and after the 155° rotation in the fast direction (the delay time is 1.8 s).

NW-SE direction. Though this direction locally varies from E-W to NNW-SSE, considering the shape of young basins (Figs. 1 and 4: Victoria Land Basin, Central Basin, etc.), is even so in agreement with the average polarization direction. The

Transantarctic Mountains also strike in the direction NNE-SSE, but this chain has developed as a shoulder of the Ross Sea rift and is not related to any compressive stress field (BEHRENDT et al., 1991). The existing data of shear-wave splitting anisotropy for Antarctica have been primarily related to frozen deformation within the crust, due to ancient tectonic events (KUBO et al., 1996). Between them, VNDA station (Fig. 4) manifests a direction quite perpendicular to our data. This small-scale variation could be evidence that the anisotropy arises from a regional more than a continental process (HELFFRICH et al., 1994). Even on the basis of this observation, we conclude that the detected anisotropy is related to the extension of the Ross Sea, even if we are not able to note whether the old or the present (thus, APM) motion is its real origin.

Conclusions

During the 1993–1995 Italian expeditions four seismographic stations were deployed in two different zones of the Victoria Land (Antarctica). They recorded teleseismic events, utilized for different applications, devoted to the study of lithosphere-asthenosphere structure in this area. In this paper, results of the study of anisotropy by seismic wave splitting are shown. Fast polarization directions,

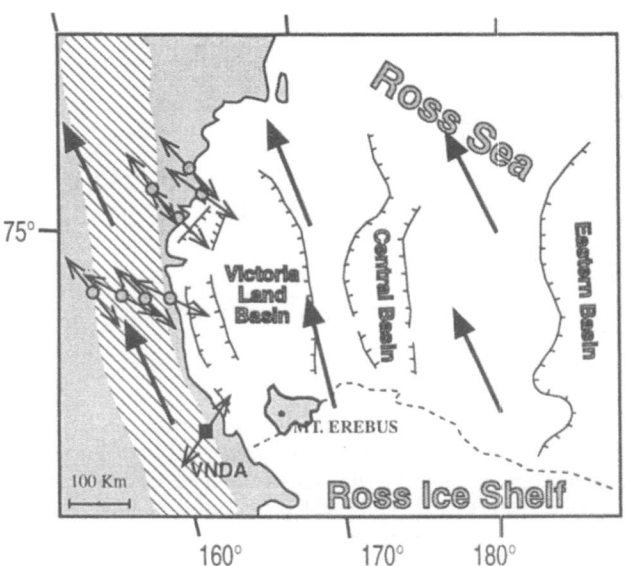

Figure 4
Map of fast polarization directions and the main tectonic features of the studied area. Circles represent our 8 sites while the square is VNDA station, studied by KUBO et al. (1996). Thicker arrows show the APM given by NUVEL1 (GRIPP and GORDON, 1990); the velocity is 1.2 cm/yr. Shaded area corresponds to the Transantarctic Mountains.

obtained by the analysis of SKS and S phases, have approximately a NNW-SSE direction, with slight differences between the two studied areas. A dispersion in the results is found, probably due to the source-side anisotropy given by some of the S phases utilized. It is still unknown whether the crust is anisotropic. At present, the data available to determine its contribution are still insufficient. The seismic phases we studied share a common path in the upper mantle and the delay times we determined are too high to be generated by crustal anisotropy. We therefore suggest that the anisotropy detected beneath the Victoria Land is located in the upper mantle.

The fast polarization direction is close to the extension direction of the Ross Sea system, but it is also close to APM. The rift developed in a Mesozoic-Cenozoic extensional regime, it is still active and, consequently, may be directly related to the relative plate motion in that area. On the basis of these observations it is really difficult, with the data presently available, to identify whether the anisotropy is drag-induced or it is caused by frozen deformation or else it is due to both of these causes. A study, to be developed in the forthcoming future, P-to-S phases recorded by the same stations, would probably better localize the anisotropy depth and thus its origin.

Acknowledgements

We would like to thank Lucia Margheriti and Concetta Nostro for providing the code employed for the calculations. Thanks to Alessandro Amato and the referees for discussions and suggestions. This work was supported by P.N.R.A.—1a.2 (Progetto Nazionale di Ricerche in Antartide).

REFERENCES

BARRUOL, G., and KERN, H. (1993), *Seismic Anisotropy and Shear-wave Splitting in Lower Crust and Upper Mantle Rocks from the Ivrea Zone–Experimental and Calculated Data*, Phys. Earth Planet. Inter. *95*, 175–194.

BEHRENDT, J. C., LeMASURIER, W. E., COOPER, A. K., TESSENSOHN, A., TREHU, A., and DAMASKE, D. (1991), *Geophysical Studies of the West Antarctic Rift System*, Tectonics *10*, 1257–1253.

CIMINI, G. B., AMATO, A., CERRONE, M., CHIAPPINI, M., DI BONA, M., and PONDRELLI, S. (1995), *Passive Seismological Studies in the Terra Nova Bay Area (Antarctica): First Results from the 1993–94 Expedition*, Terra Antarctica *2*, 81–94.

COOPER, A. K., DAVEY, F. J., and HINZ, K., *Crustal extension and origin of sedimentary basins beneath the Ross Sea and the Ross Ice Shelf, Antarctica*. In *Geological Evolution of Antarctica* (eds. Thompson, M. R. A., Crame, J. A., and Thomson, J. W.) (Cambridge Univ. Press, Cambridge, New York 1989).

CRAMPIN, S., and BOOTH, D. C. (1985), *Shear-wave Polarizations near the North Anatolian Fault, II. Interpretation in Terms of Crack-induced Anisotropy*, Geophys. J. R. Astron. Soc. *83*, 75–92.

DELLA VEDOVA, B., PELLIS, G., TREY, H., ZHANG, J., COOPER, A. K., MAKRIS, J., and ACRUP WORKING GROUP (1997), *Crustal Structure of the Transantarctic Mountains, Western Ross Sea*, I.S.A.E.S. 1995 Proceedings, Terra Antarctica, in press.

DE METS, C., GORDON, R. G., ARGUS, D. F., and STEIN, S. (1990), *Current Plate Motions*, Geophys. J. Int. *101*, 425–478.

DI BONA, M., AMATO, A., AZZARA, R., CIMINI, G. B., COLOMBO, D., and PONDRELLI, S. (1997), *Constraints on the Lithospheric Structure beneath the Terra Nova Bay Area from Teleseismic P-to-S Conversions*, I.S.A.E.S. 1995 Proceedings, Terra Antarctica, in press.

DE METS, C., GORDON, R. G., ARGUS, D. F., and STEIN, S. (1990), *Current Plate Motions*, Geophys. J. Int. *101*, 425–478.

DI BONA, M., AMATO, A., AZZARA, R., CIMINI, G. B., COLOMBO, D., and PONDRELLI, S. (1997), *Constraints on the Lithospheric Structure beneath the Terra Nova Bay Area from Teleseismic P-to-S Conversions*, I.S.A.E.S. 1995 Proceedings, Terra Antartctica, in press.

GRIPP, A. E., and GORDON, R. G. (1990), *Current Plate Velocities Relative to the Hot Spots Incorporating the NUVEL-1 Global Plate Motion Model*, Geophys Res. Lett. *17*, 1109–1112.

HELFFRICH, G., SILVER, P., and GIVEN, H. (1994), *Shear-wave Splitting Variation over Short Spatial Scales on Continents*, Geophys. J. Int. *119*, 561–573.

KUBO, A., HIRAMATSU, Y., and KANAO, M. (1996), *Shear-wave Anisotropy by SKS Splitting in Antarctica*, 30th International Geological Congress, Beijing, Abstract volume.

MAINPRICE, D., and SILVER, P. G. (1993), *Interpretation of SKS Waves Using Samples from the Subcontinental Lithosphere*, Phys. Earth Planet. Int. *78*, 257–80.

MAINPRICE, D., SILVER, P. G., and NICOLAS, A. (1993), Seismic Anisotropy in the Mantle from Petrofabrics, EOS, Trans. Am. Geophys. Union 74, 203.

MARGHERITI, L., NOSTRO, C., COCCO, M., and AMATO, A. (1996), *Seismic Anisotropy beneath the Northern Apennines (Italy) and its Tectonic Implications*, Geophys. Res. Lett. *23*, 2721–2724.

NICOLAS, A., and CHRISTENSEN, N. I., *Formation of anisotropy in upper mantle peridotites—A review*, In *Composition, Structure and Dynamics of the Lithosphere-Asthenosphere System* (eds. Fuchs, K., and Froidevauz, C.) (A. G. U., Washington D. C. 1987) pp. 111–123.

PLOMEROVÁ, J., SILENY, J., and BABUŠKA, V. (1996), *Joint Interpretation of Upper Mantle Anisotropy Based on Teleseismic P-travel Time Delays and Inversion of Shear-wave Splitting Parameters*, Phys. Earth Planet. Inter. *97*, 293–309.

PONDRELLI, S., AMATO, A., CHIAPPINI, M., CIMINI, G. B., COLOMBO, D., and DI BONA, M. (1997), *ACRUP1 Geotraverse: Contribution of Teleseismic Data Recorded on Land*, I.S.A.E.S. 1995 Proceedings, Terra Antarctica, in press.

ROULT, G., and ROULAND, D. (1994), *Antarctica I: Deep Structure Investigations Inferred from Seismology: A Review*, P.E.P.I. *84*, 15–32.

SAVAGE, M. K., SILVER, P. G., and MEYER, R. P. (1990), *Observations of Teleseismic Shear-wave Splitting in the Basin and Range from Portable and Permanent Stations*, Geophys. Res. Lett. *17*, 21–24.

SAVAGE, M. K., and SILVER, P. G. (1993), *Mantle Deformation and Tectonics: Constraints from Seismic Anisotropy in Western United States*, Phys. Earth Planet. Inter. *78*, 207–227.

SILVER, P. G., and CHAN, W. W. (1988), *Implications for Continental Structure and Evolution from Seismic Anisotropy*, Nature *335*, 34–39.

SILVER, P. G. (1996), *Seismic Anisotropy beneath the Continents: Probing the Depths of Geology*, Ann. Rev. Earth Planet. Sci. *24*, 385–432.

STERN, T. A., and TEN BRINK, U. S. (1989), *Flexural Uplift of the Transantarctic Mountains*, J. Geophys. Res. *96*, 16429–16454.

VINIK, L. P., MAKEYEVA, L. I., MILEV, A., and YU. USENKO, A. (1992), *Global Patterns of Azimuthal Anisotropy and Deformations in the Continental Mantle*, Geophys. J. Int. *111*, 433–447.

ZHANG, Z., and SCHWARTZ, S. Y. (1994), *Seismic Anisotropy in the Shallow Crust of the Loma Prieta Segment of the San Andreas Fault System*, J. Geophys Res. *99*, 9651–9661.

(Received October 9, 1996, revised/accepted June 20, 1997)

To access this journal online:
http://www.birkhauser.ch

Pure appl. geophys. 151 (1998) 443–462
0033–4553/98/040443–20 $ 1.50 + 0.20/0

Pure and Applied Geophysics

Anisotropy of the Yellowstone Hot Spot Wake, Eastern Snake River Plain, Idaho

DEREK SCHUTT,[1] EUGENE D. HUMPHREYS[1] and KEN DUEKER[2]

Abstract—Over the last 10 million years, the Yellowstone hot spot has passed beneath the eastern Snake River Plain, both magmatically modifying the Snake River Plain crust and creating a wider, wake-like "tectonic parabola" of seismicity and uplift. Analysis of SKS arrivals to a line array of 55 mostly broadband stations, distribution across the tectonic parabola, reveals a nearly uniform orientation of anisotropy, with an average fast axis orientation of N64E. The back azimuth of null splitting events is parallel to the measured fast axis, suggesting that anisotropic material consists of a single layer. Splitting parameters are independent of backazimuth, suggesting that anisotropy is constant beneath each station. Thus station-averaged split parameters are representative of the anisotropy beneath the station. Station-averaged split times range from 0.6–1.5 s, and define a pronounced depression in split times centered about 80 km southeast of the axis of the Snake River Plain.

Assuming the degree of anisotropy (averaged over the ray path) to be no more than 10%, the split times are far too great for the anisotropy to be confined solely to the lithosphere. The simplest way to explain the observed anisotropy structure is to attribute it to simple shear strain caused by the absolute motion of North America. Because anisotropy is different in nearby Colorado and Nevada, we hypothesize that fossil anisotropy created in past orogens and continent-building events in the Snake River Plain area has been reset or erased by the passage of the hot spot, and that subsequent strain of the hot spot-related asthenospheric wake created a uniformly oriented fast axis. If this is true, then our array constrains the minimum of the hot spot's asthenospheric wake.

Key words: Shear wave splitting, hot spot, mantle plume, anisotropy, yellowstone, shear wave anisotropy, SKS splitting.

Introduction

Several observations suggest the hot spot currently exciting Yellowstone magmatism has passed beneath the eastern Snake River Plain (eSRP) of southern Idaho. These include: 1) a series of time progressive rhyolitic calderas younging from the southwest to the northeast (SMITH and BRAILE, 1994); 2) a seismically inferred mid-crustal basaltic sill beneath the rhyolitic calderas (SPARLIN *et al.*, 1982); 3) a geoid high centered on Yellowstone that is thought to be caused by a deep low density anomaly (MILBERT, 1991); 4) He^3/He^4 values indicative of

[1] Department of Geological Sciences, 1272 University of Oregon; Eugene, OR 97403-1272, U.S.A. e-mail: schuttd@newberry.uoregon.edu.
[2] CIRES, University of Colorado at Boulder, Boulder, CO 80309-0449, U.S.A.

degassing primitive mantle (HEARN *et al.*, 1990); and 5) a parabolic-shaped region of heightened elevation, seismicity, and faulting, which is thought to have formed as hot spot mantle buoyantly flattened beneath North America (ANDERS and SLEEP, 1992; PIERCE and MORGAN, 1992; RIBE and CHRISTENSEN; SMITH and BRAILE, 1994).

To address the mantle structure and dynamics of this system, the Program for the Array Seismic Studies of the Continental Lithosphere (PASSCAL) supported the deployment of a 470 km long SW-trending line array of 55 mostly broadband stations across the width of the tectonic parabola in a line perpendicular to the path of the hot spot, crossing the eSRP at the location where the hot spot was active about 8 Ma (PIERCE and MORGAN, 1992; Fig. 1). These stations collected about 375 teleseismic events, which were used for an inte-grated *P* wave and *S* wave, and receiver function study of the upper mantle and crust in this area. Information about the mantle structure below this region can provide clues to understanding the mechanisms driving the hot spot, as well as giving insight into how variations in the physical state of the mantle have created the topography in the eSRP. This paper will be concentrated on the *S*-wave splitting component of our studies, which may be especially useful in providing information on the strain evolution of this hot spot asthenosphere.

Shear-wave Splitting Results

Data and Method

Over the 200-day life of the experiment, 59 shear-wave events were recorded. Of these, 32 had significant radial SKS energy that was not polluted by other phases. A record section of one of these SKS events is shown in Figure 2. In this case SKS phase is the first arriving, although we also measured splits on events where this was not the case.

We measured 141 sets of SKS split parameters. In addition, 30 events came in from backazimuths that produced no *SH* energy. These "null splitting" events occur when the ray travels along the fast or slow axes; these null events also constrain the orientation of the axes, although they do not provide any information on the split time.

We estimate anisotropy beneath our array by applying the tangential energy minimization technique of SILVER and CHAN (1991) to SKS arrivals. This method determines splitting parameters (split time and the horizontal projection of fast axis orientation) for each split observation by seeking transverse energy minima (Fig. 3—Top). As is standard in this analysis, transverse anisotropy and hexagonal symmetry are assumed. We show below that these assumptions are reasonable.

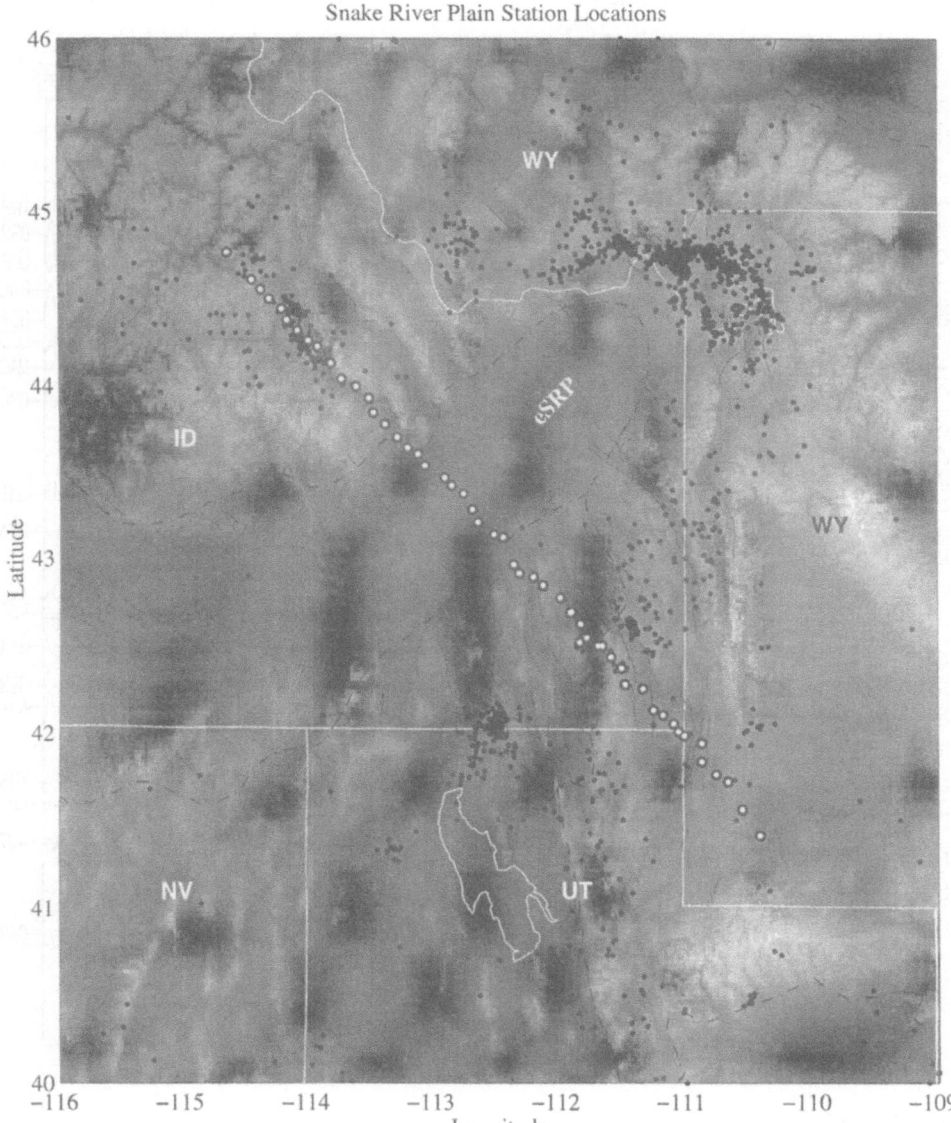

Figure 1

1993 Snake River Plain station locations. The array consisted of 55 mostly broadband seismometers stationed along a line 470 km long that runs perpendicular to the axis of the eastern Snake River Plain (eSRP), crossing the track of the hot spot near the location where magmatism was active ∼8 Ma. Earthquakes are shown with black dots, and elevation indicated with shading. The tectonic parabola is defined by the parabolic concentration of earthquakes surrounding the eSRP in a wake-like fashion, with Yellowstone lying in the seismically active NW corner of Wyoming (WY).

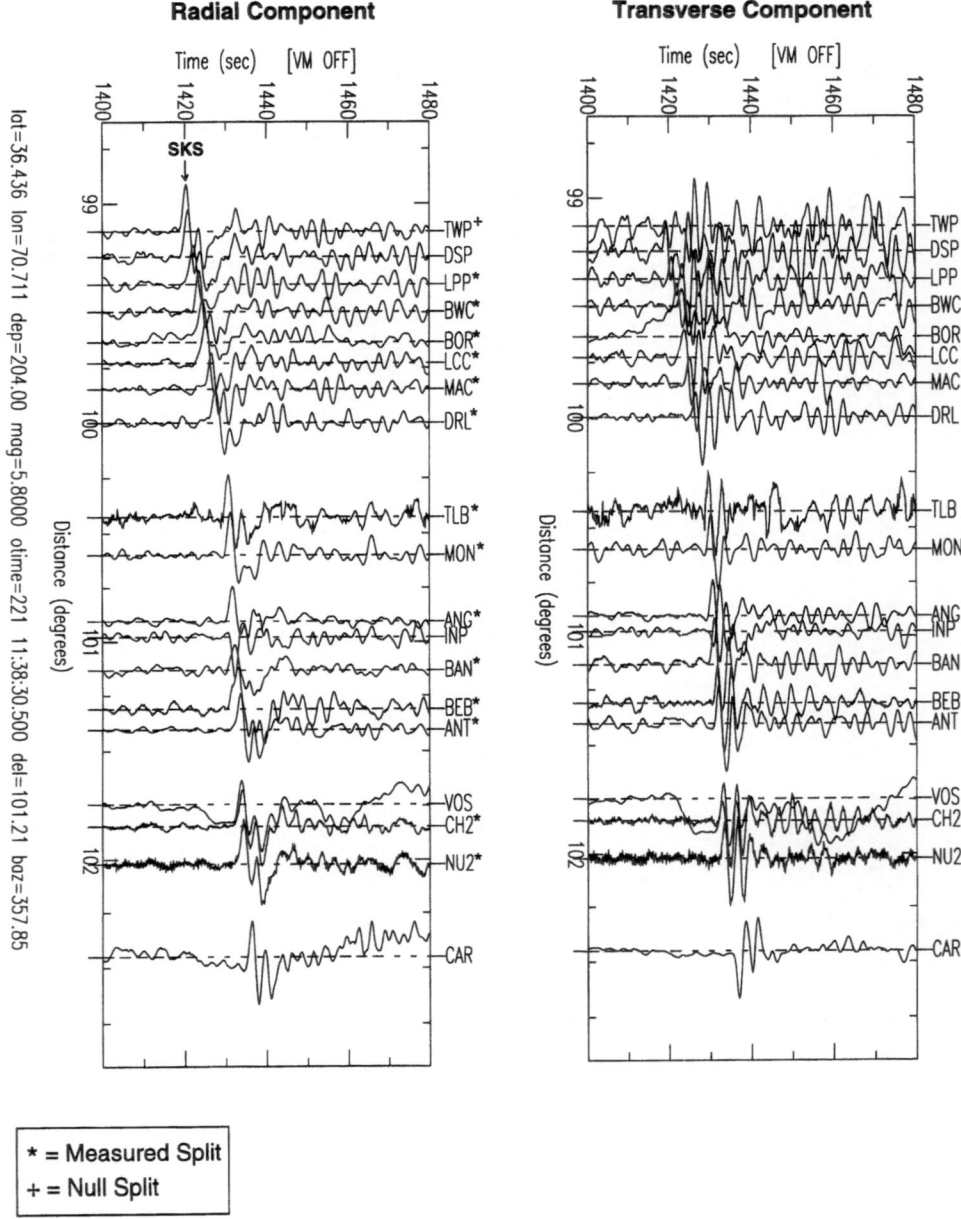

* = Measured Split
+ = Null Split

Figure 2

A record section from an average SKS event. This is certainly not the best event we received, but neither is it the worst of the 32 events with clear SKS energy. Traces are scaled individually, so that maximum amplitude displayed is the same from trace to trace. In general, for this event, the signal-to-noise ratio is lower for the transverse, by a factor of approximately 2–4. Good splits are those with split time errors of under 0.6 s. Although the single null split looks like it has energy on the transverse component, this particular trace is highly amplified. If one follows the moveout curve, there is no clear arrival at 1419 s, where one would expect to find the initial SKS energy. We also used the adjusted transverse energy plot for each station to help determine whether a split is null or not. If the transverse energy is minimum along the backazimuth of the arrival, this provides a good indication that the transverse energy seen is just noise. Of course the final determination is subjective, which motivated us to do the bulk statistical analysis which led to the station averages.

Example of Transverse Energy on a Measured Split

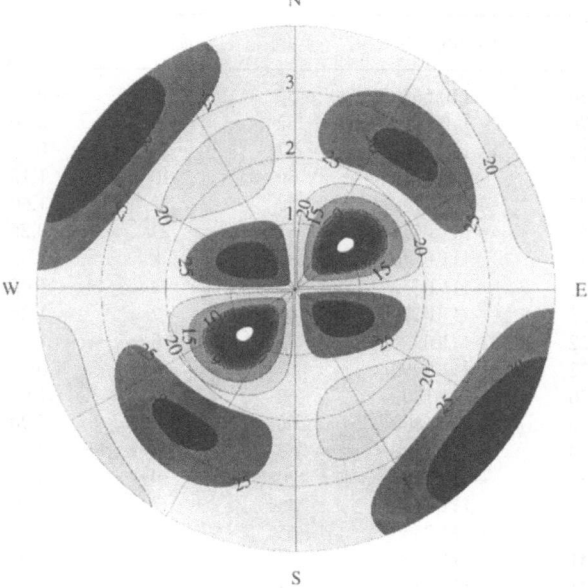

Example of Transverse Energy on a Null Split

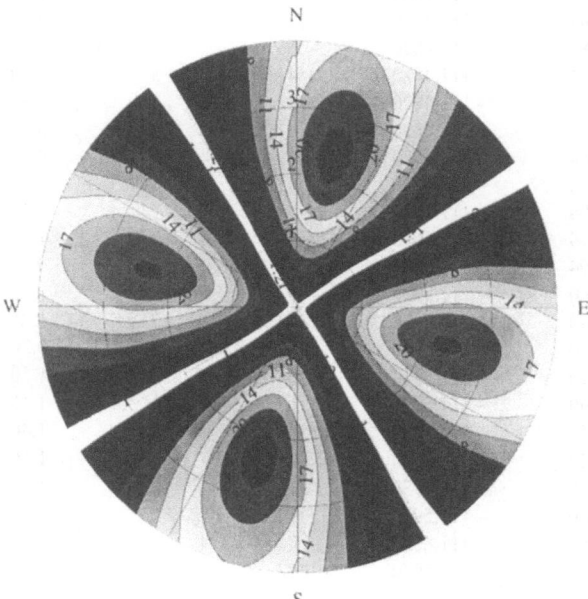

Figure 3

Plots of transverse energy as a function of fast axis orientation and split time. The minimum corrected energy gives the maximum likelihood split parameters. Radial distance is split time in seconds and angle is orientation of the fast axis. Energy levels are in relative units, normalized so that the 95% confidence bounds are indicated by the white area. (Top) Example for a split arrival. (Bottom) Example for a null split arrival.

Table 1

Raw SKS splitting data

Station	Lat. (°)	Long. (°)	ϕ (°)	ϕ_{err} (°)	δt (s)	δt_{err} (s)
CAV	41.7170	−110.6650	53	20	0	0
CAV	41.7170	−110.6650	62	16	0.70	0.20
NUG	41.8311	−110.8740	57	25	0	0
NU2	41.8319	−110.8740	50	10	1.15	0.25
LDC	41.9375	−110.8730	98	2	1.60	0.30
CH2*	42.0058	−111.0570	82	22	0.25	0.32
CH2	42.0058	−111.0570	34	8	1.05	0.12
CH2	42.0058	−111.0570	49	2	1.05	0.07
CH2	42.0058	−111.0570	41	2	1.85	0.25
DAH	42.0522	−111.0960	42	10	1.20	0.35
DAH	42.0522	−111.0960	67	9	0	0
VOS	42.1030	−111.1800	59	6	1.05	0.18
VOS	42.1030	−111.1800	66	9	0.90	0.10
VOS	42.1030	−111.1800	56	10	0.90	0.28
VOS	42.1030	−111.1800	70	30	0	0
MEM	42.1317	−111.2560	45	2	1.40	0.10
MEM	42.1317	−111.2560	65	11	0.65	0.12
AIR	42.2541	−111.3390	77	12	0.95	0.28
AIR	42.2541	−111.3390	58	18	1.25	0.50
MIC	42.2784	−111.4820	57	7	1.45	0.28
MIC	42.2784	−111.4820	49	5	1.05	0.13
MIC*	42.2784	−111.4820	123	2	2.30	0.23
MIC	42.2784	−111.4820	66	2	1.20	0.18
MIC	42.2784	−111.4820	44	5	1.30	0.30
MIC	42.2784	−111.4820	42	4	1.30	0.20
MIC	42.2784	−111.4820	41	8	0.80	0.15
MIC	42.2784	−111.4820	69	16	0	0
EMC	42.3694	−111.5090	41	6	1.30	0.30
EMC	42.3694	−111.5090	47	4	1.40	0.45
EMC	42.3694	−111.5090	47	7	1.40	0.30
EMC	42.3694	−111.5090	70	30	0	0
ANT	42.4340	−111.5930	63	4	1.30	0.40
ANT	42.4340	−111.5930	63	2	1.35	0.10
ANT	42.4340	−111.5930	69	8	1.20	0.38
ANT	42.4340	−111.5930	68	2	1.50	0.10
ANT	42.4340	−111.5930	68	11	1.15	0.25
ANT	42.4340	−111.5930	46	6	1.40	0.35
ANT	42.4340	−111.5930	55	8	1.10	0.20
ANT	42.4340	−111.5930	73	12	1.10	0.35
ANT	42.4340	−111.5930	74	4	1.60	0.28
NIT	42.4987	−111.6950	73	7	1.10	0.20
NIT	42.4987	−111.6950	73	6	1.30	0.40
NIT	42.4987	−111.6950	67	3	1.80	0.20
NIT	42.4987	−111.6950	70	10	0.90	0.20
NI2	42.4991	−111.6620	62	22	0	0
BEB	42.5201	−111.8450	49	4	1.50	0.18
BUC	42.6277	−111.8360	72	15	0	0
BUC	42.6277	−111.8360	68	6	1.00	0.10
BUC	42.6277	−111.8360	1	4	2.35	0.20

Table 1 (*continued*)

Station	Lat. (°)	Long. (°)	ϕ (°)	ϕ_{err} (°)	δt (s)	δt_{err} (s)
BA2	42.6909	−111.9190	60	19	0.30	0.15
BAN	42.6972	−111.9080	50	3	1.50	0.10
BAN	42.6972	−111.9080	77	16	1.20	0.50
BAN	42.6972	−111.9080	47	10	1.10	0.30
PEB*	42.7758	−111.9970	101	4	1.00	0.18
PEB	42.7758	−111.9970	70	4	1.25	0.12
INP	42.8457	−112.1340	34	2	2.30	0.23
MCN	42.8929	−112.2110	65	7	1.25	0.17
MCN	42.8929	−112.2110	66	4	0.80	0.10
MCN	42.8929	−112.2110	81	30	0	0
MCN	42.8929	−112.2110	81	3	1.80	0.10
MCN	42.8929	−112.2110	85	5	1.20	0.20
ANG	42.9136	−112.3240	89	12	0.70	0.17
ANG	42.9136	−112.3240	61	12	0.80	0.20
ANG	42.9136	−112.3240	68	4	1.25	0.12
ANG	42.9136	−112.3240	81	4	1.00	0.20
ANG	42.9136	−112.3240	49	6	0.95	0.18
ANG	42.9136	−112.3240	46	4	1.10	0.10
ANG	42.9136	−112.3240	57	16	0.55	0.20
ANG	42.9136	−112.3240	62	15	0	0
ANG	42.9136	−112.3240	88	7	0.90	0.20
THC	42.9618	−112.3680	78	9	1.25	0.35
THC	42.9618	−112.3680	56	10	1.15	0.20
THC	42.9618	−112.3680	65	7	1.00	0.20
FBU	42.1376	−112.5260	71	22	0	0
FBU	42.1376	−112.5260	75	4	0.95	0.10
FBU	42.1376	−112.5260	81	4	1.35	0.25
MON	43.2104	−112.6510	42	8	0.70	0.10
MON	43.2104	−112.6510	76	9	0.90	0.22
MON	43.2104	−112.6510	77	3	1.55	0.20
MON	43.2104	−112.6510	65	8	0.60	0.10
MON	43.2104	−112.6510	60	8	1.15	0.35
MON	43.2104	−112.6510	78	14	1.15	0.28
TAB	43.2869	−112.6960	59	6	1.05	0.23
TAB	43.2869	−112.6960	65	5	1.20	0.20
TAB	43.2869	−112.6960	85	10	0.60	0.20
TAB	43.2869	−112.6960	68	16	0	0
TLB	43.3791	−112.7670	45	9	1.00	0.28
TLB	43.3791	−112.7670	74	12	0.70	0.18
TLB	43.3791	−112.7670	60	5	0.90	0.15
TLB	43.3791	−112.7670	66	4	0.95	0.12
ATO	43.4259	−112.8650	57	12	1.30	0.30
ATO	43.4259	−112.8650	76	7	0	0
ILS	43.4722	−112.9220	60	7	1.80	0.40
ILS	43.4722	−112.9220	63	6	1.50	0.20
ILS	43.4722	−112.9220	79	6	1.25	0.18
ILS	43.4722	−112.9220	72	9	0	0
ILS	43.4722	−112.9220	58	6	1.35	0.18
ILS	43.4722	−112.9220	65	4	1.45	0.18
ILM	43.5417	−113.0810	67	25	0	0

Table 1 (*continued*)

Station	Lat. (°)	Long. (°)	ϕ (°)	ϕ_{err} (°)	δt (s)	δt_{err} (s)
ILM	43.5417	−113.0810	73	9	0	0
ILM	43.5417	−113.0810	67	2	1.80	0.10
ILM	43.5417	−113.0810	72	12	0	0
ILM	43.5417	−113.0810	66	3	1.40	0.12
ILN	43.6085	−113.1370	61	10	1.55	0.33
ILN	43.6085	−113.1370	63	6	1.50	0.20
ILN	43.6085	−113.1370	71	5	1.50	0.20
ARH	43.6444	−113.2190	56	2	2.00	0.18
ARH	43.6444	−113.2190	67	4	1.25	0.17
MOR	43.7050	−113.3030	69	4	1.30	0.20
MOR	43.7050	−113.3030	64	3	1.50	0.10
MOR	43.7050	−113.3030	60	2	1.75	0.12
DRL	43.7793	−113.4010	59	6	2.05	0.48
DRL	43.7793	−113.4010	60	22	0	0
DRL	43.7793	−113.4010	53	2	1.25	0.07
DRL	43.7793	−113.4010	58	5	1.10	0.15
LES	43.8451	−113.4960	67	3	1.85	0.28
LES	43.8451	−113.4960	69	2	1.75	0.18
LES	43.8451	−113.4960	51	10	1.50	0.48
LES	43.8451	−113.4960	57	14	0	0
LES	43.8451	−113.4960	70	25	0	0
LES	43.8451	−113.4960	58	14	1.15	0.35
MAC	43.9265	−113.5300	71	30	0	0
MAC	43.9265	−113.5300	64	5	1.35	0.27
MCR	43.9940	−113.6340	40	9	1.30	0.20
MCR	43.9940	−113.6340	61	5	1.75	0.30
MCR	43.9940	−113.6340	56	4	1.65	0.17
MCR	43.9940	−113.6340	47	5	1.80	0.18
MCR	43.9940	−113.6340	62	4	1.20	0.20
MCR	43.9940	−113.6340	60	14	0	0
LCC	44.0366	−113.7510	61	2	1.50	0.15
LCC	44.0366	−113.7510	62	14	0	0
LCC	44.0366	−113.7510	60	3	1.35	0.15
LCC	44.0366	−113.7510	69	9	0.80	0.10
LCC	44.0366	−113.7510	82	2	1.30	0.10
LCC	44.0366	−113.7510	78	16	0.75	0.25
LCC	44.0366	−113.7510	62	2	1.80	0.12
BOR	44.1266	−113.8370	74	6	1.50	0.20
BOR	44.1266	−113.8370	53	10	1.25	0.25
DKP	44.2220	−113.9460	68	4	1.40	0.23
DKP	44.2220	−113.9460	50	10	1.20	0.30
DKP	44.2220	−113.9460	62	8	1.00	0.22
DKP	44.2220	−113.9460	63	4	1.30	0.15
BWC	44.2556	−114.0200	72	15	0	0
BWC	44.2556	−114.0200	73	4	1.35	0.15
BWC	44.2556	−114.0200	57	10	1.40	0.30
SPC	44.3146	−114.1060	80	4	0.95	0.20
SPC	44.3146	−114.1060	66	8	0	0
LPP	44.3736	−114.1930	1	4	2.60	0.33
LPP	44.3736	−114.1930	84	5	1.60	0.20

Table 1 (*continued*)

Station	Lat. (°)	Long. (°)	ϕ (°)	ϕ_{err} (°)	δt (s)	δt_{err} (s)
LPP	44.3736	−114.1930	69	15	0	0
BIC	44.4372	−114.2340	71	5	1.10	0.23
BIC	44.4372	−114.2340	80	7	1.10	0.20
BIC	44.4372	−114.2340	63	12	0.80	0.30
BIC	44.4372	−114.2340	58	9	0.70	0.10
DSP	44.4949	−114.3330	81	5	1.00	0.10
DSP	44.4949	−114.3330	79	7	1.30	0.20
DSP	44.4949	−114.3330	48	7	1.40	0.20
DSP	44.4949	−114.3330	90	10	0	0
DSP	44.4949	−114.3330	40	20	0.60	0.27
MQR*	44.5490	−114.4020	83	4	4.00	0.05
TWP	44.6031	−114.4720	75	2	1.45	0.18
TWP	44.6031	−114.4720	62	14	0	0
TWP	44.6031	−114.4720	82	10	0	0
TWP	44.6031	−114.4720	54	22	0.55	0.25
TWP	44.6031	−114.4720	83	8	0.95	0.20
TWP	44.6031	−114.4720	76	17	1.00	0.40
TWP	44.6031	−114.4720	76	3	1.40	0.10
SDM	44.7622	−114.6660	71	22	0.45	0.18
SDM	44.7622	−114.6660	71	9	0.90	0.20
SDM	44.7622	−114.6660	85	6	1.20	0.20

Raw splitting data: all measurements with $\delta t \leq 0.5$. ϕ_{err} and δt_{err} are 95% confidence bounds. A δt of 0 indicates a null split. Based on transverse energy patterns, we believe four of our split estimates are incorrect. These are marked with an asterisk.

Results

Results are tabulated in Table 1 and shown in Figure 4. Thin lines in this figure indicate the orientation and split time of the measured split; thick lines indicate the back azimuth of the null splits (their lengths have been scaled to 1.0 s). Fast axis orientations are nearly constant across the array, with little dependence on station location or back azimuth. The angle histogram in Figure 4 emphasizes the narrow range in fast axis orientation of about 30°.

The relative consistency in splitting parameters with back azimuth allows us to discount Core Mantle Boundary (CMB) splitting as a source of the observed SKS splits (because CMB splitting would produce split parameters that vary with back azimuth) and to conclude that there is no general trend of strongly dipping fast axes across our array (also suggested by the absence of variation in split parameters with back azimuth, Fig. 5, SILVER and SAVAGE, 1994). This latter observation leads us to make the considered assumption that the fast axis is transverse in the eSRP.

The null splits, in addition to constraining the fast axis orientation into one of four orientation bands (Fig. 3—Bottom), offer important information on the simplicity of the anisotropy orientation structure. If orientation structure varies with depth, null splits would not occur (except in the case of layers of orthogonally oriented anisotropy). Thus the null splits indicate that the mantle beneath the eSRP is comprised of a single layer of transversely anisotropic material.

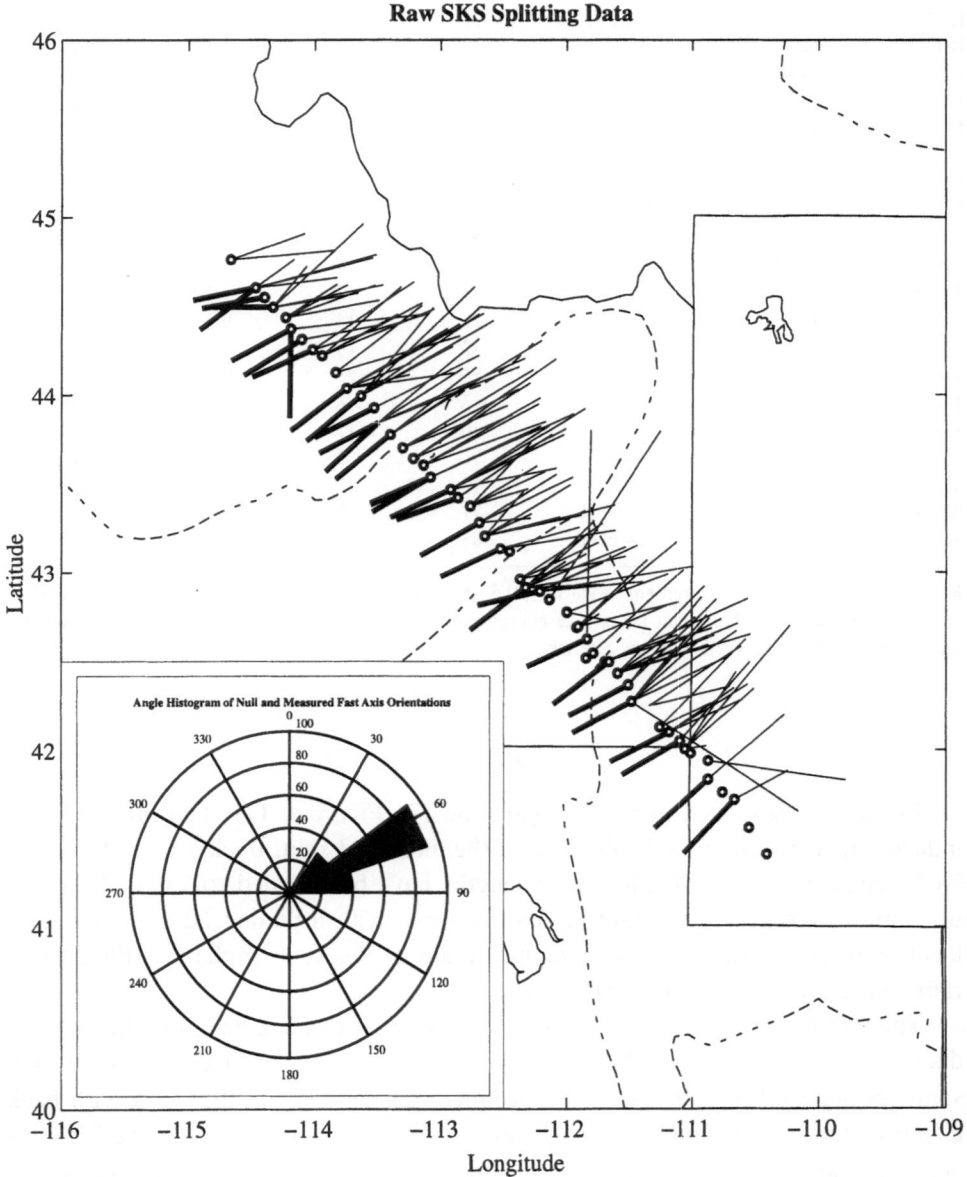

Figure 4

Raw-splitting results. Measured splits are shown with thin lines extending to the northeast, and nulls are shown with thick lines extending to the southwest. Line orientation shows the fast-axis orientation, and line length is proportional to the split time (which ranges from 0.5–1.5 s); null lengths have been set to a split time of 1 s. Both null and split measurements are limited to an azimuth of $0 < \phi < 180$. Inset is an angle histogram of measured and null orientations.

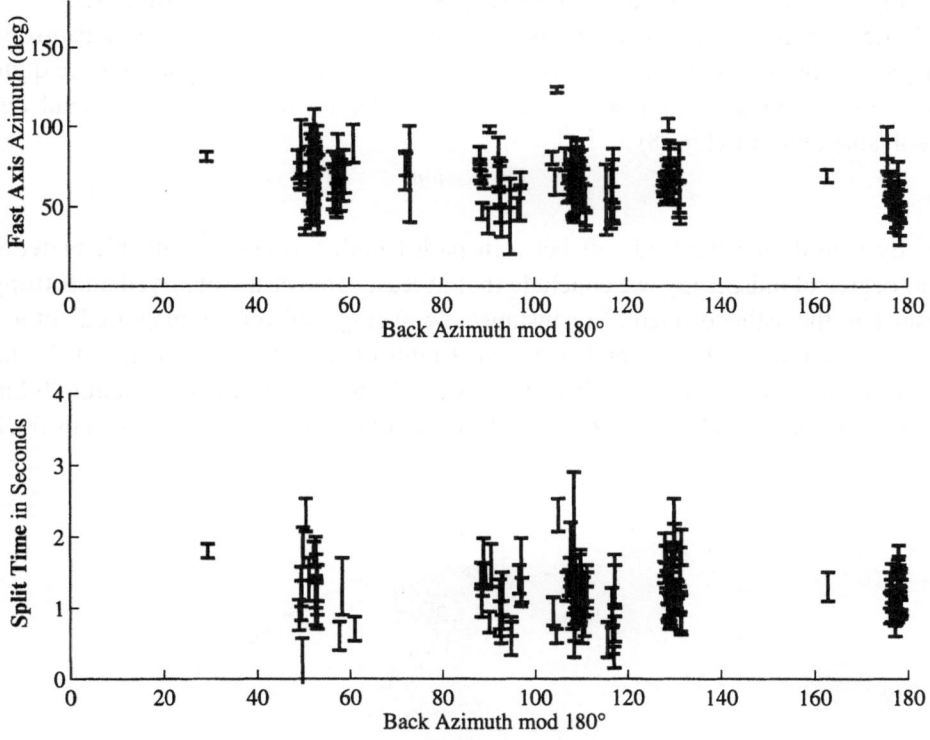

Figure 5

Split parameters with event back azimuth mod 180, using data from Table 1 ($\delta t_{err} < 0.5$ s). We have excluded the event with $\delta t = 4.0$ s. Error bars show the calculated 95% confidence intervals. We see no trend with back azimuth, suggesting that splitting occurs near our array. Actual distribution of back azimuths is void of events with back azimuths from the southeast hemisphere.

Four of our split measurements (of 171) are inconsistent with the other observations. These split determinations have transverse energy plots that appear unusual and whose results are inconsistent with our other estimates. They have still been included in our analysis, and turn out to have no significant effect on our findings.

Station Average

The independence of split parameters with back azimuth implies that anisotropy is essentially homogeneous under each station. When this is the case, it then is valid to average splitting parameters by station. SILVER and CHAN (1991) use F test statistics to determine the uncertainty estimates of split parameters. We extend this technique to station averages, which allows us to incorporate both null and observed splits into our estimates (Fig. 6 and Table 2).

Figure 7 shows station averaged split parameters. Split time averages vary from 0.5–1.5 s across the array, implying significant changes in either anisotropy magnitude or layer thickness across the array. In contrast, the orientations are quite uniform, although the northernmost stations differ in orientation by a small but resolvable amount (Fig. 8).

Discussion

By considering the trade-off between path length through anisotropic material and degree of anisotropy, we conclude that at least some of the observed anisotropy resides in the asthenosphere. For instance, an average anisotropy magnitude of 4% and a split time of 1.25 s implies a path length of nearly 150 km (Fig. 9). Using receiver functions, PENG and HUMPHREYS (1997) find the crust to be about 40 km thick and not strongly anisotropic. Unless the lithospheric mantle is 110 km thick

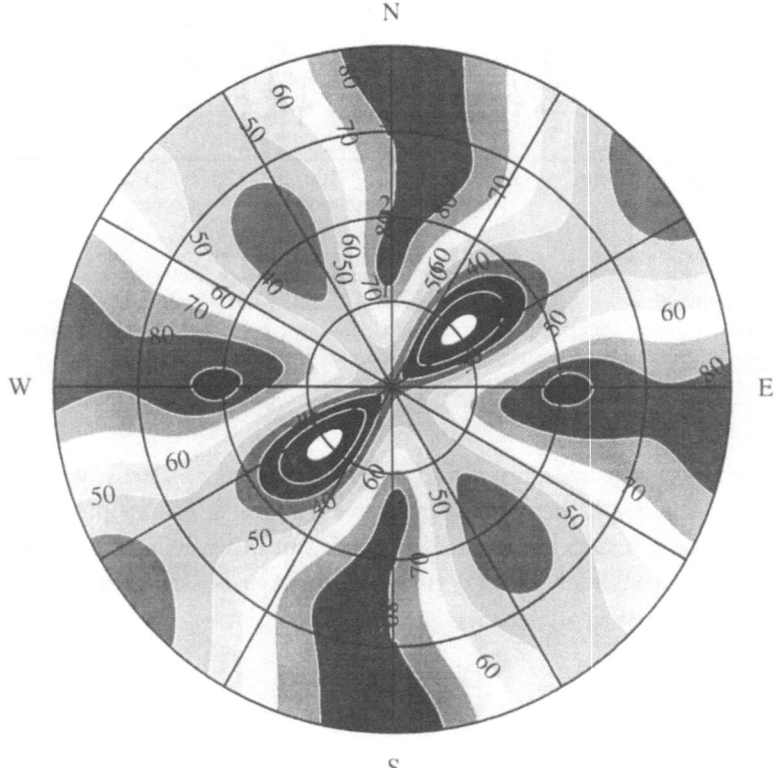

Figure 6

Plot of summed transverse energy for a selected station. Both split and null-split energies contribute to this sum. Four separate split measurements were combined to make this plot. Energy amplitude is arbitrary; an *F* test is applied to this sum to convert summed energy to probability at each point. The white region is an area of greatest likelihood, and represents roughly 95% confidence bounds.

Table 2

Station averaged splitting data

Station	Lat. (°)	Long. (°)	ϕ (°)	ϕ_{err} (°)	δt (s)	δt_{err} (s)	# of files used
CAR	41.4100	−110.4100	−70	−20 +18	2.1	−0.8 +1.0	2
HOG	41.5600	−110.5500	−56	−61 +60	1.4	−2.0 +2.0	1
CAV	41.7170	−110.6650	53	−61 +60	0.6	−1.0 +2.0	2
FOS	41.7579	−110.7600	−47	−61 +60	0.3	−0.3 +0.6	10
NUG	41.8311	−110.8740	57	−7 +07	4.0	−2.0 +2.0	1
NU2	41.8319	−110.8740	50	−20 +16	1.0	−0.4 +0.5	2
LDC	41.9375	−110.8730	10	−36 +60	0.4	−0.4 +1.1	2
CHC	41.9811	−111.0110	79	−25 +25	4.0	−2.0 +1.8	1
CH2	42.0058	−111.0570	47	−7 +07	1.1	−0.2 +0.2	4
DAH	42.0522	−111.0960	67	−22 +19	2.2	−2.0 +2.0	3
VOS	42.1030	−111.1800	58	−9 +09	0.9	−0.3 +0.3	4
MEM	42.1317	−111.2560	41	−13 +14	1.3	−0.4 +0.3	2
AIR	42.2541	−111.3390	69	−19 +28	1.1	−0.5 +0.5	2
MIC	42.2784	−111.4820	53	−13 +13	1.0	−0.4 +0.3	8
EMC	42.3694	−111.5090	52	−10 +09	0.7	−0.6 +0.6	2
ANT	42.4340	−111.5930	64	−7 +06	1.3	−0.3 +0.2	8
NIT	42.4987	−111.6950	64	−19 +19	0.9	−0.7 +0.7	2
NI2	42.4991	−111.6620	62	−13 +13	1.3	−1.5 +2.0	1
BEB	42.5201	−111.8450	49	−14 +13	1.5	−0.4 +0.4	1
BCN	42.5473	−111.7880	65	−20 +36	0.9	−0.6 +0.6	2
BUC	42.6277	−111.8360	−90	−61 +34	0.5	−0.5 +0.8	6
BA2	42.6909	−111.9190	60	−18 +40	0.3	−0.1 +0.1	1
BAN	42.6972	−111.9080	53	−20 +18	1.2	−0.4 +0.5	3
PEB	42.7758	−111.9970	71	−16 +13	1.2	−0.3 +0.2	2
INP	42.8457	−112.1340	34	−8 +06	2.3	−0.7 +0.9	1
MCN	42.8929	−112.2110	66	−14 +16	0.9	−0.3 +0.3	5
ANG	42.9136	−112.3240	66	−18 +21	0.8	−0.4 +0.3	11
THC	42.9618	−112.3680	70	−17 +17	1.0	−0.4 +0.4	4
FBU	43.1376	−112.5260	−88	−35 +38	0.7	−0.4 +0.5	4
MON	43.2104	−112.6510	71	−16 +13	0.8	−0.2 +0.4	7
TAB	43.2869	−112.6960	70	−18 +38	0.8	−0.5 +0.4	4
TLB	43.3791	−112.7670	55	−15 +14	0.8	−0.3 +0.2	3
ATO	43.4259	−112.8650	−47	−61 +60	0.2	−0.2 +0.4	2
ILS	43.4722	−112.9220	70	−10 +10	1.3	−0.4 +0.3	4
ILM	43.5417	−113.0810	67	−6 +05	1.6	−0.3 +0.2	7
ILN	43.6085	−113.1370	61	−20 +21	1.6	−0.7 +0.7	1
ARH	43.6444	−113.2190	60	−9 +07	1.6	−0.5 +0.6	3
MOR	43.7050	−113.3030	60	−7 +06	1.7	−0.3 +0.3	3
DRL	43.7793	−113.4010	57	−10 +08	1.1	−0.2 +0.3	5
LES	43.8451	−113.4960	65	−8 +07	1.7	−0.5 +0.5	8
MAC	43.9265	−113.5300	65	−13 +11	1.5	−0.7 +0.7	3
MCR	43.9940	−113.6340	60	−9 +08	1.5	−0.4 +0.3	6
LCC	44.0366	−113.7510	63	−8 +07	1.5	−0.4 +0.3	7
BOR	44.1266	−113.8370	64	−15 +13	1.5	−0.6 +0.6	3
DKP	44.2220	−113.9460	66	−12 +10	1.3	−0.4 +0.3	3
BWC	44.2556	−114.0200	66	−9 +09	1.2	−0.6 +0.6	5
SPC	44.3146	−114.1060	72	−16 +15	0.9	−0.6 +0.5	2
LPP	44.3736	−114.1930	75	−14 +14	1.3	−0.6 +0.7	4
BIC	44.4372	−114.2340	72	−12 +10	1.0	−0.3 +0.3	3

Table 2 *(continued)*

Station	Lat. (°)	Long. (°)	ϕ (°)	ϕ_{err} (°)	δt (s)	δt_{err} (s)	# of files used
DSP	44.4949	−114.3330	78	−14 +13	0.9	−0.5 +0.4	5
MQR	44.5490	−114.4020	83	−61 +60	4.0	−2.0 +2.0	1
TWP	44.6031	−114.4720	77	−12 +11	0.9	−0.4 +0.4	6
SDM	44.7622	−114.6660	78	−49 +21	0.7	−0.5 +0.5	2

Station averaged splitting data. Using the principle that the sum of χ^2 variables is itself a χ^2 variable, and following the assumption of SILVER and CHAN (1991) that the transverse energy is nearly a χ^2 variable; we sum corrected transverse energies $E_t(\phi, \delta t)$ by station. ϕ_{err} and δt_{err} are rough 95% confidence bounds. # of files used column indicates number of corrected transverse energy files (split measurements) summed to obtain the station average. We note that this technique allows us to incorporate more poorly constrained measurements than in Table 1, hence the data in Tables 1 and 2 are not 1-1. In general the data in Table 1 are a subset of the data used to make Table 2, except for six misplaced files which are in Table 1 but not in Table 2.

or more—a huge amount given that the array is situated off the craton, in a hot and extensional environment—some of the anisotropy measured (and likely most of it) must be caused by anisotropic asthenosphere.

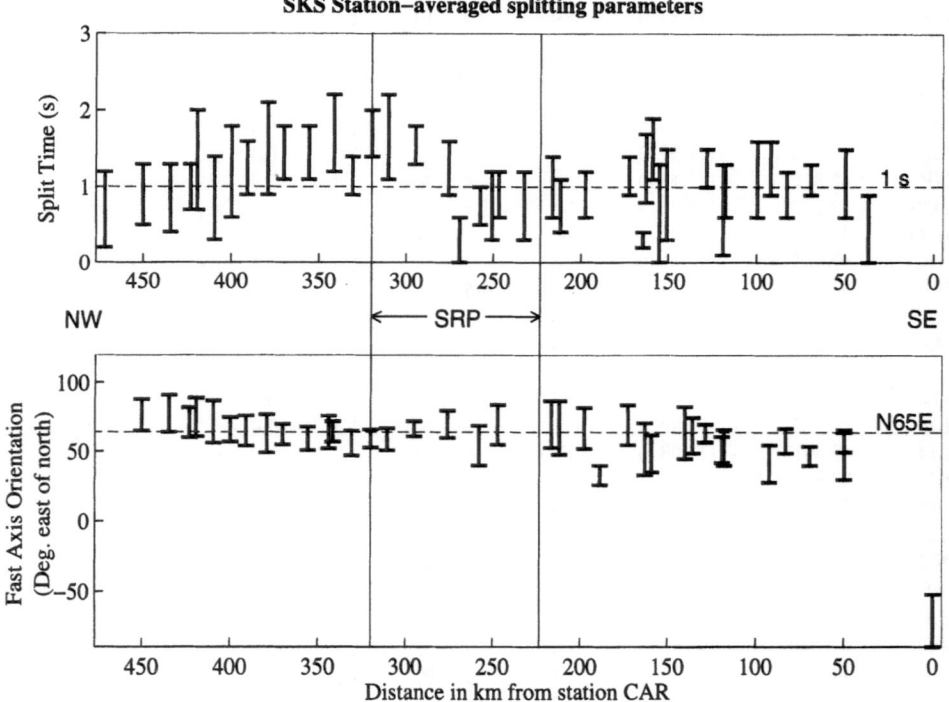

Figure 7

Station-averaged splitting results, obtained from the *F*-test results derived from plots exemplified by Figure 5. A depression in split time is seen, centered southeast of the axis of the Snake River Plain (SRP). Results with major errors have not been plotted: total $\delta t_{err} > 1.4$ s or total $\phi_{err} > 40°$ (all results tabulated in Table 2).

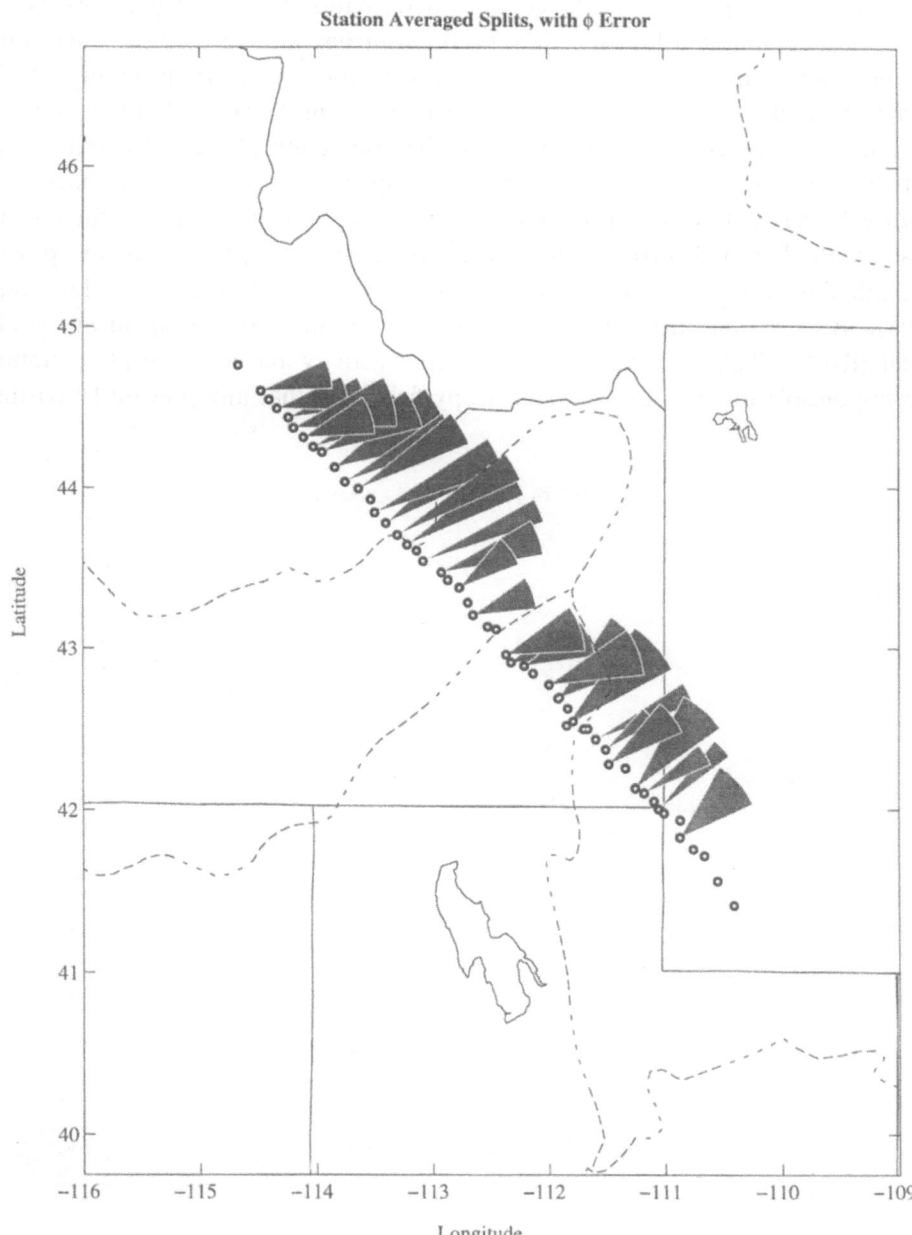

Figure 8

Station-averaged splits plotted in a map view. Width of each sector indicates approximate 95% confidence bounds in orientation of the fast axis. Split times range from 0.3–1.5 s (confidence bounds are not shown for clarity). Station averages with total $\delta t_{err} > 1.4$ s or $\phi_{err} > 40°$ are not shown. Fast axis orientations to stations in the northwest appear to be systematically rotated with respect to the other stations.

It might be expected that the complex deformation history of this area would result in a complicated lithospheric (fossil) anisotropy structure. Shear-wave splitting measurements in Colorado exhibit a very complex structure, including a wide variety of null orientations. Splitting measurements in Nevada, although simpler, have neither the degree of uniformity nor the orientation of the eSRP splits (Fig. 10). Since most (and perhaps all) of our anisotropy resides in the asthenosphere, the Lattice Preferred Orientation (LPO) inducing flow under the eSRP is thought to postdate the Laramide orogeny (when the subducted slab is thought to have passed beneath this region at a depth of ~ 100 km depth (HUMPHREYS and DUEKER, 1994)). Because anisotropy tends to align with the most recent significant strain event (RIBE, 1992), we conclude that hot spot activity and absolute plate motion are responsible for the observed anisotropy field, and that any previously existing

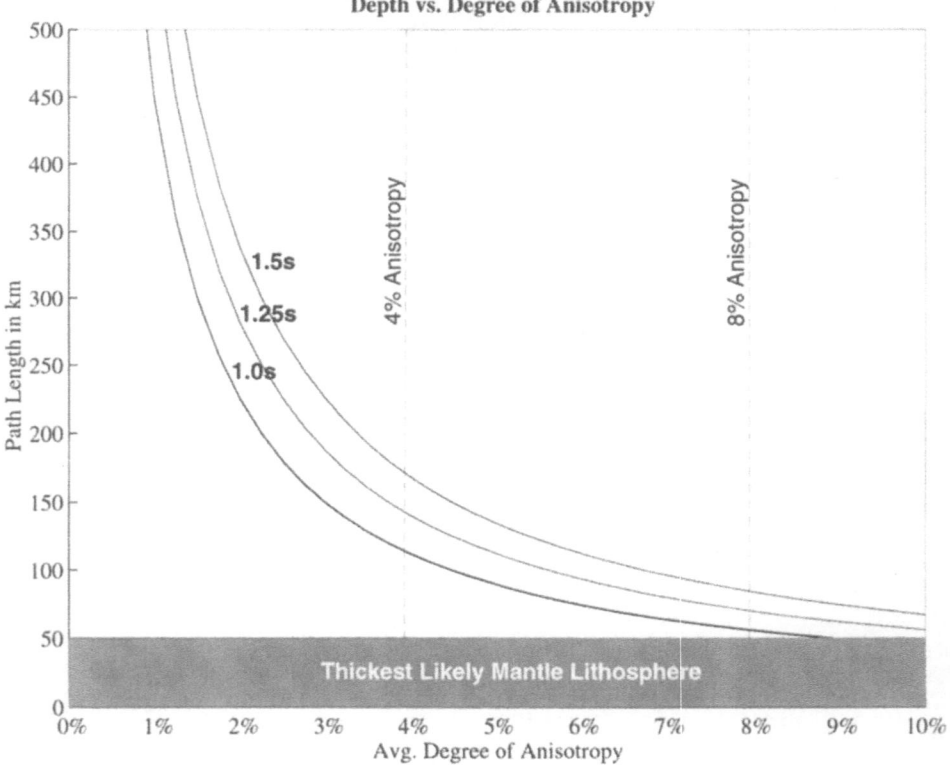

Figure 9

Plot of trade-off between path length L, anisotropy magnitude $\delta\beta$, path-averaged anisotropy magnitude $\delta\beta_0$, and split time δt, using the relation $L = \beta_0 \delta t / \delta\beta$, where β_0 is the average of fast and slow axes shear velocities (taken to be 4.5 km/s). Given an anisotropy magnitude of 4% and a split time of 1.25 s, this relation implies a path length of nearly 150 km. The 40 km thick crust is only weakly anisotropic (PENG and HUMPHREYS, 1997). As it is highly unlikely that the lithospheric mantle is 110 km thick or more, some, and probably most, of the anisotropy must lie in the asthenosphere.

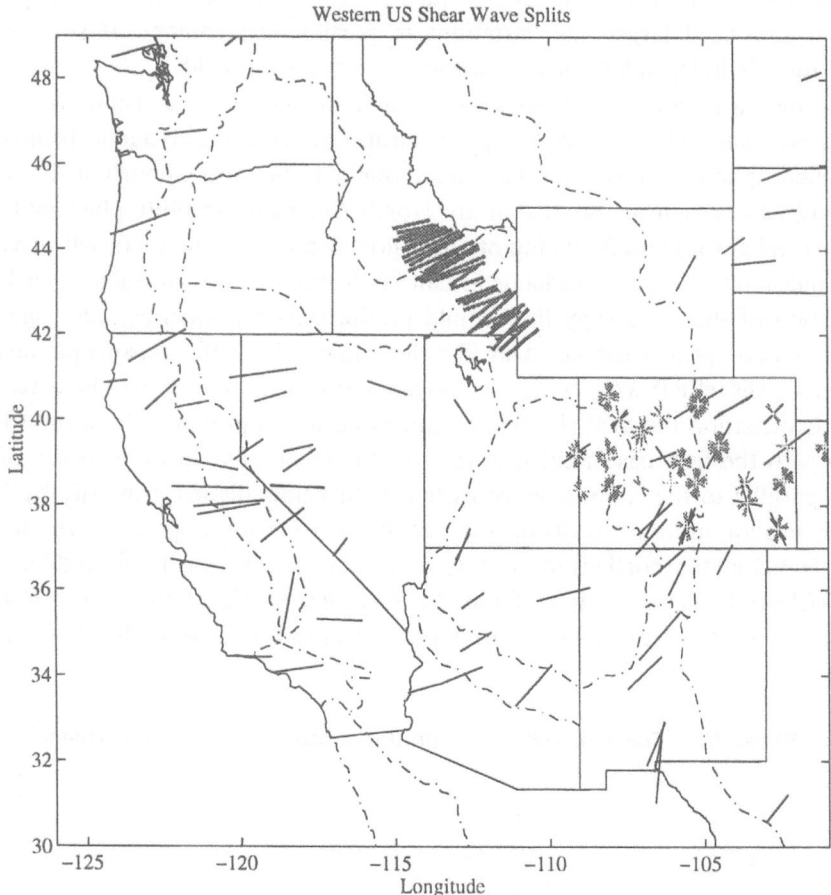

Figure 10

Eastern Snake River Plain station-averaged splits shown in regional context. Eastern Snake River Plain splits are plotted as thick dark gray lines; other measured splits are shown with thin, solid lines; and null splits (only in Colorado and Kansas) are plotted as short light gray lines. [Split data from BOSTOCK and CASSIDY, 1995; SANDVOL et al., 1992; VINNIK et al., 1992; SILVER and CHAN, 1991; HELFFRICH et al., 1994; SILVER and KANESHIMA, 1993; RUPPERT, 1992; BARRUOL et al., 1996; OZALAYBEY and SAVAGE, 1995; SAVAGE et al., 1996].

anisotropy has been reset or erased. If so, the uniform orientation structure found beneath our array lies within the asthenospheric wake of the hot spot. This is consistent with the asthenospheric wake inferred by surface deformation and uplift, but since our array does not extend beyond the tectonic parabola, the width of our investigation suggests only a minimum width of the asthenospheric wake.

The anisotropy and velocity structures provide an interesting contrast (Fig. 11). Anisotropy orientation is rather uniform across the array, whereas the S-wave velocity structure is very heterogeneous. SALTZER and HUMPHREYS (1997) argue

that the *P*-wave velocity structure—which is very similar to the *S*-velocity structure—is produced largely by variations in partial melt content. If so, the melt
distribution is independent of the anisotropy orientation field.

The measured anisotropy, which is thought to indicate the strain state of the
anisotropic material, can provide useful information on the dynamic behavior of
the asthenosphere. The two mechanisms thought to be most important in creating
asthenospheric strain in this region are North American absolute plate motion (a
passive mechanism) and flattening of anomalously buoyant mantle (a self-driven, or
active mechanism). These mechanisms can be distinguished from each other by the
orientation of the anisotropy they would produce. North American absolute plate
motion would excite a simple shear flow field that orients the anisotropic fast axis
parallel to the eSRP axis, whereas buoyancy-driven flattening would drive pure
shear deformation (vertical shortening and extension approximately normal to the
eSRP) with the fast axis oriented normal to the eSRP. The fast-axis orientation is
nearly parallel to the eSRP axis, consistent with being aligned primarily by North
American plate motion. In detail, the fast-axis orientation appears to rotate from
about N80E at the northernmost stations to about N60E near the middle of the
array (Absolute Plate Motion of the eSRP is N60 ± 20°E (GRIPP and GORDON,
1990); the most recent trend of rhyolitic volcanism is N54 ± 5°E (PIERCE and

Shear Wave Relative Velocity (% perturbations), with SKS Split Times

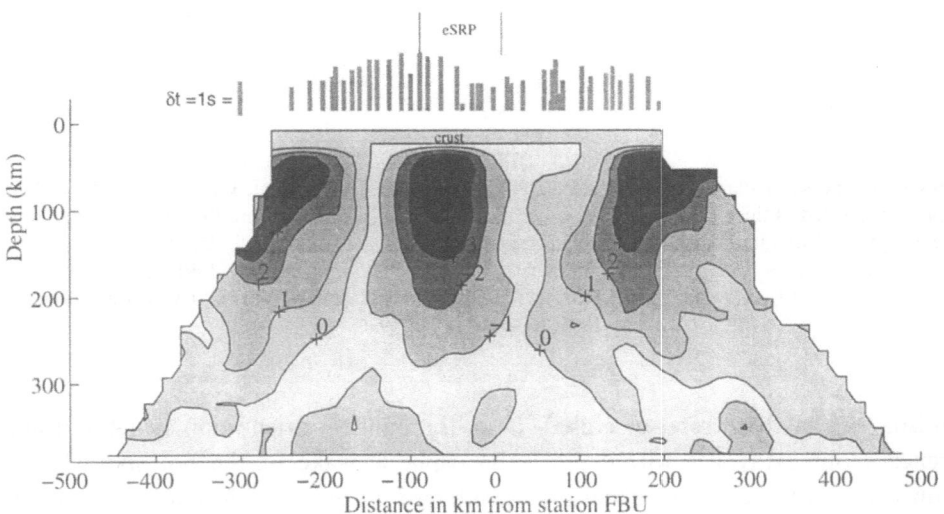

Figure 11

Upper mantle shear-wave velocity and station-averaged split times. (Split times with $\delta t_{err} > 1.4$ s are not
shown). The depression in split times appears not to correlate with velocity structure. However, because
all split determinations are derived from arrivals emanating from the NW, it is possible that the offset
pattern is a result of being projected to the SE.

Angle of Incidence of Measured SKS Splitting Events

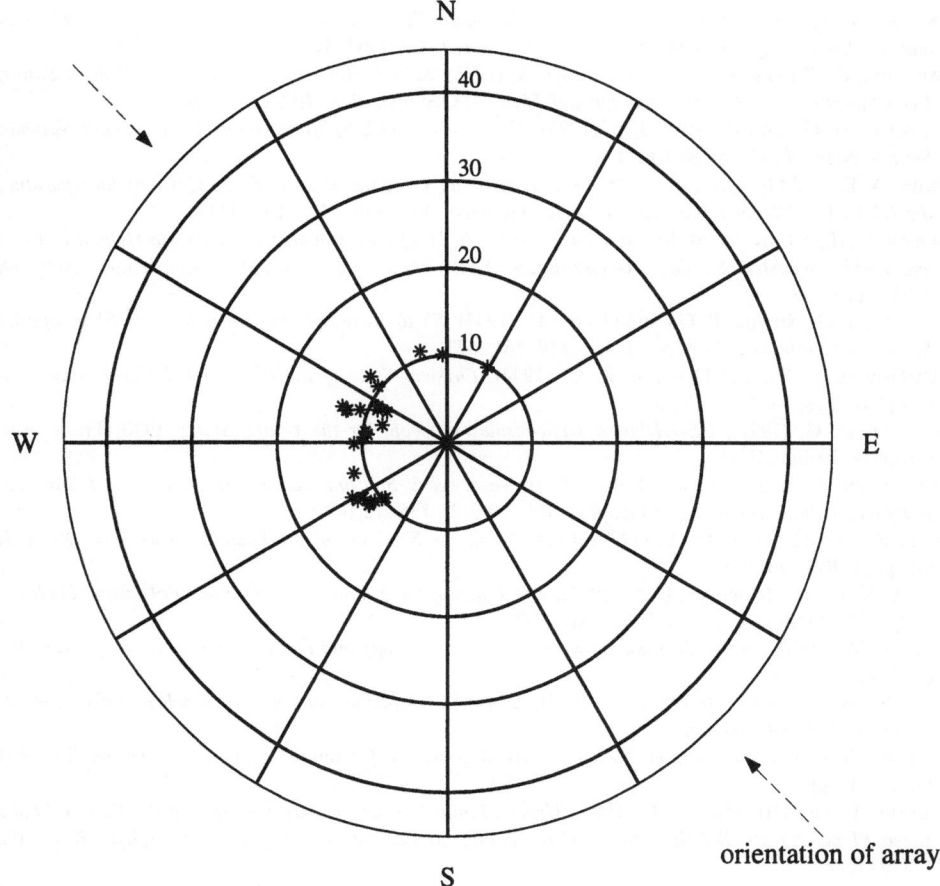

Figure 12
SKS splitting events plotted by back azimuth and angle of incidence.

MORGAN, 1992)). This trend may indicate a flattening-produced component to the integrated strain field producing the anisotropy.

Split time structure has a significant depression in station-averaged split times near the center of the array. This depression is centered about 60 km southeast of the eSRP (near the southern boundary of the eSRP). Because none of our SKS rays arrive from the southeast, the station averages are most likely projections of structure to the northwest (Fig. 12). If the anisotropy responsible for the splits were located at a depth of ~200–300 km it would be centered beneath the eSRP. The cause of the reduced anisotropy is not known. Candidate solutions include the presence of melt films oriented so as to partially compensate for the LPO, and reduced horizontal LPO development (perhaps because of upward flow or an LPO-inhibiting influence of partial melt).

References

ANDERS, M. H., and SLEEP, N. M. (1992), *Lithospheric Strengthening by Magmatic Intrusion, Thermal and Mechanical Effects of the Yellowstone Hotspot*, J. Geophys. Res. *97*, 15,379–15,394.

BARRUOL, G., SILVER, P. G., and VAUCHEZ, A. (1996), *Seismic Anisotropy in the Eastern United States; Deep Structure of a Complex Continental Plate*, J. Geophys. Res. *102*, 8329–8348.

BOSTOCK, M. G., and CASSIDY, J. F. (1995), *Variations in SKS Splitting Across the Canadian National Seismic Network*, Geophys. Res. Lett. *22*, 5–8.

GRIPP, A. E., and GORDON, R. G. (1990), *Current Plate Velocities Relative to the Hotspots Incorporating the NUVEL-1 Global Plate Motion Model*, Geophys. Res. Lett. *17*, 1109–1112.

HEARN, E. H., KENNEDY, B. M., and TREUSDELL, A. H. (1990), *Coupled Variations in Helium Isotopes and Fluid Chemistry: Shoshone Geyser Basin, Yellowstone*, Geochimica et Cosmochemica Acta *54*, 3103–3113.

HELFFRICH, G., SILVER, P. G., and GIVEN, H. (1994), *Shear-wave Splitting Variation over Short Spatial Scales on Continents*, Geophys. J. Int. *119*, 561–573.

HUMPHREYS, E. D., and DUEKER, K. G. (1994), *Physical State of the Western US Upper Mantle*, J. Geophys. Res. *99*, 9635–9650.

MILBERT, D. G. (1991), *GEOID90: A High-resolution Geoid for the United States*, EOS, Trans. Am. Geophys. Union *72*, 49.

OZALAYBEY, S., and SAVAGE, M. K. (1995), *Shear-wave Splitting beneath Western United States in Relation to Plate Tectonics*, J. Geophys. Res. *100*, 18,135–18,149.

PENG, X., and HUMPHREYS, E. (1997), *Crustal Velocity Structure of the Eastern Snake River Plain*, J. Geophys. Res., in press.

PIERCE, K. L., and MORGAN, L. A. (1992), *The Track of the Yellowstone Hot Spot: Volcanism, Faulting, and Uplift*, Memoir-Geological Soc. Am. *179*, 1–53.

RIBE, N. M. (1992), *On the Relation between Seismic Anisotropy and Finite Strain*, J. Geophys. Res. *97*, 8737–8747.

RIBE, N. M., and CHRISTENSEN, U. (1994), *Dynamical Modeling of Plume-lithosphere Interaction*, J. Geophys. Res. *99*, 669–682.

RUPPERT, S. (1992), *Tectonics of Western North America: A Teleseismic View*, Ph.D. Thesis, Stanford Univ., 216 pp.

SALTZER, R. and HUMPHREYS, E. (1997), *Upper Mantle P-wave Velocity Structure of the Eastern Snake River Plain and its Relationship to Geodynamics Models of the Region*, J. Geophys. Res. *102*, 11829–11842.

SANDVOL, E., NI, J., OZALAYBEY, S., and SCHUE, J. (1992), *Shear-wave Splitting in the Rio Grande Rift*, Geophys. Res. Lett. 2 *19*, 2337–2340.

SAVAGE, M. K., SHEEHAN, A. F., and LERNER-LAM, A. (1996), *Shear-wave Splitting across the Rocky Mountain Front*, Geophys. Res. Lett. *23*, 2267–2270.

SILVER, P. G., and CHAN, W. W. (1991), *Shear-wave Splitting and Subcontinental Mantle Deformation*, J. Geophys. Res. *96*, 16429–16454.

SILVER, P. G., and KANESHIMA, S. (1993), *Constraints on Mantle Anisotropy beneath Precambrian North America from a Transportable Teleseismic Experiment*, Geophys. Res. Lett. *20*, 1127–1130.

SILVER, P. G., and SAVAGE, M. K. (1994), *The Interpretation of Shear-wave Splitting Parameters in the Presence of Two Anisotropic Layers*, Geophys. J. Int. *119*, 949–963.

SMITH, R. B., and BRAILE, L. W. (1994), *The Yellowstone Hotspot*, J. Volc. Geotherm. Res. *61*, 121–188.

SPARLIN, M. A., BRAILE, L. W., and SMITH, R. B. (1982), *Crustal Structure of the Eastern Snake River Plain Determined from Ray Trace Modeling of Seismic Refraction Data (Idaho)*, J. Geophys. Res. *87*, 2619–2633.

VINNIK, L. P., MAKEYEVA, L. I., MILEV, A., and USENKO, Y. (1992), *Global Patterns of Azimuthal Anisotropy and Deformation in the Continental Mantle*, Geophys. J. Int. *111*, 433–447.

(Received October 14, 1996, revised April 29, 1997, accepted June 18, 1997)

Pure appl. geophys. 151 (1998) 463–475
0033–4553/98/040463–13 $ 1.50 + 0.20/0

┃Pure and Applied Geophysics

Anisotropy and Flow in Pacific Subduction Zone Back-arcs

KAREN M. FISCHER,[1] MATTHEW J. FOUCH,[1] DOUGLAS A. WIENS,[2] and
MARGARET S. BOETTCHER[1]

Abstract—We have obtained constraints on the strength and orientation of anisotropy in the mantle beneath the Tonga, southern Kuril, Japan, and Izu-Bonin subduction zones using shear-wave splitting in S phases from local earthquakes and in teleseismic core phases such as SKS. The observed splitting in all four subduction zones is consistent with a model in which the lower transition zone (520–660 km) and lower mantle are isotropic, and in which significant anisotropy occurs in the back-arc upper mantle. The upper transition zone (410–520 km) beneath the southern Kurils appears to contain weak anisotropy. The observed fast directions indicate that the geometry of back-arc strain in the upper mantle varies systematically across the western Pacific rim. Beneath Izu-Bonin and Tonga, fast directions are aligned with the azimuth of subducting Pacific plate motion and are parallel or sub-parallel to overriding plate extension. However, fast directions beneath the Japan Sea, western Honshu, and Sakhalin Island are highly oblique to subducting plate motion and parallel to present or past overriding plate shearing. Models of back-arc mantle flow that are driven by viscous coupling to local plate motions can reproduce the splitting observed in Tonga and Izu-Bonin, but further three-dimensional flow modeling is required to ascertain whether viscous plate coupling can explain the splitting observed in the southern Kurils and Japan. The fast directions in the southern Kurils and Japan may require strain in the back-arc mantle that is driven by regional or global patterns of mantle flow.

Key words: Anisotropy, mantle flow, subduction zones, shear-wave splitting.

Introduction

The presence of seismic anisotropy within subduction zones has been documented by numerous studies (e.g., ANDO et al., 1983; FUKAO, 1984; BOWMAN and ANDO, 1987; SHIH et al., 1991; YU and PARK, 1994; FISCHER and YANG, 1994; RUSSO and SILVER, 1994; YANG et al., 1995; KANESHIMA and SILVER, 1995). However, recent shear-wave splitting measurements have added new constraints on the strength and geometry of anisotropy within the back-arc mantle regions of several Pacific subduction zones: Tonga, the southern Kurils, Japan, and Izu-Bonin (FISCHER and WIENS, 1996; FOUCH and FISCHER, 1996). Shear-wave splitting

---

[1] Department of Geological Sciences, Brown University, Providence, Rhode Island, 02912, U.S.A.
E-mail: karen@emma.geo.brown.edu, Fax: 401-863-2058, Tel.: 401-863-1360.
[2] Department of Earth and Planetary Sciences, Washington University, St. Louis, Missouri, 63130, U.S.A.

parameters (fast directions and splitting times) were obtained for S phases from local earthquakes and SKS, $SKKS$, and PKS phases from teleseismic events recorded at subduction zone stations. These splitting measurements yielded two primary results: a striking variation in fast direction orientation between different back-arc regions, and patterns of splitting time with source location that constrain the distribution of anisotropy with depth in the mantle. The strength of anisotropy as a function of depth was extensively modeled in FISCHER and WIENS (1996) and FOUCH and FISCHER (1996), and in this paper we only briefly summarize these results. Rather, this paper focuses on the geometry of back-arc flow implied by the observed splitting and explores the resulting implications for coupling of the subducting slab and overriding plate to the back-arc asthenosphere.

Observed Shear-wave Splitting

In the Tonga subduction zone, shear-wave splitting parameters were obtained for S and SKS phases recorded at three stations of the southwest Pacific PASSCAL experiment (WIENS *et al.*, 1995) and an IRIS/Global Seismic Network station (Fig. 1). Data in the Izu-Bonin, Japan, and southern Kuril subduction zones were collected from the Japanese POSEIDON network, IRIS/Global Seismic Network stations, and one GEOSCOPE station (Fig. 2). Analysis was restricted to paths with free surface incidence angles of 35° or less to avoid contamination of particle motions by converted phases and phase shifts from crustal discontinuities and the free surface. In Tonga, event depths range from 380 km to 660 km. The number of splitting measurements displayed in Figure 1 is about 50% larger than the FISCHER and WIENS (1996) data set, although the splitting parameter patterns are virtually identical. Among the northwest Pacific splitting measurements of FOUCH and FISCHER (1996), event depths range from 50 km to 600 km. However, to keep Figure 2 both compact and readable, we display shear-wave splitting parameters only for events deeper than 200 km.

We parameterized shear-phase splitting by finding the fast direction, ϕ, and splitting time, δt, which most completely describe ellipticity in the shear phase particle motions. Splitting values were determined by computing the covariance matrix for the horizontal seismogram components rotated to azimuths of ϕ and $\phi + \pi/2$ and time-shifted by δt, over a grid of ϕ and δt values. Splitting parameters for a given phase were defined as those values of ϕ and δt which minimized the smaller of the two covariance matrix eigenvalues. Error bars for individual splitting parameters correspond to the 95% F-test confidence regions for $\varphi_2/\varphi_{2\,min}$, where φ_2 is the smaller of the two covariance matrix eigenvalues at any point on the ϕ and δt grid and $\varphi_{2\,min}$ is the minimum value of φ_2 that corresponds to the best-fitting ϕ and δt.

Shear-wave splitting fast directions in these subduction zones display a fairly simple pattern. For earthquakes in Tonga and in Izu-Bonin, S and SKS fast directions are roughly parallel to absolute Pacific plate motion. For earthquakes in western Honshu and the Japan Sea, S and SKS fast directions are roughly parallel to the strike of the trench, and for earthquakes in the southern Kurils, fast directions are N–S to NNE–SSW (Figs. 1 and 2). The distribution of local S phase splitting times from Tonga and Izu-Bonin also differs from that in Japan and the southern Kurils (Fig. 3). In Tonga and Izu-Bonin, local S splitting times are fairly uniform with source depth, particularly at 95% confidence, whereas in Japan and the southern Kurils, local S splitting times manifest a significant increase with source depth. In the two regions where consistent estimates of splitting in SKS and other teleseismic core phases were obtained (Tonga and the southern Kurils), the teleseismic splitting times overlap with the S splitting times from the deepest local events.

The Distribution of Anisotropy with Depth

The observed splitting time patterns qualitatively suggest that anisotropy is confined to the upper mantle in Tonga and Izu-Bonin, and that no evidence exists in any region for splitting due to lower mantle anisotropy. However, in order to fully account for differences in path length and angle between individual phases, shear-wave splitting parameters were modeled using a method that predicts shear-wave splitting on individual phase paths and determines the strength, orientation and depth extent of anisotropy that best fits the complete data set of observed splitting times and fast directions (FISCHER and WIENS, 1996; FOUCH and FISCHER, 1996). For example, the distribution of paths in the Tonga back-arc is shown in Figure 4.

Modeling results indicate that contributions to observed shear-wave splitting from the lower mantle are not required in any of these subduction zones and are ruled out beneath the southern Kurils and Tonga. This result is most plausibly explained by an absence of significant anisotropy widely distributed within the lower mantle. Assuming that the strength and orientation of anisotropy are roughly uniform throughout the back-arc mantle, anisotropy is ruled out beneath 400 km in Izu-Bonin and beneath 435 km in Tonga. In contrast, the increase of observed splitting times over transition zone source depths in the southern Kurils (Fig. 3b) requires anisotropy on the order of 0.5% to depths of at least 500 km. However, the strength of transition zone anisotropy could be weaker if the strength of anisotropy was allowed to vary with depth. Data from all the subduction zones are consistent with a model in which the lower transition zone (520–660 km) and lower mantle are largely isotropic, and the upper transition zone (410–520 km) contains weak anisotropy (possibly due to preferred orientation of β spinel) in some regions but not others. Significant anisotropy in the back-arc upper mantle is ubiquitously required.

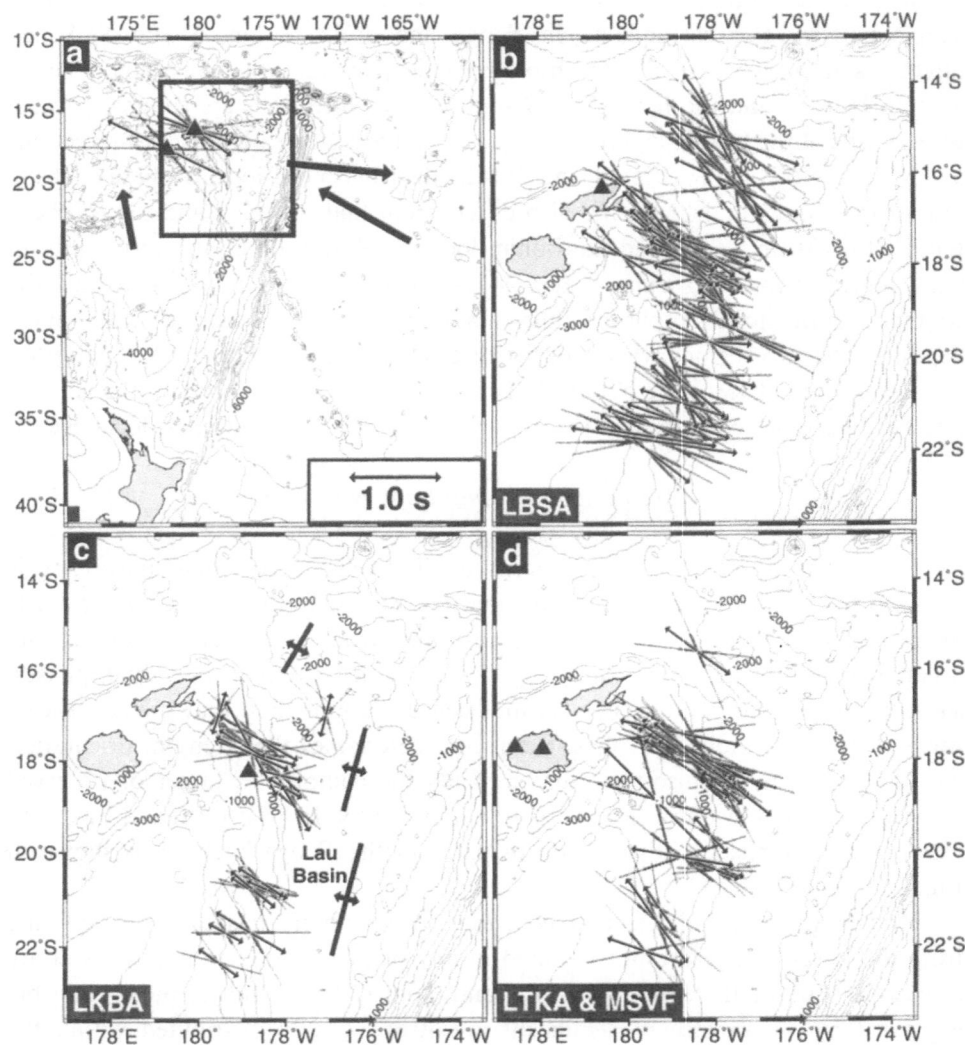

Figure 1

Shear-wave splitting in the Tonga subduction zone observed at 3 stations of the southwest Pacific PASSCAL experiment (LBSA, LKBA, and LTKA) and one IRIS/GSN permanent station (MSVF) (triangles). a) *SKS* splitting parameters at LBSA and LTKA, plotted at the stations. Bold arrows denote the direction and relative rates of absolute Pacific plate motion (PPM), absolute Australian plate motion, and absolute Tonga arc motion (GRIPP and GORDON, 1990; BEVIS *et al.*, 1995). Box outlines area magnified in b–d. b–d) Local *S* splitting measurements plotted at the earthquake source. The location and orientation of active back-arc spreading are indicated by the thicker arrows in (c). Splitting parameters are shown as vectors parallel to ϕ (fast direction) with lengths scaled to δt (splitting time); corresponding 95% confidence limits are represented by thick and thin line segments. A 1 s reference vector is shown at bottom right, and bathymetry is contoured in 1000 m intervals. Fast directions for paths in Tonga are roughly parallel to PPM.

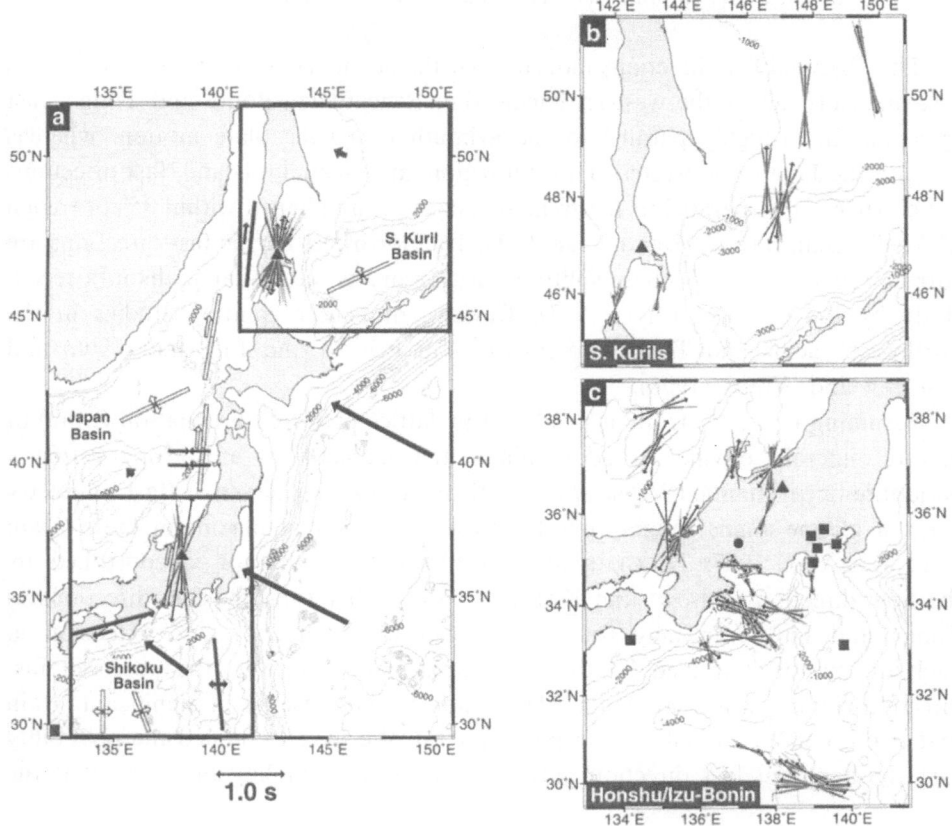

Figure 2

Shear-wave splitting observed in Japan and the southern Kurils at IRIS/GSN stations YSS (Sakhalin Island) and MAJO (Honshu) (triangles), GEOSCOPE station INU (solid circle), and 8 POSEIDON network stations on Honshu and in the Izu-Bonin arc, HKY, TKO, SGN, YCU, JIZ, HCH, MRM, and OGS (squares, OGS out of box). (To make this condensed figure clearer, we have omitted a few of the splitting measurements and stations shown in FOUCH and FISCHER (1996).) a) *SKS* splitting parameters at MAJO and YSS, plotted at the stations. Bold arrows denote the direction and relative rates of absolute Pacific plate motion (PPM), and the absolute motions of the Philippine, Eurasian and North American plates (GRIPP and GORDON, 1990). Smaller solid arrows indicate present-day rifting, shearing, and compression. Smaller outlined arrows indicate past and now inactive back-arc spreading and shearing. Boxes outline areas magnified in b–c. b–c) Local *S* splitting measurements plotted at the earthquake source. Splitting parameters are shown as vectors parallel to ϕ (fast direction) with lengths scaled to δt (splitting time); corresponding 95% confidence limits are represented by thick and thin line segments. A 1 s reference vector is shown at bottom right, and bathymetry is contoured in 1000 m intervals. Fast directions for paths in Izu-Bonin are roughly parallel to PPM, but beneath the Japan Sea, western Honshu region and Sakhalin Island, fast directions are N–S to NNE–SSW.

Implications for the Geometry and Dynamics of Back-arc Flow

The observed fast directions indicate that the geometry of back-arc strain varies systematically across the western Pacific rim. Beneath Izu-Bonin and Tonga, fast directions are roughly parallel to the azimuth of Pacific plate motion, whereas beneath the Japan Sea, western Honshu region, and Sakhalin Island, fast directions are either roughly parallel to the trench (Japan) or are aligned within 35° of trench (Sakhalin Island) (Figs. 1 and 2, Table 1). The northwest Pacific fast directions are corroborated by local *S*-phase splitting measurements on similar paths in a recent study by SANDVOL and NI (1997). Reviews of earlier splitting studies in the northwest Pacific and Tonga are provided in FOUCH and FISCHER (1996) and FISCHER and WIENS (1996).

Assuming that anisotropy is produced by lattice preferred orientation (LPO) of mantle minerals, olivine should dominate the geometry of anisotropy within a peridotite upper mantle (WENK *et al.*, 1991; RIBE, 1992). The *a* axis (fast symmetry axis) of olivine aligns roughly parallel to the direction of maximum finite strain (e.g., KARATO, 1987; NICOLAS and CHRISTENSEN, 1987) or is controlled by the flow direction (ZHANG and KARATO, 1995). Therefore, the fast directions in Tonga and Izu-Bonin may be easily explained by shearing and extension in back-arc mantle entrained by the subducting slab. However, the observed fast directions in Japan and Sakhalin Island indicate the presence of significant strain that is not coupled to subducting plate motion. We have examined the possibility that the apparent fast direction patterns are due to sampling bias, i.e., that the

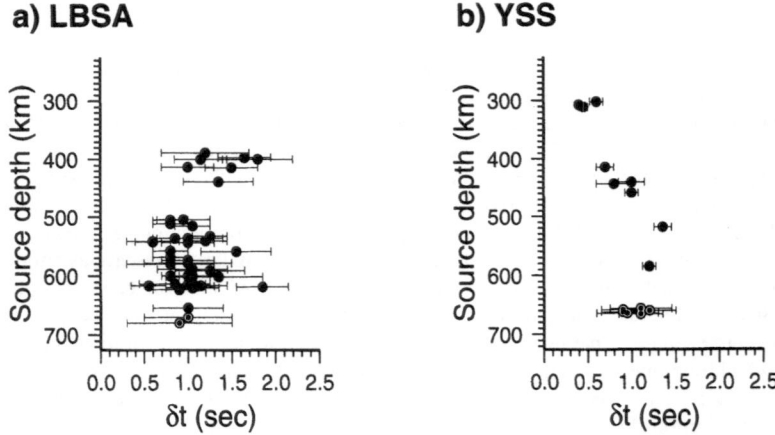

Figure 3
Observed splitting times for local *S* phases (solid circles) and their 95% confidence limits shown as a function of earthquake source depth. Splitting times for *SKS* phases (bullseyes) are plotted near 670 km depth for comparative purposes. a) Splitting times for station LBSA in the Tonga back-arc are roughly uniform with source depth, particularly at 95% confidence. b) Splitting times for station YSS on Sakhalin Island in the southern Kuril back-arc show a significant increase with source depth.

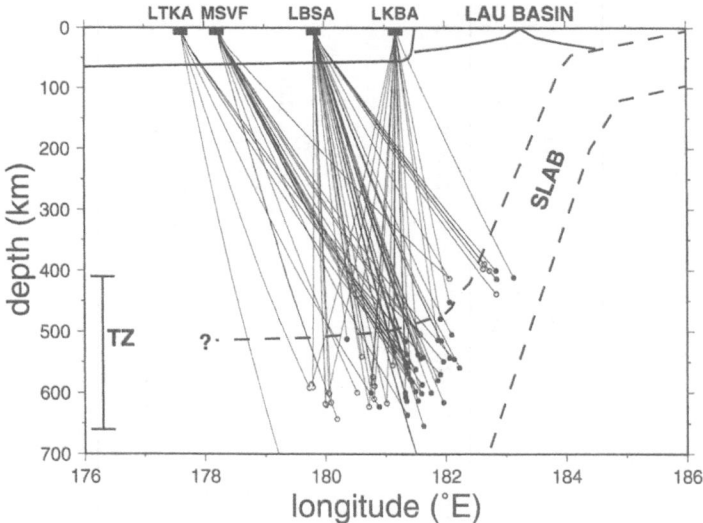

Figure 4

S and *SKS* raypaths through the Tonga slab and back-arc. Bold dashed contours show the approximate outline of the high velocity subducting lithosphere from recent inversions for *P*-wave velocity (VAN DER HILST, 1995; ZHAO *et al.*, 1995). Bold solid contours show the probable shape of the overriding plate. The profile passes through station LKBA at an azimuth of 80°. Solid dots show events located within 2° of the profile; open dots are from more distant arc segments where slab shape may differ from the given outline. Raypaths are relative to the IASP91 radial velocity model.

back-arc strain patterns in the four subduction zones are similar but the *S* paths in Tonga and Izu-Bonin sample a different region of the back-arc than those in Japan and the southern Kurils. However, although *S* paths from local earthquakes in Izu-Bonin in general lie closer to the strike of the subducting slab than local *S* paths in the other subduction zones, no systematic difference in path distribution can explain the observed fast direction patterns.

What factors control the apparent differences in back-arc strain patterns between Tonga and Izu-Bonin and Japan and the southern Kurils? An examination of absolute plate motions indicates little difference in the properties of the subducting plate between these subduction zones, but demonstrates fundamental variations in the rate of overriding plate motion and in the style and geometry of back-arc tectonism. The subducting Pacific plate enters the trench in all four subduction zones with roughly the same azimuth and rate of absolute plate motion (GRIPP and GORDON, 1990) (Table 1). However, the total rates of overriding plate motion (V_{up}, Table 1) and the components of overriding plate motion normal to the trench are much larger in Tonga and Izu-Bonin than in Japan and the southern Kurils (Figs. 1 and 2). In terms of back-arc tectonism, Tonga and Izu-Bonin are strongly extensional environments, while the southern Kuril and Japan back-arcs are dominated by a combination of N–S shearing and E–W compression (Table 1).

Table 1

Shear-wave splitting, plate motions, and back-arc tectonism

Region	φ average	PPM azimuth[a]	PPM rate[a] (cm/y)	V_{up} (cm/y)	Dominant mode of overriding plate deformation	Rate of overriding plate deformation	Minimum overriding plate age
Tonga	~300°	~300°	10.5	Australian: 5.0[a] Tonga arc: >10[b]	extension at 280°–300°[b,c]	10–16 cm/y[b,c]	present[c]
Izu-Bonin	~290°	~293°	10.7	Philippine: 5.0[a]	extension at 255°–20°[d]	?	Shikoku Basin: 15 Ma[d] Sumisu Rift: present[d] 18 Ma[f]
Japan	~20°	~294°	10.5	Eurasian: 0.9[a]	compression at ~90° after shear at ~0°[e]	1–2 cm/y or less[e]	
S. Kurils	~10°	~299°	9.8	N. American: 1.3[a]	shear at ~0°[e]	1–2 cm/y[e]	~20 Ma[e,g]

[a] GRIPP and GORDON (1990).
[b] BEVIS et al. (1995).
[c] WEISSEL (1977), HAMBURGER and ISACKS (1988), HAWKINS (1995).
[d] HASTON and FULLER (1991), TAYLOR et al. (1991), KOBAYASHI et al. (1995), CLIFT (1995).
[e] FOURNIER et al. (1994, 1995), JOLIVET et al. (1994, 1995).
[f] TAMAKI (1995).
[g] GNIBIDENKO et al. (1995).

In Tonga, earthquake source mechanisms (HAMBURGER and ISACKS, 1988) and back-arc spreading in the Lau Basin (active since 6 Ma) (WEISSEL, 1977; HAWKINS, 1995) indicate extension at an azimuth that is roughly normal to the trench and parallel to the observed fast directions (Fig. 1). In Izu-Bonin, back-arc spreading occurred in the Shikoku Basin from 29 to 15 Ma and has continued in the Sumisu Rift from 2 Ma to the present (TAYLOR et al., 1991; CLIFT, 1995; KOBAYASHI et al., 1995). The initial orientation of back-arc spreading in the Shikoku Basin was roughly ENE–WSW, but later spreading had an azimuth closer to E (KOBAYASHI et al., 1995) (Fig. 2). The orientation of normal faults in the Sumisu Rift (TAYLOR et al., 1991) and source mechanisms from the Harvard Centroid Moment Tensor Catalog for Sumisu Rift earthquakes are consistent with extension at a roughly E–W azimuth that is normal to the trench and lies within 30° of most observed fast directions. In the southern Kuril and Japan back-arcs, paleomagnetic data suggest that the overriding plates deformed along a roughly N–S dextral shear system during the Miocene. This shear zone appears to have remained active in the southern Kurils through the present, but has been replaced by E–W compression to the west of Honshu (Fig. 2, Table 1) (FOURNIER et al., 1994, 1995; JOLIVET et al., 1994, 1995). The N–S shearing is roughly parallel to the observed fast directions in both the southern Kurils and Japan. In contrast to the active rifting in the Tonga and Izu-Bonin back-arcs, spreading in the Japan Sea and southern Kuril basins was due to pull-apart tectonism and ended prior to 18 Ma (TAMAKI, 1995; GNIBIDENKO et al., 1995). Finally, the Izu-Bonin and Tonga trenches have undergone much greater seaward trench migration over the last 25 Ma than have the trenches in Japan and the southern Kurils (HASTON and FULLER, 1991; VAN DER HILST and SENO, 1993; FOURNIER et al., 1994; VAN DER HILST, 1995). We are therefore left with an intriguing picture in which fast directions in Tonga and Izu-Bonin are parallel to subducting plate motion and parallel or subparallel to overriding plate deformation, but fast directions in the southern Kurils and Japan are highly oblique to subducting plate motion and parallel to present or past overriding plate shearing.

How might these differences in back-arc tectonics influence the back-arc strain patterns responsible for the apparent anisotropy? One possibility worth considering is that overriding plate anisotropy with the geometry of recent overriding plate deformation makes a large contribution to the observed splitting and is particularly dominant in Japan and the southern Kurils where the overriding plate is thicker due to the absence of recent back-arc spreading. (Even though the N–S shear system no longer appears to be active in the Japan Sea (FOURNIER et al., 1994, 1995; JOLIVET et al., 1994, 1995), the recent E–W compression could produce E–W olivine b axes and N–S fast directions.) However, while lithospheric anisotropy may have some influence on the observed fast directions, it is insufficient to completely explain their variation. The observed increase of splitting times with source depth from 200–600 km in the southern Kurils (Fig. 3b) and Japan indicates that anisotropy with a N–S to NNE–SSW fast direction must persist deep into the asthenosphere. In addition, the portions of the overriding plate sampled in Tonga and Izu-Bonin largely lie outside the Lau Basin and the Sumisu Rift where the youngest and thinnest lithosphere occurs (Fig. 4), so the differences in sampled lithospheric thickness between subduction zones may not be significant.

Two types of models might explain the apparent differences in deep back-arc mantle strain geometry between Tonga/Izu-Bonin and the southern Kurils/Japan: large-scale flow and local viscous plate coupling. In the first type of model, regional to global patterns in mantle buoyancy and flow dictate back-arc strain patterns, and overriding plate deformation either passively reflects or is decoupled from these deeper processes. For instance, larger-scale flow patterns could drive predominantly N–S flow along the Kuril and Japan back-arcs, producing the N–S to NNE–SSW fast directions. In Tonga and Izu-Bonin, larger-scale flow might either be negligible, allowing viscous coupling to surface plate motions to dominate, or have an azimuth roughly in the direction of absolute Pacific plate motion.

In the second type of model, viscous coupling to local plate motion dominates back-arc flow. For instance, McKENZIE (1979) and RIBE (1989) show that a subducting slab entrains surrounding mantle material, aligning olivine *a* axes with the slab-mantle interface and driving circulation throughout the back-arc wedge. This type of model works well in Izu-Bonin where Pacific plate motion. Philippine plate motion, and the probable orientation of recent back-arc extension are all either parallel or subparallel to the observed fast directions (Fig. 2, Table 1) (STINE *et al.*, 1996). It is also easily applicable to Tonga where Pacific plate motion and back-arc extension are parallel to the observed fast directions and have much higher rates than the more northerly motion of the Australian plate (Fig. 1, Table 1) (HALL *et al.*, 1997). However, the viscous plate coupling model is less obviously consistent with the N–S to NNE–SSW fast directions in the southern Kurils and Japan. Although the roughly N–S geometry of shearing in the overriding plate near Sakhalin Island (Fig. 2) aligns with the trend of the fast directions observed in the southern Kurils, its rate is much less than the rate of subducting plate motion. Moreover, tectonic reconstructions suggest that similar ~N–S shearing in Honshu and the Sea of Japan ended at 12–15 Ma (JOLIVET *et al.*, 1995). The ability of viscous plate coupling to explain fast directions observed in the southern Kurils and Japan is now under investigation using three-dimensional mantle flow calculations (HALL *et al.*, 1997).

The viability of the viscous plate coupling model in the southern Kurils and Japan could be enhanced if the subducting slab was partially decoupled from back-arc mantle flow, allowing overriding plate motions to dominate shallow back-arc strain. Two-dimensional flow calculations indicate that the presence of a thin low viscosity layer at the slab-mantle interface can reduce entrainment of the surrounding mantle by the slab, but would not prevent viscous plate coupling from modeling the shear-wave splitting in Tonga and Izu-Bonin (STINE *et al.*, 1996).

Because the viscosity of olivine dramatically decreases with increasing water content (CHOPRA and PATERSON, 1984; HIRTH and KOHLSTEDT, 1996), de-watering of slab minerals and hydration of the adjacent mantle (DAVIES and STEVENSON, 1992; PEACOCK, 1996), could produce such a low viscosity zone, at least to depths of roughly 200 km. Subducted sediments could also contribute to low viscosities at the slab surface, perhaps to even greater depths (PLANK and LANGMUIR, 1993). Finally, more complex viscosity structures, including viscosity variations due to partial melting widely distributed within the back-arc mantle (DAVIES and STEVENSON, 1992; PEACOCK, 1996), would also affect the distribution of back-arc strain and should be investigated.

In summary, shear-wave splitting observed in the Tonga, Izu-Bonin, Japan and southern Kuril subduction zones requires significant anisotropy in the upper mantle beneath their back-arcs. Fast directions in Tonga and Izu-Bonin are parallel to the azimuth of subducting Pacific plate velocity and are easily modeled by mantle strain coupled to local plate motion. In contrast, fast directions in the southern Kurils and Japan are highly oblique to the azimuth of subducting Pacific plate velocity. Future three-dimensional back-arc flow modeling will address whether these fast directions can be explained by viscous plate coupling, or whether they require strain in the back-arc mantle that is driven by larger-scale mantle flow.

Acknowledgements

We thank Seiji Tsuboi of the Earthquake Research Institute in Tokyo, Japan for his assistance with the Japanese POSEIDON data, and the IRIS Data Management System for access to IRIS/GSN data. Thanks to Marc Parmentier, Chad Hall, Chris Kincaid, Greg Hirth, and Marc Spiegelman for helpful discussions, and to D. Schutt and an anonymous reviewer for good suggestions. Figures were created using the GMT software package of WESSEL and SMITH (1995). This research was funded by the National Science Foundation under award EAR-9506502.

REFERENCES

ANDO, M., ISHIKAWA, Y., and YAMAZAKI, F. (1983), *Shear Wave Polarization Anisotropy in the Upper Mantle beneath Honshu, Japan,* J. Geophys. Res. *88,* 5850–5864.

BEVIS, M., TAYLOR, F. W., SCHUTZ, B. E., RECY, J., ISACKS, B. L., HELU, S., SINGH, R., KENDRICK, E., STOWELL, J., TAYLOR, B., and CALMANT, S. (1995), *Geodetic Observations of Very Rapid Convergence and Back-arc Extension at the Tonga Arc,* Nature 374, 249–251.

BOWMAN, J. R., and ANDO, M. (1987), *Shear-wave Splitting in the Upper-mantle Wedge above the Tonga Subduction Zone,* Geophys. J. R. Astr. Soc. *88,* 25–41.

CHOPRA, P. N., and PATERSON, M. S. (1984), *The Role of Water in the Deformation of Dunite,* J. Geophys. Res. *89,* 7861–7876.

CLIFT, P. D., *Volcaniclastic sedimentation and volcanism during the rifting of western Pacific backarc basins*. In *Active Margins and Marginal Basins of the Western Pacific* (eds. Taylor, B., and Natland, J.) (American Geophysical Union, Washington 1995) pp. 67–96.

DAVIES, J. H., and STEVENSON, D. J. (1992), *Physical Model of Source Region of Subduction Zone Volcanics*, J. Geophys. Res. *97*, 2037–2070.

FISCHER, K. M., and WIENS, D. A. (1996), *The Depth Distribution of Mantle Anisotropy beneath the Tonga Subduction Zone*, Earth Planet. Sci. Lett. *142*, 253–260.

FISCHER, K. M., and YANG, X. (1994), *Anisotropy in Kuril-Kamchatka Subduction Zone Structure*, Geophys. Res. Lett. *21*, 5–8.

FOUCH, M. J., and FISCHER, K. M. (1996), *Mantle Anisotropy beneath Northwest Pacific Subduction Zones*, J. Geophys. Res. *101*, 15987–16002.

FOURNIER, M., JOLIVET, L., HUCHON, P., SERGEYEV, K. F., and OSCORBIN, L. S. (1994), *Neogene Strike-slip Faulting in Sakhalin and the Japan Sea Opening*, J. Geophys. Res. *99*, 2701–2725.

FOURNIER, M., JOLIVET, L., and FABBRI, O. (1995), *Neogene Stress Field in SW Japan and Mechanism of Deformation during the Sea of Japan Opening*, J. Geophys. Res. *100*, 24295–24314.

FUKAO, Y. (1984), *Evidence from Core-reflected Shear Waves for Anisotropy in the Earth's Mantle*, Nature *309*, 695–698.

GNIBIDENKO, H. S., HILDE, T. W. C., GRETSKAYA, E. V., and ANDREYEV, A. A., KURIL, *Kuril (South Okhotsk) Backarc Basin*, In *Backarc Basins: Tectonics and Magmatism* (ed. Taylor, B.) (Plenum Press, New York 1995) pp. 441–449.

GRIPP, A. E., and GORDON, R. G. (1990), *Current Plate Velocities Relative to the Hotspots Incorporating the NUVEL-1 Global Plate Motion Model*, Geophys. Res. Lett. *17*, 1109–1112.

HALL, C. E., PARMENTIER, E. M., and FISCHER, K. M. (1997), *Implications of Seismic Anisotropy for Three-dimensional Mantle Flow Structure in Plate Convergent Zones*, EOS Trans. AGU 78(17), Spring Meet. Suppl. S323.

HAMBURGER, M. W., and ISACKS, B. L. (1988), *Diffuse Back-arc Deformation in the Southwestern Pacific*, Nature *332*, 599–604.

HASTON, R. B., and FULLER, M. (1991), *Paleomagnetic Data from the Philippine Sea Plate and their Tectonic Significance*, J. Geophys. Res. *96*, 6073–6098.

HAWKINS, JR, J. W., *The geology of the Lau Basin*. In *Backarc Basins: Tectonics and Magmatism* (ed. Taylor, B.) (Plenum Press, New York 1995) pp. 63–138.

HIRTH, G., and KOHLSTEDT, D. L. (1996), *Water in the Oceanic Upper Mantle: Implications for Rheology, Melt Extraction and the Evolution of the Lithosphere*, Earth Planet. Sci. Lett. *144*, 93–108.

JOLIVET, L., TAMAKI, K., and TOURNIER, M. (1994), *Japan Sea, Opening History and Mechanism: A Synthesis*, J. Geophys. Res. *99*, 22237–22259.

JOLIVET, L., SHIBUYA, H., and FOURNIER, M. (1995), *Paleomagnetic rotations and the Japan Sea opening*. In *Active Margins and Marginal Basins of the Western Pacific* (eds. Taylor, B., and Natland, J.) (American Geophysical Union, Washington 1995) pp. 355–369.

KANESHIMA, S., and SILVER, P. G. (1995), *Anisotropic Loci in the Mantle beneath Central Peru*, Phys. Earth Planet. Inter. *88*, 257–272.

KARATO, S. (1987), *Seismic anisotropy due to lattice preferred orientation of minerals: Kinematic or dynamic?* In *High-pressure Research in Mineral Physics* (eds. Manghnani, M. H., and Syono, S.) (TERRAPUB, Tokyo 1987) pp. 455–471.

KOBAYASHI, K., KASUGA, S., and OKINO, K., *Shikoku Basin and its margins*. In *Backarc Basins: Tectonics and Magmatism* (ed. Taylor, B.) (Plenum Press, New York 1995) pp. 381–405.

MCKENZIE, D. (1979), *Finite Deformation during Fluid Flow*, Geophys. J. R. Astr. Soc. *58*, 687–715.

NICOLAS, A., and CHRISTENSEN, N. I. (1987), *Formation of Anisotropy in Upper Mantle Peridotites—A Review*, Rev. Geophys. *25*, 11–123.

PEACOCK, S. M., *Thermal petrologic structure of subduction zones*. In *Subduction: Top to Bottom* (eds. DeBout, G. E., Scholl, D. W., Kirby, S. H., and Platt, J. P.) (American Geophysical Union, Washington 1996) pp. 119–133.

PLANK, T., and LANGMUIR, C. H. (1993), *Tracing Trace Elements from Sediment Input to Volcanic Output at Subduction Zones*, Nature *362*, 739–743.

RIBE, N. M. (1989), *Seismic Anisotropy and Mantle Flow*, J. Geophys. Res. *94*, 4213–4223.

RIBE, N. M. (1992), *On the Relation between Seismic Anisotropy and Finite Strain*, J. Geophys. Res. *97*, 8737–8747.

RUSSO, R. M., and SILVER, P. G. (1994), *Trench-parallel Flow beneath the Nazca Plate from Seismic Anisotropy*, Science *263*, 1105–1111.

SANDVOL, E., and NI, J. (1997), *Deep Azimuthal Anisotropy in the Southern Kurile and Japan Subduction Zones*, J. Geophys. Res. *102*, 9911–9922.

SHIH, X. R., SCHNEIDER, J. F., and MEYER, R. P. (1991), *Polarization of P and S Waves, and Shear-wave Splitting Observed from the Bucaramanga Nest, Columbia*, J. Geophys. Res. *96*, 12069–12082.

STINE, A. R., FISCHER, K. M., and PARMENTIER, E. M. (1996), *Anisotropy and Shear-wave Splitting Inferred from Back-arc Flow Models*, EOS Trans. AGU *77*(46), Fall Meet. Suppl. F733.

TAMAKI, K., *Opening tectonics of the Japan sea*. In *Backarc Basins: Tectonics and Magmatism* (ed. Taylor, B.) (Plenum Press, New York 1995) pp. 407–420.

TAYLOR, B., KLAUS, A., BROWN, G. R., MOORE, G. F., OKAMURA, Y., and MURAKAMI, F. (1991), *Structural Development of Sumisu Rift, Izu-Bonin Arc*, J. Geophys. Res. *96*, 16113–16129.

VAN DER HILST, R., and SENO, T. (1993), *Effects of Relative Plate Motion on the Deep Structure and Penetration Depth of Slabs below the Izu-Bonin and Mariana Island Arcs*, Earth Planet. Sci. Lett. *120*, 395–407.

VAN DER HILST, R. (1995), *Complex Morphology of Subducted Lithosphere in the Mantle beneath the Tonga Trench*, Nature *374*, 154–157.

WIENS, D. A., SHORE, P. J., McGUIRE, J. J., ROTH, E., BEVIS, M. G., and DRAUNIDALO, K. (1995), *The Southwest Pacific Seismic Experiment*, IRIS Newsletter *14*, 1–4.

WEISSEL, J. K., *Evolution of the Lau basin by the growth of small plates*. In *Island Arcs, Deep Sea Trenches, and Back-arc Basins* (eds. Talwani, M., and Pitman III, W. C.) (American Geophysical Union, Washington 1977) pp. 429–436.

WENK, H.-R., BENNETT, K., CANOVA, G. R., and MOLINARI, A. (1991), *Modeling Plastic Deformation of Peridotite with the Self-consistent Theory*, J. Geophys. Res. *96*, 8337–8349.

WESSEL, P., and SMITH, W. H. F. (1995), *New Version of the Generic Mapping Tools Released*, EOS Trans. AGU *76*, 329.

YANG, X., FISCHER, K. M., and ABERS, G. A. (1995), *Seismic Anisotropy beneath the Shumagin Islands Segment of the Aleutian-Alaska Subduction Zone*, J. Geophys. Res. *100*, 18165–18177.

YU, Y., and PARK, J. (1994), *Hunting for Azimuthal Anisotropy beneath the Pacific Ocean Region*, J. Geophys. Res. *99*, 15399–15421.

ZHANG, S., and KARATO, S. (1995), *Lattice Preferred Orientation of Olivine Aggregates Deformed in Simple Shear*, Nature *375*, 774–777.

ZHAO, D., WIENS, D. A., DORMAN, L., HILDEBRAND, J., WEBB, S., and McDONALD, M. (1995), *High Resolution Seismic Tomography of the Tonga Slab and Lau Back-arc Basin*, EOS Trans. AGU *76*(46), Fall Meet. Suppl. F392.

(Received December 10, 1996, revised/accepted June 16, 1997)

 To access this journal online:
http://www.birkhauser.ch

II. Mantle Heterogeneity vs. Anisotropy— 3-D Velocity and Density Structures and Inferences on Mantle Dynamics

Pure appl. geophys. 151 (1998) 479–493
0033–4553/98/040479–15 $ 1.50 + 0.20/0

© Birkhäuser Verlag, Basel, 1998

Pure and Applied Geophysics

Passive Seismology and Deep Structure in Central Italy

A. AMATO,[1] L. MARGHERITI,[1] R. AZZARA,[1] A. BASILI,[1] C. CHIARABBA,[1]
M. G. CIACCIO,[1] G. B. CIMINI,[1] M. DI BONA,[1] A. FREPOLI,[1] F. P. LUCENTE,[1]
C. NOSTRO[1] and G. SELVAGGI[1]

Abstract—In the last decade temporary teleseismic transects have become a powerful tool for investigating the crustal and upper mantle structure. In order to gain a clearer picture of the lithosphere-asthenosphere structure in peninsular Italy, between 1994 and 1996, we have deployed three teleseismic transects in northern, central, and southern Apennines, in the framework of the project *GeoModAp* (European Community contract EV5V-CT94–0464). Some hundreds of teleseisms were recorded at each deployment which lasted between 3 and 4 months. Although many analyses are still in progress, the availability of this high quality data allowed us to refine tomographic images of the lithosphere-asthenosphere structure with an improved resolution in the northern and central Apennines, and to study the deformation of the upper mantle looking at seismic anisotropy through shear-wave splitting analysis. Also, a study of the depth and geometry of the Moho through the receiver function technique is in progress. Tomographic results from the northernmost 1994 and the central 1995 teleseismic experiments confirm that a high-velocity anomaly (HVA) does exist in the upper 200–250 km and is confined to the northern Apenninic arc. This HVA, already interpreted as a fragment of subducted lithosphere is better defined by the new temporary data, compared to previous works, based only on data from permanent stations. No clear high-velocity anomalies are detected in the upper 250 km below the central Apennines, suggesting either a slab window due to a detachment below southern peninsular Italy, or a thinner, perhaps continental slab of Adriatic lithosphere not detectable by standard tomography. We found clear evidence of seismic anisotropy in the uppermost mantle, related to the main tectonic processes which affected the studied regions, either NE–SW compressional deformation of the lithosphere beneath the mountain belt, or arc-parallel asthenospheric flow (both giving NW–SE fast polarization direction), and successive extensional deformation (~E–W trending) in the back-arc basin of northern Tyrrhenian and Tuscany. Preliminary results of receiver function studies in the northern Apennines show that the Moho depth is well defined in the Tyrrhenian and Adriatic regions while its geometry underneath the mountain belt is not yet well constrained, due to the observed high complexity.

Key words: Passive seismology, central Mediterranean, Italy, seismic tomography, seismic anisotropy, receiver function, upper mantle structure, geodynamics.

Introduction

The evolution of the Apennines-Tyrrhenian region is still poorly understood due to a limited knowledge of the deep structure. Evidence of lithospheric subduction

[1] Istituto Nazionale di Geofisica, Via di Vigna Murata, 605; 00143, Roma, Italy. Tel: 0039-6-51860414, 0039-6-51860486; Fax: 0039-6-5041181; e-mail: amato@ing750.ingrm.it, margheriti@ing750.ingrm.it

during the Neogene formation of the Apenninic belt comes from geological, petrological, and seismological evidence (MALINVERNO and RYAN, 1986; ROYDEN *et al.*, 1987; SERRI *et al.*, 1993; SELVAGGI and AMATO, 1992), although there is open debate on whether a subduction model is applicable to the whole region or not (see e.g., LAVECCHIA, 1988). Moreover, there are many uncertainties on the slab geometry, extension, and nature (if continental or oceanic) (SPAKMAN *et al.*, 1993; AMATO *et al.*, 1993). Hereafter in the introduction, we first describe the tectonic setting of peninsular Italy and the related open problems, then we focus on the seismological studies carried our during GeoModAp (a multidisciplinary project funded by the E.C.). Finally, we present preliminary results of tomography and seismic anisotropy studies achieved to date.

The Apennines Mountain Belt and the Tyrrhenian Basin

The present-day geologic setting of the Italian peninsula is very complex due to the interaction of different geodynamic processes. Italy is located between the African and the European plates, which are presently converging approximately in a N–S direction at a rate of less than 1 cm/year (ARGUS *et al.*, 1989; DEWEY *et al.*, 1989; DE METS *et al.*, 1990). Along the northern Apennines from Upper Miocene to Quaternary the contemporaneous existence of extension in the Tyrrhenian area and along the inner margin of the Apenninic belt, and compression along the outer eastern arcs of the peninsula, is commonly observed (PATACCA and SCANDONE, 1989; FREPOLI and AMATO, 1997).

The present configuration of the Apennines consists of two major arcs, the northern Apenninic and the Calabrian arcs (Fig. 1), separated in central-southern Apennines by a ~N–S trending fault zone that according to some authors is related to a deep lithospheric discontinuity (PATACCA and SCANDONE, 1989; AMATO *et al.*, 1993). Subcrustal seismicity is detected only below the northern

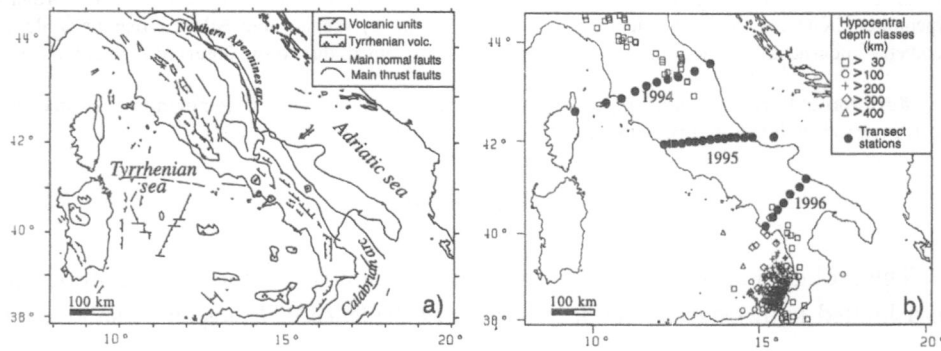

Figure 1
(a) Simplified structural map of Italy. (b) Subcrustal seismicity sorted by depth (see legend in the box) and approximate location of the three transects seismic stations (solid circles).

Apennines (SELVAGGI and AMATO, 1992), and the Calabrian arc (ANDERSON and JACKSON, 1987), with a gap in between (Fig. 1). It is widely accepted that the southern Tyrrhenian Sea formed as a back-arc basin related to subduction of the Ionian lithosphere beneath the Calabrian arc (MALINVERNO and RYAN, 1986). This subduction process is well documented by the intermediate and deep seismicity located down to 450 km (Fig. 1) in the southern Tyrrhenian Sea (ANDERSON and JACKSON, 1987; FREPOLI et al., 1996), by seismic tomography (SPAKMAN et al., 1993; AMATO et al., 1993; SELVAGGI and CHIARABBA, 1995) and by the presence of calc-alkaline volcanism in the Aeolian Islands (BARBERI et al., 1973), even if it is probably limited to the effect of a passively sinking slab (PATACCA and SCANDONE, 1989). The northern Apenninic arc developed with a series of thrusting episodes over the Adriatic continental plate, which flexed below the foredeep for as much as 8 km since Pliocene (ROYDEN et al., 1987). The presence of subcrustal seismicity down to 90 km depth (Fig. 1) (SELVAGGI and AMATO, 1992) suggests that the arc has developed due to westward subduction of the Adriatic plate. The depth extent of the subducted lithosphere, observed by tomography, and its composition and geometry (including presence and extent of slab detachments) are still a matter of debate.

Studies of seismic anisotropy carried out in the Italian peninsula, besides those obtained by the GeoModAp and described in this paper, include a study of P-wave residuals by PLOMEROVA (1997), who found that underneath the northern Apennines, fast direction within the lithosphere plunge to the WSW, approximately parallel to the dip direction of the subducting slab. Anisotropy obtained using Pn phases (MELE et al., 1995) shows NW–SE fast directions (parallel to the surface structural trends) at the crust mantle boundary beneath the northern Apenninic arc.

The Moho depth is not well defined everywhere beneath the Italian peninsula due to an insufficient coverage of good deep reflection and refraction surveys. The results from CROP (CROsta Profonda = deep crust) deep seismic reflection surveys for the southern (CROP 04), the northern (CROP 03) and the central Apennines (CROP11) are becoming available, but in many regions the deep crust and the Moho geometry are not well resolved. By now the existence of different types of Moho on either side of the Apennines is well established—in the Tyrrhenian area Moho depth is around 20–25 km, whereas in the Adriatic plate it is about 35 km (GEISS, 1987).

GeoModAp Seismological Studies of the Deep Structure

The main goal of the project GeoModAp (GEOdynamic MODeling of an active region of the Mediterranean: the APennines) is the definition of a geodynamic model of the Apennines-Tyrrhenian system through the collection of multidisciplinary data specifically designed to provide tighter constraints for modeling. The main fields of activities of GeoModAp are summarized in Table 1. Relevant papers

Table 1

GeoModAp main fields of activities

Geodesy	Collection and processing of new GPS data. Analysis of old triangulation data.
Seismology	Compilation of a new catalogue of seismic moments. Tomographic studies. Deployment of three temporary seismological transects across the Apennines to infer Moho depth and dip from teleseismic receiver functions, and to investigate anisotropic structures in the upper mantle.
Present-day stress	Compilation of a database containing borehole breakout measurements and fault plane solutions.
Structural geology	Paleomagnetic analyses and anisotropy of magnetic susceptibility. Field observations in both the brittle and the ductile deformation regime.
Modeling	Two and three-dimensional laboratory models. Numerical modeling based on finite elements codes.

published by GeoModAp include: AMATO *et al.* (1995); FACCENNA *et al.* (1996); GIUNCHI *et al.* (1996); SPERANZA *et al.* (1996); MARGHERITI *et al.* (1996); BASSI *et al.* (1997); FREPOLI and AMATO (1997).

Concerning the seismological studies carried out to constrain the geometry of the deep structures, we have obtained an improved three-dimensional velocity model of the lithosphere-asthenosphere system beneath the Apennines. We have used new data collected during temporary deployments, consisting of three linear arrays, that are the first experiments of this kind ever carried out in Italy. There are three main objectives of these temporary arrays, namely (a) refining the tomographic images, as already mentioned, (b) studying the intensity and direction of anisotropy in the crust and upper mantle; and (c) studying the Moho geometry through the determination and analysis of teleseismic receiver functions.

The Teleseismic Transects

The three transects we have deployed (Fig. 1) are linear arrays crossing the northern, central, and southern Apennines (AMATO *et al.*, 1995, 1996). The stations were equipped with 24-bit RefTek digitizers, in continuous recording mode at 40 sps or 20 sps, and with Lennartz 5 s or broad-band Guralp sensors (CMG3, CMG4 and CMG40). About 40–50 Gbytes of data were recorded during each experiments, which lasted for 3–4 months.

The first transect was installed in the northern Apennines during the summer of 1994, with ten stations from Corsica to M. Conero. This transect was 340 km long, and average station spacing was about 35 km in the Italian peninsula. During its three months of operation, 150 teleseismic events were recorded by at least one station of the temporary array, and 70 by at least 5 stations. For all these events the digital seismograms recorded by about 20 permanent seismic stations of the ING National Network were also selected in order to obtain a larger data set for the northern Apennines.

The second transect was deployed in the spring of 1995 in the central Apennines with 15 stations from the Tyrrhenian coast north of Rome to the Tremiti Islands in the Adriatic Sea. It was about 250 km long and average stations spacing across the Apennines was about 20 km. During four months of deployment about 120 teleseismic events were recorded, 60 of which were by at least 9 stations.

The third transect was deployed across the southern Apennines in the summer of 1996 with 7 stations from the Cilento to the Apulian region and a dense 2D array of 12 stations around the town of Potenza, in collaboration with the University of Grenoble. The transect is about 200 km long and average stations spacing is about 30 km; data processing is in progress.

Preliminary Results

The results we present in this paper were derived from the analysis of the teleseismic data collected during the 1994 northern Apennines and the 1995 central Apennines experiments.

Seismic Tomography

Teleseismic transects improved the resolution of tomographic images by adding new recording sites, though only for a few months, in regions characterized by a paucity of seismic stations (Fig. 2a). We applied the well established teleseismic tomography technique ACH (AKI et al., 1977) to an increased data set of teleseismic arrival times, accurately picked from digital wave forms recorded by the ING National Seismic Network (RSNC) and by the temporary linear arrays. At present, we computed two separate tomographic models beneath the Apenninic regions crossed by the northern and central transects, including in both inversions data recorded at the stations of the RSNC located in the surroundings of our temporary stations (Fig. 2a).

In the northern Apennines our improved tomographic images confirm the main features outlined by previous reconstruction (e.g., AMATO et al., 1993), however the increased resolution allows us to depict some details better than before. The most relevant result in the northern Apennines (Fig. 2) is that the shallowest portion of the slab, imaged by seismic tomography as a high velocity anomaly, appears to be laterally heterogeneous (down to 100–130 km) and has an arcued shape, while its deepest part is more linear, suggesting that the slab curvature was acquired in the late stage of subduction. We find no evidence of slab detachment in the upper 200 km below the northern Apenninic arc, like the one evidenced by SPAKMAN *et al.* (1993). It must be reminded that the vertical resolution of teleseismic tomography is poor, and therefore a thin low velocity layer (due for instance to a slab detachment) throughout the region would not be recognized. However, it must be also noticed that the high velocity anomaly must be shallow (<200 km), because of the strongly azimuthal variability of the relative residuals in a profile which is only 300 km long. As a result, the tomograms reveal that the high velocity anomaly (with *Vp* perturbations up to +5–7%) is limited to the upper 200 km, and is less intense below this depth, suggesting either that the total slab length is about 200 km, or that the deepest portion has been thermally assimilated or detached.

The results of the tomographic inversion beneath the central Apennines (between 35 and 250 km, Fig. 2) do not show the presence of significant high velocity anomalies (larger than +2%); this region of the mantle appears as a zone of separation between the northern Apennines and the Calabrian arcs, both character-ized by clear evidence of subducted lithosphere. On the other hand, some low velocities spots (as low as −6%) are evident along the Tyrrhenian coast. The

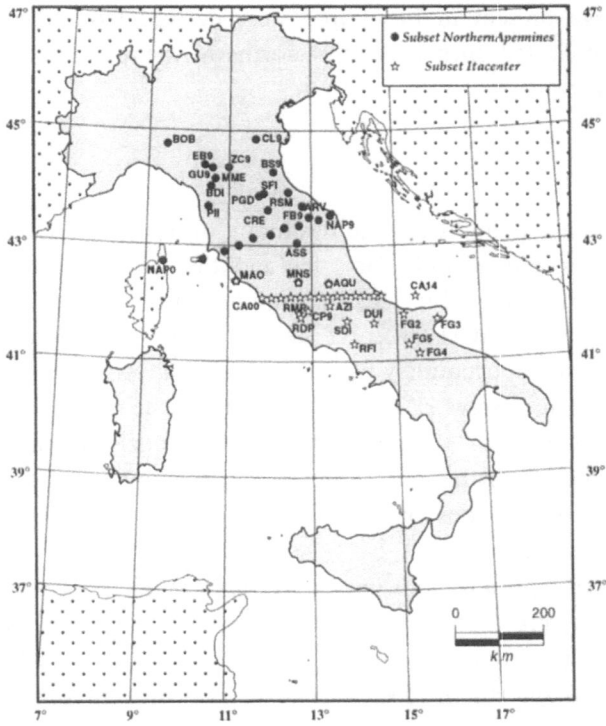

Figure 2 (a)

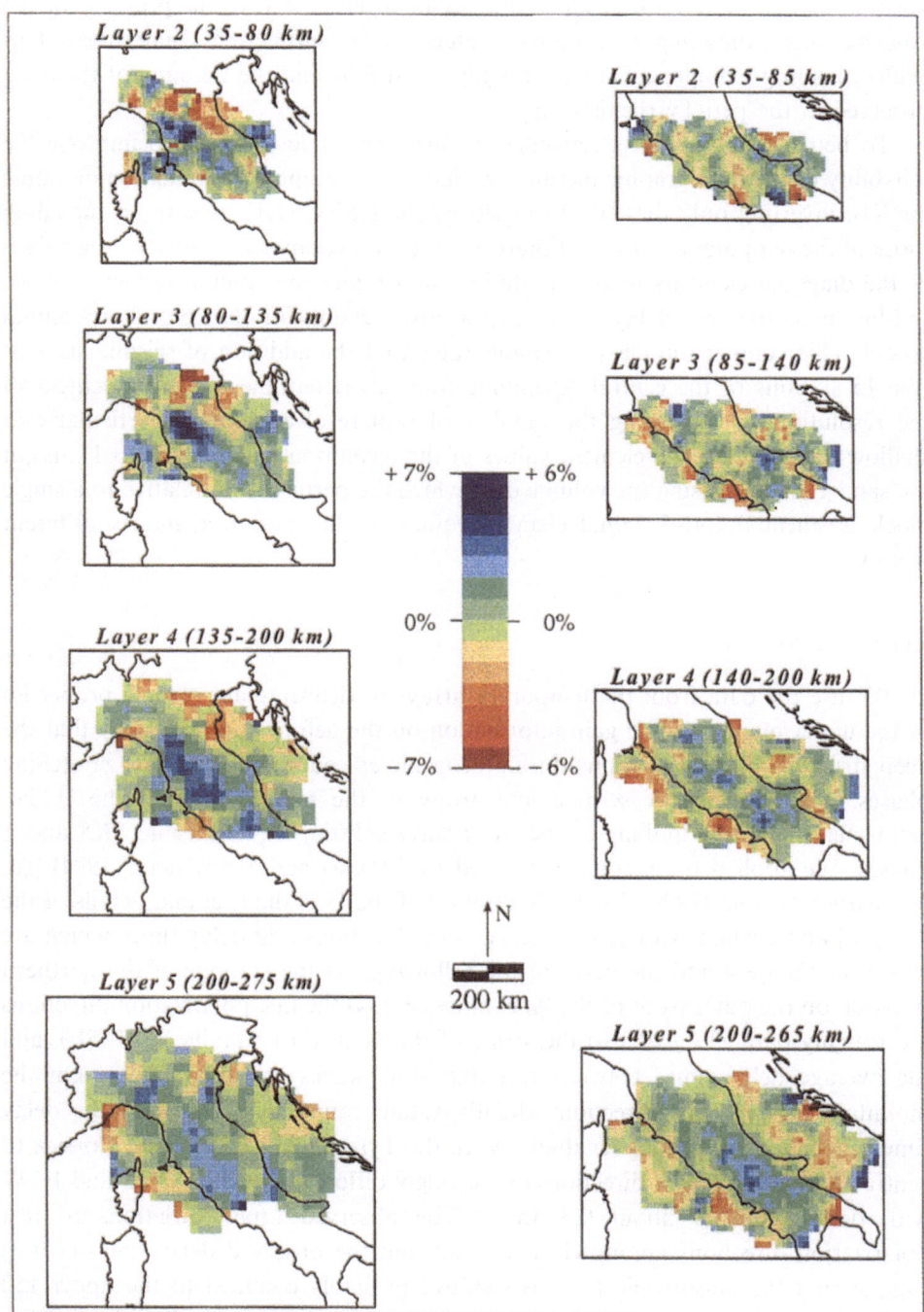

Figure 2
Tomography results for the northern and central Apennines: (a) Stations used in the two inversions
(solid circles: northern Apennines model, stars: central Apennines model). Velocity perturbation of *P*
wave in the upper mantle of the northern Apennines (b) and of the central Apennines (c). Note the
different range of the computed anomalies, particularly for the shallowest layers, as high as 7% in the
northern Apennines, lower than 2–3% in the central Apennines.

position of these low velocity anomalies seems to be related to the presence of the volcanic belt, consisting of eruptive centers active during the Quaternary. This evidence is in good agreement with the high heat flow and the thinning of the crust observed in the peri-Tyrrhenian area.

To better evaluate the effectiveness of the teleseismic transects to improve the reliability of the tomographic pictures, we have also computed parallel tomographic models, inverting only the data recorded by the RSNC stations without including those of the temporary array. In Figure 3 we show a comparison between the values of the diagonal elements of the resolution matrix and the column elements of the resolution matrix into a layer and into a cross section for the central Apennines model. This comparison clearly demonstrates that the addition of seismic stations (the 15 stations of the central Apenninic transect) determines an improvement of the resolution by increasing the number of well resolved blocks, particularly at shallow depths (diagonal element values of the resolution matrix close to 1) and at the same time decreasing the volume over which the perturbation relative to a single block is smeared (off-diagonal element values of the resolution matrix different from 0).

Anisotropy Studies

We use the data from the temporary arrays to delineate anisotropic properties of the upper mantle and to gain information on the deformational history that the deep structures have undergone during orogenic episodes and successive stretching phases. The presence of seismic anisotropy in the mantle beneath the Italian peninsula is detected looking at the shear-wave splitting of teleseismic *SKS* and *S* phases. We applied the method suggested by ZHANG and SCHWARTZ (1994) [see MARGHERITI *et al.* (1996) for the description of the 1994 data set and details of the analysis] and evaluated weighted average fast directions and delay times which are shown in Figure 4 and discussed in the following. At the stations of the northern transect, on the outer front of the Apennines (Fig. 4) the fast polarization directions are roughly parallel ($\pm 20°$) to the strike of the main thrust faults (NW–SE), and the average delay times between fast and slow waves are about 0.5 s. On the mountain belt the fast directions slightly rotate counterclockwise and the delay times increase at values higher than 1 s. In the Tyrrhenian region, from Corsica to central Tuscany, the fast directions are strongly different, between ENE and E–W with delay times of about 0.5–1.0 s. The observed rapid variations of fast polarization directions among close stations and the observed delay times (~ 1 s) suggest that the anisotropic layer is shallow, probably confined to the upper 150 km. The variability of ϕ and δt at the same station for different earthquakes implies a 3D structure more complex than a single anisotropic flat layer, which we will try to investigate carefully in the future looking at azimuthal and distance dependence. For the central Apennines transect only a few *SKS* phases suitable for shear-wave

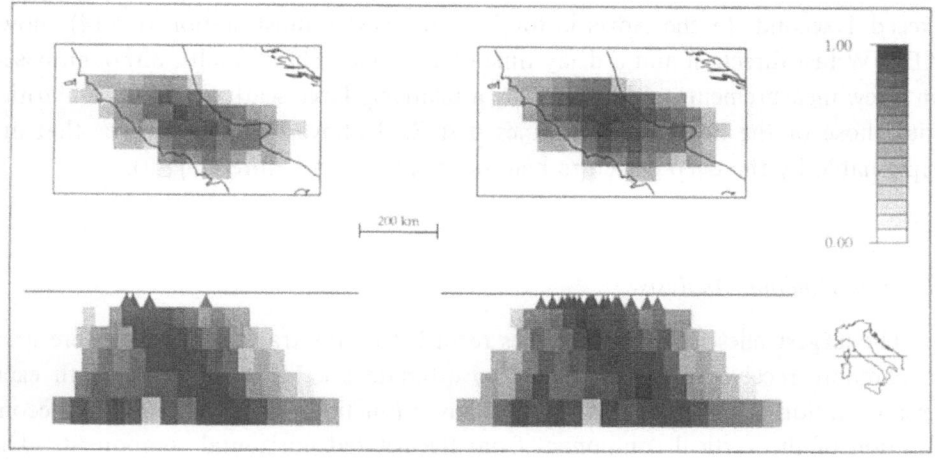

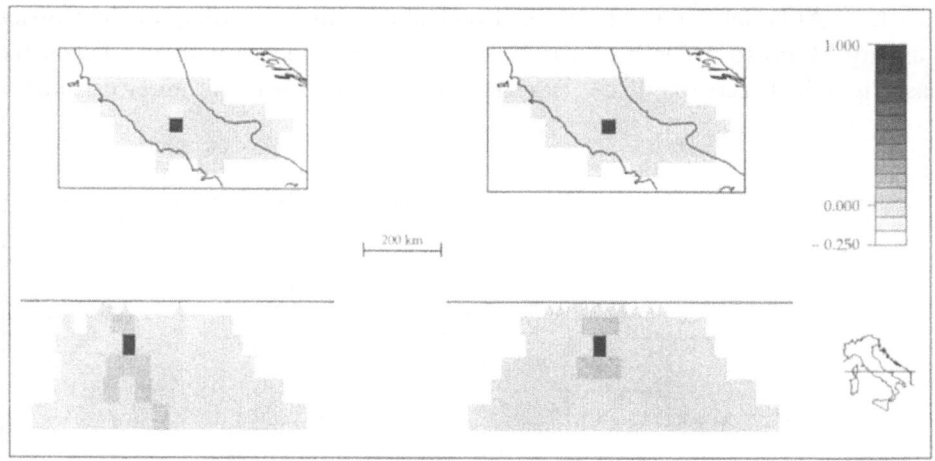

Figure 3
Upper part, comparison into a layer and into a cross section between the resolution matrix diagonal elements of the model including (on the right) and not including (on the left) the data recorded by the central Apenninic transect stations. Lower part, comparison between the resolution matrix column elements, computed for the same block in both the models.

splitting analysis were recorded at each station. Work is in progress to constrain the anisotropic parameters. In a preliminary analysis (Fig. 4) we find E–W to SW–NE fast directions and delay times of about 1 second in the Tyrrhenian side of the

Apennines (the 8 western stations). Two stations on the central section of the chain show NW–SE fast direction that would be consistent with a NE–SW compressional deformation of the lithosphere, whereas the easternmost stations on the Adriatic side have scattered fast directions. The delay times in this area generally exceed 1 second. In the Adriatic foreland the easternmost station (CA14) shows NE–SW fast direction and a delay time of 1 second. These results, although based on a few measurements and affected by a relatively large scatter, exhibit similarities with those of the northern Apennines described above, with differences that are explainable by the deep structure heterogeneity shown before (Fig. 2).

Receiver Function Analysis

The largest teleseismic earthquakes recorded by the transect ($M > 6$) were used to compute receiver functions useful to estimate the Moho depth beneath each seismic station. Tangential and radial receiver functions were computed by deconvolution of the vertical component from the rotated horizontal components. The interpretation of the receiver functions in terms of homogeneous layers models, in progress at the moment, will help to constrain the crustal and upper mantle structure. Although the results are still preliminary, the upwarding of the Corsica-Tuscany Moho to a depth of 20–25 km in correspondence with the Tyrrhenian basin is well detected (Fig. 5). Beneath the Apenninic belt, the receiver function

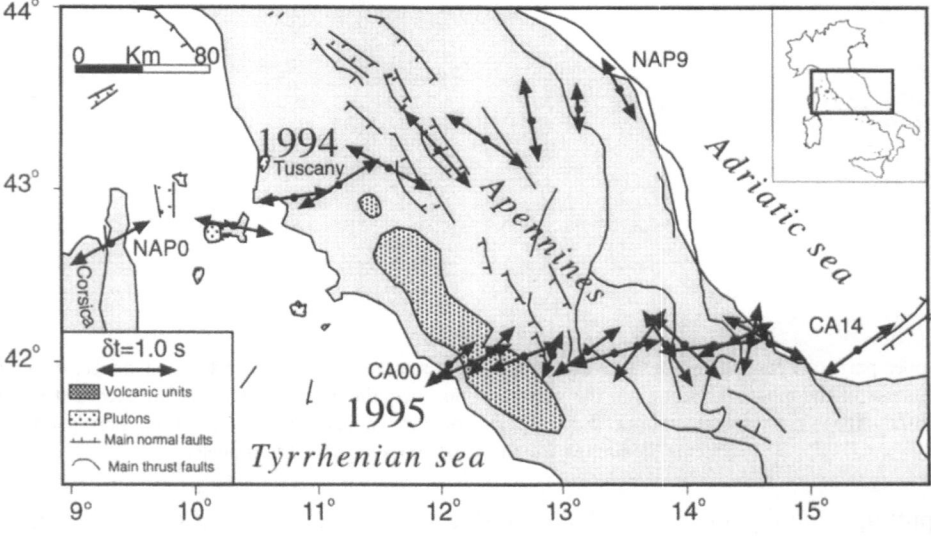

Figure 4
Splitting results for the study of seismic anisotropy in the northern and central Apennines. Arrows show average fast directions, their length is proportional to the delay times.

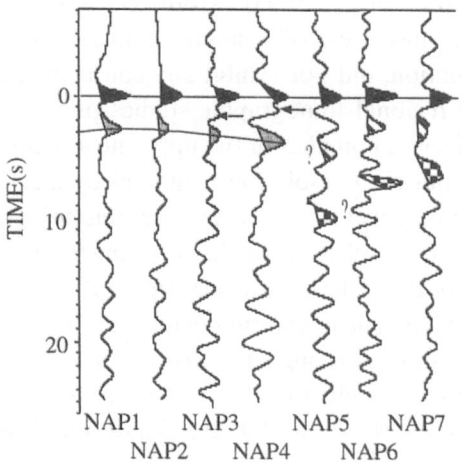

Figure 5

Example of radial receiver functions computed for a M 6.4 earthquake at $\Delta = 87°$. In the western station from Elba to eastern Tuscany (NAP1-NAP4) a *P*-to-*S* conversion attributed to the crust-mantle boundary, is well evident at 3–3.5 s after the direct *P* pulse, corresponding to a Moho depth of 20–25 km. The minimum Moho depth (provided that the velocity in the crust are the same) is below Elba Island (NAP1) and the Tuscany coast, and an eastward dipping is also evident. Below the Apennines (NAP5–NAP7) the structure appears to be more complex, with many converted phases after the direct arrival, probably due to multiple reverberations.

complexity strongly increases and large azimuthal variations are observed. Work is in progress to model the observed *P*-to-*S* conversions with synthetic ray-tracing techniques.

Discussion and Conclusive Remarks

Tomographic pictures of the Italian peninsula produced in the last decade have shown different patterns of the high velocity features that are thought to depict the slabs at depth. In the Calabrian arc, tomographic images of the upper mantle, computed using various techniques, identify the presence of a deep high velocity body interpreted as a descending slab, continuously deepening to about 500 km toward NW with a dip angle of about 70° (SELVAGGI and CHIARABBA, 1995; CIMINI and DE GORI, 1997), in good agreement with the distribution of the intermediate and deep seismicity (FREPOLI et al., 1996). In the northern Apennines, teleseismic tomography enhanced the presence of a continuous high velocity zone from the surface down to 200–300 km (AMATO et al., 1993), whereas the joint inversion of bulletin arrival times of regional and teleseismic events revealed the existence of a slab detachment in the upper 150–200 km beneath the entire

Apenninic belt (SPAKMAN *et al.*, 1993). More recent work by PIROMALLO and
MORELLI (1997) which makes use of a similar data set and parameterization, but
with an increased resolution, did not exhibit any continuous detachment below the
Italian region. All the regional tomographic studies previously mentioned did not
enhance significant velocity anomalies in the uppermost mantle (<250 km) beneath
the central Apennines but their resolution is limited by insufficient station spacing
and inhomogeneous ray coverage. Our passive seismological studies described
above, although not yet complete, provide new constraints in understanding the
geodynamic evolution of the Italian peninsula. In the deep structure of the northern
Apennines no clear detachment in the subducted slab is revealed by the improved
tomographic reconstruction. Although the vertical resolution of the tomographic
images would not allow us to identify thin (a few tens of km) horizontal features,
the seismicity distribution down to 90 km depth indicates that the proposed
detachment should be located at greater depth (~100–200 km), where we find the
largest high velocity anomaly (Fig. 2). It is interesting to note that modeling long
wavelength gravity anomalies, ROYDEN (1993) suggests the existence of a dense
slab between 50 and 200 km depth below Tuscany, slightly to the west of our
tomographic slab. The steep dipping of the tomographic slab suggests that the
present configuration was reached through a progressive slab rollback, determined
either by older (deeper) oceanic lithosphere drawing from below, or alternatively by
an eclogitized continental lithosphere. If the former case applies, we may think that
the sinking process is close to its end, otherwise it may be a self-sustained process
that lasts until the entire continental lithosphere is consumed. The apparent
increase of the slab curvature toward shallow depth, which fits well with geological
and paleomagnetic results (SPERANZA *et al.*, 1996), could be due to the recent
(during/after Pliocene) subduction of a laterally heterogeneous lithosphere in the
northern Apenninic arc (for instance a central, denser oceanic part with lighter
portions of continental lithosphere at one or both sides).

The crucial point to understand this heterogeneous subduction-sinking process
along the Italian peninsula could be the nature and the role played by the thicker
Adriatic lithosphere in the central-southern Apennines, where a clear-cut separation
between the northern Apenninic and the southern Tyrrhenian arcs is evident. In
fact no subcrustal seismicity is detected in the central-southern Apennines (Fig. 1b)
and our improved tomographic images do not depict significant velocity anomalies
(>2%) in the uppermost mantle beneath the central Apennines as if there is a slab
window between the two arcs.

The study of deep seismic anisotropy in the northern Apennines has revealed
that the mantle lithosphere is deformed in the Adriatic foreland and in the belt,
whereas in the Tyrrhenian side the anisotropic layer probably extends well below the
lithosphere. The consistence of deep anisotropy directions with surface geology and
tomographic images demonstrates that both the compressional orogenic process and
the extensional post-orogenic evolution are deep events which affect at least the

entire lithosphere and not just the crust. Deep deformation of the Adriatic lithosphere generates NW–SE fast directions. We do not know whether this deformation is inherited by NE compression acquired in an early stage of Apenninic subduction, or if it is due to active compressional stress acting today in the lithosphere. If the latter hypothesis is true, since the compressional belt in the brittle crust is confined to a narrow region on the easternmost front of the belt (FREPOLI and AMATO, 1997) there must be a decoupling (in the lower crust?) between an extending upper crust and the underlying Adriatic lithosphere beneath NAP4-NAP7.

The strongly different, mostly ENE trending, anisotropy direction in the Tyrrhenian region, associated with the small lithosphere thickness (SUHADOLC and PANZA, 1989), and the correspondence with surface structural trends (JOLIVET *et al.*, 1997) suggest that the Miocene to present extension in this region was driven by deep asthenospheric flow and that asthenospheric uplift did not occur with a radial symmetry, but rather with a well-defined orientation. Since this orientation is roughly parallel to the migration direction of the orogenic activity, we propose that the E–W asthenospheric flow was triggered by the slab retreat. This process is possibly at a further stage in the central Apennines where the proposed slabless window down to 250 km depth has allowed the 'Tyrrhenian asthenosphere' to flow underneath the mountain belt, determining a more diffuse pattern of trench-normal fast polarization directions. The region over which this asthenospheric flow is hypothesized (the central and southern Apennines) is characterized by widespread extension in the crust (AMATO *et al.*, 1995) and general uplifting (WESTWAY, 1993), suggesting that the present-day deformation is driven by deep processes.

REFERENCES

AKI, K., CHRISTOFFERSON, A., and HUSEBYE, E. (1977), *Determination of Three-dimensional Seismic Structure of Lithosphere*, J. Geophys. Res. 82, 277–296.

AMATO, A., ALESSANDRINI, B., CIMINI, G. B., FREPOLI, A., and SELVAGGI, G. (1993), *Active and Remnant Subducted Slabs beneath Italy: Evidence from Seismic Tomography and Seismicity*, Annali di Geofisica 26, 201–214.

AMATO, A., MONTONE, P., and CESARO, M. (1995), *State of Stress in Southern Italy from Borehole Breakout and Focal Mechanism Data*, Geophys. Res. Lett. 22, 3119–3122.

AMATO, A., AZZARA, R., BASILI, A., CHIARABBA, C., CIMINI, G. B., DI BONA, M., MARGHERITI, L., NOSTRO, C., and SELVAGGI, G. (1995), *The Northern Apennines Teleseismic Transect (1994)*, Technical report, Pubbl. of the Istituto Nazionale di Geofisica, no. 568.

AMATO, A., AZZARA, R., BASILI, A., CHIARABBA, C., CIMINI, G. B., DI BONA, M., LUCENTE, F. P., MARGHERITI, L., NOSTRO, C., SELVAGGI, G., ACERRA, C., BACCHESCHI, S., BATTELLI, P., FRANCESCHELLI, G., MARCHETTI, A., MODICA, G., MONDIALI, L., PICCOLINI, L., PIRRO, F., PIRRO, M., TARDINI, R., and VALLOCCHIA, M. (1996), *The Central Apennines Teleseismic Transect (1995)*, Technical Report, Pubbl. of the Istituto Nazionale di Geofisica, no. 581.

ANDERSON, H. J., and JACKSON, J. (1987), *The Deep Seismicity of the Tyrrhenian Sea*, Geophys. J. R. Astr. Soc. London 91, 613–637.

ARGUS, D. F., GORDON, R. G., DEMETS, C., and STEIN, S. (1989), *Closure of the Africa-Eurasia-North America Plate Motion Circuit and Tectonics of the Gloria Fault*, J. Geophys. Res. *94*, B5, 5585–5602.

BARBERI, F., GASPARINI, P., INNOCENTI, F., and VILLARI, L. (1973), *Volcanism of the Southern Tyrrhenian Sea and its Geodynamical Implications*, J. Geophys. Res. *78*, 5221–5232.

BASSI, G., and SABADINI, R. (1994), *The Importance of Subduction for the Modern Stress Field in the Tyrrhenian Area*, Geoph. Res. Lett. *21*, 329–332.

BASSI, G., SABADINI, R., and REBAÏ, S. (1997), *Modern Tectonic Regime in the Tyrrhenian Area: Observations and Models*, Geophys. J. Int. *129*, 330–346.

CIMINI, G. B., and DE GORI, P. (1997), *Upper Mantle Velocity Structure beneath Italy from Direct and Secondary P-wave Teleseismic Tomography*, Annali di Geofisica *XL*, 175–194.

DE METS, G., GORDON, R. G., ARGUS, D. F., and STEIN, S. (1990), *Current Plate Motions*, Geophys. J. Int. *101*, 425–478.

DEWEY, J. F., HELMAN, M. L., TURCO, E., HUTTON, D. H. W., and KNOTT, S. D., *Kinematics of the western Mediterranean*. In *Alpine Tectonics* (eds. Coward, M. P., Dietrich, D. and Park, R. G.) (Geol. Soc. Sp. Pub., London 1989) pp. 265–283.

FACCENNA, C., DAVY, P., BRUN, J. P., FUNICIELLO, R., GIARDINI, D., MATTEI, M., and NALPAS, T. (1996), *The Dynamics of Back-arc Extension: An Experimental Approach to the Opening of the Tyrrhenian Sea*, Geophys. J. Int. *126*, 781–795.

FREPOLI, A., SELVAGGI, G., CHIARABBA, C., AMATO, A. (1996), *State of Stress in the Southern Tyrrenian Subduction Zone from Fault-plane Solution*, Geophys. J. Int. *125*, 879–891.

FREPOLI, A., and AMATO, A. (1997), *Contemporaneous Extension and Compression in the Northern Apennines from Earthquake Fault Plane Solutions*, Geophys. J. Int. *129*, 368–388.

GEISS, E. (1987), *A New Compilation of Crustal Thickness in the Mediterranean Area*, Ann. Geophys. *5B*(6), 623–630.

GIUNCHI, C., SABADINI, R., BOSCHI, E., and GASPERINI, P. (1996), *Dynamic Models of Subduction: Geophysical and Geological Evidence in the Tyrrhenian Sea*, Geophys. J. Int. *126*, 555–578.

JOLIVET, L., FACCENNA, C., BRUNET, C., CADET, J. P., FUNICIELLO, R., MATTEI, M., ROSSETTI, F., STORTI, F., GOFFE', B., and THEYE, T. (1997), *Mid-crustal Shear Zones in Post-orogenic Extension: The Northern Tyrrhenian Sea Case*, J. Geophys. Res., submitted.

LAVECCHIA, G. (1988), *The Tyrrhenian-Apennines System: Structural Setting and Seismotectogenesis*, Tectonophysics *147*, 263–296.

MALINVERNO, A., and RYAN, W. B. F. (1986), *Extension in the Tyrrhenian Sea and Shortening in the Apennines as Results of Arc Migration Driven by Sinking of the Lithosphere*, Tectonics *5*, 227–245.

MARGHERITI, L., NOSTRO, C., COCCO, M., and AMATO, A. (1996), *Seismic Anisotropy Beneath the Northern Apennines (Italy) and its Tectonic Implication*, Geophys. Res. Lett. *23*, 2721–2724.

MELE, G., ROVELLI, A., SEBER, D., BARAZANGI, M., and HEARN, T. (1995), *High-Frequency Seismic Wave Propagation in the Uppermost Mantle Beneath Italy*, IUGG XXI General Assembly Boulder (Colorado), 3–14 July 1995.

PATACCA, E., and SCANDONE, P., *Post-Tortonian mountain building in the Apennines. The role of the passive sinking of a relic lithospheric slab*. In *The Lithosphere in Italy* (eds. Boriani, A., Bonafede, M., Picardo, G. B., and Vai, G. G.) (Accademia dei Lincei, Roma 1989) pp. 157–176.

PIROMALLO, C., and MORELLI, A. (1997), *Modelling Mediterranean Upper Mantle by Travel-time Tomography: Preliminary Results*, Annali di Geofisica, submitted.

PLOMEROVA, J. (1997), *Seismic Anisotropy in Tomographic Studies of the Upper Mantle Beneath Southern Europe*. Annali di Geofisica *XL*, 11–121.

ROYDEN, L., PATACCA, E., and SCANDONE, P. (1987), *Segmentation and Configuration of Subducted Lithosphere in Italy: An Important Control on Trust-belt and Foredeep-basin Evolution*, Geology *15*, 714–717.

ROYDEN, L. (1993), *The Tectonic Expression Slab Pull at Continental Convergent Boundaries*, Tectonics *12*(2), 303–325.

SELVAGGI, G., and AMATO, A. (1992), *Intermediate-depth Earthquakes in the Northern Apennines (Italy): Evidence for a Still Active Subduction?* Geophys. Res. Lett. *19*, 2127–2130.

SELVAGGI, G., and CHIARABBA, C. (1995), *Seismicity and P-wave Velocity Image of the Southern Tyrrhenian Subduction Zone*, Geophys. J. Int. *121*, 818–826.

SERRI, G. F., INNOCENTI, F., and MANETTI, P. (1993), *Geochemical and Petrological Evidence of the Subduction of Delaminated Adriatic Continental Lithosphere in the Genesis of the Neogene-Quaternary Magmatism of Central Italy*, Tectonophysics *223*, 117–147.

SPAKMAN, W., VAN DER LEE, S., VAN DER HILST, R. (1993), *Travel-time Tomography of the European-Mediterranean Mantle Down to 1400 km*, Phys. Earth Plan. Int. *79*, 3–74.

SPERANZA, F., SAGNOTTI, L., and MATTEI, M. (1996), *Tectonics of the Umbria-Marche-Romagna Arc (Central-northern Apennines, Italy): New Paleomagnetic Constraints*, J. Geophys. Res. *102*, 3153–3166.

SUHADOLC, P., and PANZA, G. F. (1995), *Physical properties of the lithosphere-asthenosphere system in Europe from geophysical data.* In *The Lithosphere in Italy. Advances in Earth Science Research* (eds. Boriani, A., Bonafede, M., Piccardo, G. B., and Vai, G. B.) Atti Convegno Lincei *80*, 15–40.

WESTAWAY, R. (1993), *Quaternary Uplift of Southern Italy*, J. Geophys., Res. *98*, 21741–21772.

ZHANG, Z., and SCHWARTZ, S. Y. (1994), *Seismic Anisotropy in the Shallow Crust of the Loma Prieta Segment of the San Andreas Fault System*, J. Geophys. Res. *99*, 9651–9661.

(Received October 23, 1996, revised June, 1997, accepted June 16, 1997)

To access this journal online:
http://www.birkhauser.ch

Pure appl. geophys. 151 (1998) 495–502
0033–4553/98/040495–08 $ 1.50 + 0.20/0

▌Pure and Applied Geophysics

Pn Anisotropy in the Northern Apennine Chain (Italy)

GIULIANA MELE[1]

Abstract — *Pn* travel times recorded by the stations of the Italian seismic network have been used to image the azimuthal variations of seismic velocity in the uppermost mantle beneath the northern part of the Apennine chain. The azimuthal variation of *Pn* velocity is interpreted here in terms of seismic anisotropy. We have found that about 5% anisotropy characterizes the uppermost mantle of the studied area, and that the fastest direction of *Pn* follows the arcuate trend of the chain. This suggests that seismic anisotropy is strongly related to the tectonic regime that originated the northern Apennine arc.

Key words: Seismic anisotropy, *Pn* waves, northern Apennines.

Introduction

The Apennine chain is an orogen of Tertiary age lying along the peninsular part of Italy. This chain formed since the late Oligocene-early Miocene following the westward subduction of lithosphere beneath the Corsica-Sardinia continental margin (e.g., BECCALUVA *et al.*, 1989, and references therein). The Apennine chain then rotated counterclockwise, reaching its present NW–SE setting (e.g., CIVETTA *et al.*, 1978). During the Messinian, extensional deformations affected the Tyrrhenian margin of the Italian peninsula and then migrated toward the internal units of the Apennines, where normal faulting is still active along a NW–SE trend (e.g., ELTER *et al.*, 1975; LAVECCHIA, 1988). The present setting of the chain is characterized by two major arcs merging in central Italy along the Ortona-Roccamonfina transverse zone (Fig. 1). Evolutionary models of the area can be found in MALINVERNO and RYAN (1986), LAUBSCHER (1988), LOCARDI (1988), PATACCA *et al.* (1990), DOGLIONI (1991), among many others.

In this paper the results of a tomography study of the uppermost mantle beneath the northern Apennines are discussed. This work is part of a more extended study on the *Pn* velocity structure beneath Italy and surrounding regions (MELE *et al.*, 1997). The *Pn* phase travels mainly in the uppermost part of the mantle, along and beneath the Moho discontinuity, with velocities ranging from

[1] Istituto Nazionale di Geofisica, Via di Vigna Murata 605, 00143 Rome, Italy.

7.7 to 8.4 km/s depending on the rheological characteristics of the lithosphere (e.g., CHUNG, 1977; BLACK and BRAILE, 1982), and for its propagation characteristics it has been used to investigate the velocity structure at sub-Moho depth in many regions of the world (e.g., HEARN, 1984; BEGHOUL and BARAZANGI, 1990; HEARN et al., 1991; HEARN and NI, 1994; HEARN, 1996).

Data and Methodology

The results of this study are obtained by applying the anisotropic tomography technique of HEARN (1996) to *Pn* travel times recorded in the Italian area from 1988 to 1991 (ING, 1996). First arrival times have been selected for epicentral distance (between 200 and 1000 km), event depth (less than 30 km), event size

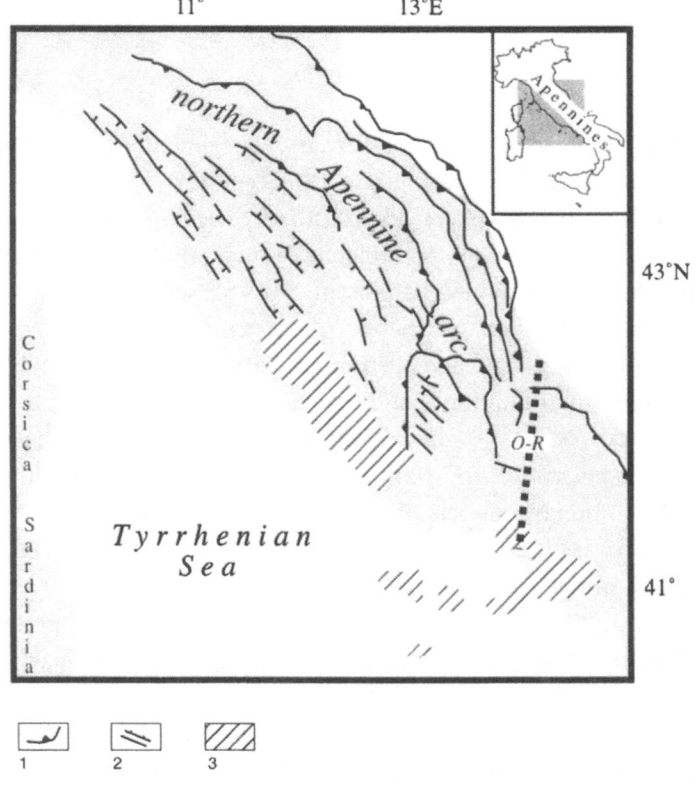

Figure 1

Structural sketch of the northern Apennine area. 1: main thrusts; 2: normal and vertical faults; 3: Plio-Quaternary volcanics; *O-R*: Ortona-Roccamonfina transverse zone.

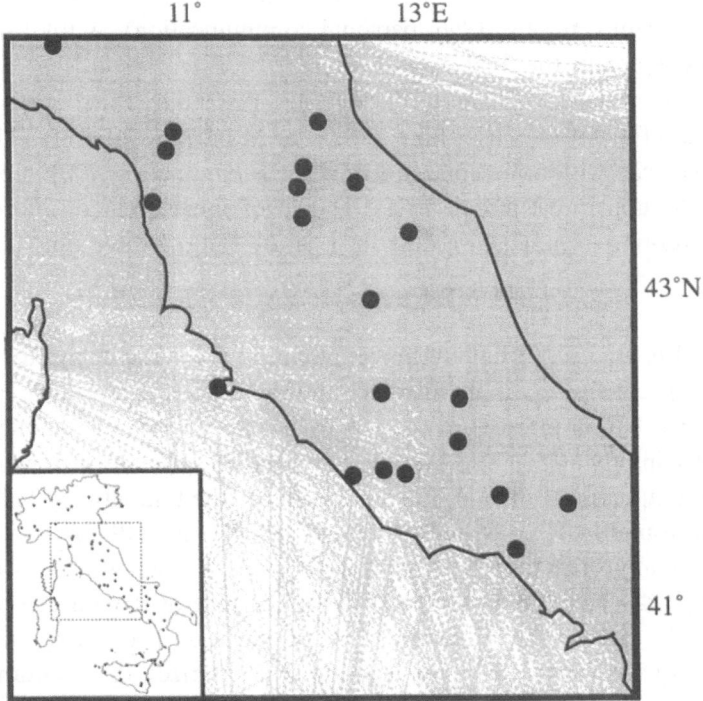

Figure 2
All (11,880) *Pn* ray paths used in this study for the tomographic inversion. Dots represent seismic stations.

(more than 10 recordings per event), residuals (less than 9 s). A straight line was fit to the travel times and residuals were recalculated with respect to this best fit. The maximum residual allowed was then 6 s. The average *Pn* velocity resulting from the straight-line fit to the data set is 8.2 km/s over the Italian area (MELE *et al.*, 1997).

The *Pn* ray paths used in this study range mostly between 200 and 600 km of length and sample the uppermost mantle beneath the northern Apennines from different azimuths (Fig. 2). The study of seismic velocity in the topmost mantle could benefit from the use of short-distance refracted waves, i.e., the so-called *Pd* phase. This phase develops between 80 and 250 km from the source, being refracted just below the Moho. However, significant errors can be made in reading the *Pd* arrival times, due to the contamination of the first crustal *Pg* arrival within 200 km of epicentral distance. For this reason, the use of the "teleseismic" refracted phase *Pn*, which represents the first arrival on the seismogram in the distance range of 200 to 1000 km, has been preferred. As for the longest rays we used, it is important to point out that, due to the earth's sphericity, ray paths as long as 1000 km can dive to 30 km beneath the Moho in the center of the path, and even more if a large mantle gradient exists. However, the *Pn* travel times are corrected for the earth's sphericity, and travel-time variations due to large gradients in the mantle are found to be insignificant (HEARN, 1996, personal communication). A total of 11,880 *Pn* travel times represent the final dataset.

Within the epicentral distances considered in this study, the *Pn* ray paths can be modelled as refracted rays travelling along the Moho discontinuity. The variation of seismic velocity within the uppermost mantle is parameterized by subdividing the surface of the uppermost mantle in a 2-D grid of square cells of appropriate size. The *Pn* travel-time residuals are described as the sum of three time terms:

$$t_{ij} = a_i + b_j + \Sigma d_{ijk}(s_k + A_k \cos 2\phi + B_k \sin 2\phi)$$

where a_i and b_j are crustal time terms representing the static delays for event i and station j, respectively, d_{ijk} is the distance the ray travels in the k-th cell, s_k is the slowness perturbation into cell k, A_k and B_k are the anisotropic coefficients and ϕ is the backazimuth angle. The sum is run over all cells through which the ray travels in the uppermost mantle. Further details of the tomography technique used here can be found in HEARN and NI (1994) and HEARN (1996).

As a first approximation, seismic anisotropy in the mantle is assumed to be described by a 2ϕ azimuthal variation. The magnitude of the anisotropy is defined as $(A_k^2 + B_k^2)^{1/2}$ and the fast direction of *Pn* propagation is given by $1/2 \arctan(B_k/A_k)$.' A set of Laplacian damping equations regularizes the solution, and two damping constants are separately applied to the unknown slowness and anisotropic coefficients. A proper pair of damping constants is chosen to balance the error size and the resolution width. In this approach the trade-off between velocity and anisotropy variations is a crucial issue, which has been checked by using different combinations of damping parameters for both velocity and anisotropy. In this study the main features of the velocity and anisotropy fields are observed to be significantly stable, even though the extent of the velocity anomalies and the amount of anisotropy vary slightly for different combinations of the damping constants.

Observations and Results

We have found that seismic anisotropy is an important feature in the uppermost mantle beneath the northern Apennine area (Fig. 3); here, the azimuthal variations of *Pn* velocity have a maximum deviation of ± 0.2 km/s with respect to the *Pn* velocity locally found by MELE *et al.* (1997). The error resulting from the inversion procedure for the *Pn* velocity is ± 0.05 km/s.

The percentage of *Pn* anisotropy shown in Figure 3 is defined as:

$$\frac{V_{max}^{Pn} - V_{min}^{Pn}}{V_{av}^{Pn}}$$

where V_{max}^{Pn} and V_{min}^{Pn} are the maximum and minimum value of *Pn* velocity, and V_{av}^{Pn} is the average *Pn* velocity.

Laboratory and field experiments have demonstrated that seismic anisotropy represents a powerful tool to investigate the state of strain in the upper mantle (e.g., BROTHERS, 1959; HESS, 1964; RAITT et al., 1971; FUCHS, 1977; MJELDE and SELLEVOL, 1993). In fact, during deformation the fast axes of upper mantle anisotropic minerals (mainly olivine) tend to align with the longest axis of the ellipsoid strain which is perpendicular to the direction of maximum compression in the pure shear deformation regime (e.g., MCKENZIE, 1979; RIBE, 1992). In orogenic belts, where pure shear deformations are likely to occur, the fastest direction of *Pn* velocity would thus be perpendicular to the direction of orogenic compression (e.g., SILVER and CHAN, 1988; KARATO, 1989).

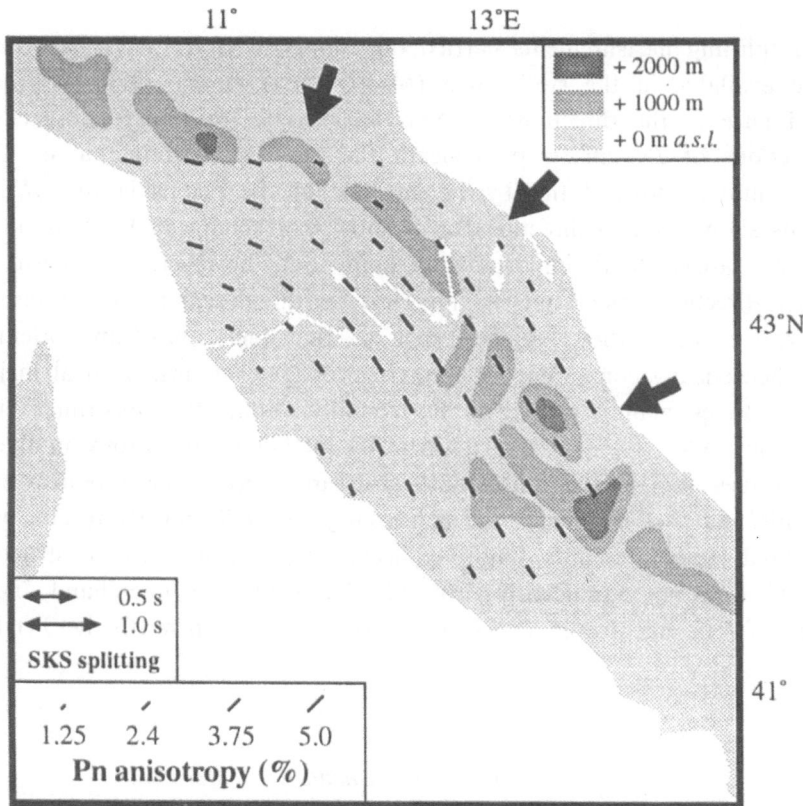

Figure 3

Pn anisotropy in the uppermost mantle beneath the northern Apennines (re-drawn after MELE et al., 1997). Black lines indicate the fast direction of *Pn* velocity. The length of the lines is proportional to the deviation from the average *Pn* velocity. Lines with arrowheads represent the direction of anisotropy inferred from *SKS* splitting measurements by MARGHERITI et al. (1996); their lengths indicate the average delay time. Thick arrows indicate the direction of the arc compression in the northern Apennines. Elevation is represented by the grey scale.

In the uppermost mantle beneath the northern Apennines the pattern of *Pn* anisotropy appears to be strongly related to the regional tectonic trend, following the arc-structure of the chain. Beneath the northern Apennine arc we consistently find that the fastest direction of *P*-wave velocity is perpendicular to the direction of the arc compression (Fig. 3). The *Pn* anisotropy pattern suggests that the arcuate setting of the northern Apennines originated from deep processes involving at least the uppermost mantle. On the other hand, *Pn* anisotropy coincides with the direction of the Neogene and present-day extension beneath the internal units of the chain and the Tyrrhenian margin of the studied area, suggesting that such deformation might have controlled a preferred orientation of minerals in the uppermost mantle. Similar results were found by BEGHOUL and BARAZANGI (1990) beneath the Basin and Range province.

SKS splitting measurements carried out along an ENE–WSW seismic transect are available in the study area (MARGHERITI *et al.*, 1996). As one can see in Figure 3, the directions of *SKS* anisotropy are rather consistent with the directions of *Pn* anisotropy beneath the external Apennine units. On the contrary, moving toward the Tyrrhenian side of the peninsula the *SKS* measurements show rotating directions of anisotropy, reaching an E–W trend along the coastal areas. It is important to point out, however, that even if the *SKS* fast direction orientation is the same as the direction of the fastest *P*-wave propagation (CHRISTENSEN, 1984), so that their anisotropy orientations can be theoretically compared, *SKS* waves propagate at near vertical incidence, while *Pn* waves mainly propagate horizontally within the uppermost mantle. Thus, while *SKS* waves measure a vertically integrated anisotropy in the mantle, *Pn* waves sample the anisotropic structures within the topmost part of the mantle. As a consequence, the consistency of *SKS* and *Pn* results beneath the external Apennine units simply indicates that the anisotropic structure in the mantle does not significantly vary with depth. On the other hand, the presence of different anisotropic layers are inferred at depth beneath the Tyrrhenian coast.

Acknowledgements

This work is part of a more extended tomography study in which the contribution of Thomas Hearn has been significant; the author wishes to thank Muawia Barazangi for encouraging this study. I also thank Dogan Seber and Antonio Rovelli for the critical review of the manuscript. Thanks are due to Michel Granet, Jarka Plomerova and an anonymous reviewer for their criticism and helpful suggestions.

REFERENCES

BECCALUVA, L., BROTZU, P., MORBIDELLI, L., SERRI, G., and TRAVERSA, G., *Cenozoic tectono-magmatic evolution and inferred mantle sources in the Sardo-Tyrrhenian area*. In *The Lithosphere in Italy: Advances in Earth Science Research* (eds. Boriani, A., Bonafede, M., Piccardo, G. G., and Vai, G. B.) (Accademia Nazionale dei Lincei, Roma 1989) pp. 229–248.

BEGHOUL, N., and BARAZANGI, M. (1990), *Azimuthal Anisotropy of Velocity in the Mantle Lid beneath the Basin and Range Province*, Nature *348*, 536–538.

BLACK, P. R., and BRAILE, L. W. (1982), *Pn Velocity and Cooling of the Continental Lithosphere*, J. Geophys. Res. *87*, B13, 10557–10568.

BROTHERS, R. N. (1959), *Flow Orientation of Olivine*, Am. J. Sci. *257*, 574–584.

CHRISTENSEN, N. I. (1984), *The Magnitude, Symmetry and Origin of Upper Mantle Anisotropy Based on Fabric Analyses of Ultramafic Tectonites*, Geophys. J. R. Astr. Soc. *76*, 89–111.

CHUNG, D. H. (1977), *Pn Velocity and Partial Melting, Discussion*, Tectonophysics *42*, T35–T42.

CIVETTA, L., ORSI, G., and SCANDONE, P. (1978), *Eastwards Migration of the Tuscan Anatectic Magmatism due to the Anticlockwise Rotation of the Apennines*, Nature *276*, 604–606.

DOGLIONI, C. (1991), *A Proposal for Kinematic Modelling of W-dipping Subductions—Possible Applications to the Tyrrhenian-Apennines System*, Terra Nova *3*, 423–434.

ELTER, P., GIGLIA, G., TONGIORGI, M., and TREVISAN, L. (1975), *Tensional and Compressional Areas in the Recent (Tortonian to present) Evolution of the Northern Apennines*, Boll. Geofis. Teor. Appl. *27*, 3–18.

FUCHS, K. (1977), *Seismic Anisotropy of the Subcrustal Lithosphere as Evidence for Dynamical Processes in the Upper Mantle*, Geophys. J. R. Astr. Soc. *49*, 167–179.

HEARN, T. M. (1984), *Pn Travel Times in Southern California*, J. Geophys. Res. *89*, 1843–1855.

HEARN, T. M. (1996), *Anisotropic Pn Tomography in the Western United States*, J. Geophys. Res. *101*, 8403–8414.

HEARN, T. M., BEGHOUL, N., and BARAZANGI, M. (1991), *Tomography of the Western United States from Regional Arrival Times*, J. Geophys. Res. *96*, 16369–16381.

HEARN, T. M., and NI, J. F. (1994), *Pn Velocities beneath Continental Collision Zones: The Turkish-Iran Plateau*, Geophys. J. Int. *117*, 273–283.

HESS, H. H. (1964), *Seismic Anisotropy of the Uppermost Mantle under Oceans*, Nature *203*, 629–631.

ING (Istituto Nazionale di Geofisica) (1996) *Seismic Bulletin*, Roma.

KARATO, S. I., *Seismic anisotropy: mechanisms and tectonic implications*. In *Rheology of Solids and of the Earth* (eds. Karato, S. I., and Toriumi, M.) (Oxford Science Publications 1989) pp. 393–422.

LAUBSCHER, H. P. (1988), *The Arcs of the Western Alps and the Northern Apennines: An Updated View*, Tectonophysics *146*, 67–78.

LAVECCHIA, G. (1988), *The Tyrrhenian-Apennines System: Structural Setting and Seismotectogenesis*, Tectonophysics *147*, 263–296.

LOCARDI, E. (1988), *The Origin of the Apenninic Arcs*, Tectonophysics *146*, 105–123.

MALINVERNO, A., and RYAN, W. B. F. (1986), *Extension in the Tyrrhenian Sea and Shortening in the Apennines as Result of Arc Migration Driven by Sinking of the Lithosphere*, Tectonics *5*, 227–245.

MARGHERITI, L., NOSTRO, C., COCCO, M., and AMATO, A. (1996), *Seismic Anisotropy beneath the Northern Apennines (Italy) and its Tectonic Implications*, Geophys. Res. Lett. *23*(20), 2721–2724.

MCKENZIE, D. (1979), *Finite Deformation during Fluid Flow*, Geophys. J. R. Astr. Soc. *58*, 689–715.

MELE, G., ROVELLI, A., SEBER, D., HEARN, T., and BARAZANGI, M. (1997), *Compressional Velocity Structure and Anisotropy in the Uppermost Mantle beneath Italy and Surrounding Regions*, J. Geophys. Res., accepted.

MJELDE, R., and SELLEVOL, M. A. (1993), *Seismic Anisotropy Inferred from Wide-angle Reflections off Lofoten, Norway, Indicative of Shear-aligned Minerals in the Upper Mantle*, Tectonophysics *222*, 21–32.

PATACCA, E., SARTORI, R., and SCANDONE, P. (1990), *Tyrrhenian Basin and Apenninic Arcs: Kinematic Relations since late Tortonian Times*, Mem. Soc. Geol. It. *45*, 425–451.

RAITT, R. W., SHOR, G. G., MORRIS, G. B., and KIRK, H. K. (1971), *Mantle Anisotropy in the Pacific Ocean*, Tectonophysics *12*, 173–186.

RIBE, N. M. (1992), *On the Relation between Seismic Anisotropy and Finite Strain*, J. Geophys. Res. *97*, 8737–8747.

SILVER, P. G., and CHAN, W. W. (1988), *Implications for Continental Structure and Evolution from Seismic Anisotropy*, Nature *335*, 34–39.

(Received November 19, 1996, revised/accepted June 26, 1997)

 To access this journal online:
http://www.birkhauser.ch

Pure appl. geophys. 151 (1998) 503–525
0033–4553/98/040503–23 $ 1.50 + 0.20/0

© Birkhäuser Verlag, Basel, 1998

❙ Pure and Applied Geophysics

New Perspectives on Mantle Dynamics from High-resolution Seismic Tomographic Model P1200

ONDŘEJ ČADEK,[1] DAVID A. YUEN,[2] HANA ČÍŽKOVÁ,[1] MOTOYUKI KIDO,[3]
HUA-WEI ZHOU,[4] DAVID BRUNET[5] and PHILIPPE MACHETEL[5]

Abstract—Recently a high-resolution tomographic model, the *P1200*, based on *P*-wave travel times was developed, which allowed for detailed imaging of the top 1200 km of the mantle. This model was used in diverse ways to study mantle viscosity structure and geodynamical processes. In the spatial domain there are lateral variations in the transition zone, suggesting interaction between the lower-mantle plumes and the region from 600 km to 1000 km. Some examples shown here include the continental region underneath Manchuria, Ukraine and South Africa, where horizontal structures lie above or below the 660 km discontinuity. The blockage of upwelling is observed under central Africa and the interaction between the upwelling and the transition zone under the slow Icelandic region appears to be complex. An expansion of the aspherical seismic velocities has been taken out to spherical harmonics of degree 60. For degrees exceeding around 10, the spectra at various depths decay with a power-law like dependence on the degree, with the logarithmic slopes in the asymptotic portion of the spectra containing values between 2 and 2.6. These spectral results may suggest the time-dependent nature of mantle convection. Details of the viscosity structure in the top 1200 km of the mantle have been inferred both from global and regional geoid data and from the high-resolution tomographic model. We have considered only the intermediate degrees ($l = 12–25$) in the nonlinear inversion with a genetic algorithm approach. Several families of acceptable viscosity profiles are found for both oceanic and global data. The families of solutions for the two data sets have different characteristics. Most of the solutions asociated with the global geoid data show the presence of asthenosphere below the lithosphere. In other families a low viscosity zone between 400 and 600 km depth is found to lie atop a viscosity jump. Other families evidence a viscosity decrease across the 660 km discontinuity. Solutions from oceanic geoid show basically two low viscosity zones: one lying right below the lithosphere; the other right under 660-km depth. All of these results bespeak clearly the plausible existence of strong vertical viscosity stratification in the top 1000 km of the mantle. The presence of the second asthenosphere may have important dynamical ramifications on issues pertaining to layered mantle convection. Numerical modelling of mantle convection with two phase transitions and a realistic temperature- and pressure-dependent viscosity demonstrates that a low viscosity region under the endothermic phase transition can indeed be generated self-consistently in time-dependent situations involving a partially layered configuration in an axisymmetric spherical-shell model.

Key words: Geodynamics, seismic tomography, spectral analysis, inferences of viscosity from geoid, mantle convection.

[1] Department of Geophysics, Faculty of Mathematics and Physics, Charles University, V Holešovičkách 2, 18000 Prague, Czech Republic.
[2] Department of Geology and Geophysics and Minnesota Supercomputer Institute, University of Minnesota, Minneapolis, MN 55415–1227, U.S.A.
[3] Ocean Research Institute, University of Tokyo, 1-15-1, Minami-dai, Nakano-ku, Tokyo 164, Japan.
[4] Department of Geological Sciences, University of Houston, Houston, TX 77204–5503, U.S.A.
[5] UMR5566, GRGS/CNRS/CNES/UPS, 18, Av. E. Belin, 31401 Toulouse Cedex 4, France.

1. Introduction

In the last dozen years a variety of inferences pertaining to the dynamical processes of the earth's mantle has been drawn from utilizing the three-dimensional seismic tomographic models (MASTERS et al., 1982; DZIEWONSKI, 1984; WOOD-HOUSE and DZIEWONSKI, 1984; INOUE et al., 1990; TANIMOTO, 1990; ZHANG and TANIMOTO, 1991; MONTAGNER and TANIMOTO, 1991; FUKAO et al., 1992; SU et al., 1994; ROMANOWICZ, 1995) together with the linear viscous theory (e.g., RICHARDS and HAGER, 1984; RICARD et al., 1984; HAGER and CLAYTON, 1989; RICARD and BAI, 1991; FORTE and PELTIER, 1991; KING and MASTERS, 1992; ZHANG and CHRISTENSEN, 1993; MORGAN and SHEARER, 1993; WOODWARD et al., 1993; KING, 1995; THORAVAL et al., 1995; PARI and PELTIER, 1995; FORTE and MITROVICA, 1996; ČADEK et al., 1997). These studies mainly concerned answering questions related to long wavelength features, up to degree 8 in the spherical harmonic expansion. Such wavelengths would be hard-pressed to address outstanding issues in geodynamics, such as the degree of mantle flow across the transition zone. In fact, it has been shown from fluid dynamical studies that the mass flux through the phase change regions depends critically on the horizontal wavelength of the local convective cell (TACKLEY, 1995; YUEN et al., 1995; BUTLER and PELTIER, 1997). Thus the question of whether mantle flow is layered or not cannot be answered in such a straightforward fashion from linear harmonic analysis (e.g., MORGAN and SHEARER, 1993), since the convective processes in the mantle, especially in the transition zone, are highly nonlinear in nature. Results from recent numerical simulations of mantle convection with phase transitions have revealed the very complex character of flow modulated by the phase transitions (e.g., MACHETEL and WEBER, 1991; ZHAO et al., 1992; WEINSTEIN, 1993; TACK-LEY et al., 1993, 1994; HONDA et al., 1993; SOLHEIM and PELTIER, 1994; KING and ITA, 1995; STEINBACH and YUEN, 1995; YUEN et al., 1994; CSEREPES and YUEN, 1997).

With the recent availability of high-resolution global tomographic models from both surface waves (TRAMPERT and WOODHOUSE, 1995, 1996) and body wave data (GRAND, 1994; ZHOU, 1996; VAN DER HILST et al., 1997), a new venue has appeared for imaging the interior of the earth and for providing new constraints on the viscosity structure of the mantle from the geoid. In this study we will show how the usage of this new tomographic model *P1200* (ZHOU, 1996), based on *P* waves, can lead to new and interesting findings in the viscosity structure of the upper 1000 km of the mantle. The region of profound interest upon which we will focus lies between the transition zone and 1000 km depth. Recent studies, e.g., by KAWAKATSU and NIU (1994), KÝVALOVÁ et al. (1995), ČADEK et al. (1995), WEN and ANDERSON (1995), FORTE and MITROVICA (1996), PELTIER (1996) and ČÍŽKOVÁ et al. (1996), have argued for a more complex character of the lower mantle region immediately below the 660-km seismic discontinuity.

The plan of this paper is as follows. First, we briefly describe the *P1200* model and show some interesting vertical cross sections of the top 1200 km of the mantle. These views will present a convincing picture of the complicated nature of this region near the transition zone under oceans and continents. Next we will show the spectral dependencies of the seismic heterogeneities for several depths, since it will suggest the nonlinear character of the mantle flow processes by the power-law dependence these spectra are shown to display. The usage of the slope of the geoid spectrum is then proposed as a means of employing the intermediate wavelength portion of the geoid to determine the viscosity structure in the top 1000 km of the mantle. Both global and regional geoid data will be employed in the nonlinear inversion algorithms. A remarkable outcome of the inversion will be the revelation of a second asthenosphere right below the spinel to perovskite transition. We will then demonstrate by way of a numerical simulation of mantle convection with phase transitions and a temperature- and pressure-dependent Newtonian rheology, how low viscosity regions can be developed in mantle convection for realistic convection conditions. In the final section we will summarize the results and discuss their geophysical implications.

The main aim of the paper is to draw attention to the new generation of high-resolution tomographic models and to demonstrate their possible applications in geodynamics. Our approach, by emphasizing the methodological aspects, will be qualitative rather than quantitative, and the results presented in this paper thus should be considered as preliminary. We believe that, in spite of certain problematic features, the effects discussed here can inspire future, more sophisticated investigation, which will lead to a deeper understanding of mantle dynamics.

2. High-resolution Tomographic Model P1200 and its Spectral Characteristics

The global tomographic model *P1200* used here has been obtained by applying a multi-cell inversion technique (ZHOU, 1996) to *P*-wave travel times. The data-processing has included both computing delay times with respect to the *iasp91* model (KENNETT and ENGDAHL, 1991) and hypocentral redetermination (ZHOU and WANG,-1994). The *P1200* model resolution, which refers to the smallest cell size of the model, is $1° \times 1° \times$ about 50 km for the top 1200 km of the mantle. On the other hand, the data resolution, which measures the smallest features retrievable by the ray coverage, is about 5° to 15° laterally at places of poor ray coverage, such as some regions under the Pacific and Antarctica. At places with the best ray coverage, such as around the major subduction zone, the data resolution is close to 1° laterally. The model has undergone several tests against different reference models and the influence of deep mantle contributions below 1200 km depth. In general, the long wavelength features of the *P1200* model correlate well with previous *P*-wave and surface-wave models (INOUE *et al.*, 1990; PULLIAM *et al.*,

1993; TRAMPERT and WOODHOUSE, 1995) and the short wavelength components are in accord with regional studies (ZHOU and CLAYTON, 1990; VAN DER HILST et al., 1991; FUKAO et al., 1992). The details can be found in ZHOU (1996).

Figure 1 shows the vertical sections of the *P1200* model across three long paths traversing the Eurasian continent (A–A1), Africa (B–B1) and the Atlantic and North America (C–C1). Earthquakes are shown in dots. One notices that there are distinct differences across the 660-km discontinuity (dotted line) beneath both continental and oceanic regions. Below Manchuria in A–A1 there is fast anomaly associated with the Japanese subducting slabs which are trapped above the 660-km discontinuity. Close to this pool of fast material appears some slow anomaly coming from the western Pacific. In contrast to this fast anomaly we see that there is a pool of slow material trapped in the transition zone directly under the cratonic region of the Ukraine. Under southern Africa (B) there is fast material directly beneath the craton. The megaplume does not make a direct connection from the lower mantle under South Africa to the slow region under Central Africa. Beneath the Afar triangle (B1) there appears a channel of slow matter below the 660-km discontinuity. Underneath Iceland (C1) we can observe that there is a complex interaction between the slow material in the lower mantle and the plume-like structure in the upper-mantle under Iceland. In the western part of North America (C) there exists a large province with a slow but concentrated anomaly at 660 km depth. This slow anomaly lies adjacent to the fast anomaly under the Canadian shield, which extends to the transition zone under eastern Canada. These large-scale images of the top 1200 km of the mantle reveal the existence of complex heterogeneities in the transition zone, particularly in the region between 660 km and 1000 km depth. The lack of a clear connectivity of the fast and slow anomalies across the 660-km discontinuity would suggest that the mantle flow is rather complex in the top part of the lower mantle and the transition zone does in some sense modulate the local mass transfer.

We have carried out a spectral analysis of the *P1200* model up to the spherical harmonic degree 60. The spherical harmonic coefficients of the relative aspherical velocity $f = 100 \times dV/V$ have been estimated by the method of a least-squares adjustment of model values given in a regular grid of 180×360 points (MARTINEC, 1991). The cut-off degree $l_{max} = 60$ has been chosen at one third of the Nyquist frequency, which guarantees that the resultant spectra are not biased by the effects of aliasing. The degree 60 corresponds to a resolution of 6 degrees laterally, which is a reasonable estimate of the data resolution in most places. The effect of a limited data resolution in some parts of the mantle on the spectra has been carefully tested. We have compared the spectra, obtained from the least-squares analysis of the complete *P1200* model, with the spectra computed by integrating over the well resolved areas only. The agreement between the two groups of spectra is rather good for degrees greater than 10 but only modest for the lower degrees. This indicates that the tomographic spectra may be, in general, more biased by the data

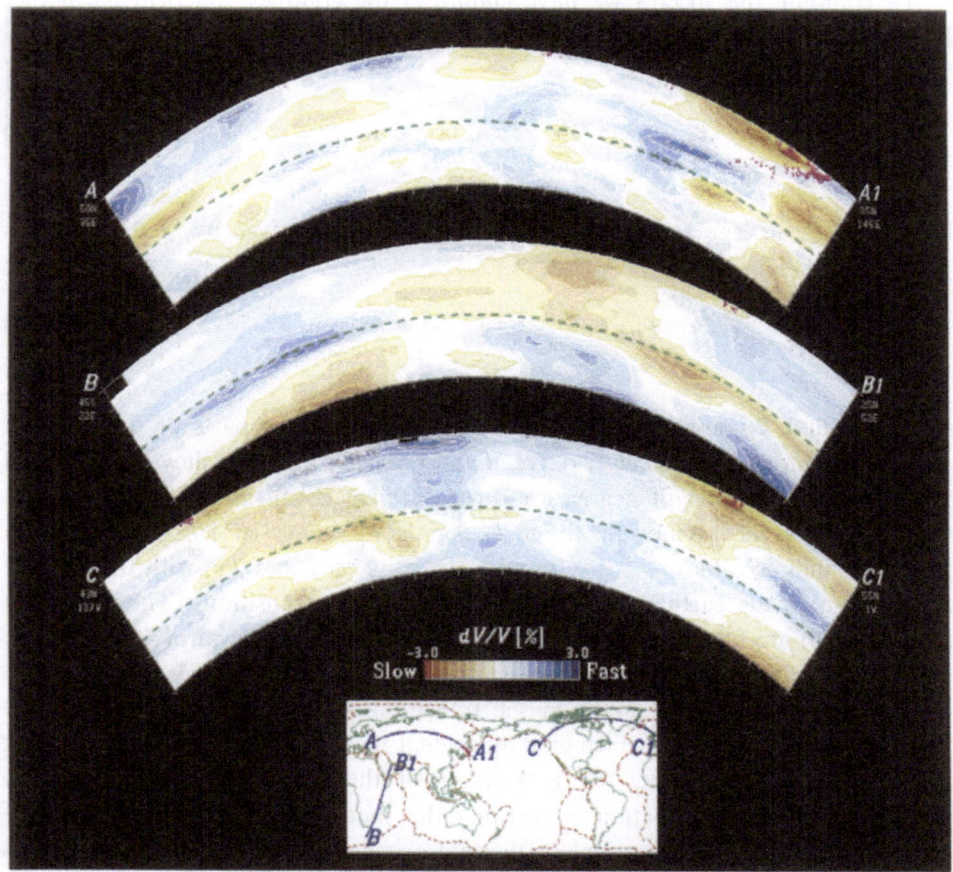

Figure 1

Vertical cross sections of the *P1200* model along three paths. The top 1200 km of the mantle is shown. The three paths are shown along the great circle trajectories shown below in the world map. Blue and brown represent respectively the fast and slow relative *P*-wave anomalies obtained from inversion. Earthquakes are portrayed by the purple dots. For additional information, the reader is referred to ZHOU (1996).

gaps at low degrees than at the higher degree harmonics. Another possible source of the bias, which could mainly influence the harmonic contributions above degree 20, is noise due to poor resolution of the body-wave seismic tomography below oceanic areas, which cover two thirds of the earth. It is rather difficult to assess this effect quantitatively and thus certain caution is necessary in interpreting the spectra relative to the flow processes in the mantle. It should be noted that the *P1200* model does not include the influence of anisotropy (MONTAGNER and TANIMOTO, 1990, 1991) and anelasticity (ROMANOWICZ, 1994, 1995) which may also be important.

To compute the spectra we have employed the complex spherical harmonics $Y_{lm}(\theta, \phi)$ (JONES, 1985), normalized such that

$$\int_0^\pi \int_0^{2\pi} Y_{l_1 m_1}(\theta, \phi) Y_{l_2 m_2}^*(\theta, \phi) \sin\theta \, d\phi \, d\theta = \delta_{l_1 l_2} \delta_{m_1 m_2}. \tag{1}$$

The power P_l at degree l in the layer at radius r is then given by

$$P_l(r) = \sum_{m=-l}^{l} [A_{lm}^2(r) + B_{lm}^2(r)], \tag{2}$$

where A_{lm} and B_{lm} are respectively the real and imaginary parts of the complex spherical harmonic coefficient at degree l and order m. Figure 2 shows the spectral power P_l as a function of degree for three different depth intervals. The spectra suggest that in the asymptotic (large l) regime P_l decays as $1/l^\beta$, where β is a power-law index. Values of β are found to lie between 2 and 2.6, with the extreme values close to the earth's surface and the two-phase transitions. The maximum values (~ 2.6) are found in the lithosphere and the depth interval of 410–460 km, while very low values (~ 2.1) are characteristic for the region between 250 and 410 km depth. The maximum power in the transition zone is usually found close to degree 2 and 6 (see Fig. 3) which is in agreement with previous analyses (MASTERS *et al.*, 1982; MONTAGNER, 1994). Analysis of the depth variations of β may provide important information concerning the time-dependent nature of mantle dynamics as well as the chemical or thermal origins of the mantle heterogeneities (YUEN *et al.*, 1996). It should be mentioned that usage of slopes in the asymptotic part of the spectra is quite common in fluid dynamics where it is commonly used as a diagnostic technique for time-dependent processes (MONIN and YAGLOM, 1975).

Figure 3 depicts the seismic heterogeneity spectrum as a function of depth. The spectrum is plotted in a logarithmic scale and normalized to the maximum at each depth. The same scale, between 0 and 1, is used at each degree to illustrate the relative differences between the high-degree parts of the spectra at various depths. The power at higher degrees significantly varies with depth, marking the interfaces at 50, 200, 400, 660 and 820 km. Dark regions, indicating a more broadband spectrum, may be associated with a complicated chemical structure (crust) and/or with the increased chaotic nature of the convective heterogeneities (asthenosphere and a layer between the 660 and 820 km depths). In contrast, the light layers found between 50 and 200 km (lithosphere) and in the phase transition region should be associated with a less chaotic convective circulation, which is consistent with plate movements in the lithosphere and a high value of the viscosity expected in the transition zone (KARATO *et al.*, 1995).

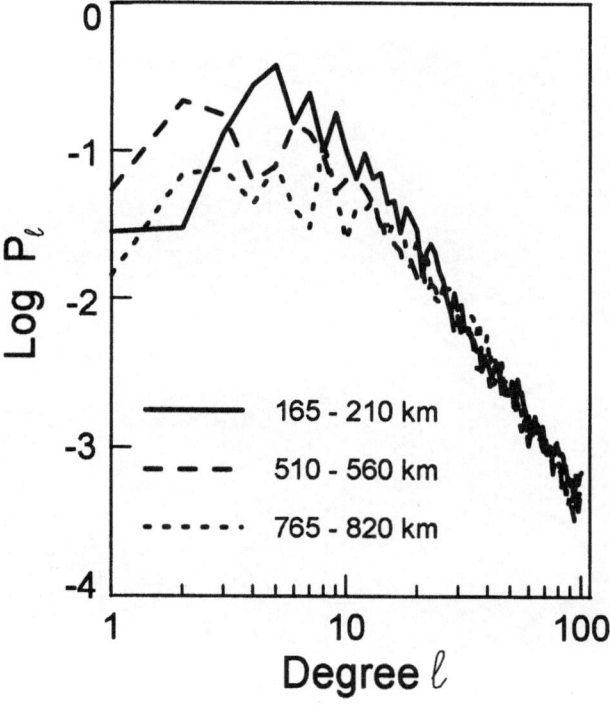

Figure 2
Log-log plot of the seismic heterogeneity power spectrum as a function of degree l of the spherical harmonics. Three depth intervals have been considered.

3. Inferences of Viscosity Profile from the Geoid

A similar power-law decay as in the case of the seismic heterogeneities (see Fig. 2) was also found for the geoid (KAULA, 1967). The logarithmic slope of a recent geoid model by RAPP and PAULIS (1990) is -3.1 for l ranging between 10 and 25. Around degree 30 the slope suddenly changes to a value close to -2 which characterizes well the decay in the degree range $l = 30-100$. The change of the spectral slope around 30° has not yet been properly studied; it is, however, probable that it is associated with a change of physical mechanism of the geoid generation. If this is the case, the change of the slope may be interpreted as a transition from dynamical to isostatic compensation of the surface topography, and degree 30 represents the upper limit for the inferences of viscosity from the geoid.

The inferences of viscosity from the geoid are based on the fact that the low degree geoid reflects not only the distribution of density anomalies in the mantle but also the undulations of the surface and internal boundaries due to a mantle flow (see e.g., RICARD et al., 1984; RICHARDS and HAGER, 1984; and HAGER and CLAYTON, 1989). The convective currents, induced by the anomalous densities,

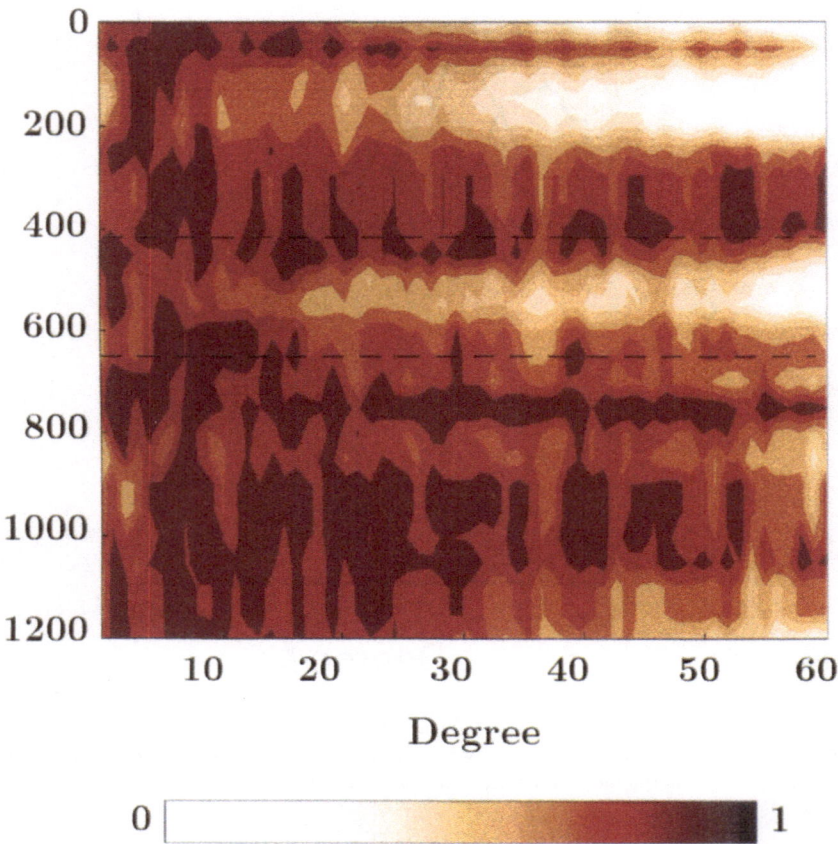

Figure 3

Seismic heterogeneity spectrum of the *P1200* model (Zhou, 1996) as a function of depth. The spectrum is plotted in a logarithmic scale and normalized to the maximum value at each depth. The same scale, between 0 and 1, is used at each degree to illustrate the relative differences between the high-degree parts of the spectra at various depths. The dashed lines mark the positions of the phase boundaries at 410 and 660 km.

deform the boundaries and the internal surfaces. These deformations, the amplitudes of which strongly depend on the mantle viscosity structure, also contribute to
the geoid anomalies. The inverse problem for viscosity is usually formulated as a
minimization of the misfit between the observed geoid and the geoid predicted for
a given viscosity structure of the mantle (e.g., King, 1995). A different approach
was used by Čížková *et al.* (1996) who employed the value of the logarithmic slope
of the intermediate degree geoid together with the *P1200* model to constrain the
viscosity structure of the top 1000 km of the mantle. By conducting over one
million runs in a Monte Carlo inversion, they found basically three families of

viscosity models which can fit the slope of the geoid spectrum. All these families are characterized by a viscosity increase in the lower mantle but significantly differ in the region between the 660 km and 1000 km depth. We present here more advanced results of this inversion obtained from the application of genetic algorithm techniques (e.g., GOLDBERG, 1989). Different from ČÍŽKOVÁ et al. (1996), we do not fit the two parameters characterizing the linear decay of the geoid power spectrum, rather we minimize the squared misfit between the observed and predicted power degree by degree:

$$S_1(\eta) = \sum_{l=12}^{25} [\log_{10} P_l^{\text{obs}} - \log_{10} P_l^{\text{pred}}(\eta)]^2 = \min, \tag{3}$$

where P_l^{obs} is the power of the observed geoid and P_l^{pred} is the power of the predicted dynamical geoid at degree l. In general, the power of the predicted geoid is a function of viscosity $\eta(r)$ and density $\rho(r, \theta, \phi)$ computed from a given seismic tomographic model (see e.g., HAGER and CLAYTON, 1989). In this study we use the P1200 model in the top 1200 km and the model by FUKAO et al. (1992) in the rest of the mantle. We note that the effects of mass anomalies located in the deep mantle on the intermediate-degree geoid are small.

The conversion of seismic anomalies to density perturbations represents a weak point in the geoid inversions. To estimate the sensitivity of the inversion to the velocity-to-density scaling, we test five models of $\partial \ln \rho / \partial \ln v$, based on recent mineral physics data (Table 1). As suggested by various authors (e.g., JORDAN, 1988; KARATO, 1993; DOIN et al., 1996), the interpretation of the seismic anomalies in terms of densities is particularly problematic in the uppermost mantle, where the effects of chemical anomalies and anelasticity have yet to be properly quantified. This accounts for why models A–D significantly differ in the top 200 km of the mantle, which allows one to estimate the role of the lithospheric and asthenospheric masses in the inversion. Below a depth of 200 km we impose the values of $\partial \ln \rho / \partial \ln v$, taken from KARATO (1993). In model E, a constant conversion parameter 0.4 is prescribed throughout the mantle.

The viscosity η is parameterized in seven layers containing boundaries at 0, 100, 200, 400, 660, 820, 1000 and 2900 km. The intermediate-degree harmonics of the geoid used in our inversion are only slightly sensitive to viscosity changes below a depth of about 1000 km and they cannot be used for determining the details of the viscosity structure in a deeper mantle. This provides the reason why the viscosity can only be roughly parameterized below this depth. A more detailed discussion of the parameterization can be found in a recent paper by KIDO and ČADEK (1997).

For determining the viscosity model which supplies the minimum value to functional $S_1(\eta)$, we have employed a genetic algorithm with similar parameters as KING (1995). This technique allows for the mapping of the minima of S_1 on a

Table 1

Models of velocity-to-density scaling ($\partial \ln \rho / \partial \ln v$) used in this study

Model	Depth range (km)			
	0–100	100–200	200–660	660–2890
A	0.6	0.3	0.45–0.55	KARATO (1993)
B	0.6	0.1	0.45–0.55	KARATO (1993)
C	0.0	0.3	0.45–0.55	KARATO (1993)
D	0.0	0.1	0.45–0.55	KARATO (1993)
E	0.4	0.4	0.4	0.4

six-dimensional model space with an efficiency, which is higher by two orders of magnitude than for a Monte Carlo search or a systematic exploration of the model space.

Figure 4 shows the viscosity profiles which can best satisfy the constraint given by eq. (3). Two viscosity profiles are plotted for each of the velocity-to-density scaling models A–E defined in Table 1. The top panels are for profiles with a viscosity increase at 410 km, while the models with a decrease at this depth are shown in the bottom set. The fit of the predicted geoid spectra with the spectrum of the observed geoid is illustrated in Figure 5.

The viscosity models can roughly be classified into three families. Models in the first family (models A–D in the bottom set) exhibit a low viscosity zone between 410 and 660 km and viscosity peak between 660 and 1000 km. The models of the second family (models A and D in the top set, and model E in the bottom set) evidence ony small changes in viscosity between 100 and 660 km and a viscosity increase by 1–2 orders in the lower mantle. The third family, represented by models B, C and E in the top set, is characterized by two low viscosity zones. The first, located somewhere between 100 and 410 km, can be identified with the traditional asthenosphere, while the other low viscosity zone is found below the 660-km discontinuity. All the models shown in Figure 4 are characterized by a high viscosity lithosphere and higher values of viscosity below a depth of 1000 km than in the upper mantle. Models of the first family have been described by ČÍŽKOVÁ *et al.* (1996) and they have certain common features with the previous viscosity estimates by FORTE and PELTIER (1987, 1991), KING (1995), and PARI and PELTIER (1995) as well. The second family is close to the traditional estimates of mantle viscosity from the lowest degree geoid (e.g., RICARD and BAI, 1991). The models of the third family are rather similar to the viscosity profiles inferred from the lowest degree geoid by ZHANG and CHRISTENSEN (1993) and KING (1995).

These results strongly indicate that the viscosity structure between 660 and 1000 km depth may be rather complex. It is also possible that the viscosity at this depth

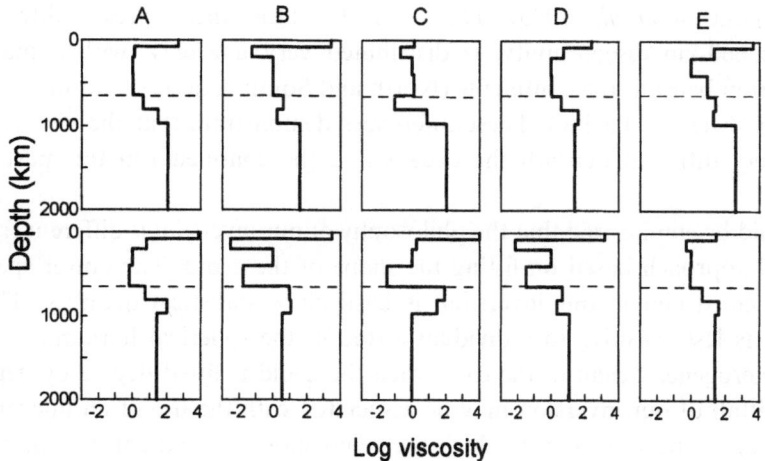

Figure 4

Viscosity profiles which provide the best fit of the geoid spectrum at intermediate degree range ($l = 12$–25). Two viscosity profiles are plotted for each of velocity-to-density scaling models A–E, defined in Table 1. The top panels are for profiles with a viscosity increase at 410 km, while the profiles with a decrease at this depth are shown in the bottom set. The dashed line marks the position of the 660-km boundary. Since the geoid is insensitive to the absolute value of viscosity, only relative variations of viscosity can be inferred from these models.

varies laterally and the different minima of S_1 correspond to different regions of the earth. This view is supported by a significant correlation found between the seismic image at this depth and subduction in the last 120 Myr (WEN and ANDERSON,

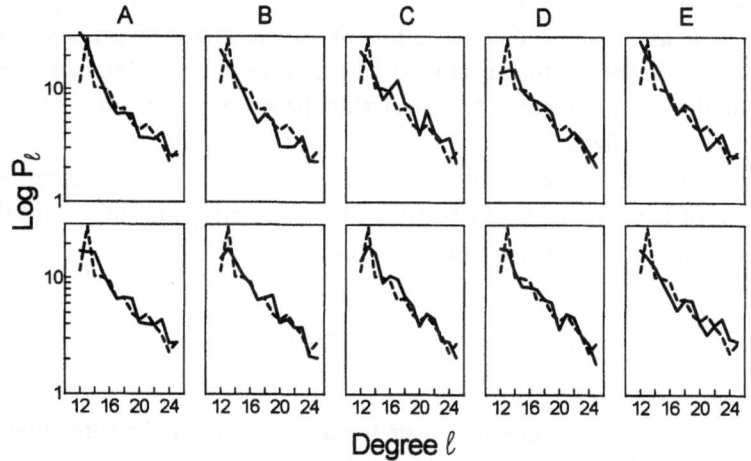

Figure 5

Comparison between the power spectrum of the observed geoid (RAPP and PAULIS, 1990, dashed line) and the spectra computed from the profiles shown in Figure 4 (thick lines) for degree $l = 12$–25.

1995; KÝVALOVÁ *et al.*, 1995). The subducted lithosphere, presumably located below the 660-km discontinuity, is distributed very unevenly, with a maximum concentration beneath the continents (North and South America) and the continental margins (western Pacific). These inferences demonstrate that the viscosity field may be very different beneath the oceans and the continents in this part of the mantle.

It should be emphasized that the philosophy of our inversion is different from the traditional approach based on fitting the shape of the geoid. The power spectrum, used as a constraint in our inversion, is basically a statistical quantity. Thus, its prediction is less sensitive to individual errors in the spherical harmonic series of seismic heterogeneity than in the case when the geoid is fitted degree by degree. A possible pitfall of our inversion may be associated with the use of an intermediate-degree range of the spectrum in which the geoid shows a good correlation with the continental crustal thickness (LESTUNFF and RICARD, 1995). The results in Figure 4, illustrating the proposed method, should thus be considered as very preliminary. Further analysis of the *P1200* model, especially in a very shallow depth, together with an isostatic correction of the geoid, will be needed for more reliable information concerning the intermediate-degree part of the geoid spectrum.

The viscosity profiles shown in Figure 4 have been derived from a global distribution of the seismic anomalies. In the inversion scheme, these anomalies are assumed to be of a thermal nature, which allows their easy conversion into densities. However, this assumption may not be valid in a large portion of the mantle; the shallow mantle under the continents and the subduction zones being particularly problematic (JORDAN, 1988; DOIN *et al.*, 1996). The new high-resolution tomographic models allow the relationship between the geoid and the seismic heterogeneity pattern to be addressed only in regions where the role of thermal heterogeneities may be dominant (KIDO and ČADEK, 1997). Moreover, the effects associated with an isostatic compensation of the surface topography, which are rather important in the continents (LESTUNFF and RICARD, 1995), can probably be neglected in the oceans where the correlation between the geoid and the crustal thickness is less significant.

We will follow the traditional formulation of the inverse problem based on minimization of the squared misfit between the observed geoid δN_{obs} and the geoid $\delta N_{\text{pred}}'$ predicted for a given viscosity structure η:

$$S_2(\eta) = \left\| \delta N^{\text{obs}} - \delta N^{\text{pred}}(\eta) \right\|_{L_2} = \min, \qquad (4)$$

where the norm has been computed over the area of one of the three main oceans (Atlantic, Indian and Pacific). To suppress the gravity signal from the problematic masses associated with the continents and the subduction zones, we have employed only the intermediate-degree part of the spectrum ($l = 12\text{--}25$). The viscosity η is

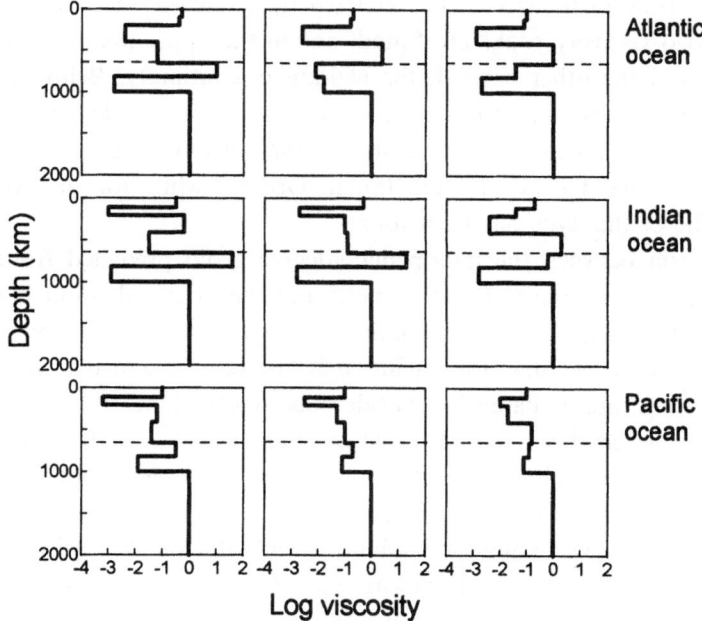

Figure 6

Viscosity profiles which give the best fit of the oceanic geoid observed at degrees $l = 12$–25. The profiles are depicted separately for each ocean. The dashed line marks the position of the 660-km boundary. Since the geoid is insensitive to the absolute value of viscosity, only relative variations of viscosity can be inferred from these models.

parameterized in the same way as in the previous case. The sensitivity of the results to various velocity-to-density scalings has been tested and is found to be relatively small, which agrees with the previous inferences (KING, 1995; KIDO and ČADEK, 1997). The genetic algorithm, including a penalization of wildly oscillating viscosity models, is again employed to find the minimum value of functional $S_2(\eta)$.

The viscosity models providing the best fit of the oceanic geoid observed at degrees 12–25 are plotted in Figure 6. The models can be divided into two groups. The models with a viscosity increase at 410 km are characterized by the existence of two low viscosity zones. The first zone is located somewhere between 100 and 410 km, while the other is found either below the 660-km discontinuity or at 820 km. The two zones are separated by a high viscosity hill peaking somewhere between 410 and 820 km. The viscosity decrease in the second low viscosity zone is comparable to the asthenosphere in the Atlantic and the Indian Oceans although it is rather weak in the Pacific. We note that the resolution of our inversion is insufficient for the viscosity to be retrieved with a higher accuracy. The models of this group are rather similar to some of models obtained from the inversion of the geoid spectrum (cf. models in the top panels in Fig. 4).

The other group of models is characterized by a viscosity decrease at 410 km. In this case, two viscosity peaks are found, one in the upper mantle, between 200 and 410 km, and the other beneath the 660-km discontinuity. Below a depth of 820 km the viscosity significantly decreases, forming a similar low viscosity zone as in models of the first group. We emphasize that this viscosity pattern has only been obtained in the Pacific and the Indian Oceans, while for the Atlantic no viscosity profile of this type has been found.

A comparison between the intermediate-degree geoid predicted from the viscosity profiles given in Figure 6 (left panels) and the observed geoid is shown in Figure 7. The two fields are rather similar in shape, which is also quantified by correlation coefficients of 0.42 for Pacific, 0.52 for Indian, and 0.53 for Atlantic. These correlations are apparently considerably lower than those obtained by lower-degree geoid analyses (e.g. RICHARDS and HAGER, 1989), however their confidence levels are higher than 99%. We note that the amplitudes of the predicted geoid are almost two times smaller in comparison with the observed geoid. In general, a lower amplitude solution yields the smallest misfit, when the correlation is lower. It is worth noting that KIDO and ČADEK (1997) have obtained similar viscosity profiles, however the resultant geoid amplitudes are comparable to the observation, by using inversion in terms of maximizing the correlation coefficient for the same degree range and velocity-to-density scaling parameters.

4. Numerical Modelling

Numerical modelling can also shed light on the issues of the development of low viscosity zones beneath 660-km depth. Physically one can explain the potential existence of a low viscosity zone under the endothermic phase change by bringing into play temperature- and pressure-dependent viscosity. If the horizon at the endothermic phase transition coincides with the internal boundary in a layered or partially layered system and there is no jump in the creep law across the phase change, then one would expect a viscosity minimum somewhere below 660-km depth, simply from the trade-off between the viscosity decrease due to temperature under the internal boundary layer at 660-km depth and the increase due to the pressure-dependence of mantle rheology. In a layered or partially layered mantle the hotter material from a hot plume head will locally cause a low viscosity zone by impinging the 660-km discontinuity from below.

In Figure 8 we depict the temperature and viscosity fields taken from a two-dimensional numerical simulation of convection in an axisymmetric spherical shell in which a Newtonian rheology with both temperature- and pressure-dependent viscosity is employed. These results come from the model by BRUNET and

MACHETEL (1997) in which both the olivine to spinel and the spinel to perovskite transitions are accounted for with a self-consistent treatment of both compressibility and the distortion of the phase boundaries by lateral temperature gradients. The rheological and physical parameters are given in Table 2. There is no viscosity

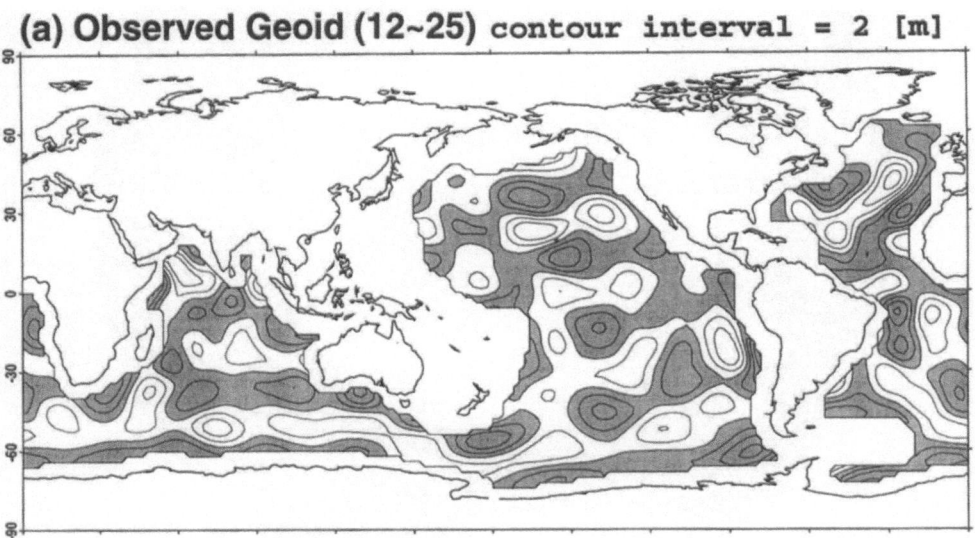

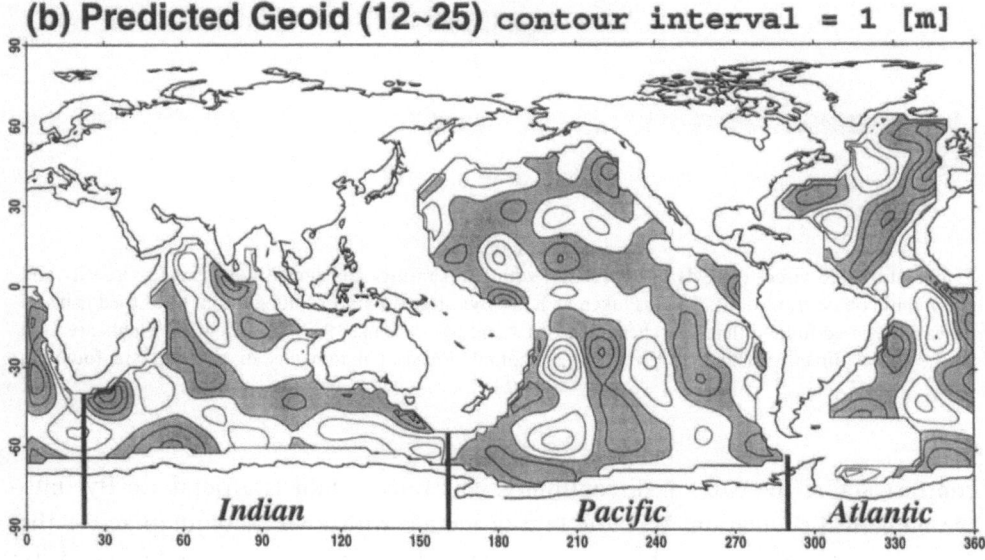

Figure 7
Comparison between the observed geoid (a) and the geoid predicted for the viscosity profiles which give the best fit with observations in each of the oceanic regions (b). The shading marks regions of negative values.

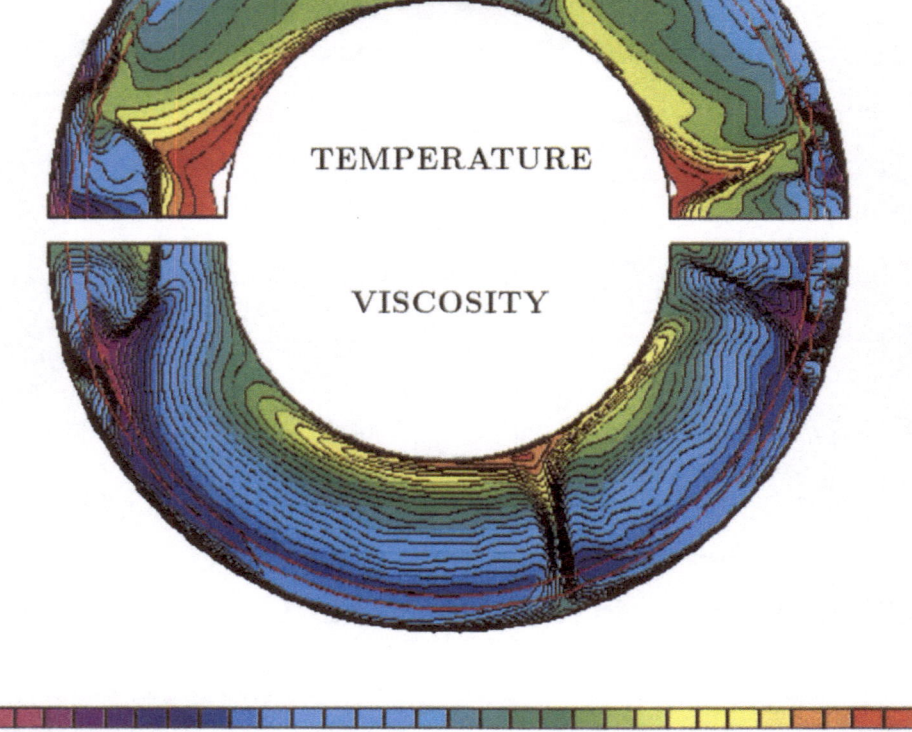

Figure 8
Temperature and viscosity fields of convection with temperature- and pressure-dependent viscosity and two mantle phase transitions. Time is taken at 1.558 Byr after the start of integration. Grid used is based on variable-sized finite-difference scheme (BRUNET and MACHETEL, 1997), in which 200 points are used in the radial direction and 510 along the horizontal. Physical parameters of this run are found in Table 2.

contrast across the 670-km discontinuity. The time instant is around 1.6 Byr into the numerical simulation. We note that in models with a viscosity jump across the 660 km, as in BRUNET and MACHETEL (1997), the tendency to form a low viscosity zone is decreased. Nevertheless, in the viscosity profiles from BRUNET and MA-CHETEL (1997) one can notice a small decrease in the viscosity below 660-km discontinuity. Since the activation energy used is minimal, only 70 kJ/mole, a larger viscosity contrast is to be expected for more realistic activation energies, around 400–500 kJ/mole, expected for transition zone minerals.

Figure 8 illustrates that indeed a low viscosity region (dark color) may exist in this partially layered convective system. We note that this low viscosity channel extends only a few hundred kilometers below the endothermic phase change. The thermal anomalies also evidence that there is complex interaction between the hot plume head and the transition zone, very similar to the tomographic images from *P1200* shown in Figure 1. The white regions in the thermal field represent temperatures some 200 degrees above the temperature at the CMB, which is a consequence of intense viscous dissipation at the base of a big plume, where there is severe deformation from the stagnation point in the flow pattern. Figure 9 displays the corresponding laterally averaged temperature and viscosity profile. From the thermal profile one recognizes that there are pronounced nonadiabatic features in the transition zone, as revealed by the bump in the transition zone. Contrastingly, the presence of an asthenosphere below 660 km appears quite clearly in the viscosity profile. The viscosity rises steeply in the lower mantle, until it reaches a maximum directly near 150 km above the CMB. Thus we have demonstrated from a direct numerical simulation how a low viscosity region may be produced in a self-consistent fashion in convection with phase transitions in a temperature- and pressure-dependent viscous mantle for a model with no viscosity jump across the 670-km discontinuity.

5. Summary and Conclusions

We have studied the geodynamics of the earth's mantle by several new approaches which have been stimulated by recent progress made in seismic tomogra-

Table 2

Physical parameters used in numerical modelling shown in Figures 8 and 9

Surface radius	6291 km
Surface temperature	1400 K
Density at surface	3370 kg/m^3
Average mantle density	4850 kg/m^3
Core radius	3480 km
Temperature at CMB	3000 K
Thermal diffusivity	0.8×10^{-6} m^2/s
Reference viscosity of lower mantle phase	10^{22} Pa · s
Activation energy throughout the mantle	100 kJ/mole
Activation volume of upper mantle	5×10^{-7} m^3/mole
Activation volume of lower mantle	9×10^{-7} m^3/mole
Clapeyron slope coefficient at 670 km	−2.8 MPa/K
Clapeyron slope coefficient at 400 km	3.0 MPa/K
Change of density at 670 km	390 kg/m^3
Change of density at 400 km	180 kg/m^3

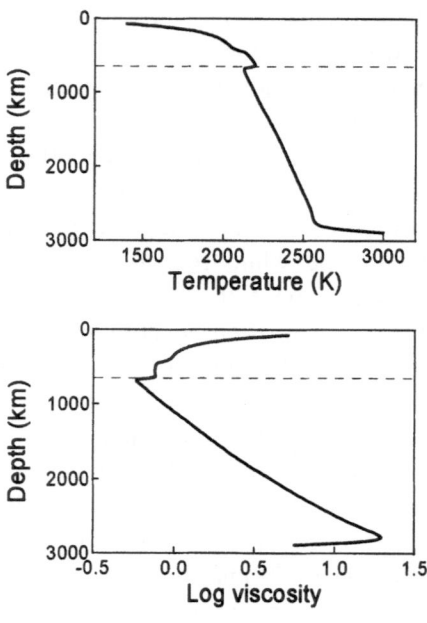

Figure 9
Horizontally averaged viscosity and temperature profiles of the two-dimensional fields depicted in Figure 8. The dashed line marks the position of the 660-km boundary.

phy. Through the window afforded by the high-resolution *P1200* tomographic model (ZHOU, 1996), we have glimpsed the interior heterogeneities of the upper 1200 km of the mantle to the extent that certain tantalizing features can be observed, such as slabs interacting with the top part of the lower mantle and the deflection of megaplume structures under Africa and the complex interaction of the lower-mantle plume beneath Iceland. Global tomographic models (ZHOU, 1996) can now approach the resolution of a couple to a few degrees characteristic of regional models (VAN DER HILST *et al.*, 1991; VAN DER HILST, 1995). The *P1200* model has also revealed the complicated nature of the region below the 660-km discontinuity, where many horizontally lying structures can be discerned. The spectra of the seismic heterogeneities also reveal striking features involving the power-law dependence with the degree. Such a dependence in the spectrum is indicative of underlying nonlinear processes in mantle dynamics. The power-law dependence of the spatial variations of the mantle heterogeneities also sheds light on the nature of time-dependence in mantle convection by the use of Taylor's hypothesis (TAYLOR, 1938; CHILLA *et al.*, 1993), which connects the properties of the spatial spectra with those from temporal spectra in turbulent flows.

The fine resolution made feasible by the *P1200* model has also prompted us to look into the inferences of mantle viscosity from geoid anomalies from a different vantage point. We have employed a new method, in which the slope of the geoid spectrum between $l = 12$ and 25 is used for inferring the viscosity profile in the top

1200 km of the mantle. By adopting a genetic algorithm approach, we have unveiled several interesting features of the viscosity structure in the top 1000 km of the mantle. Most interesting is the possible existence of a low viscosity zone under the 660-km discontinuity which contains a clearer expression from the inversion of the oceanic geoid data. The existence of a low viscosity layer in the top part of the lower mantle is consistent with recent developments in mineral physics (LI et al., 1996; TSCHAUNER et al., 1996), seismology (KAWAKATSU and NIU, 1994; SAKURAI, 1996; MONTAGNER and KENNETT, 1996) and geochemistry (GALER et al., 1989). This layer provides a natural explanation for the significant correlation found between the positions of subduction zones in the past 120 Myr and the seismic structure at a depth of about 1000 km (WEN and ANDERSON, 1995; KÝVALOVÁ et al., 1995). Such a low viscosity zone can potentially have important consequences for enforcing layered convection (CSEREPES and YUEN, 1997). We have demonstrated from numerical modelling how such a low viscosity channel can be produced under the 660-km phase change by means of a realistic mantle convection model with a strong temperature- and pressure-dependent viscosity and two major phase transitions, in which there is no viscosity contrast between the upper and lower mantles. There exist several families of solutions from these nonlinear inversions and most of them are found to be endowed with this low viscosity channel. On the other hand, the inversion of global geoid data reveals a possible trade-off between a low-viscosity channel above the 660-km discontinuity and a hard layer directly below the discontinuity. Another interesting outcome is the presence of the traditional asthenosphere right below the lithosphere. Such narrow features were unresolvable by means of longer wavelength geoid and long wavelength tomographic models, used previously.

We expect that this trend of utilizing higher resolution tomographic models, both regionally and globally, will bring about a new era in the way geodynamical questions are posed. What is now clear is that the days of using simple two- or three-layer earth models for mantle viscosity interpretation are over. These models have been educational for the past 12 years and now it is time to march forward with more sophisticated models, which should concurrently incorporate more testable mineral physical constraints (e.g., KARATO, 1996).

Acknowledgments

We are grateful for discussions with Shun Karato, Reini Boehler, and László Cserepes regarding many topics in mantle rheology and dynamics. Critical comments and suggestions from Jean-Paul Montagner and an anonymous referee are gratefully acknowledged. This research has been supported by the Humboldt Foundation, geophysics program of the National Science Foundation, U.S.–Japan

International Program, French INSU, and Czech national grants GAUK 144, GAUK 228 and GAČR 212.

References

BRUNET, D., and MACHETEL, PH. (1997), *Large-scale Tectonic Features Induced by Mantle Avalanches with Phase, Temperature and Pressure Dependent Viscosity*, J. Geophys. Res., in press.

BUTLER, S., and PELTIER, W. R. (1997), *Internal Thermal Boundary Layer Stability in Phase Transition Influenced Mantle Mixing*, J. Geophys. Res., 2731–2749.

ČADEK, O., KÝVALOVÁ, H., and YUEN, D. A. (1995), *Geodynamical Implications from the Correlation of Surface Geology and Seismic Tomographic Structure*, Earth Planet. Sci. Lett. *136*, 615–627.

ČADEK, O., ČÍŽKOVÁ, H., and YUEN, D. A. (1997), *Can Long-wavelength Dynamical Signatures be Compatible with Layered Mantle Convection?* Geophys. Res. Lett. *24*, 2091–2094.

CHILLA, F., CILIBERTO, S., INNOCENTI, C., and PAMPALONI, E. (1993), *Boundary Layer and Scaling Properties in Turbulent Thermal Convection*, Il Nuovo Cimento *15D*, 1229–1249.

ČÍŽKOVÁ, H., ČADEK, O., YUEN, D. A., and ZHOU, H. (1996), *Slope of the Geoid Spectrum and Constraints on Mantle Viscosity Stratification*, Geophys. Res. Lett. *23*, 3063–3066.

CSEREPES, L., and YUEN, D. A. (1997), *Dynamical Consequences of Mid-mantle Viscosity Stratification on Mantle Flows with an Endothermic Phase Transition*, Geophys. Res. Lett. *24*, 181–184.

DOIN, M. P., FLEITOUT, L., and McKENZIE, D. (1996), *Geoid Anomalies and the Structure of Continental and Oceanic Lithospheres*, J. Geophys. Res. *101*, 16119–16135.

DZIEWONSKI, A. M. (1984), *Mapping the Lower Mantle: Determination of Lateral Heterogeneities in P Velocity up to Degree and Order 6*, J. Geophys. Res. *89*, 5929–5952.

FORTE, A. M., and PELTIER, W. R. (1987), *Plate Tectonics and Aspherical Earth Structure: The Importance of Poloidal-toroidal Coupling*, J. Geophys. Res. *92*, 3645–3679.

FORTE, A. M., and PELTIER, W. R. (1991), *Viscous Flow Models of Global Geophysical Observables, 1. Forward Problems*, J. Geophys. Res. *96*, 20131–20159.

FORTE, A. M., and MITROVICA, J. X. (1996), *New Inferences of Mantle Viscosity from Joint Inversion of Long-wavelength Mantle Convection and Postglacial Rebound Data*, Geophys. Res. Lett. *23*, 1147–1150.

FUKAO, Y., OBAYASHI, M., INOUE, H., and NENBAI, M. (1992), *Subducted Slabs Stagnant in the Mantle Transition Zone*, J. Geophys. Res. *97*, 4809–4822.

GALER, S. J. G., GOLDSTEIN, S. L., and O'NIONS, R. K. (1989), *Limits on Chemical and Convective Isolation in the Earth's Interior*, Chemical Geology *75*, 257–290.

GOLDBERG, D. E., *Genetic Algorithms in Search, Optimization, and Machine Learning* (Addison-Wesley Publishing Company, Inc. 1989).

GRAND, S. P. (1994), *Mantle Shear Structure beneath the Americas and Surrounding Oceans*, J. Geophys. Res. *99*, 11591–11621.

HAGER, B. H., and CLAYTON, R. W., *Constraints on the structure of mantle convection using seismic observations, flow models and the geoid*. In *Mantle Convection, Plate Tectonics and Global Dynamics* (ed. Peltier, W. R.) (Gordon and Breach Scientific Publishers 1989) pp. 657–763.

HONDA, S., YUEN, D. A., BALACHANDAR, S., and REUTELER, D. (1993), *Three-dimensional Instabilities of Mantle Convection with Multiple Phase Transitions*, Science *259*, 1308–1311.

INOUE, H., FUKAO, Y., TANABE, K., and OGATA, Y. (1990), *Whole Mantle P-wave Travel Time Tomography*, Phys. Earth Planet. Inter. *59*, 294–328.

JONES, M. N., *Spherical Harmonics and Tensors for Classical Field Theory* (Research Studies Press Ltd., Letchworth 1985).

JORDAN, T. H. (1988), *Structure and Formation of the Continental Tectosphere*, J. Petrology, *Special Lithospheric Issue*, 11–37.

KARATO, S. (1993), *Importance of Anelasticity in the Interpretation of Seismic Tomography*, Geophys. Res. Lett. *20*, 1623–1626.

KARATO, S., WANG, Z., LIU, B., and FUJINO, K. (1995), *Plastic Deformation of Garnets: Systematics and Implications for the Rheology of the Mantle Transition Zone*, Earth Planet. Sci. Lett. *130*, 13–30.

KARATO, S., *Phase transformations and rheological properties of mantle materials.* In *Earth's Deep Interior* (eds. Crossley D., and Soward, A. M.) (Gordon and Breach Press, 1996) Chapter 8.

KAULA, W. M. (1967), *Theory of Statistical Analysis of Data Distributed over a Sphere*, Rev. Geophys. *5*, 83–107.

KAWAKATSU, H., and NIU, F. (1994), *Seismic Evidence for a 920-km Discontinuity in the Mantle*, Nature *371*, 301–305.

KENNETT, B. L. N., and ENGDAHL, E. R. (1991), *Traveltimes for Global Earthquake Location and Phase Identification*, Geophys. J. Int. *105*, 429–465.

KIDO, M., and ČADEK, O. (1997), *Inferences of Viscosity from the Oceanic Geoid: Indication of a Low Viscosity Zone below the 660-km Discontinuity*, Earth Planet. Sci. Lett. *151*, 125–138.

KING, S. D. (1995), *Radial Models of Mantle Viscosity: Results from a Genetic Algorithm*, Geophys. J. Int. *122*, 725–734.

KING, S. D., and ITA, J. J. (1995), *Effect of Slab Rheology on Mass Transport Across a Phase Transition Boundary*, J. Geophys. Res. *100*, 20211–20222.

KING, S. D., and MASTERS, G. (1992), *An Inversion for Radial Viscosity Structure Using Seismic Tomography*, Geophys. Res. Lett. *19*, 1551–1554.

KÝVALOVÁ, H., ČADEK, O., and YUEN, D. A. (1995), *Correlation Analysis between Subduction in the Last 180 Myr and Lateral Seismic Structure in the Lower Mantle*, Geophys. Res. Lett. *22*, 1281–1284.

LESTUNFF, Y., and RICARD, Y. (1995), *Topography and Geoid due to Lithospheric Mass Anomalies*, Geophys. J. Int. *122*, 982–990.

LI, P., KARATO, S., and WANG, Z. (1996), *High-temperature Creep in Fine-grained Polycrystalline $CaTiO_3$, an Analogue Material of $(Mg, Fe)SiO_3$ Perovskite*, Phys. Earth Planet. Inter. *95*, 19–36.

MACHETEL, P., and WEBER, P. (1991), *Intermittent Layered Convection in a Model Mantle with an Endothermic Phase Change at 670 km*, Nature *350*, 55–57.

MARTINEC, Z. (1991), *Program to Calculate the Least-squares Estimates of the Spherical Harmonic Expansion Coefficients of an Equally Angular-gridded Scalar Field*, Comput. Phys. Commun. *64*, 140–148.

MASTERS, G., JORDAN, T. H., SILVER, P. G., and GILBERT, F. (1982), *Aspherical Earth Structure from Fundamental Spheroidal Mode Data*, Nature *298*, 609–613.

MONIN, A., and YAGLOM, A. M., *Statistical Fluid Mechanics, vol. 2* (M.I.T. Press, Cambridge, MA 1975).

MONTAGNER, J.-P., and TANIMOTO, T. (1990), *Global Anisotropy in the Upper Mantle Inferred from the Regionalization of Phase Velocities*, J. Geophys. Res. *95*, 4797–4819.

MONTAGNER, J.-P., and TANIMOTO, T. (1991), *Global Upper Mantle Tomography of Seismic Velocities and Anisotropies*, J. Geophys. Res. *96*, 20337–20351.

MONTAGNER, J.-P. (1994), *Can Seismology Tell us Anything about Convection in the Mantle?* Rev. Geophys. *32*, 115–138.

MONTAGNER, J.-P., and KENNETT, B. L. N. (1996), *How to Reconcile Body-wave and Normal-mode Reference Earth Models*, Geophys. J. Int. *125*, 229–248.

MORGAN, J. P., and SHEARER, P. M. (1993), *Seismic Constraints on Mantle Flow and Topography of the 660-km Discontinuity: Evidence for Whole Mantle Convection*, Nature *365*, 506–511.

PARI, G., and PELTIER, W. R. (1995), *The Heat Flow Constraint on Mantle Tomography-based Convection Models: Towards a Geodynamically Self-consistent Inference of Mantle Viscosity*, J. Geophys. Res. *100*, 12731–12752.

PELTIER, W. R. (1996), *Mantle Viscosity and Ice-age Ice Sheet Topography*, Science *273*, 1359–1364.

PULLIAM, R. J., VASCO, D. W., and JOHNSON, L. R. (1993), *Tomographic Inversions for Mantle P-wave Velocity Structure Based on the Minimization of l_2 and l_1 Norms of International Seismological Centre Travel-time Residuals*, J. Geophys. Res. *98*, 699–734.

RAPP, R. H., and PAULIS, N. K. (1990), *The Development and Analysis of Geopotential Coefficient Models to Spherical Harmonic Degree 360*, J. Geophys. Res. *95*, 21885–21911.

RICARD, Y., FLEITOUT, L., and FROIDEVAUX, C. (1984), *Geoid Heights and Lithospheric Stress for a Dynamic Earth*, Ann. Geophys. *2*, 267–286.

RICARD, Y., and BAI WUMING (1991), *Inferring the Viscosity and the 3-D Density Structure of the Mantle from Geoid, Topography and Plate Velocities*, Geophys. J. Int. *105*, 561–571.

RICHARDS, M. A., and HAGER, B. H. (1984), *Geoid Anomalies in a Dynamic Earth*, J. Geophys. Res. *89*, 5987–6002.

ROMANOWICZ, B. (1994), *Anelastic Tomography: A New Perspective on Upper Mantle Thermal Structure*, Earth Planet. Sci. Lett. *128*, 113–122.

ROMANOWICZ, B. (1995), *A Global Tomographic Model of Shear Attenuation in the Upper Mantle*, J. Geophys. Res. *100*, 12375–12394.

SAKURAI, T., *Whole-mantle P-wave Tomography and Differential PP-P Time Measurement* (Master Thesis, University of Tokyo 1996).

SOLHEIM, L. P., and PELTIER, W. R. (1994), *Avalanche Effects in Phase Transition Modulated Thermal Convection: A Model of Earth's Mantle*, J. Geophys. Res. *99*, 6997–7018.

SU, W.-J., WOODWARD, R. L., and DZIEWONSKI, A. M. (1994), *Degree 12 Model of Shear-velocity Heterogeneity in the Mantle*, J. Geophys. Res. *99*, 6945–6980.

STEINBACH, V., and YUEN, D. A. (1995), *The Effects of Temperature-dependent Viscosity on Mantle Circulation with Two Major Phase Transitions*, Phys. Earth Planet. Int. *90*, 13–36.

TACKLEY, P. J., STEVENSON, D. J., GLATZMAIER, G. A., and SCHUBERT, G. (1993), *Effects of an Endothermic Phase Transition at 670-km Depth on Spherical Mantle Convection*, Nature *361*, 699–704.

TACKLEY, P. J., STEVENSON, D. J., GLATZMAIER, G. A., and SCHUBERT, G. (1994), *Effects of Multiple Phase Transitions in a Three-dimensional Spherical Model of Convection in Earth's Mantle*, J. Geophys. Res. *99*, 15877–15902.

TACKLEY, P. J. (1995), *On the Penetration of an Endothermic Phase Transition by Upwellings and Downwellings*, J. Geophys. Res. *100*, 15477–15488.

TANIMOTO, T. (1990), *Long Wavelength S-wave Velocity Structure Throughout the Mantle*, Geophys. J. Inter. *100*, 327–336.

TAYLOR, G. I. (1938), *The Spectrum of Turbulence*, Proc. Roy. Soc. London *A164*, 476–488.

THORAVAL, C., MACHETEL, PH., and CAZENAVE, A. (1995), *Locally Layered Convection Inferred from Dynamic Models of the Earth's Mantle*, Nature *375*, 777–780.

TRAMPERT, J., and WOODHOUSE, J. H. (1995), *Global Phase Velocity Maps of Love and Rayleigh Waves between 40 and 150 Seconds*, Geophys. J. Int. *122*, 675–690.

TRAMPERT, J., and WOODHOUSE, J. H. (1996), *High-resolution Global Phase Velocity Distributions*, Geophys. Res. Lett. *23*, 21–24.

TSCHAUNER, O., BOEHLER, R., SPECHT, S., ZERR, A., and PALME, H. (1996), *Probable Phase Transition in Mg and Fe-perovskite under Lower Mantle Conditions*, AGU Abstract, Fall Meeting 1996.

VAN DER HILST, R., ENGDAHL, R., SPAKMAN, W., and NOLET, G. (1991), *Tomographic Imaging of Subducted Lithosphere below Northwest Pacific Island Arcs*, Nature *353*, 733–739.

VAN DER HILST, R. (1995), *Complex Morphology of Subducted Lithosphere in the Mantle beneath the Tonga Trench*, Nature *374*, 154–157.

VAN DER HILST, R. D., WIDIYANTORO, S., and ENGDAHL, E. R. (1997), *Evidence for Deep Mantle Circulation from Global Tomography*, Nature *386*, 578–584.

WEINSTEIN, S. A. (1993), *Catastrophic Overturn in the Earth's Mantle Driven by Multiple Phase Changes and Internal Heat Generation*, Geophys. Res. Lett. *20*, 101–104.

WEN, L., and ANDERSON, D. L. (1995), *The Fate of Slabs Inferred from 130 Million Years of Subduction and Seismic Tomography*, Earth Planet. Sci. Lett. *133*, 185–198.

WOODHOUSE, J. H., and DZIEWONSKI, A. M. (1984), *Mapping the Upper Mantle: Three-dimensional Modelling of Earth Structure by Inversion of Seismic Waveforms*, J. Geophys. Res. *89*, 5953–5986.

WOODWARD, R. L., FORTE, A. M., SU, W.-J., and DZIEWONSKI, A. M., *Constraints on the large-scale structure of the earth's mantle*. In *Evolution of the Earth and Planets* (eds. Takahashi, E., *et al.*) (Geophysical Monograph 74, American Geophysical Union, 1993) pp. 89–110.

YUEN, D. A., REUTELER, D. M., BALACHANDAR, S., STEINBACH, V., MALEVSKY, A. V., and SMEDSMO, J. L. (1994), *Various Influences on Three-dimensional Mantle Convection with Phase Transitions*, Phys. Earth Planet. Inter. *86*, 185–203.

YUEN, D. A., BALACHANDAR, S., STEINBACH, V., HONDA, S., REUTELER, D. M., SMEDSMO, J. L., and LAUER, G. S. (1995), *Nonequilibrium Effects of Core Cooling and Time-dependent Internal Heating on Mantle Flush Events*, Nonlinear Processes in Geophysics *2*, 206–221.

YUEN, D. A., ČADEK, O., VAN KEKEN, P., REUTELER, D. M., KÝVALOVÁ, H., and SCHROEDER, B. A., *Combined results from mineral physics, tomography and mantle convection and their implications on global geodynamics*. In *Seismic Modelling of the Earth's Structure* (eds. Morelli A., and Ekström, G.) (Editrice Compositori, Bologna, Italy 1996) pp. 463–506.

ZHANG, S., and CHRISTENSEN, U. R. (1993), *Some Effects of Lateral Viscosity Variations on Geoid and Surface Velocities Induced by Density Anomalies in the Mantle*, Geophys. J. Int. *114*, 531–547.

ZHANG, Y.-S., and TANIMOTO, T. (1991), *Global Love Wave Phase Velocity Variation and its Significance to Plate Tectonics*, Phys. Earth Planet. Inter. *66*, 160–202.

ZHAO, W., YUEN, D. A., and HONDA, S. (1992), *Multiple Phase Transitions and the Style of Mantle Convection*, Phys. Earth Planet. Inter. *72*, 185–210.

ZHOU, H., and CLAYTON, R. W. (1990), *P- and S-wave Travel-time Inversions for Subducting Slab under the Island Arcs of the Northwest Pacific*, J. Geophys. Res. *95*, 6829–6852.

ZHOU, H., and WANG, H. (1994), *A Revisit to P-wave Travel-time Statistics at Teleseismic Stations*, J. Geophys. Res. *99*, 17849–17861.

ZHOU, H. (1996), *A High Resolution P-wave Model for the Top 1200 km of the Mantle*, J. Geophys. Res. *101*, 27791–27810.

(Received November 5, 1996, revised July 15, 1997, accepted July 17, 1997)

 To access this journal online:
http://www.birkhauser.ch

Pure appl. geophys. 151 (1998) 527–537
0033–4553/98/040527–11 $ 1.50 + 0.20/0

Pure and Applied Geophysics

Regional Correlation Analysis between Seismic Heterogeneity in the Lower Mantle and Subduction in the Last 180 Myr: Implications for Mantle Dynamics and Rheology

HANA ČÍŽKOVÁ,[1] ONDŘEJ ČADEK[1] and ALICE SLANCOVÁ[2]

Abstract—We have carried out a regional correlation analysis between the seismic structure of the lower mantle and the reconstructions of subduction sites in the past 180 Myr with the aim of estimating individual styles of slab motion over different parts of the earth. The correlation patterns obtained for three subduction branches (West Pacific, East Pacific and Alpine-Himalayan) are remarkably different. In the West Pacific, the subducting slabs tend to be stagnant beneath the 660-km discontinuity, while basically no subducted lithosphere has been detected below the depth of 1000 km. In contrast, the lithosphere subducted beneath the Americas seems to penetrate through the lower mantle continuously, showing correlation peaks at depth intervals of 800–1100 km and 1900–2500 km. In the Alpine-Himalayan region, significant correlation has been found below the 660-km discontinuity for recent subduction and in the mid-mantle for subduction younger than 120 Myr. An increase in the correlation close to the core-mantle boundary nevertheless indicates that, under certain circumstances, the slabs can reach the bottom of the mantle in the West Pacific and in the Alpine-Himalayan regions as well. The correlation peak at a depth of around 1000 km is common to all the subduction branches. However, its depth rather varies for different subduction zones and, thus, it is not clear whether this correlation maximum may be associated with a global mid-mantle discontinuity.

Key words: Correlation analysis, seismic velocity heterogeneities, subduction history.

1. Introduction

Over the last two decades, various techniques have been developed to aid in understanding the relationship between the geological features at the surface of the earth and the dynamic processes in the mantle. Among these techniques, one of the most popular is correlation analysis because of its relative simplicity and its ability

[1] Department of Geophysics, Faculty of Mathematics and Physics, Charles University, V Holešovičkách 2, 18000 Praha 8, Czech Republic.
[2] Geophysical Institute of the Academy of Sciences of the Czech Republic, BočníII/1401, 14131 Praha 4, Czech Republic.

to compare data coming from different branches of the geosciences. The increasing resolution of seismic tomography together with new information about past surface tectonic features provide a fruitful area where correlation analysis can be effectively applied.

Several recent papers (e.g., SCRIVNER and ANDERSON, 1992; ČADEK et al., 1994; WEN and ANDERSON, 1995) have shown that seismically fast regions at a depth of about 1000 km significantly correlate with reconstructions of subduction sites in the past 120 Myr. For subduction older than 120 Myr, another two less significant correlation maxima have been found at around 2000 km and close to the core-mantle boundary (KÝVALOVÁ et al., 1995). Between 1200 km and 2000 km a gap exists where no significant correlation has been found in any time. This *global* correlation pattern, however, does not seem to be fully valid on a *regional* scale. Tomographic analysis of the mantle shear structure beneath the Americas (GRAND, 1994) indicates that the style of subduction in this region is different from the global pattern in that the subducted lithosphere penetrates through the mantle continuously, without exhibiting any gap in the mid-mantle. In contrast, the mid-mantle gap is quite obvious in the West Pacific region where no remnants of the old lithosphere have been detected below the depth of about 1100 km (KAWAKATSU and NIU, 1996). The existence of the regional differences in subduction have been predicted by numerical simulations of the subducting plate (CHRISTENSEN, 1996) and confirmed by both regional and global high-resolution tomographic studies (SAKURAI, 1996; ZHOU, 1996; VAN DER HILST et al., 1997).

In last few years, two possible styles of slab motion in the lower mantle, the steady state (slabs penetrating through the mantle continuously) and the catastrophic (discontinuous avalanches), have often been discussed by both the convection modellers and the geodynamicists endeavouring to explain the long-wavelength geoid. The concept of slabs passing through the mantle in a steady-state regime is favoured, especially due to strikingly easy prediction of the dynamic geoid based on the reconstructions of subduction history in the last 180 Myr (RICARD et al., 1993). In contrast, numerical simulations of the thermal convection with phase transitions indicate that the catastrophic style of slab motion may also occur during certain periods of the mantle's development (see, e.g., MACHETEL and WEBER, 1991; STEINBACH and YUEN, 1992; PELTIER and SOLHEIM, 1992; HONDA et al., 1993; TACKLEY et al., 1993; WEINSTEIN, 1993). The main limitation of the present-day thermal convection models is an unrealistic rheological and geochemical description of the lithospheric plates. Recent two-dimensional numerical studies involving more realistic characteristics of subducted slabs have indicated that the catastrophic scenario may not be appropriate for the mantle. CHRISTENSEN (1996) shows that the behavior of the subducted lithosphere may be rather complex and strongly dependent on trench migration. The critical value of trench retreat is between 2 and 4 cm/year. If trench migration is faster than the critical value, the descending slab flattens above the phase boundary. At slower rates the slab penetrates to the lower mantle, although its morphology may be rather complex. A broad variety of styles in slab behavior is also compatible with the microphysical analysis of slab material under phase-transition conditions (RIEDEL and KARATO, 1997).

As this topic is of great importance to the further development of geophysical investigation, we have carried out a correlation analysis between the seismic structure of the lower mantle and the reconstructions of subduction sites in three different regions (West Pacific, East Pacific, and Alpine-Himalayan). The aim of our effort, covering 180 Myr of subduction history, was to shed light on the following problems: (i) Is the correlation peak at a depth of 1000 km, found by global correlation analysis, really common to all the subducted branches? (ii) Does each of the subduction branches have its own subduction style? (iii) If yes, can the style be classified according to the trench migration rate?

2. Data

In recent years, several whole mantle tomographic models have been produced for both P and S waves (e.g., TANIMOTO, 1990; INOUE et al., 1990; MASTERS et al., 1992; VASCO et al., 1993; SU et al., 1994). In the present study, we used the model of shear velocity heterogeneity by SU et al. (1994) which is based on an analysis of both body and surface waves and which provides a good lateral resolution below most of the recent subduction sites. The model is parameterized laterally by spherical harmonics up to degree 12 and by Legendre polynomials up to degree 13 in the radial direction. This corresponds to a lateral resolution of 3500 km and a radial resolution of about 200 km.

For the correlation analysis between the seismic pattern and the subduction, we used the average locations of subduction zones in the six time intervals covering the last 180 Myr, digitized from the figure in RICHARDS and ENGEBRETSON (1992). The dipping plate effect is taken into account by projecting the subduction zone positions 700 km horizontally in the direction of the overriding plate (see RICHARDS and ENGEBRETSON (1992) for details). The subduction sites were treated as simple lines (two-dimensional delta functions) which have the same weight everywhere. Differences in the subduction rate were not considered. For each time interval, the subduction was divided into three branches roughly corresponding to the East Pacific subduction (EP), the West Pacific subduction (WP), and the Alpine-Himalayan subduction (AH) considered together with the subduction in southeast Asia (Fig. 1). Subduction zones located in regions of deteriorated seismic resolution (South Pacific) and uncertain subduction zones were not taken into account, which explains the differences between Figure 1 and the original figure in RICHARDS and ENGEBRETSON (1992).

3. Correlation Technique

Statistical correlations of geophysical fields are usually computed in the spectral domain. This approach, very simple from the formal point of view, has many pitfalls which can be avoided through the use of spatial domain correlations (RAY and ANDERSON, 1994). For reasons explained in previous papers (KÝVALOVÁ *et al.*, 1995; ČADEK *et al.*, 1994), we prefer the spatial correlation formula based on integrating the seismic heterogeneity $t(d, \theta, \phi)$ at a depth d along the subduction line L:

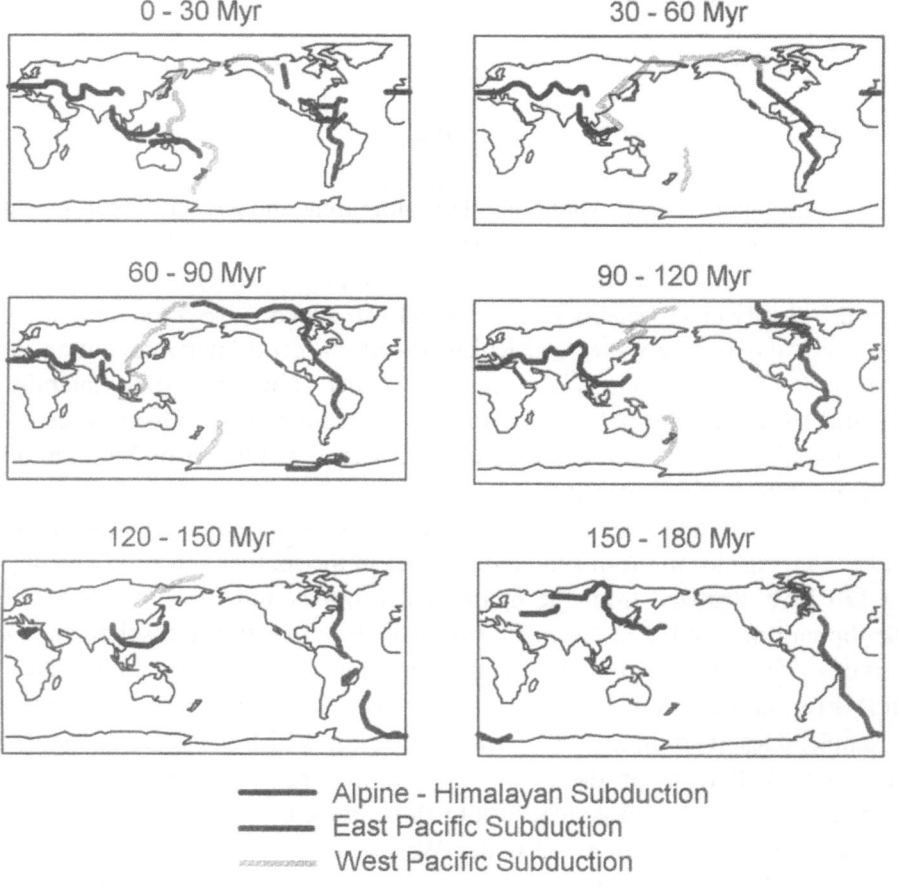

Figure 1
Average location of subduction zones used in this paper. The data were digitized from the figure in RICHARDS and ENGEBRETSON (1992). For each time interval, the subduction has been divided into three branches roughly corresponding to the East Pacific subduction, the West Pacific subduction, and the Alpine-Himalayan subduction considered together with the subduction in southeast Asia.

$$c(d) = \frac{\int_L t(d, \theta, \phi)\, dL}{t_m \int_L dL}.$$ (1)

The normalization factor $t_m \int_L dL$, where t_m is the extreme value of t at depth d, was chosen so that c was equal to 1 if the subduction line perfectly fits the extreme of t. It should be mentioned that in our case the correlation coefficient c was always significantly smaller than 1. A relatively small value of c, usually not exceeding 0.3, does not, however, necessarily mean that the two fields are not intrinsically related. The purpose of the correlation analysis is to reveal a linkage that is not obvious at first sight but which is hidden in noise and overprinted by other effects. If the uncertainties in the data are large, and this is likely the case for the heterogeneity pattern given by seismic tomography, then it is natural for the correlation coefficient to be small. In order to quantify the significance of the correlation, we evaluated the confidence level corresponding to correlation coefficient c. There is no general formula for evaluating the confidence level in the case of spatial domain correlations. The simplest technique for determining the confidence level is the Monte Carlo test (see, e.g., RAY and ANDERSON, 1994) which is based on the statistical evaluation of a large set of correlation coefficients obtained for randomly rotated functions L and t. One of the functions (t in our case) was randomly rotated and the correlation between subduction L and the rotated function t was computed from eq. (1). This was repeated for a large number of rotations (50,000 in the present paper). By constructing a histogram of correlation values, one can determine the confidence level as a probability of the occurrence of a correlation coefficient smaller than c. The correlation is usually judged to be significant if the confidence level is greater than 95%.

4. Results and Discussion

The results of our correlation analysis, translated into appropriate confidence levels, are shown in Figure 2. The correlation patterns obtained for individual subduction branches are remarkably different. In the Alpine-Himalayan region (left panel) a significant correlation was found below the 660-km discontinuity for recent subduction and in the mid-mantle for subduction younger than 120 Myr. The lithosphere subducted beneath the Americas (middle panel) seems to penetrate through the lower mantle continuously, showing correlation peaks at depth intervals of 800–1100 km and 1900–2500 km (cf. GRAND, 1994 and VAN DER HILST et al., 1997). In contrast, the slabs subducted in the West Pacific (right panel) tend to be stagnant below the 660-km discontinuity, and slight subducted lithosphere is detected below a depth of 1000 km. An increase in correlation close to the core-mantle boundary nevertheless indicates that, under certain circumstances, the slabs in the West Pacific and the Alpine-Himalayan region can also reach the

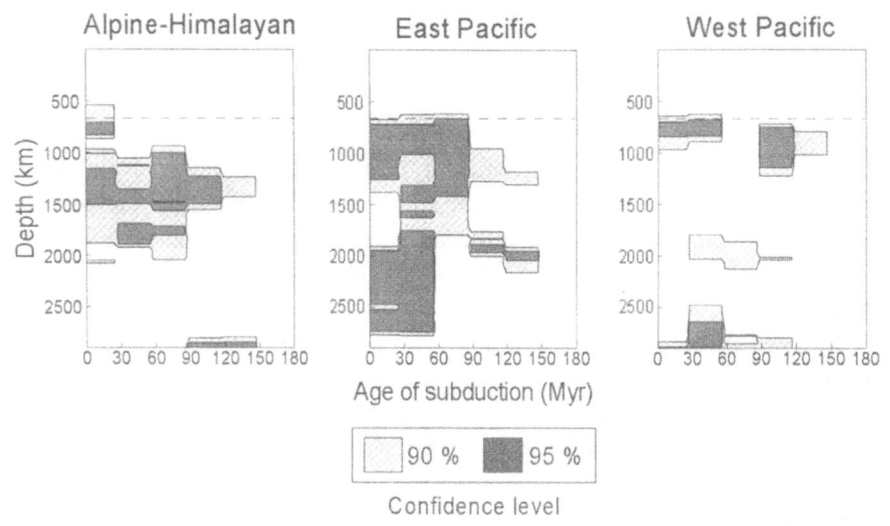

Figure 2

Regions of statistically significant correlation between the subduction and the distribution of fast seismic velocity anomalies in the mantle, depicted as a function of depth and age of subduction for three different subduction branches. Only areas where the confidence level of correlation exceeds 90% are shown. The dashed line marks the position of the 660-km boundary.

bottom of the mantle. This result is in good agreement with recent seismic studies by KENDALL and SILVER (1996), VAN DER HILST *et al.* (1997) and WYSESSION (1996).

The results of the correlation analysis are illustrated in Figure 3, where seismic structures beneath the three subduction zones are plotted on two-dimensional sections passing along the subduction lines. The subduction zone positions are considered within the time interval of 30–60 Myr, and their exact locations are given in Figure 4. Such a refolding of the subduction offers a new view of seismic structures below the subduction zone and allows us to discern details which otherwise would have been missed with standard projection methods. In the Alpine-Himalayan region (top panel), the subducted slabs are almost invisible in the top 1000 km. The faster-than-average seismic velocities, however, prevail in the mid-mantle and, in certain regions, continue down to the core-mantle boundary. The seismic heterogeneity pattern is very different beneath the East Pacific subduction zone (middle panel) where seismically fast material dominates throughout the mantle, covering more than 90% of the area of the section. The only significant region of seismically slow material is found in the mid-mantle. This anomaly is pronounced especially in the northern part of the section, but its weak continuation can also be recognized below South America. A similar, but broader and less deflected mid-mantle anomaly is found in the West Pacific (bottom panel) where it separates seismically fast material in the top 1200 km from a seismically more complex lowermost mantle. In the bottom 1000 km of the mantle, the fast seismic material prevails at a depth of around 2000 km and close to the core-mantle boundary.

Our results indicate that the ability of the cold lithosphere in the West Pacific to penetrate to the lower mantle is limited. In contrast, the younger and, thus, warmer lithosphere subducted beneath the Americas penetrates through the lower mantle freely. This difference in behaviour of the young and old subducting slabs is a challenging task for both the mineral physicists and the convection modellers. The numerical simulations of subduction (CHRISTENSEN, 1996) indicate that the behaviour strongly depends on the velocity of trench migration. If trench retreat is faster than 2–4 cm/year, then the descending slab flattens above the phase boundary. At slower rates the slab penetrates to the lower mantle. The characteristic rates of trench retreat in the Pacific subduction system are between 1 and 2.5 cm/year in the last 60 Myr (Fig. 5), thus close to the critical values. The migration rates in the

Alpine-Himalayan Subduction

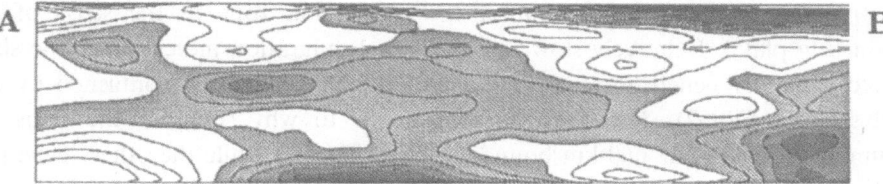

East Pacific Subduction

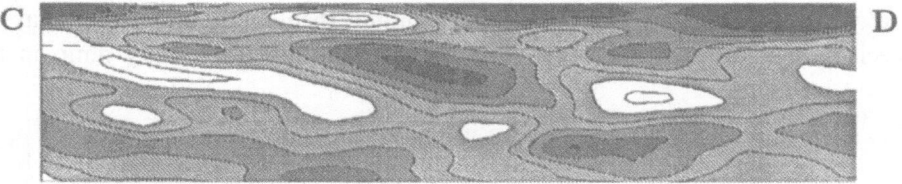

West Pacific Subduction

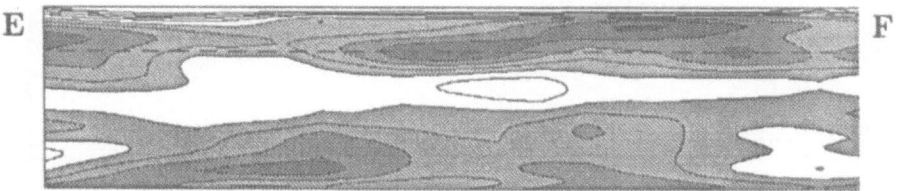

Figure 3
Seismic anomalies after Su *et al.* (1994) plotted on sections passing along the three subduction zones. Subduction zones are considered within the time interval of 30–60 Myr, and their exact locations are given in Figure 4. Seismic anomalies are given in percent of PREM (DZIEWONSKI and ANDERSON, 1981). Isoline interval is 0.3%. Regions with faster-than-average shear velocities are shaded. The dashed line marks the position of the 660-km boundary.

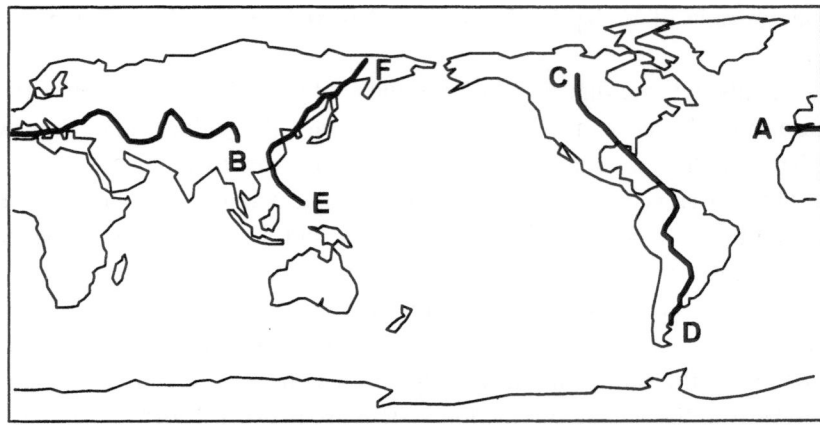

Figure 4
Positions of the subduction zones used in Figure 3.

Alpine-Himalayan system are even smaller, between −2.4 and 0 cm/year (the negative sign denotes the motion of the subduction zone in the sense of the subducting plate). Relatively slow rates of trench migration indicate that the slabs should generally penetrate to the lower mantle. This is indeed confirmed by our analysis (see Fig. 2). The question remains as to why some of the slabs are accumulated below the 660-km boundary (West Pacific) while the others penetrate to the deeper mantle (East Pacific, Alpine-Himalayan region). Recent investigation of the grain-size evolution in the subducted oceanic lithosphere (RIEDEL and KARATO, 1997) indicates that the strength of the slabs depends on the temperature in a very unusual way, and that the resultant rheological weakening may be especially significant for cold and fast slabs. If the major viscosity jump in the

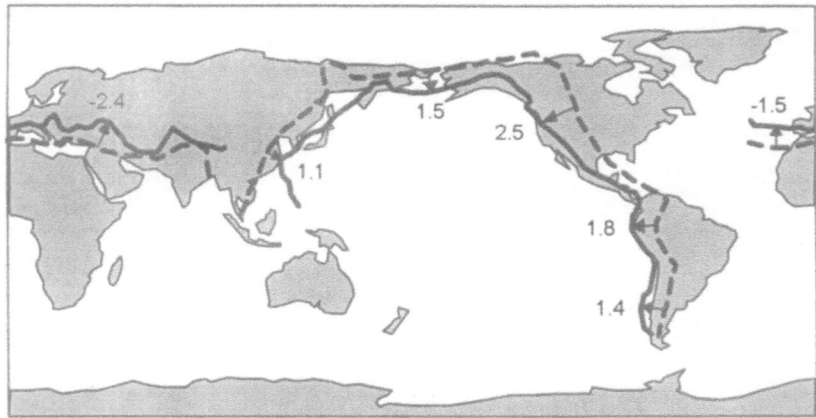

Figure 5
Average location of the Circum-Pacific and Alpine-Himalayan subduction zones in the time intervals of 0–10 Myr (full lines) and 56–64 Myr (dashed lines) plotted together with characteristic rates of the trench retreat in cm/year. Positions of plate boundaries are taken from GORDON and JURDY (1986).

mantle is not associated with the 660-km discontinuity but is located somewhat deeper in the lower mantle (FORTE and PELTIER, 1991; PARI and PELTIER, 1995; FORTE and MITROVICA, 1996; ČÍŽKOVÁ et al., 1996), then it is quite natural that the slabs in the West Pacific, which are fast and cold, are not able to penetrate through the viscosity interface and accumulate above it. In contrast, the fast slabs in the East Pacific penetrate to the mid-mantle because they are warmer and, thus, less affected by the rheological weakening. The differences in behavior of subducted slabs may also be associated with a low viscosity zone that has developed in certain regions below the 660-km discontinuity. ČADEK et al. (1997) demonstrate that such a low viscosity zone is indeed compatible with an intermediate-wavelength geoid in the oceanic regions. Numerical simulations of the thermal convection with strongly pressure- and temperature-dependent viscosity show that Rayleigh-Taylor instabilities can develop in this low viscosity channel. This leads to the spreading of cold material, which is trapped for a long time at depths between 670 km and 1000 km (CSEREPES and YUEN, 1997).

Results from the regional correlation analysis differ from those obtained on a global scale (KÝVALOVÁ et al., 1995). The only feature common to all the subduction branches is the correlation peak at a depth of around 1000 km. As already mentioned by KÝVALOVÁ et al. (1995), this peak may be associated with a global discontinuity expected in the mid-mantle by some seismologists and mineral physicists. Recent seismic array analysis (KAWAKATSU and NIU, 1994) has indicated that this discontinuity should be strongly deflected (depth variations 900–1100 km) which is in agreement with our findings. It is possible, however, that the very good correlation found at this depth only marks a mid-mantle viscosity increase suggested by an analysis of the non-hydrostatic geoid (FORTE and PELTIER, 1991; PARI and PELTIER, 1995; ČÍŽKOVÁ et al., 1996).

The above correlation analysis and accompanied discussion have been based on the premise that the fast seismic anomalies in the mantle are of a slab origin. It is evident, however, that even if no slabs penetrate to the lower mantle, the fast seismic region will exist there and some of them will correlate with subduction sites at the surface, either by accident, or due to a thermal coupling between the upper and lower mantle. Further investigations employing high-resolution tomographic models in connection with numerical simulation of subduction will be necessary for the fate of slabs to be fully understood.

Acknowledgements

The authors thank Shun Karato for inspiring discussions. Critical comments and suggestions of two anonymous reviewers are gratefully acknowledged. The work was supported by the Czech National Grant No. 205/96/0212 and the Charles University Grant Nos. 144 and 228.

REFERENCES

ČADEK, O., YUEN, D. A., ČÍŽKOVÁ, H., KIDO, M., ZHOU, H., BRUNET, D., and MACHETEL, PH. (1997), *New Perspectives on Mantle Dynamics from High-resolution Seismic Tomographic Model P1200*, Pure appl. geophys. *151*, 503–525.

ČADEK, O., YUEN, D. A., STEINBACH, V., CHOPELAS, A., and MATYSKA, C. (1994), *Lower Mantle Thermal Structure Deduced from Seismic Tomography, Mineral Physics, and Numerical Modelling*, Earth Planet. Sci. Lett. *121*, 385–402.

CHRISTENSEN, U. R. (1996), *The Influence of Trench Migration on Slab Penetration into the Lower Mantle*, Earth Planet. Sci. Lett. *140*, 27–39.

ČÍŽKOVÁ, H., ČADEK, O., YUEN, D. A., and ZHOU, H. (1996), *Slope of the Geoid Spectrum and Constraints on Mantle Viscosity Stratification*, Geophys. Res. Lett. *23*, 3063–3066.

CSEREPES, L., and YUEN, D. A. (1997), *Dynamical Consequences of Mid-mantle Viscosity Stratification on Mantle Flows with an Endothermic Phase Transition*, Geophys. Res. Lett. *24*, 181–184.

DZIEWONSKI, A. M., and ANDERSON, D. L. (1981), *Preliminary Reference Earth Model (PREM)*, Phys. Earth Planet. Int. *25*, 297–356.

FORTE, A. M., and MITROVICA, X. J. (1996), *New Inferences of Mantle Viscosity from Joint Inversion of Long-wavelength Mantle Convection and Postglacial Rebound Data*, Geophys. Res. Lett. *23*, 1147–1150.

FORTE, A. M., and PELTIER, W. R. (1991), *Viscous Flow Models of Global Geophysical Observables, 1. Forward Problems*, J. Geophys. Res. *96*, 20131–20159.

GORDON, R. G., and JURDY, D. M. (1986), *Cenozoic Plate Motion*, J. Geophys. Res. *91*, 1049–1057.

GRAND, S. P. (1994), *Mantle Shear Structure Beneath the Americas and Surrounding Oceans*, J. Geophys. Res. *99*, 11591–11621.

HONDA, S., YUEN, D. A., BALACHANDAR, S., and REUTELER, D. (1993), *Three-dimensional Instabilities of Mantle Convection with Multiple Phase Transitions*, Science *259*, 1308–1311.

INOUE, H., FUKAO, Y., TANABE, K., and OGATA, Y. (1990), *Whole Mantle P-wave Travel-time Tomography*, Phys. Earth Planet. Int. *59*, 294–328.

KAWAKATSU, H., and NIU, F. (1994), *Seismic Evidence of a 920-km Discontinuity in the Mantle*, Nature *371*, 301–305.

KAWAKATSU, H., and NIU, F. (1996), *Depth Variations of the "920-km Discontinuity" in the Mid-mantle*, Annales Geophysicae *14*, C43.

KENDALL, J. M., and SILVER, P. G. (1996), *Constraints from Seismic Anisotropy on the Nature of the Lowermost Mantle*, Nature *381*, 409–412.

KÝVALOVÁ, H., ČADEK, O., and YUEN, D. A. (1995), *Correlation Analysis between Subduction in the Last 180 Myr and Lateral Seismic Structure in the Lower Mantle*, Geophys. Res. Lett. *22*, 1281–1284.

MACHETEL, P., and WEBER, P. (1991), *Intermittent Layered Convection in a Model Mantel with an Endothermic Phase Change at 670 km*, Nature *350*, 55–57.

MASTERS, G., BOLTON, H., and SHEARER, P. (1992), *Large-scale 3-dimensional Structure of the Mantle*, E.O.S., American Geophys. Union Trans. *73* (14), 201.

PARI, G., and PELTIER, W. R. (1995), *The Heat Flow Constraint on Mantle Tomography-based Convection Models: Towards a Geodynamically Self-consistent Inference of Mantle Viscosity*, J. Geophys. Res. *100*, 12731–12752.

PELTIER, W. R., and SOLHEIM, L. P. (1992), *Mantle Phase Transitions and Layered Chaotic Convection*, Geophys. Res. Lett. *19*, 321–324.

RAY, T. W., and ANDERSON, D. L. (1994), *Spherical Disharmonics in the Earth Sciences and the Spatial Solution: Ridges, Hotspots, Slabs, Chemistry, and Tomography Correlations*, J. Geophys. Res. *99*, 9605–9614.

RICARD, Y., RICHARDS, M., LITHGOW-BERTELLONI, C., and LE STUNFF, Y. (1993), *A Geodynamic Model of Mantle Density Heterogeneity*, J. Geophys. Res. *98*, 21895–21909.

RICHARDS, M. A., and ENGEBRETSON, D. C. (1992), *Large-scale Mantle Convection and the History of Subduction*, Nature *355*, 437–440.

RIEDEL, M. R., and KARATO, S. (1997), *Grain-size Evolution in Subducted Lithosphere Associated with the Olivine-spinel Transformation and its Effect on Rheology*, Earth Planet. Sci. Lett., in press.

SAKURAI, T., *Whole Mantle P-wave Tomography and Differential PP-P Time Measurement*, Master Thesis (University of Tokyo, 1996).

SCRIVNER, C., and ANDERSON, D. L. (1992), *The Effect of Post Pangia Introduction on Global Mantle Tomography and Convection*, Geophys. Res. Lett. *19*, 1053–1056.

STEINBACH, V., and YUEN, D. A. (1992), *The Effects of Multiple Phase Transitions on Venusian Mantle Convection*, Geophys. Res. Lett. *19*, 2243–2246.

SU, W.-J., WOODWARD, R. L., and DZIEWONSKI, A. M. (1994), *Degree 12 Model of Shear Velocity Heterogeneity in the Mantle*, J. Geophys. Res. *99*, 6945–6980.

TACKLEY, P. J., STEVENSON, D. J., GLATZMAIER, G. A., and SCHUBERT, G. (1993), *Effects of an Endothermic Phase Transition at 670 km Depth on Spherical Mantle Convection*, Nature *361*, 699–704.

TANIMOTO, T. (1990), *Long Wavelength S-wave Velocity Structure Throughout the Mantle*, Geophys. J. Int. *100*, 327–336.

VAN DER HILST, R. D., WIDIYANTORO, S., and ENGDAHL, E. R. (1997), *Evidence for Deep Mantle Circulation from Global Tomography*, Nature *386*, 578–584.

VASCO, D. W., JOHNSON, L. R., PULLIAM, R. J., and EARLE, P. S. (1993), *Robust Inversion of IASP91 Travel-time Residuals for Mantle P- and S-velocity Structure, Earthquake Mislocation, and Station Correction*, J. Geophys. Res. *99*, 13727–13755.

WEINSTEIN, S. A. (1993), *Catastrophic Overturn in the Earth's Mantle Driven by Multiple Phase Changes and Internal Heat Generation*, Geophys. Res. Lett. *20*, 101–104.

WEN, L., and ANDERSON, D. L. (1995), *The Fate of Slabs Inferred from 130 Million Years of Subduction and Seismic Tomography*, Earth Planet. Sci. Lett. *133*, 185–198.

WYSESSION, M. E. (1996), *Large-scale Structure of the Core-mantle Boundary from Diffracted Waves*, Nature *382*, 244–248.

ZHOU, H. (1996), *A High Resolution P-wave Model for the Top 1200 km of the Mantle*, J. Geophys. Res. *101*, 27791–27810.

(Received October 17, 1996, revised June 23, 1997, accepted June 24, 1997)

 To access this journal online:
http://www.birkhauser.ch

Pure appl. geophys. 151 (1998) 539–548
0033–4553/98/040539–10 $ 1.50 + 0.20/0

❙Pure and Applied Geophysics

Velocity and Density Heterogeneities of the Tien-Shan Lithosphere

T. M. Sabitova,[1] O. M. Lesik[2] and A. A. Adamova[1]

Abstract—The Tien-Shan orogene is a region in which the earth's crust undergoes considerable thickening and tangential compression. Under these conditions the lithosphere heterogeneities (composition, rheological) create the prerequisites for the development of various phenomena of tectonic layering (lateral shearing, different deformation of layers). To study the distribution of velocity, density and other elastic parameters, the results from a seismic tomography study on *P*-wave as well as *S*-wave velocities were used. Using empirical as well as theoretical formulas on the relationship between velocity, density and silica content in rocks, their distribution in the Tien-Shan's lithosphere has been calculated. In addition, other elastic parameters, such as Young's modulus, shear modulus, Poisson's ratio and coefficient of general compressions have been determined. Zoning of different types of crust was carried out for the region investigated. The characteristics of the "crust-mantle" transition have been investigated. Large blocks with different types of the earth's crust were distinguished. Layers with inverse values of velocity, density and shear and Young modulus are revealed in the Tien-Shan lithosphere. All of the above described features open new ways to solve geodynamics problems.

Key words: Seismic zoning, crust, silica content of rocks, elastic parameters.

Introduction

The Tien-Shan is the most prominent mountain building (Fig. 1), in which the orogene processes such as tangential compression of crust, the forming of linear folds, upthrusts, thrusts, overthrust sheets, crustal thickening and high seismicity have been intensively occurring since the second half of the Eocene. The important problem in these geodynamic conditions is to reveal the lithosphere heterogeneities that cause its nonadequate response to the actual stress. This problem is solved by different geophysical methods. Especially important are these layers, where an inversion of physical parameters like velocities, densities, shear modulus and Young modulus are observed.

The inversion layers function as detachment horizons, which in turn provide the freedom of relative movement during deformation events. The purpose of our

[1] Institute of Seismology, NAS, Asanbai 52/1, 720060, Bishkek, Kyrgyzstan.
[2] Experiment—Methodical Expedition of IVTAN RAS, 720049, Bishkek-49, Kyrgyzstan.

research is to distinguish velocity, density and other geophysical heterogeneities in the Tien-Shan lithosphere in order to make a zoning of different crustal types and to study the origin of the heterogeneities by joint analysis of geological and geophysical data.

Method

To solve the problem as introduced above, we have interpreted the three-dimensional (3-D) velocity model of the Tien-Shan of ROECKER *et al.* (1993) and calculated the elastic parameters and the silica content in this model. Reliability of the velocity model was verified by the forward problem resolution for the source with parameters known (explosion) and correlated with the velocities obtained by the other geophysical methods in this region.

The initial data were *P*- and *S*-wave velocities for the crustal and the uppermost mantle layers as obtained by ROECKER *et al.* (1993), assuming uncertainties as given by the authors. The systematization of the different velocity section types was carried out on the basis of the following indicators: presence of the waveguides (low-velocity zones), the high velocity bodies and a special type of velocity-depth function. Analysis of the *P*-wave velocity distribution allowed us to reveal three main types of the velocity pattern: 1—a layered crust with one or several waveguides; 2—a low velocity crust without waveguides ($v_P = 5.6$–6.0 km/s are observed here down to 35 km and even to 50 km); 3—a normal crust without waveguides (v_P increases with depth, Fig. 2a). There are several distinguished subtypes of the velocity patterns in every main type.

The *P*- and *S*-wave velocity data were also used to calculate the rock density and silica contents, employing the empiric formula of KHALEVIN *et al.* (1986).

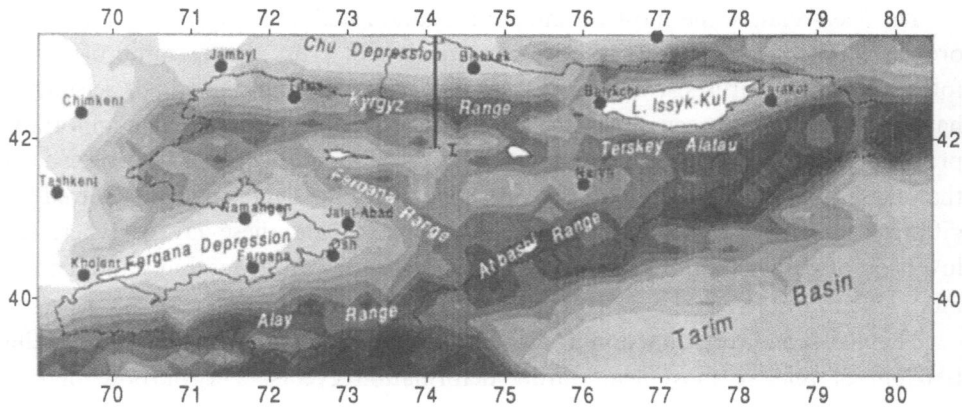

Figure 1
Map showing the major geographic features of the Tien-Shan; I—profile "Kendyktas-Djumgol."

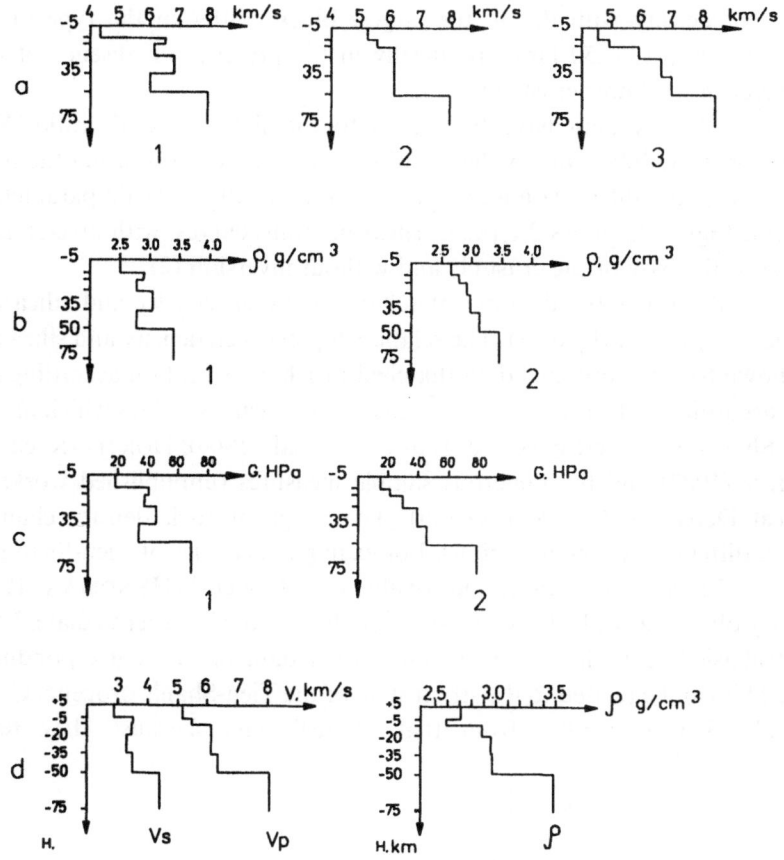

Figure 2

Types of the velocity, density and shear modulus section (a)—velocity section types: 1—layering crust with waveguides (low-velocity zones) in depths $H = 10–20$ km and $H = 35–50$ km; 2—low velocity crust; 3—normal crust; (b)—density section types: 1—with density inversion in the lower crust ($H = 35–50$ km); 2—without density inversion in the lower crust; (c)—shear modulus (G) section types: 1—with inversion of G in the lower crust; 2—without inversion of G in the lower crust; (d)—relations of the P- and S-wave velocities and rock density (ρ).

Calculated densities had been correlated with recalculated maps of the anomaly gravity field, which are characterized by density heterogeneities in corresponding layers. A strong correlation of the anomaly areas (LESIK, 1996) is observed. The reliability of the calculated density increases due to joint use of both P- and S-wave velocities. Quite frequently, one can observe that v_P and v_S change not accordingly in the Tien-Shan, which apparently is related to the fact that P-wave velocities are being determined mainly by the mineral composition of the rocks, while the S-wave velocities are being stipulated mostly by their physical state. Consequently, a rather complicated pattern of relations between v_P, v_S and density can be observed (Fig. 2d). One can distinguish different types of the earth's crust according to the variations of density with depth; the largest blocks differ in the type of "crust-mantle" transition (35–50 km), specifically in the presence or absence of the low density layer in the lower crust (Fig. 2b).

The rock density data have been used to calculate Poisson's ratio, Young's modulus, shear modulus and coefficient of general compression, using the formulas by DORTMAN, ed. (1984). Zoning on sections type of these elastic parameters was carried out. Figure 2c shows 2 types of shear modulus curves: with inversion of this parameter at the base of the crust (1) and without inversion (2).

Figure 3 illustrates the diagram of relation between density and silica content calculated on v_P and v_S (by dots). The relationship between density and silica content is also shown for the published data (hatched): rock classification according to SiO_2 content, according to BOGATIKOV *et al.*, eds. (1981), density values which are typical for Tien-Shan's rocks were collected from CLARK, ed. (1966); DORTMAN, ed. (1984); YUDAKHIN (1983) and by numerous sample measures (unpublished works of the Geological Department of Kyrgyzstan). The range of rock density changes for Tien-Shan differs for different regions. For example, densities of Tien-Shan granites are 2.60–2.62 g/cm^3 on average, and a value of 2.80 g/cm^3 (HYNDMAN, 1972) has never been observed. Calculated densities of the "acidic," "intermediate," "basic" and "ultrabasic" rocks have higher values than data on the corresponding rock samples. This can be related to the strain state of the Tien-Shan's orogene. Certainly, the calculated values differ from that of real data, because they represent

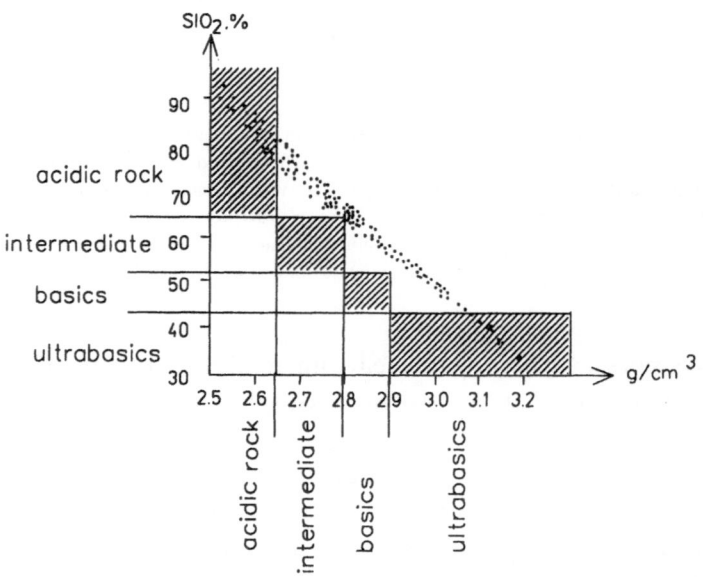

Figure 3

Relation of density versus silica content. Values of the parameters, computed for empiric formulas (KHALEVIN *et al.*, 1986) are shown by dots; values of the parameters collected from (BOGATIKOV *et al.*, eds., 1981; CLARK, ed., 1966; DORTMAN, ed., 1984; YUDAKHIN, 1983) are hatched.

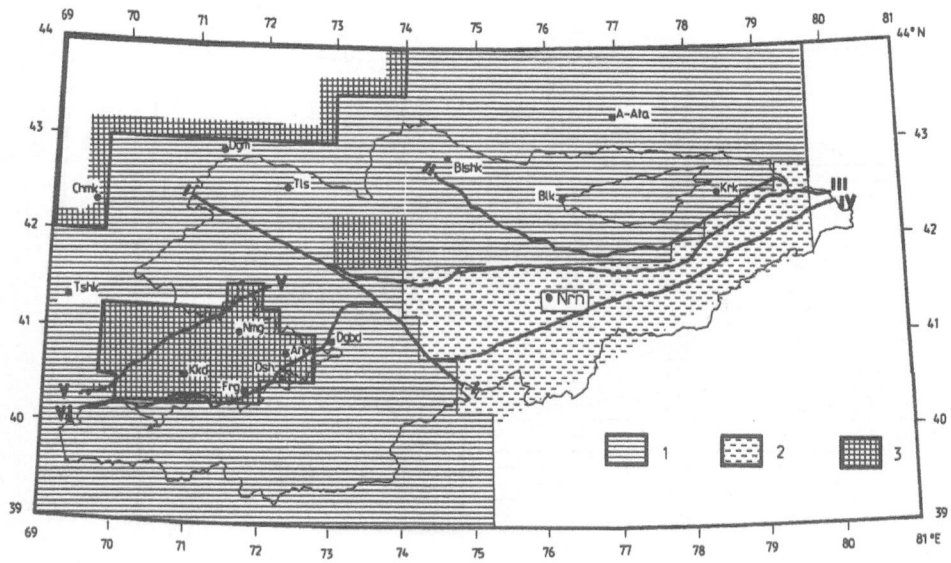

Figure 4

Scheme of seismic zoning for the Tien-Shan's crust on velocity section types: 1—the layered crust with one or several waveguides; 2—the low-velocity crust without waveguides; 3—the normal crust without waveguides. Faults: I—Talaso-Fergana, II—Central Terskei, III—Nikolaev line, IV—Atbashi-Enylchek, V—North-Fergana, VI—South-Fergana.

average densities of thick crustal layers and, moreover, they could be affected by data errors. Nonetheless, we can use these densities to distinguish large density heterogeneities in the crust.

Results

Here only the major features, i.e. the largest blocks with the main types of velocity sections (Fig. 4) and also the density and shear modulus maps characterizing the zone of the "crust-mantle" transition (35–50 km) (Figs. 5 and 6) are shown.

Moreover, an example of composition determined from relations between density and silica content, calculated from P- and S-wave velocities along the profile "Kendyktas-Djumgol", is presented (Figs. 7 and 8). This profile is chosen across the strike of the anomalous velocity zones (Fig. 1). There is a waveguide (V_p) observed in the middle crust in the southern part of the profile, but in the Kyrgyz and Chu tectonic blocks it is in the upper and lower crusts (Fig. 7). High velocity bodies are distinguished in the upper and middle crusts which are underlain with the waveguides. Analysis of the relation of density and silica content has allowed us to determine the crustal composition for this profile. The bed body of "ultrabasic" composition in the 20–35 km depth range is of interest in the Chu tectonic block

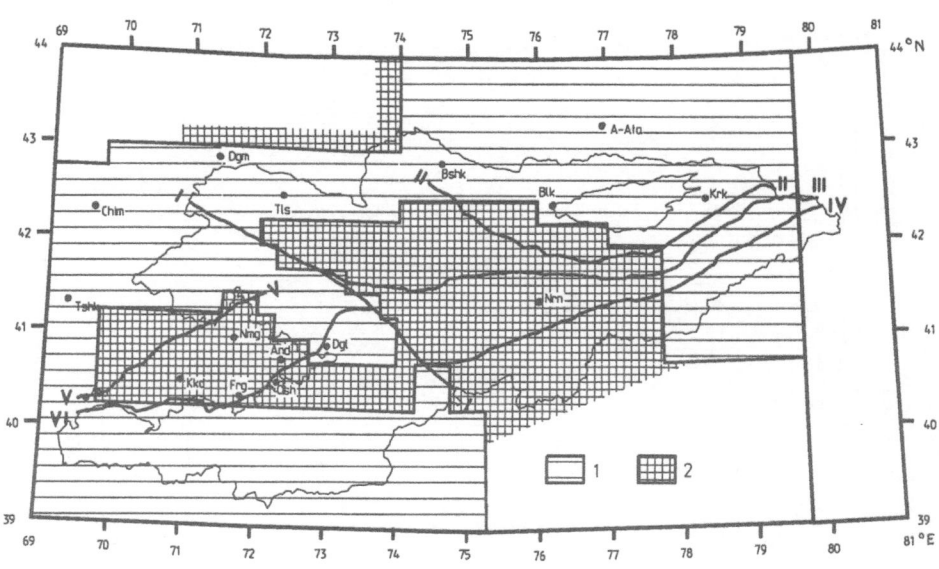

Figure 5
Scheme of density distribution in the lower crust ($H = 35-50$ km). 1—areas where the density inversion is observed; 2—areas without the density inversion.

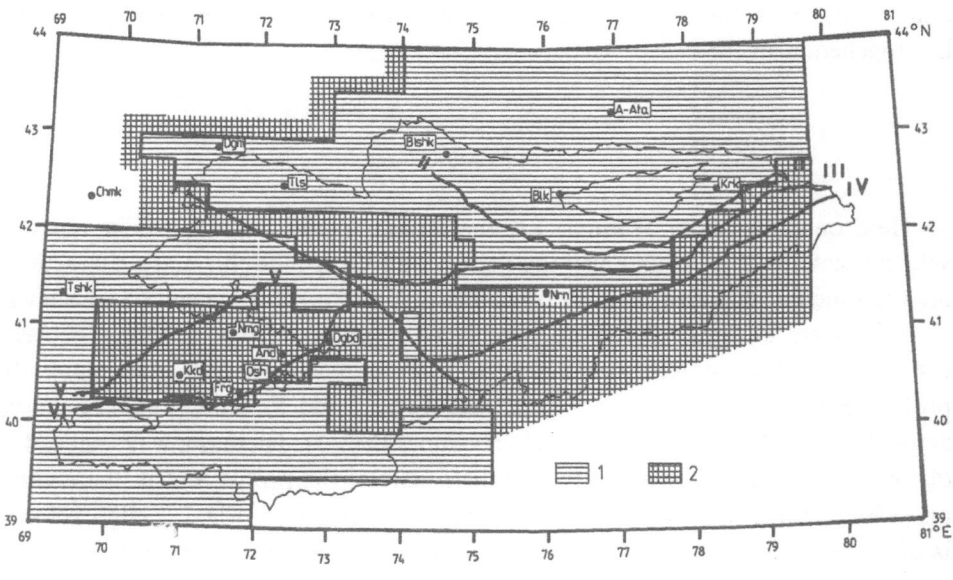

Figure 6
Scheme of shear modulus distribution in the lower crust ($H = 35-50$ km). 1—areas with shear modulus inversion; 2—areas without shear modulus inversion.

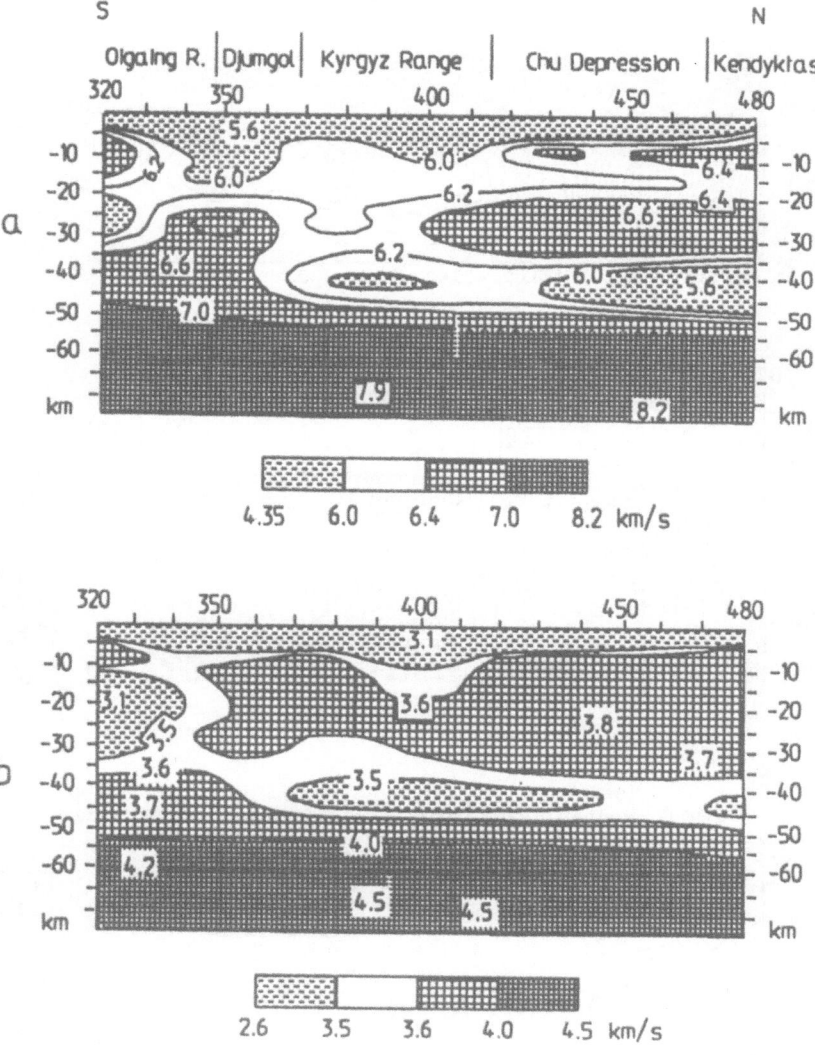

Figure 7
Velocity sections v_P (a) and v_S (b) on "Kendyktas-Djumgol" profile.

(Fig. 8c): the top of this layer is likely to be the remainder of the Moho boundary of the platform of the preorogene stage of the Tien-Shan development (crust thickness in the preorogene stage was 30–35 km, amplitude of neotectonics motions is about 10 km). Depending on the degree of layering and composition, the tectonic blocks with allochtone (Chu and Djumgol) and autochtone (Kyrgyz) types of crust are distinguished. Obviously, they are blocks with a different development history.

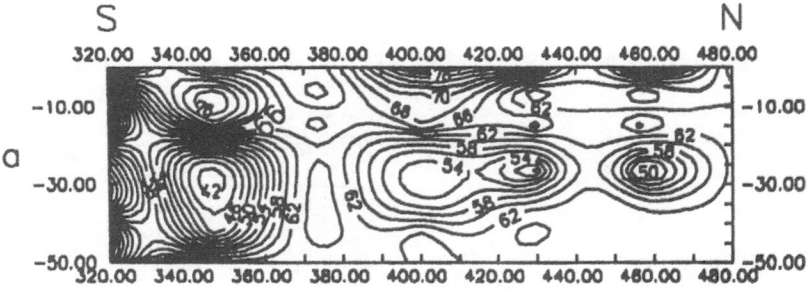

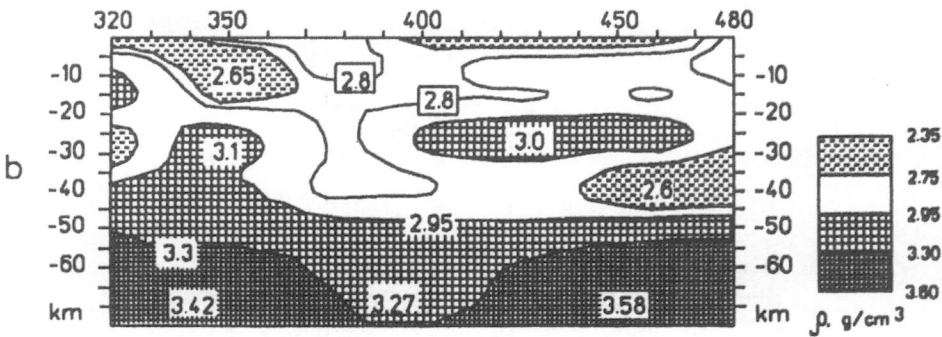

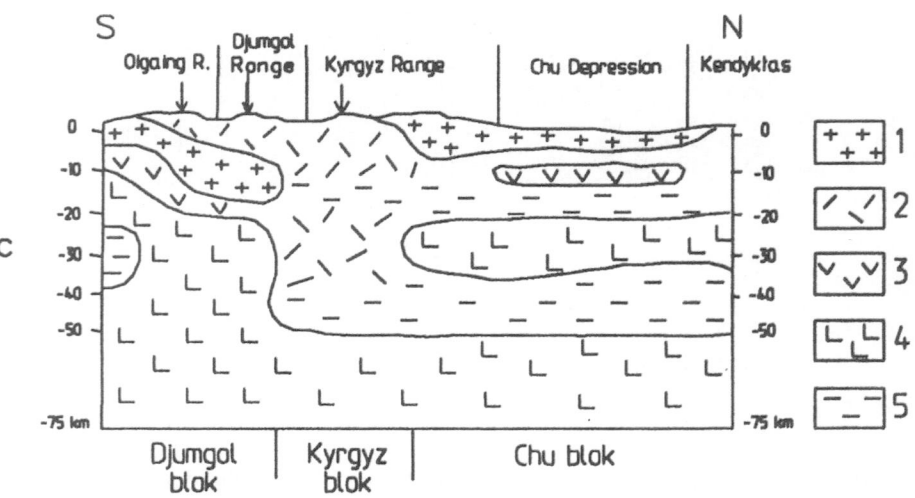

Figure 8

Profile "Kendyktas-Djumgol": a—distribution of the silica content, %; b—density section; c—composition section, rocks: 1—acidics, 2—intermediate, 3—basics, 4—ultrabasics.

Discussion

The crust in the Tien-Shan region has experienced various degrees of development (Fig. 4). It is characterized by an alternation of high- and low-velocity layers. The origin of crustal waveguides is not simple: the velocities of seismic waves depend on the composition of the rocks and density, thermodynamical conditions, fluidization and the mechanical and physical state of material. Our explanation of the origin of the waveguides follows the ideas of BAKIROV et al. (1996). In that published investigation, based on complex geology-geophysical data, analysis revealed that the high-temperature metamorphism and the magma forming have occurred in the Tien-Shan's crust recently.

The lower crust waveguide is identified as the layer of partial melting of material, where high-temperature subfacies of amphibolite facies dominate, and migma (melt consisted of remaining unmelted material) is forming. We think that the upper isolated low-velocity bodies are extruded magmatic chambers. The migma layer existence facilitates upper crust deformation, constrains faulting to certain depth levels and introduces important corrections into geotectonic notions (BAKIROV et al., 1996). The inverse layers with lower density values and shear modulus (Figs. 5 and 6), existing in the Tien-Shan's lithosphere in conditions of tangential compression (CHEDIA, 1986), are determined by its response to the recent tectonical strains. Existence of such a layer at the base of the crust provides some of its autonomy during deformation. Presence of several layers with low density causes the gravity of the non-equilibrated system, in which more dense layers overlay light layers (Fig. 2b-1). A system of this kind is very sensitive to every outward influence, like the recent tectonical strain (FADEEV, 1993). The above-mentioned type of crust should be taken into account in the study of recent processes from the position of nonlinear geodynamics, research actively developing in recent years (PUSHCHAROVSKY, 1993).

In areas, where inverse layers are absent, both in the middle crust and at its base, the crust is strongly related to the upper mantle, forming a lithosphere block, where such tectonic elements as warping, overthrusting and accretion are absent. Such a block has been distinguished in the Fergana depression (Figs. 4, 5 and 6), where v_P and v_S increase with depth, the lower part of the crust has high velocity and density ($v_P = 7.0$ km/s, $\rho = 3.41$ g/cm^3), which is associated with ultrabasic rocks. Quite another crustal type is observed in the central part of the Tien-Shan, where extremely low P-wave velocities ($v_P = 5.9$–6.1 km/s) are noted to 35–50 km depth. Our preliminary research has shown that there are conditions for high-temperature metamorphism and magma generation in the Tien-Shan's crust (BAKIROV et al., 1996). Velocities 5.9–6.1 km/s in the lower part of the crust are obviously stipulated with migma (partly melted rocks). Calculated densities for these velocities are about 2.80 g/cm^3 (Fig. 2b). Analog low velocities at the lower crust can be observed in the other young folded systems also (NERSESOV, ed., 1987). Density smoothly increases with depth. Other elastic parameters are also weakly differentiated. The inverse layers are absent in the lower part of the crust.

Such sharp distinctions in physical properties of crustal types are stipulated by the previous history of geological development of these large blocks. These large blocks consist of smaller size blocks with various heterogeneities. Interpretation of such heterogeneities is possible only by means of complex geology-geophysical data analysis.

The heterogeneities of physical parameters revealed in the Tien-Shan's crust lithosphere open new approaches to solution of problems of nonlinear geodynamics, which has been actively developing during recent years.

REFERENCES

BAKIROV, A. B., LESIK, O. M., LOBANCHENKO, A. N., and SABITOVA, T. M. (1996), *Indication of Modern Magmatism in Tien-Shan*, Geologia Geofiz., Novosibirsk *37* (12), 42–53 (in Russian).

CHEDIA, O. K., *Morphostructures and Recent Tectogenesis of Tien-Shan* (Ilim, Frunze 1986).

Classification and Nomenclature of the Magmatic Rocks (eds. BOGATIKOV, O. A. *et al.*) (Moscow 1981).

Deep Structure of Low Seismic Activity Regions of USSR (ed. NERSESOV, I. L.) (Nauka, Moscow 1987).

FADEEV, V. E. (1993), *Non-Linear Phenomena and its Role in Geotectonics*, Geotectonics *1*, 7–12 (in Russian).

Handbook of Physical Constants (ed. CLARK, C.) (Yale Univ., New Haven, Connecticut 1966).

HYNDMAN, D. W., *Petrology of Igneous and Metamorphic Rocks* (McGraw-Hill, Inc. 1972).

KHALEVIN, N. I., ALEINIKOV, A. L., KOLUPAJEVA, E. N., TIUNOVA, A. M., and YUNUSOV, F. F. (1986), *On Joint Use of Longitudinal and Transverse Waves in Deep Seismic Sounding*, Geologia Geofiz., Novosibirsk *10*, 94–98 (in Russian).

LESIK, O. M., *Use of the seismic tomography data for determination rock's composition and elastic parameters of the Earth's crust*. In *Neotectonics and Recent Geodynamics: Continents and Oceans* (Theses of XXIX tectonic workshop, Moscow 1996) pp. 79–80.

PUSHCHAROVSKY, YU. M. (1993), *Non-linear Geodynamics (author's credo)*, Geotectonics *1*, 3–7.

Physical Properties of the Rocks and Minerals (Petrophysics) (ed. DORTMAN, N. B.) (Nedra, Moscow 1984).

ROECKER, S. W., SABITOVA, T. M., VINNIK, L. P., BURMAKOV, Y. A., GOLOVANOV, M. I., MAMATKANOVA, R., and MUNIROVA, L. (1993), *Three-dimensional Elastic Wave Structure of the Western and Central Tien-Shan*, J. Geophys. Res. *98*, 15779–15795.

SABITOVA, T. M., LESIK, O. M., ADAMOVA, A. A., and MAMATKANOVA, R., *Seismotomography model of Tien-Shan Earth crust in connection with seismicity*, European Geophysical Society (Katlenburg-Lindau, Germany 1995), Annales Geophys. Part 1, p. 55.

YUDAKHIN, F. N., *Geophysical Fields, Deep Structure and Seismicity of the Tien-Shan* (Ilim, Frunze 1983).

(Received December 30, 1996, revised August 12, 1997, accepted September 17, 1997)

To access this journal online:
http://www.birkhauser.ch

Pure appl. geophys. 151 (1998) 549–561
0033–4553/98/040549–13 $ 1.50 + 0.20/0

Pure and Applied Geophysics

Large-scale 3-D Gravity Analysis of the Inhomogeneities in the European-Mediterranean Upper Mantle

T. P. YEGOROVA,[1] V. I. STAROSTENKO,[1] and V. G. KOZLENKO[1]

Abstract—Methods and the results of estimating the anomalies characterising the density inhomogeneities in the European-Mediterranean upper mantle are described. These anomalies were obtained by subtracting the gravity effect of a crustal density model derived from seismic velocities from the observed gravity field averaging over an area of $1° \times 1°$. The 3-D density model of the study region comprises two regional layers of varying thickness with lateral variation of average density: the sedimentary cover and the crystalline crust. The average densities for model layers were evaluated by using a velocity/density conversion function and taking into account sediment consolidation with depth. Clear correlation between residual gravity anomalies and both velocity heterogeneities and thermal regime data of the upper mantle has been revealed. An agreement of positive anomalies over the Alps, the Adriatic plate and the Calabrian Arc with high velocity domains in the upper mantle and reduced temperatures at the subcrustal layer are caused by lithospheric "roots" and thickened lithosphere below these structures. Gravity residual lows, revealed over the Western Mediterranean Basin and Pannonian Basin, are in correspondence with both low velocities and high temperatures in the upper mantle. These anomalies are the result of the presence of asthenosphere in shallow near-Moho depths below these basins.

Key words: 3-D gravity modelling, European-Mediterranean, lithosphere and asthenosphere, mantle density heterogeneities.

Introduction

Data on subcrustal density heterogeneities are important information commonly used with data from other sources for the better understanding of endogenic processes operating in the upper mantle, its composition, possible state and evolution. Such subcrustal density inhomogeneities, especially in young active tectonic regions, cause gravity effects reaching several hundreds of mGal. As a rule, these density anomalies of the upper mantle are not noticed in the observed gravity field (Fig. 1) because they are compensated isostatically by crustal inhomogeneities. To estimate subcrustal density heterogeneities, residual anomalies obtained by subtracting the gravity influence of the earth's crust from the observed gravity field, were used (ARTEMJEV *et al.*, 1994; YEGOROVA *et al.*, 1995). The correct method for estimating subcrustal density heterogeneities is large-scale gravity modelling based on the 3-D mass distribution of a simplified crustal model.

[1] Institute of Geophysics, National Academy of Sciences of Ukraine, 32 Pr. Palladina, Kiev 252680, Ukraine.

Structure of the Model and Initial Data

Our model for the crust of the European continent consists of two regional layers of varying thickness with lateral variation of average density: the sedimentary cover and the crystalline crust; offshore, the model is supplemented by a sea-water layer. This 3-D density model is based on a simplified velocity model taken from maps of the main seismic horizons of the European crust—the "seismic" basement and the Moho boundary—and from a map of the average *P*-wave velocity in the crystalline crust (GIESE and PAVLENKOVA, 1988; HURTIG *et al.*, 1992). For the Alpine-Mediterranean region we have also taken into account recent seismic data on the Moho topography obtained from investigations on the EGT profile (BLUNDELL *et al.*, 1992). The accepted 3-D crustal model for the European-Mediterranean region is shown in Figure 2.

OBSERVED GRAVITY FIELD

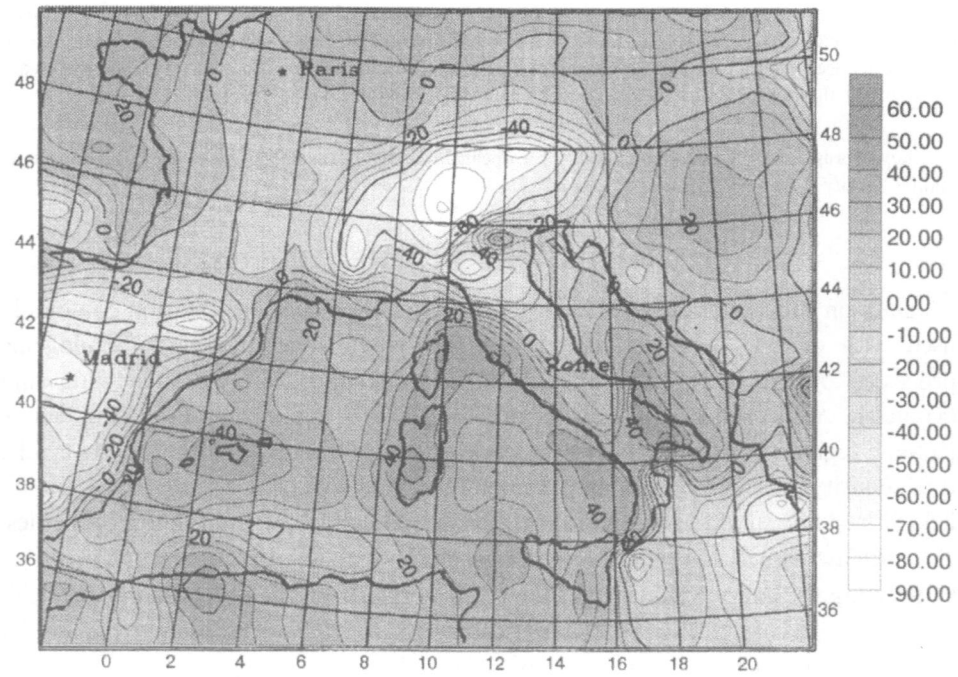

Figure 1
Generalized map of the observed gravity field of the European-Mediterranean region. The map uses Bouguer anomalies on land and free-air anomalies offshore, averaged over a $1° \times 1°$ grid. Portion of the map of Europe and North Atlantic (YEGOROVA *et al.*, 1995). Contour interval 10×10^{-5} m s^{-2}.

Density values were derived by using the following conversion functions between P-wave velocity V_p [km s^{-1}] and density ρ [10^3 kg m^{-3}]. For the crystalline crust we have used the relation $\rho = 2.7 + 0.27(V_p - 6.0)$. A specific type of the $\rho(V_p)$ relation was used for the sediments of the Western Mediterranean area (YEGOROVA et al., 1997), and for the rest of the study area the average density was determined taking into account the relation of sediment consolidation with depth (Fig. 3). The distribution of average density in the two model layers is shown in Figure 4. Offshore the model was supplemented by a sea-water layer with density $\rho = 1.03 \times 10^3$ kg m^{-3}.

Method of Modelling

The gravity effect of the model, approximated by $1° \times 1°$ planar parallelepipeds, was calculated by a program designed for solving the 3-D gravity problem considering the spherical configuration of the earth (STAROSTENKO and MANUKYAN, 1987). The gravity effect of each of the model layers (sea-water layer g_w (Fig. 5a); sedimentary cover g_{sd} (Fig. 5b); crystalline crust g_{cc} (Fig. 5c)) was calculated at the center of each block. For gravity calculations we have used anomalous densities of the layers obtained by subtracting the upper mantle density (taken to be 3.3×10^3 kg m^{-3}) from the average density of the layer ρ_1: $\Delta\rho = \rho_1 - 3.3 \times 10^3$ kg m^{-3}. The total gravity effect of the model g_{cr} (Fig. 5d) is the sum of the effects of all the model layers: $g_{cr} = g_w + g_{sd} + g_{cc}$.

The residual anomalies Δg_r (Fig. 6) were obtained by subtracting the calculated effect of the model g_{cr} and a constant from the observed gravity field g_{ob} (Fig. 1): $\Delta g_r = g_{ob} - g_{cr} - 790 \times 10^{-5}$ m s^{-2}. The values of -790×10^{-5} m s^{-2}, considered as the "normal" g_{cr} field, is the average gravity effect of the model for the crust of the stable East-European Platform.

Since our model has been constructed on the base of a simplified velocity model for the crust of the European continent, we are obliged to make accuracy evaluations of our modelling by choosing an optimal isoline interval for maps of gravity effects from the model layers. The sedimentary cover, with an average thickness of about 2 km (contour interval is 1 km) and with 0.05×10^3 kg m^{-3} isoline interval of average density, should be presented by a gravity effect map with a contour interval not less than 50×10^{-5} m s^{-2}. The crystalline crust of the study area, described by an average thickness of 30–35 km (with 4 km isoline interval) and by 0.1 km s^{-1} isoline interval of average velocity and correspondingly by that of 0.025×10^3 kg m^{-3} of average density, is featured by a gravity effect map no less than 35×10^{-5} m s^{-2}. Therefore, we have defined the main isoline interval for our gravity maps as 50×10^{-5} m s^{-2} which is in correspondence with a rms of about $20–25 \times 10^{-5}$ m s^{-2}.

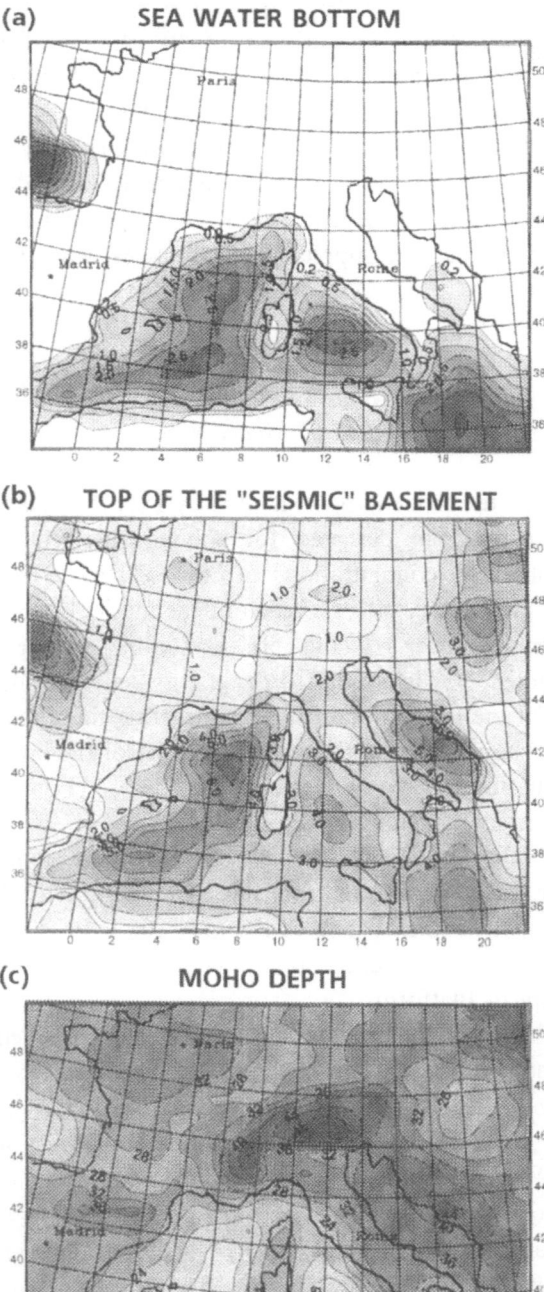

Figure 2

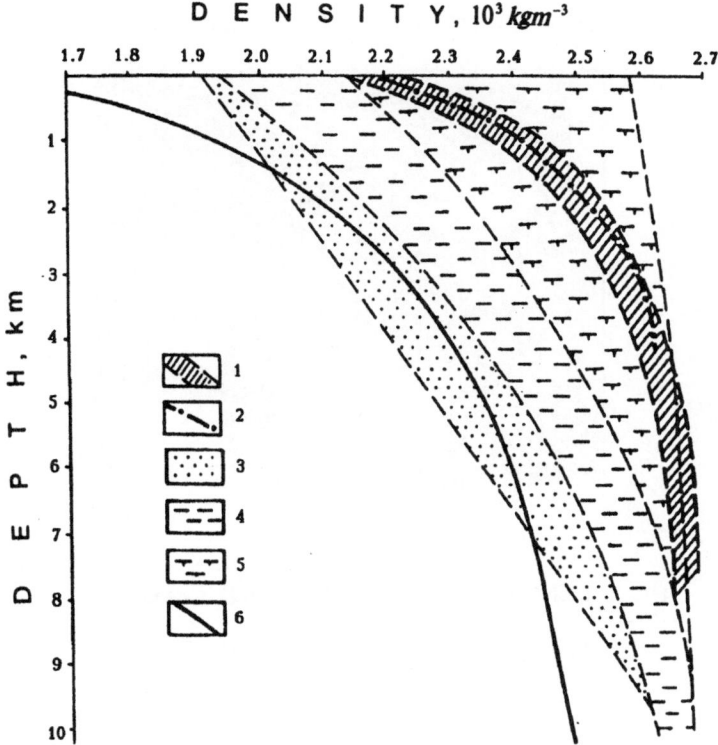

Figure 3

Dependence of sediment density on depth. 1: distribution of average density with depth for sediments of the East-European Platform (PODOBA and OZERSKAYA, 1975), accepted as the basic distribution for gravity modelling in this study; 2: function used by GRANSER (1987) for the Pannonian Basin; 3–5: areas of density variations obtained by KINSMAN (1975) for sandstones (3), clays (4), sandstones and limestones (5); 6-sediment density distribution for the Bay of Biscay basin, according to RUSAKOV (1990).

Modelling Results

Most of the study region has Δg_r anomalies of ca. -100×10^{-5} m s^{-2} amplitude. The maximal Δg_r anomalies of more than -200×10^{-5} m s^{-2} are found in the Tyrrhenian Sea, the South Balearic Basin and the Iberian Peninsula. Gravity lows in excess of -150×10^{-5} m s^{-2} were obtained over the Rhinegraben and the Pannonian Basin. Over the Alpine orogeny (Alps and Adriatic plate/Dinarides)

Figure 2

3-D density model for the crust of the European-Mediterranean region. Data averaged over a 1° × 1° grid. The model consists of a sea-water layer with $\rho = 1.03 \times 10^3$ kg m^{-3} and two regional layers with varying average density: the sedimentary cover and the crystalline crust. a: sea-bottom topography, km; b: depth to the bottom of the sedimentary layer (GIESE and PAVLENKOVA, 1988), km; c: depth to the bottom of the crust (Moho), km (GIESE and PAVLENKOVA, 1988; BLUNDELL et al., 1992; HURTIG et al., 1992).

(a) DENSITY OF THE SEDIMENTARY LAYER

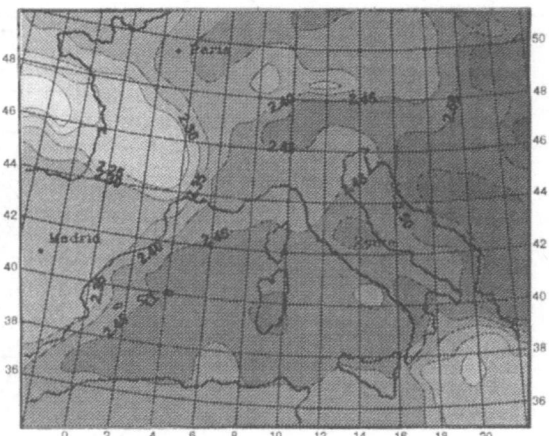

(b) DENSITY OF THE CRYSTALLINE CRUST

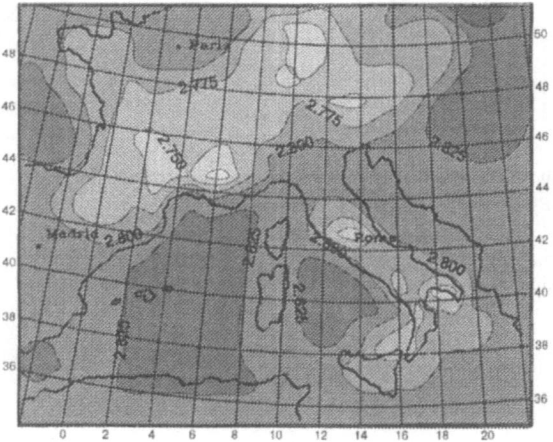

Figure 4

Distribution of average density in the model layers, 10^3 kg m^{-3}: a: in the sedimentary layer, $\bar{\rho}$ was obtained by using the dependence shown by pattern number 1 in Figure 3 and a velocity/density conversion function $\bar{\rho} = 1.88 - 0.09\bar{V} + 0.11\bar{V}^2 - 0.01\bar{V}^3$ (YEGOROVA *et al.*, 1997); b: in the crystalline crust, $\bar{\rho}$ values were calculated from $\bar{V}$ data (GIESE and PAVLENKOVA, 1988), using the conversion function $\rho = 2.7 + 0.27\,(V_p - 6.0)$.

there are relative Δg_r highs with more than $+50 \times 10^{-5}$ m s^{-2} amplitude. Local highs are also observed above the Calabrian Arc and the Pyrenees.

It is evident that the residual anomalies are caused mainly by both mantle density heterogeneities and the effect of crustal heterogeneities that were not taken into account by the modelling. The latter factor in general is due to: (1) inaccuracy of the initial velocity model because of the availability of sparse and irregular set of the DSS profiles and (2) use of a unique linear conversion function between velocity and density. However, the effect caused by these crustal factors creates small local anomalies. We therefore presume that large residual anomalies are caused mainly by mantle heterogeneities (YEGOROVA *et al.*, 1995).

Comparison of Mantle Gravity Anomalies with Other Data

At present, large amounts of data on structure, composition and state of the upper mantle are available. In our analysis we consider the correlation between Δg_r

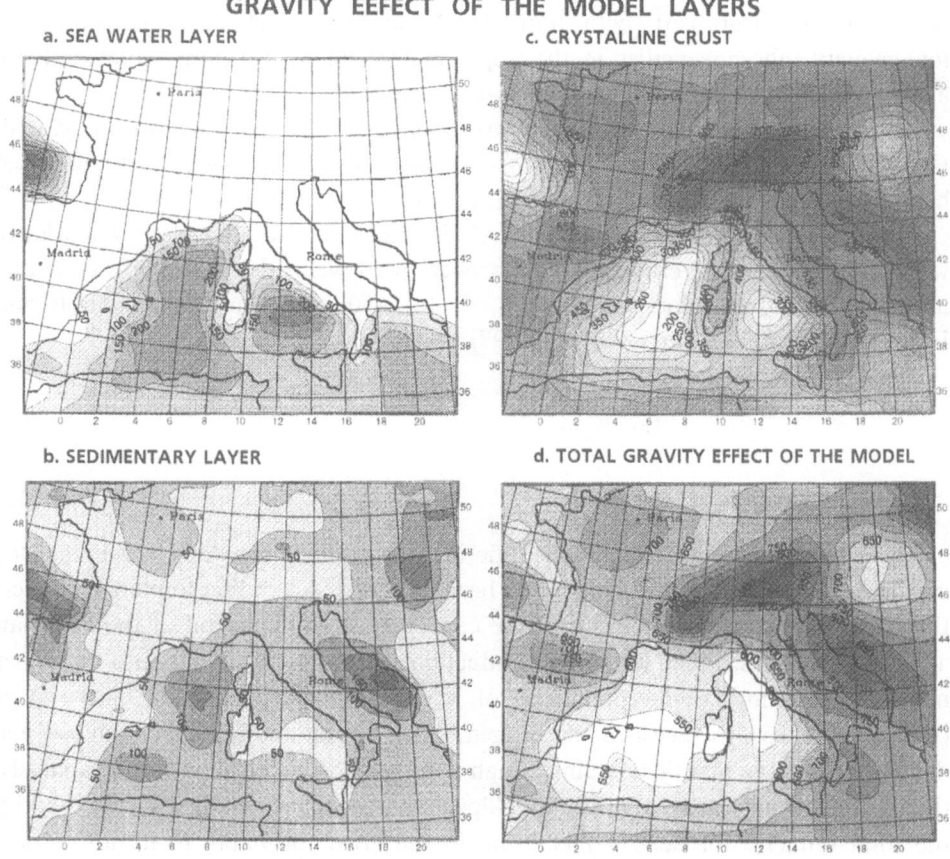

Figure 5
Gravity effect of the layers of the 3-D density model (Figs. 2 and 4), derived by solving the direct 3-D gravimetric problem (STAROSTENKO and MANUKYAN, 1987) over a 1° × 1° grid: a: sea-water layer; b: sedimentary cover; c: crystalline crust; d: total gravity effect of all (a–c) layers. Contour interval 50×10^{-5} m s^{-2}.

anomalies and representative seismotomographic and geothermic data. The comparison was made by invoking data on lithosphere structure provided by interpretations of earthquake S waves (MUELLER and PANZA, 1984; SNIEDER, 1988; PANZA and SUHADOLC, 1990) and P waves (BABUŠKA *et al.*, 1988; GRANET and TRAMPERT, 1989; BABUŠKA and PLOMEROVA, 1992; SPAKMAN *et al.*, 1993). This comparative analysis led to the conclusion that the highest gravity effect is produced by subcrustal inhomogeneities to 150 km depth, with a maximal influence from anomalous masses at approximately 95 km. Also considered were terrestrial heat-flow density data (HURTIG *et al.*, 1992), calculated temperatures at Moho level (ČERMAK, 1993).

By generalising the above data and invoking the concept of endogenic regimes by BELOUSOV (1990), one can outline the following series of tectonic units.

1) Taphrogenic and rift regime (Western Mediterranean region, the Pannonian Basin, the Rhinegraben), with thin (15–25 km) crust, shallow asthenosphere, low-velocity inhomogeneities in the upper mantle, high heat-flow values and Δg_r anomalies of -150 to -200×10^{-5} m s^{-2} amplitude;

2) Areas of the Hercynian orogeny (Bohemian Massif, Armorican Massif, French Massif Central, Iberian Peninsula), with 30–35 km thick crust, increased heat-flow values and $\Delta g_r = -50$ to -150×10^{-5} m s^{-2};

3) Orogenic regime (Alps, Pyrenees, Calabrian Arc, Adriatic plate, Dinarides), with thick (35–50 km) crust, thick lithosphere, high velocities in the upper mantle, moderate heat-flow values, and $\Delta g_r = -50$ to $+50 \times 10^{-5}$ m s^{-2} (see Table and Figs. 8 and 9 in YEGOROVA *et al.*, 1997).

Discussion

As a result of large-scale geoscientific investigations in the Alps, a general idea on the deep structure of the orogen has been obtained. According to plate tectonics, the Alps were formed as a result of the convergence and collision of the Eurasian and African plates, which led to the subduction of a lithospheric slab into the upper mantle. According to interpretations of seismic observations made in the Alps (KISSLING, 1993), the crust of the orogen either has an anomalously high average density and/or the mantle contains a high density body. According to our analysis, the residual gravity anomaly over the Alps is of lithospheric nature, i.e., caused by the total influence of density increasing in the crust, probably in its lower part (poorly elucidated by seismic methods), and high density body occurrence in the subcrustal lithosphere. A mantle component is confirmed by many results of seismological investigations. Our estimates of density in the subcrustal layer, made under the assumption that this layer contains main density heterogeneities causing the residual gravity anomaly, evidence that, accepting a lower limit at $H = 200$ km

RESIDUAL GRAVITY ANOMALIES

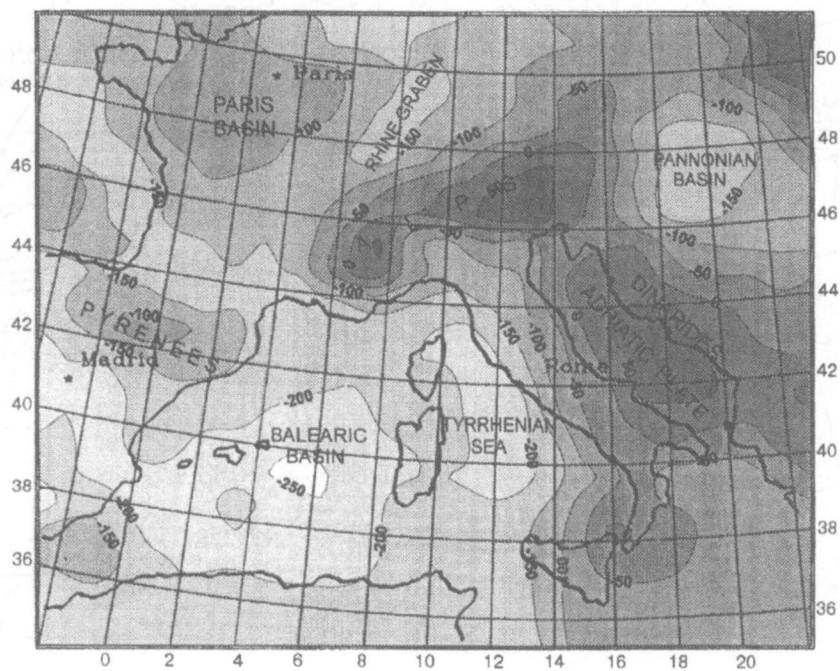

Figure 6

Residual gravity anomalies Δg_r of European-Mediterranean area obtained by subtracting the calculated gravity effect (g_{cr}) in Figure 5(d) from the observed gravity field (g_{ob}) in Figure 1, accepting -790×10^{-5} m s^{-2} as the average crustal gravity effect over the East-European platform: $\Delta g_r = g_{ob} - g_{cr} - 790 \times 10^{-5}$ m s^{-2} (contour interval 50×10^{-5} m s^{-2}).

(the upper one is the Moho), there is a density increase of $\sim 0.045 \times 10^3$ kg m^{-3}, relative to a conventional upper mantle density of 3.3×10^3 kg m^{-3}. The choice of the lower depth constraint is based on the results of seismotomographic investigations (SPAKMAN et al., 1993), showing that the strongest ΔV anomalies are located above this level. It is interesting to note, that the lithospheric "roots" of the Western and Eastern Alps are disconnected (Fig. 6), which completely agrees with seismological data interpretation by BABUŠKA et al. (1990) (see Fig. 7).

The main part of the Δg_r high over the Adriatic plate is explained by thickening of the "cold" lithosphere, the thickness of which is estimated to be about 150 km. In the regions of the Apennines and Dinarides adjacent to the Adriatic plate, two high-velocity inhomogeneities, associated with subduction zones, have been found at $H = 200\text{–}400$ km (SPAKMAN et al., 1993; WORTEL and SPAKMAN, 1996). These domains contribute to the mantle gravity anomaly formation. A similar nature is suggested by GRANET and TRAMPERT (1989) for the relatively high Δg_r over the Calabrian Arc.

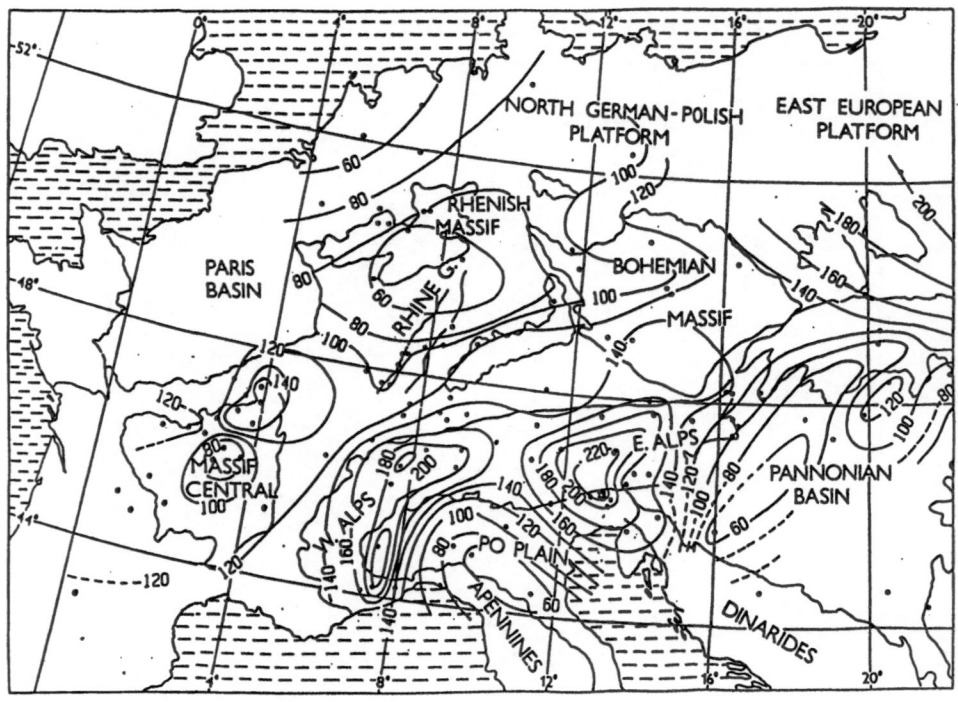

Figure 7
Model of the lithosphere thickness (isothickness in km) derived from *P*-wave residuals (BABUŠKA and PLOMEROVÁ, 1992).

The negative Δg_r anomalies over the Pannonian Basin and the Western Mediterranean Basin are due to an asthenospheric uprise to $H = 60$ km depth under the former (BABUŠKA and PLOMEROVÁ 1992; LILLIE *et al.*, 1994), and to $H = 30$ km under the latter (PANZA and SUHADOLC, 1990). Here, the seismotomographic data (SPAKMAN *et al.*, 1993) suggest the asthenosphere to be well pronounced between 95 km and 200 km depth. These distinguished domains of velocity decrease correlate with high values of calculated mantle temperatures: $T = 1000°–1200°$ at $H = 40$ km (ČERMAK, 1993). If the density decrease due to mantle heating is assumed to occur below the Moho to a depth of 200 km, the anomalies with -150 to -200×10^{-5} m s^{-2} amplitude over the Western Mediterranean Basin and Pannonian Basin are caused by a density decrease of -0.045×10^3 kg m^{-3}. The density decrease thus corresponds to a calculated thermal density decrease values of -0.01×10^3 kg m^{-3} for each 100°C. For this reason we assume the negative mantle anomalies of the Pannonian and the Western Mediterranean basins to be due to heated (up to partially molten) asthenosphere, the top of which is elevated up to near-crustal depths.

Conclusion

Progress in studying the earth's deep structure, using 3-D regional gravity modelling, depends on solving two main problems: (1) the availability of a regional interpretation of seismic data (Deep Seismic Sounding) for setting up a 3-D velocity model for the crust and (2) development of an appropriate software for gravity effect calculation from approximated mass distribution. At present, both problems are essentially solved. As a base for a 3-D density model of the earth's crust of the European continent, approximated by two regional layers—sedimentary cover and the consolidated crust (an offshore sea-water layer was added), we have used a simplified velocity model (GIESE and PAVLENKOVA, 1988; HURTIG *et al.*, 1992). Also taken into account were recent data on the Moho topography for the part of Alpine-Mediterranean region obtained as a result of seismic studies along the EGT profile (BLUNDELL *et al.*, 1992). The gravity effects of this model were calculated with the program developed by STAROSTENKO and MANUKYAN (1987) designed to solve the 3-D direct problem taking into account the spherical form of the earth. Residual gravity anomalies, obtained by subtracting the gravity influence of the crustal model from the observed field, reach amplitudes of a few hundreds of mGals. By using the mentioned program, the density anomaly distribution in the subcrustal layer, responsible for that residual gravity field, was calculated (see Fig. 7 in YEGOROVA *et al.*, 1997).

The mantle origin of the main part of the residual gravity anomalies is supported by their clear correlation with both velocity heterogeneities of the subcrustal upper mantle obtained by seismic tomography and geothermal data. A pronounced feature of mantle gravity and density anomalies of the study region is the positive anomalies below the Alps, the Adriatic plate and the Calabrian Arc. High velocity domains in the upper mantle below these structures and reduced temperatures at subcrustal levels suggest that these anomalies are caused by cold and thickened (up to 150 km) lithosphere of the Adriatic plate, and the lithospheric "roots" beneath the Alps and the Calabrian Arc.

Gravity residual lows of -200 to $-200 \times 250 \times 10^{-5}$ m s^{-2} amplitude, revealed over the Pannonian Basin and the Western Mediterranean Basin, agree with low-velocity mantle domains, corresponding to high temperatures in the upper mantle. These anomalies are due to thermal expansion of the material in the asthenosphere, the top of which is locally elevated close to the Moho boundary. For the rest of the area, low-amplitude, negative residual anomalies characterise the generally lower density of the upper mantle beneath Western Europe, compared with that below the East-European platform.

The work carried out provides the first experience in setting up regional crustal gravity models of this kind. Its main result is that the residual gravity anomalies obtained are verified by available data of other geophysical methods. These data (especially seismic tomography) suggest that the mantle of the study region is notably differentiated in lateral and vertical directions. These mantle heterogeneities will be considered and studied in our further regional gravity modelling of Europe's lithosphere.

Acknowledgements

We gratefully acknowledge to the reviewers R. Lillie and H. Zeyen for their insightful comments and English assistance. Partial support for this work was provided by INTAS-project No. 1946.

REFERENCES

ARTEMJEV, M. E., KABAN, M. K., KUCHERINENKO, V. A., DEMYANOV, G. V., and TARANOV, V. A. (1994), *Subcrustal Density Inhomogeneities of Northern Eurasia as Derived from the Gravity Data and Isostatic Models of the Lithosphere*, Tectonophysics *240*, 249–280.

BABUŠKA, V., PLOMEROVÁ, J., and PAJDUŠAK, P., *Lithosphere-asthenosphere in central Europe: Model derived from P residuals*. In *Proc. of the Fourth EGT Workshop: The Upper Mantle* (eds. Nolet, G., and Dost, B.) (European Science Foundation, Strasbourg 1988) pp. 37–48.

BABUŠKA, V., PLOMEROVÁ, J., and GRANET, M. (1990), *The Deep Lithosphere in the Alps: A Model Inferred from P Residuals*, Tectonophysics *176*, 137–165.

BABUŠKA, V., and PLOMEROVÁ, J. (1992), *The Lithosphere in Central Europe—Seismological and Petrological Aspects*, Tectonophysics *207*, 141–163.

BELOUSOV, V. V. (1990), *Tectonosphere of the Earth: Upper Mantle and Crust Interaction*, Tectonophysics *180*, 139–183.

BLUNDELL, D., FREEMAN, R., and MUELLER, St. (eds.), *A Continent Revealed* (Cambridge University Press, Cambridge 1992).

ČERMAK, V. (1993), *Lithopheric Thermal Regime in Europe*, Phys. Earth Planet. Inter. *79*, 179–193.

GIESE, P., and PAVLENKOVA, N. I. (1988), *Structural Maps of the Earth's Crust in Europe*, Izv. AN SSSR, Fizika Zemli *10*, 3–14 (in Russian).

GRANET, M., and TRAMPERT, J. (1989), *Large-scale P-velocity Structures in the Euro-Mediterranean Area*, Geophys. J. Int. *99*, 583–594.

GRANSER, H. (1987), *Three-dimensional Interpretation of Gravity Data from Sedimentary Basins Using an Exponential Density-depth Function*, Geophys. Prospect. *35*, 1030–1041.

HURTIG, E., ČERMAK, V., HAENEL, R., and ZUI, V. (eds), *Geothermal Atlas of Europe* (Hermann Haack Verlagsgesellschaft GmbH, Geographisch-Kartographische Anstalt Gotha 1992).

KINSMANN, D. J. J., *Rift basins and peculiarities of sediment accumulation in the conditions of sagged margins of continents*. In *Petroleum and Global Tectonics* (eds. Fischer, G., and Judson, S.) (Princeton University Press 1975) pp. 61–91.

KISSLING, E. (1993). *Deep Structure of the Alps—What Do We Really Know?* Phys. Earth Planet. Inter. *79*, 87–112.

LILLIE, R. J., BIELIK, M., BABUŠKA, V., and PLOMEROVÁ, J. (1994), *Gravity Modelling of the Lithosphere in the Eastern Alpine-Western Carpathian-Pannonian Basin Region*, Tectonophysics *231*, 215–235.

MUELLER, St., and PANZA, G. F., *The lithosphere-asthenosphere system in Europe*. In *1st EGT Workshop, the Northern Segment* (eds. Galson, D. A., and Mueller, St.) (European Science Foundation, Strasbourg 1984) pp. 23–26.

PODOBA, N. V., and OZERSKAYA, M. A. (eds.), *Physical Properties of Sedimentary Cover of East-European Platform* (Nauka, Moscow 1975) (in Russian).

PANZA, G. F., and SUHADOLC, P. (1990), *Properties of the Lithosphere in Collisional Belts in the Mediterranean—A Review*, Tectonophysics *182*, 39–46.

RUSAKOV, O. M. (1990), *The Thickness and Density of Sedimentary Cover in the Indian Ocean*, Geophys. J. *3* (8), 390–399.

SNIEDER, R. (1988), *Large-scale Waveform Inversions of Surface Waves for Lateral Heterogeneity—II: Application to Surface Waves in Europe and Mediterranean*, J. Geophys. Res. *93*, 12067–12080.

SPAKMAN, W., VAN DER LEE, S., and VAN DER HILST, R. (1993), *Travel-time Tomography of the European-Mediterranean Mantle Down to 1400 km*, Phys. Earth Planet. Inter. *79*, 3–74.

STAROSTENKO, V. I., and MANUKYAN, A. G. (1987), *Gravimetric Problems for Spherical Celestial Bodies and their Application to the Study of the Nectaris Basin Mascon on the Moon*, Boll. Geofis. Teor. Appl. *29* (115), 201–219.

WORTEL, R., and SPAKMAN, W. (1996), *The Geodynamic Evolution of the Alpine-Mediterranean Region: From Structure to Dynamics*, Mitteilungen der Gesellschaft der Geologie- und Bergbaustudenten in Österreich *41*, 142–143.

YEGOROVA, T. P., KOZLENKO, V. G., PAVLENKOVA, N. I., and STAROSTENKO, V. I. (1995), *3-D Density Model for the Lithosphere of Europe: Construction Method and Preliminary Results*, Geophys. J. Int. *121*, 873–892.

YEGOROVA, T. P., STAROSTENKO, V. I., KOZLENKO, V. G., and PAVLENKOVA, N. I. (1997), *Three-dimensional Gravity Modelling the European-Mediterranean Lithosphere*, Geophys. J. Int. *129*, 355–367.

(Received October 14, 1996, revised/accepted August 12, 1997)

 To access this journal online:
http://www.birkhauser.ch

III. Mineral and Rock Physics Studies

Pure appl. geophys. 151 (1998) 565–587
0033–4553/98/040565–23 $ 1.50 + 0.20/0

▌Pure and Applied Geophysics

Seismic Anisotropy in the Deep Mantle, Boundary Layers and the Geometry of Mantle Convection

SHUN-ICHIRO KARATO[1]

Abstract—An attempt is made to explore the geodynamical significance of seismic anisotropy in the deep mantle on the basis of mineral physics. The mineral physics observations used include the effects of deformation mechanisms on lattice and shape preferred orientation, the effects of pressure on elastic anisotropy and the nature of lattice preferred orientation in deep mantle minerals in dislocation creep regime. Many of these issues are still poorly constrained, but a review of recent results shows that it is possible to interpret deep mantle seismic anisotropy in a unified fashion, based on the solid state processes without invoking partial melting. The key notions are (i) the likely regional variation in the magnitude of anisotropy as deformation mechanisms change from dislocation to diffusion creep (or superplasticity), associated with a change in the stress level and/or grain-size in the convecting mantle with a high Rayleigh number, and (ii) the change in elastic anisotropy with pressure in major mantle minerals, particularly in (Mg, Fe)O. The results provide the following constraints on the style of mantle convection: (i) the $SH > SV$ anisotropy in the bottom transition zone and the $SV > SH$ anisotropy in the top lower mantle can be attributed to anisotropy structures (lattice preferred orientation and/or laminated structures) caused by the horizontal flow in this depth range, suggesting the presence of a mid-mantle boundary layer due to (partially) layered convection, (ii) the observed no significant seismic anisotropy in the deep mantle near subduction zones implies that deformation associated with subducting slabs is due mostly to diffusion creep (or superplasticity) and therefore slabs are weak in the deep mantle and hence easily deformed when encountered with resistance forces, and (iii) the $SH > SV$ anisotropy in the cold thick portions of the D'' layer is likely to be due to horizontally aligned shape preferred orientation in perovskite plus magnesiowüstite aggregates formed by strong horizontal shear motion in the recent past.

Key words: Seismic anisotropy, mantle convection, boundary layer, lattice preferred orientation, shape preferred orientation, deformation mechanism map, superplasticity.

Introduction

The nature of seismic anisotropy is sensitive to the geometry of convection and the mechanisms of deformation and hence seismic anisotropy may provide useful constraints on the style of mantle convection (e.g., KARATO, 1989). Earlier attempts to infer the geometry of convection from seismic anisotropy include TANIMOTO and ANDERSON (1984) and NATAF *et al.* (1986) in which the regional variation of azimuthal and polarization anisotropy of surface waves in the upper mantle was interpreted in terms of the geometry of convection. They used the relation between

---

[1] University of Minnesota, Department of Geology and Geophysics, Minneapolis, MN 55455, U.S.A.

the azimuthal and the polarization anisotropy of surface waves and the geometry of flow found in shallow upper mantle peridotites (e.g., NICOLAS and CHRISTENSEN, 1987; KARATO, 1989). Similar works include MONTAGNER (1994) and SILVER (1996).

Although these studies provided useful insights into the tectonics of the upper mantle, perhaps a more critical issue in global dynamics is the style of convection in the deep mantle (the transition zone and the lower mantle). More specifically, the degree and the geometry of slab penetration into the lower mantle, the extent to which convection in the upper mantle is related to the convection in the lower mantle and the role of the D'' layer on convection are critical but still highly controversial (e.g., YOUNG and LAY, 1987; SILVER *et al.*, 1988; JORDAN *et al.*, 1989, 1993; DAVIES and RICHARDS, 1992; PELTIER, 1996; KENDALL and SILVER, 1996; GARNERO and LAY, 1997). These issues are closely related to the nature of boundary layers in mantle convection, namely (i) subducting slabs, (ii) a possible boundary layer between the upper and lower mantle and (iii) the bottom boundary layer (the D'' layer). The deformation mechanisms and deformation-induced fabrics in deep mantle minerals must be understood before these issues can be investigated using seismic anisotropy. It will be shown in this paper that observations on seismic anisotropy can provide useful constraints on the nature of these boundary layers and hence global dynamics of the deep mantle, when they are interpreted on the basis of appropriate mineral physics. It should be emphasized, however, that the mineral physics study related to such an exercise is still in its infancy (for review see HEMLEY and COHEN, 1996; KARATO, 1997a) and many of the conclusions reached in this paper could be subject to revisions when more experimental data become available. The main purpose of this paper is not to present any definite conclusions, but rather to illustrate the potential usefulness of this approach in mantle dynamics and to highlight important issues on which more work is needed.

Seismic Anisotropy and Mantle Convection

The Rayleigh number appropriate for mantle convection ranges from $\sim 10^5$ to $\sim 10^7$ which exceeds significantly the critical Rayleigh number for convective instability ($\sim 10^3$; e.g., JARVIS and PELTIER, 1989). At such a high Rayleigh number, convection is vigorous, and the driving forces for convection and much of the deformation are concentrated in boundary layers (e.g., BUSSE, 1989; JARVIS and PELTIER, 1989). Thus, the style of convection is largely controlled by the distribution of buoyancy forces and resistance forces in or near boundary layers. JORDAN *et al.* (1989) used the observed velocity heterogeneities to infer the structure of boundary layers. The use of seismic anisotropy will significantly contribute to a better understanding of boundary layers and hence the dynamics of convection for several reasons. First, observations of seismic anisotropy provide constraints on the deformation mechanisms and hence some insights into the creep strength in the boundary layers (KARATO and WU, 1993; KARATO et al., 1995). In addition, seismic anisotropy provides important constraints on the geometry of boundary layers. A notable example is the inference of vertical flow in the upper mantle beneath mid-ocean ridges and in old oceanic upper mantle from the observation of SH/SV polarization anisotropy (NATAF et al., 1986; MONTAGNER and TANIMOTO, 1991). A similar approach will provide useful constraints on the geometry of deep mantle flow, if seismological observations on the deep mantle are interpreted on the basis of properties of its constituent materials.

The following questions may be addressed as to the nature of mantle convection that could be constrained from seismic anisotropy.

1. Is There a Mid-mantle Horizontal Boundary Layer?

This is a key question regarding the style of mantle convection. If convection is more or less whole mantle wide (e.g., DAVIES and RICHARDS, 1992; JORDAN et al., 1993), then no horizontal boundary layer is expected in the mid-mantle. If, on the other hand, convection is layered to some extent, then a horizontal boundary layer must occur in between the upper and lower mantle (e.g., PELTIER, 1996). A critical question then is how to infer the presence or the absence of a mid-mantle boundary layer from observations? High resolution seismic tomography showing velocity heterogeneities in the mantle suggests that subducting slabs are deflected and convection current is to some extent layered (e.g., VAN DER HILST et al., 1991; FUKAO et al., 1992; CADEK et al., 1997), but the degree to which convection is layered is still a highly controversial issue (e.g., JORDAN et al., 1993; PELTIER, 1996). The presence of a mid-mantle horizontal boundary layer is testable from the observation of seismic anisotropy. A horizontal flow in a certain depth region will produce anisotropic structures if stress associated with it is high enough (Fig. 1). If the constituent materials have large elastic anisotropy (in the case of lattice preferred orientation, Fig. 1(a), or a large contrast in elastic constants (in the case of laminated structures, Fig. 1(b), then such a flow will be detectable from seismic anisotropy.

2. How Strong are the Subducting Slabs?

Numerical modeling of mantle convection incorporating the effects of density changes associated with mantle phase transformations and temperature-dependent rheology suggest that convection is intermittently layered when slabs have creep strength similar to the surrounding mantle, but whole mantle convection will result if slabs have significantly higher creep strength than the surrounding mantle (DAVIES, 1995). Thus, the strength of slabs appears to play a decisive role in determining the style of mantle convection. KARATO et al. (1995) showed that observations on seismic anisotropy can be used to identify the deformation mechanisms in the mantle and hence indirectly provide insights into the strength of materials.

3. How Do Upwelling Plumes Interact with the 660 km Discontinuity?

The interaction of upwelling plumes with phase transformation boundaries depends on a number of parameters and is not well understood (e.g., STEINBACH and YUEN, 1995; NAKAKUKI *et al.*, 1994). Plumes may go through these boundaries or interact with them to spread their heads near these boundaries. If the latter mode of interaction occurs, and if the stress level in the plume head is high enough in these regions, then such a flow will show characteristic signature in seismic anisotropy.

4. Is the D'' Layer Partially Molten? What is the Nature of Deformation in the D'' Layer?

The D'' layer at the base of the mantle is considered to be a thermal and chemical boundary layer (e.g., YOUNG and LAY, 1987). There will be some temperature increase in the D'' layer, however since the D'' layer is at the bottom of the convecting mantle, large chemical heterogeneities can also be expected. Chemical reaction with the core might also change its chemistry. In fact there is a growing consensus that the D'' layer is highly heterogeneous (e.g., YOUNG and LAY, 1987; SYLVANDER and SOURIAU, 1996; WYSESSION, 1996; GARNERO and LAY, 1997).

(A)   (B)

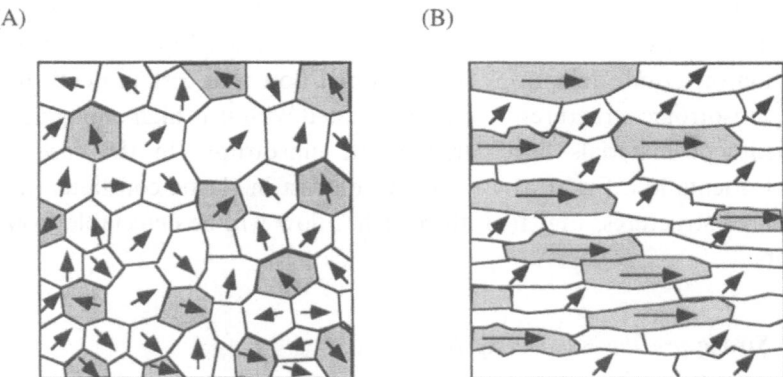

Figure 1
Schematic diagrams showing microstructures in polycrystalline (two-phase) materials during deformation (e.g., KARATO *et al.*, 1995). White and shade grains indicate two different phases and arrows show crystallographic orientation. (A) Microstructure formed by deformation via superplastic creep. Both lattice preferred orientation (LPO) and shape preferred orientation (SPO) are absent in this regime. (B) Microstructures characteristic of deformation via dislocation creep. Strong LPO and SPO are developed. A similar laminated structure will also occur at a larger scale when a mixture of two materials is sheared (ALLÉGRE and TURCOTTE, 1986).

KENDALL and SILVER (1996) argued that the observed $SH > SV$ (SH: velocity of SH wave, SV: velocity of SV wave) anisotropy in the D'' layer is caused by partial melting. However, a physical basis for this hypothesis is rather weak and the observed correlation of strongly anisotropic areas and the thick and cold portions of the D'' layer appears to be inconsistent with this model. As will be shown in this paper, solid state mechanisms can naturally explain the observed seismic properties in which case the nature of anisotropy is likely to be related to the deformation mechanisms in the D'' layer, where subducted slabs are likely to be accumulated.

Seismic Anisotropy in the Deep Mantle

Various sources of seismological observations can provide constraints on the anisotropic structure of the deep mantle. They include: (i) shift (or splitting) of peaks of free oscillations (MONTAGNER and KENNETT, 1996), (ii) shear-wave splitting (SKS, ScS etc. KANESHIMA and SILVER, 1995; FISCHER and WIENS, 1996; FOUCH and FISCHER, 1996; VINNIK and MONTAGNER, 1996), (iii) anomalous diffraction of body waves in the D'' layer (MAUPIN, 1994; VINNIK et al., 1995), and (iv) anomalies in P to S conversion at some discontinuities (VINNIK and MONTAGNER, 1996). These studies have provided the following conclusions as to the anisotropic structure of the deep mantle:
(1) Compared to the anisotropy of the shallow lithosphere (less than ~ 100 km), the amplitude of anisotropy in the deep mantle (deeper than ~ 300 km) is significantly smaller (MONTAGNER and TANIMOTO, 1991; MONTAGNER and KENNET, 1996).
(2) In particular, the lower mantle of the earth is devoid of any significant anisotropy (KANESHIMA and SILVER, 1995; MEADE et al., 1995a) except for the D'' layer and presumably the topmost lower mantle (MONTAGNER and KENNETT, 1996).
(3) The D'' layer, however, has significant $SH > SV$ anisotropy, and the magnitude of anisotropy changes laterally (VINNIK et al., 1995; KENDALL and SILVER, 1996; MATZEL et al., 1996; GARNERO and LAY, 1997).
(4) The layers near the 660 km discontinuity appear to have a weak but detectable anisotropy (MONTAGNER and KENNETT, 1996; VINNIK and MONTAGNER, 1996). According to MONTAGNER and KENNETT (1996), the layer just above the 660 km discontinuity has $SH > SV$ anisotropy whereas the layer just below has $SV > SH$ anisotropy.
It must be noted that among the four items above, (1) through (3) are robust, but the item (4), namely the anisotropy around the 660 km discontinuity is highly controversial. There are several studies in which significant anisotropy in this depth range or deeper portions of the mantle is ruled out (e.g., FOUCH and FISCHER, 1996; FISCHER and WIENS, 1996). However, these studies investigate

different aspects of anisotropy, namely the SH/SV polarization anisotropy in MONTAGNER and KENNETT (1996) whereas the shear-wave splitting in FOUCH and FISCHER (1996) and FISCHER and WIENS (1996), hence the different conclusions reached in these studies are not necessarily mutually contradictory.

In the following, I will provide geodynamical interpretations of these observations on the basis of mineral physics.

Seismic Anisotropy and Deformation in the Deep Mantle

There are two classes of anisotropic structures that could cause seismic anisotropy in the earth's mantle (Fig. 1). One is the lattice preferred orientation (LPO) of anisotropic minerals and the other is the shape preferred orientation (SPO) of secondary phases. SPO may occur at larger scales as laminated structures (e.g., ALLÉGRE and TURCOTTE, 1986). In this case, when the scale of laminated structure is significantly smaller than the wavelength of seismic waves, a laminated material will behave like a homogeneous but anisotropic material. The latter effect is important only when the contrast in elastic moduli in coexisting phases is large. In most of the mantle, contrasts in elastic constants in co-existing minerals are less than $\sim 10\%$ and the anisotropy due to SPO or laminated structures is usually small. An exception is partially molten rocks and, as will be shown later, the case of the deep lower mantle where perovskite and magnesiowüstite have large contrasts in shear moduli and in the deep transition zone where garnetite (previous oceanic crust) and spinel-rich assembly may occur as a laminated structure, as a result of the separation of a garnetite layer from subducted oceanic lithosphere (RINGWOOD and IRIFUNE, 1988; KARATO, 1997b). Since partial melting is unlikely in most of the deep mantle (e.g., BOEHLER, 1996), I will consider seismic anisotropy due to LPO of minerals in most cases but the effects of SPO or laminated structure in the D'' layer and in the deep transition zone will also be examined.

Anisotropy due to SPO causes $SH > SV$ polarization anisotropy but no azimuthal anisotropy when the laminated structure shows horizontal layering. Anisotropy due to LPO depends both on geometry of flow and on the slip systems, and causes both polarization and azimuthal anisotropy.

Two issues are important in interpreting seismic anisotropy in terms of LPO. First, the LPO will develop in deformed materials when deformation is due to dislocation creep (or twinning), although deformation due to diffusion creep or superplasticity does not result in LPO. Such a notion has long been well established in materials science (e.g., EDINGTON et al., 1976) and KARATO (1988) demonstrated this for olivine even when significant grain boundary migration occurs in diffusion creep. Similarly, SPO develops in dislocation creep regime, but not in superplasticity. This is because most of the strain in superplastic deformation is taken by grain-boundary sliding and not by the shape change in each grain (e.g., EDINGTON

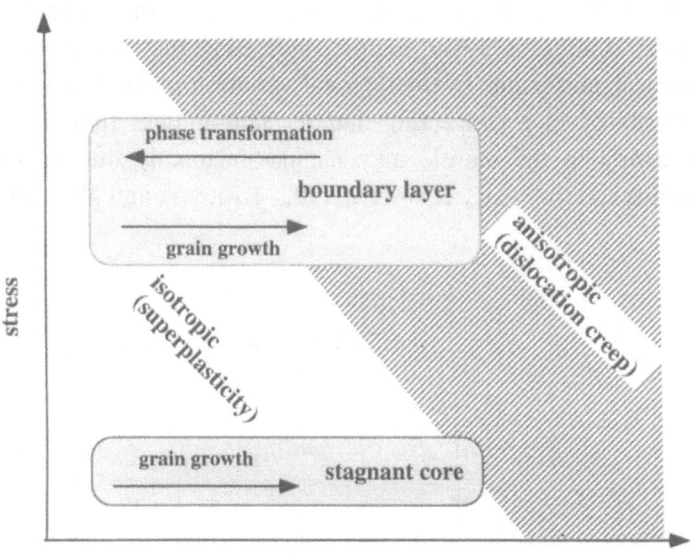

grain size

Figure 2

A schematic deformation mechanism map. Dislocation creep dominates at high stress and/or coarse grain size and results in anisotropic structures whereas superplastic creep dominates at low stress and/or small grain size and results in isotropic structures. Stresses in boundary layers are higher than those in stagnant cores. Therefore, dislocation creep is favored in boundary layers, if grain size is large. Phase transformations in subducting slabs could reduce grain size to promote superplastic creep, until grain growth brings grain size large enough for dislocation creep.

et al., 1976). Thus seismic anisotropy due to SPO is also expected only in dislocation creep regime.

Figure 2 shows a schematic deformation mechanism map illustrating that anisotropic structure is formed when deformation occurs at relatively high stress and/or large grain size, whereas structure will be isotropic when deformation occurs at relatively low stress and/or small grain-size. Stress level in a boundary layer is usually higher than that in a stagnant core (I refer to the central portion of a convecting mantle as "stagnant core" because deformation there is significantly less than that in boundary layers. However, even in "stagnant core," strain due to mantle convection in geological time scales is extensive, particularly when convection current is not steady state (e.g., McKENZIE, 1979).) Therefore dislocation creep is generally favored in the boundary layers unless grain size is very small. Small grain size is, however, expected in subducting slabs as a result of phase transformations (e.g., ITO and SATO, 1990; RIEDEL and KARATO, 1997). In these cases, deformation mechanisms might change to superplastic flow (or diffusion creep) until grain size enlarges due to grain growth. Deformation mechanism maps for MgO and perovskite (Fig. 3) illustrate that the conditions for transition between superplasticity (diffusion creep) and dislocation creep in these mantle minerals are close to typical conditions in the mantle, indicating that a regional variation in deformation mechanisms due to the regional variation in stress and/or grain size is a distinct possibility in a convecting mantle. This implies that the amplitude of seismic anisotropy in the mantle may change regionally due to the change in deformation mechanisms (e.g., KARATO, 1992; KARATO and WU, 1993).

(A)

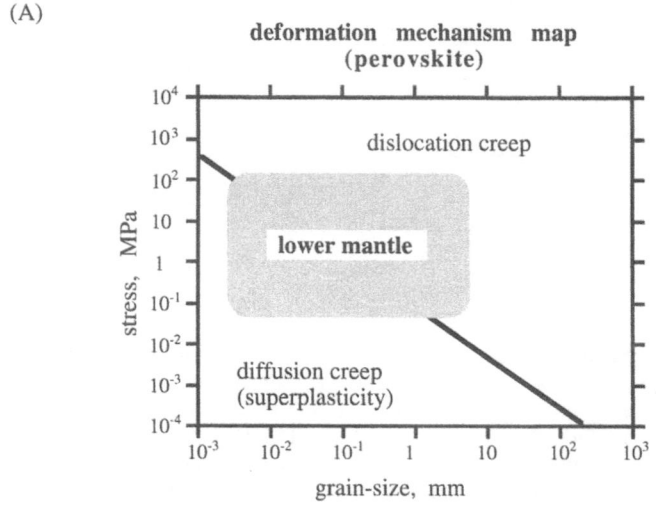

(B)

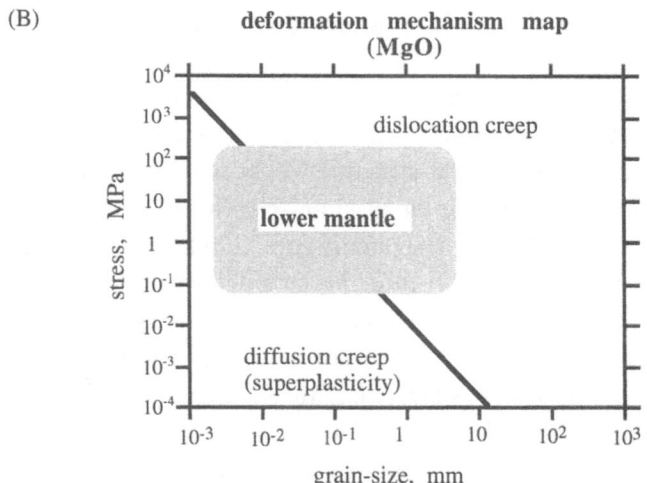

Figure 3
Deformation mechanism maps for (A) perovskite (modified from LI *et al.*, 1996) and (B) for MgO (modified from FROST and ASHBY, 1982). A range of deformation conditions in the lower mantle is also shown.

The second issue is that in dislocation creep regime where LPO will develop, the nature of LPO is controlled by the nature of slip systems and deformation geometry (VAN HOUTTE and WAGNER, 1985). When the effects of dynamic recrystallization are not large as in the case of most upper mantle minerals and in crustal minerals where dynamic recrystallization due to subgrain rotation is the dominant mechanism of dynamic recrystallization (KARATO, 1988), then the controlling factor for the LPO is still the slip systems. The slip systems in a given crystal are controlled by the crystal structure and the nature of chemical bonding (for general discussion, see e.g., FROST and ASHBY, 1982). This implies that the dominant slip systems will be the same for materials with the same crystal structure and similar chemical bonding (FROST and ASHBY, 1982). Therefore the nature of LPO may be inferred from those of analogue materials. In this paper, the LPO of deep mantle minerals are inferred from those in analogue materials except for MgO where direct measurements of LPO are available (RICE, 1970). I consider LPO corresponding to simple shear deformation rather than uni-axial compression or tension, because simple shear is a likely deformation geometry in most of the mantle, especially in and around boundary layers.

The following minerals are considered: spinel in the transition zone and perovskite and magnesiowüstite in the lower mantle. Garnet (majorite) is an important constituent mineral in the transition zone (volume fraction is $\sim 40-80\%$; e.g., RINGWOOD, 1991), yet since garnets have nearly isotropic elastic properties, their effects on seismic anisotropy through LPO are only to dilute the magnitude of anisotropy and not very important for the present discussion, although its presence might contribute to anisotropy when laminated structures such as Fig. 1(b) are present. Modified spinel is an important mineral in the shallow transition zone that might contribute to seismic anisotropy there (e.g., SHARP et al., 1994; TONG et al., 1994). However, the present paper focuses on boundary layers, namely deep transition zone, and I will not consider the modified spinel in this paper. The lower mantle is considered to be composed of $\sim 70-80\%$ perovskite and $\sim 20-30\%$ magnesiowüstite (RINGWOOD, 1991), although the exact mineralogy is controversial.

In magnesiowüstite, two types of slip systems are activated, namely $\{110\}\langle1\bar{1}0\rangle$ and $\{100\}\langle011\rangle$ (CHIN and MAMMEL, 1973), the activity of which depends on temperature. At high temperatures (say $T/T_m > 0.5$ (T_m: melting temperature)), both are equally activated, however the slip along the $\{110\}$ plane dominates and hence the $\{110\}$ planes become subparallel to the shear plane in simple shear flow (e.g., FRANSSEN, 1993). These slip systems are common to most of the ionic crystals with the NaCl structure, irrespective of the nature of elastic anisotropy (FROST and ASHBY, 1982), hence it is expected that the preferred orientation of magnesiowüstite due to dislocation creep will have a similar pattern at a similar T/T_m. Thus I assume that the $\{110\}$ planes are parallel to shear plane and the $\langle111\rangle$ direction is parallel to the shear direction.

The LPO in perovskites have been studied by KARATO et al. (1995) (see also ZHANG and KARATO, 1997) and MEADE et al. (1995a). KARATO et al. (1995) and ZHANG and KARATO (1997) investigated the LPO in $CaTiO_3$ (a perovskite that assumes crystal structure similar to (Mg, Fe)SiO$_3$ perovskite) at high T/T_m (~ 0.7– 0.8) in simple shear deformation geometry and found that a strong LPO is produced when perovskite is deformed in dislocation creep regime, but no strong LPO is formed in superplastic regime. In dislocation creep regime, the (010) plane becomes subparallel to the shear plane, and the [100] orientation becomes subparallel to the flow direction. In contrast, MEADE et al. (1995a) deformed $MgSiO_3$ perovskite at room temperature and high pressures ($T/T_m \sim 0.1$) and found no significant LPO. The reason for this difference is not clear although deformation mechanisms other than dislocation creep are likely to operate at room temperature. Since the earth's mantle is under high T/T_m, I will assume the results by KARATO et al. (1995) and ZHANG and KARATO (1997) in this paper.

The LPO in spinel was studied by FUJIMURA (1984) using an analogue material ($NiFe_2O_4$) with uni-axial compression or tension. He found an LPO consistent with dislocation creep by the $\{111\}\langle 1\bar{1}0 \rangle$ slip system under some conditions. The LPO in simple shear deformation in crystals with the same slip system (e.g., fcc metals) has been studied by WILLIAMS (1962) who found that the $\{111\}$ planes become subparallel to the shear plane, although another peak of $\{100\}$ poles is observed parallel to the shear plane. The $\langle 110 \rangle$ orientation becomes subparallel to the shear direction. The dominant slip system in spinels is $\{111\}\langle 1\bar{1}0 \rangle$, although another slip plane is activated when the degree of non-stoichiometry increases (MITCHELL et al., 1976). The slip systems identified in germanate and silicate spinel (MADON and POIRIER, 1980; DUPAS et al., 1997) agree with those in nearly stoichiometric spinel. Considering these uncertainties I assume that the $\{111\}$ or the $\{100\}$ planes become subparallel to the shear plane and that the $\langle 110 \rangle$ direction is subparallel to the shear direction.

Elasticity of Deep Mantle Minerals and Seismic Anisotropy due to Lattice Preferred Orientation

Given LPO, the nature of seismic anisotropy is determined by the elastic anisotropy of constituent minerals and mineralogy. Elastic constants of deep mantle minerals (spinel, perovskite and magnesiowüstite) have been determined only at relatively low pressures and temperatures (e.g., BASS, 1995), and hence the elastic constants under lower mantle conditions can only be estimated by theoretical calculations. Elastic constants will change with pressure and temperature (and possibly with frequency of elastic waves). However, I will consider only the effect of pressure in this paper, because the effect of pressure is by far more important than other effects (the incorporation of the temperature effects estimated by ISAAK *et al.* (1990) reduces the magnitude of calculated anisotropy but does not significantly change the conclusions). I will use the results of theoretical calculations for MgO (ISAAK *et al.*, 1990; KARKI *et al.*, 1997) and MgSiO$_3$ (WENTZCOVITCH, KARKI *et al.*, private communication) neglecting the effect of Fe. For spinel, no theoretical nor experimental studies are available where the pressure effects of individual elastic constant have been investigated, although the effects of pressure on seismic wave velocities in isotropic aggregates have been measured up to 3 GPa (RIGDEN *et al.*, 1991). Since the pressure effects on anisotropy are usually minor at the relatively low pressure conditions (see the study on olivine, CHEN *et al.*, 1996), I will use the anisotropic elastic constants of spinel determined at room pressure (WEIDNER *et al.*, 1984).

The calculation of seismic anisotropy caused by LPO is made as follows. In perovskite, where the LPO caused by shear flow is known, we used the orientation of individual grains and assumed elastic constants of MgSiO$_3$ perovskite (at a given depth) to calculate the elastic constant matrix of a deformed aggregate. For magnesiowüstite or spinel for which no detailed data on LPO are available, we calculated the elastic constant matrix corresponding to a perfect LPO consistent with known slip systems in these crystals. Therefore the latter results should be considered to give the upper limit for the magnitude of seismic anisotropy. In a typical situation, the magnitude of anisotropy due to LPO of a polycrystalline aggregate is about 1/3–1/2 of that of a perfectly oriented single crystal (e.g., KARATO, 1989).

Given an elastic constant matrix, the *SH/SV* polarization anisotropy, the azimuthal anisotropy of *S* waves and *PH/PV* (*PH*: velocity of horizontally propagating *P* wave, *PV*: velocity of vertically propagating *P* wave) anisotropy can be calculated using a scheme described by MONTAGNER and NATAF (1986). The results are summarized in Table 1 and in Figure 4. When a comparison of these results with seismological observations is made, appropriate averaging must be made about the likely mixing effect. The estimated anisotropy for pyrolite composition is also given. For pyrolite composition, I assumed 50:50% spinel and majorite in the transition zone and 70:30% perovskite and magnesiowüstite, and the strength of anisotropy in an aggregate is assumed to be 1/2 of that of a single crystal.

For anisotropy due to SPO or laminated structure, I used the theory of BACKUS (1962) which predicts *SH* > *SV* anisotropy and no azimuthal anisotropy for horizontal structure.

Interpretation of Deep Mantle Seismic Anisotropy

1. Mid-mantle Boundary Layer

The above calculation shows that a possible horizontal flow near the 660 km will manifest *SH* > *SV* anisotropy in the bottom part of the transition zone

Table 1

Seismic anisotropy due to lattice preferred orientation in some deep mantle minerals associated with horizontal flow

Mineral	Depth		$(SH\text{-}SV)/SV$ (%)	$\Delta V_s/V_s$ (%)		$(PH\text{-}PV)/PV$ (%)
spinel*						
$\{111\}\langle 1\bar{1}0\rangle$	~ 600 km		$+2.7$	0.3	(//)	-0.9
$\{100\}\langle 1\bar{1}0\rangle$			-3.7	0		$+1.4$
(pyrolite)	~ 600 km		$+0.7$	0.1	(//)	-0.2
perovskite**	~ 700 km	W	-2.5	2	($\perp$)	$+1.7$
(010)[100]		K_1	-1.6	6	($\perp$)	$+0.4$
	~ 2800 km	W	-1.2	11	($\perp$)	$+1.7$
		K_1	$+1.4$	9	($\perp$)	-2.1
MgO***	~ 700 km	I	$+1.3$	7	($\perp$)	-1.3
$\{110\}\langle 111\rangle$		K_2	-0.6	3	(//)	$+0.7$
	~ 2800 km	I	-4.3	22	(//)	$+1.4$
		K_2	-5.9	33	(//)	$+2.0$
(pyrolite)	~ 700 km	K	-1.3	3	($\perp$)	$+0.4$
	~ 2800 km	K	-0.1	1	($\perp$)	-1.1

$\Delta V_s/V_s$: velocity anisotropy of two vertically travelling shear waves
//: fast S wave parallel to flow direction
$\perp$: fast S wave perpendicular to flow direction
*: based on single crystal (perfect LPO)
**: based on LPO by KARATO et al. (1995) and ZHANG and KARATO (1997)
***: based on single crystal (perfect LPO)
W: based on elastic constants from WENTZCOVITCH (personal communication)
K_1: based on elastic constants from STIXRUDE (personal communication)
I: based on elastic constants from ISAAK et al. (1990)
K_2: based on elastic constants from KARKI et al. (1997)
pyrolite: 50:50% majorite and spinel at the base of the transition zone, and 70:30% perovskite and magnesiowüstite in the lower mantle
K: based on elastic constants from STIXRUDE (personal communication) and KARKI et al. (1997)

and $SV > SH$ anisotropy in the top lower mantle. This prediction is consistent with the results of seismological study by MONTAGNER and KENNETT (1996) which is based mostly on free oscillation data. However, the observed amplitude of the SH/SV anisotropy in the bottom transition zone is too high to attribute to LPO in spinel. This becomes more serious when the effects of secondary peak of LPO ($\{100\}$) are included (see Table 1). It is likely that contributions from laminated structures (SPO) contribute to the $SH > SV$ anisotropy in the deep transition zone. Such an anisotropy structure is possible because this zone is likely to be made of mixtures of previous oceanic crust (garnetite in the transition zone) and spinel-rich assembly (see ALLÉGRE and TURCOTTE, 1986; RINGWOOD and IRIFUNE, 1988; RINGWOOD, 1991; KARATO, 1997b) which have largely contrasting elastic moduli.

Some of the results from body waves such as the splitting of SKS or ScS waves indicate no appreciable anisotropy in this depth range (FOUCH and FISCHER, 1996; FISCHER and WIENS, 1996). It should be noted that these two sets of seismological observations use different techniques and the apparently contradictory results can be mutually compatible. The analyses of surface waves and free oscillations assume that the mantle has transverse isotropy (radial anisotropy) and investigate the

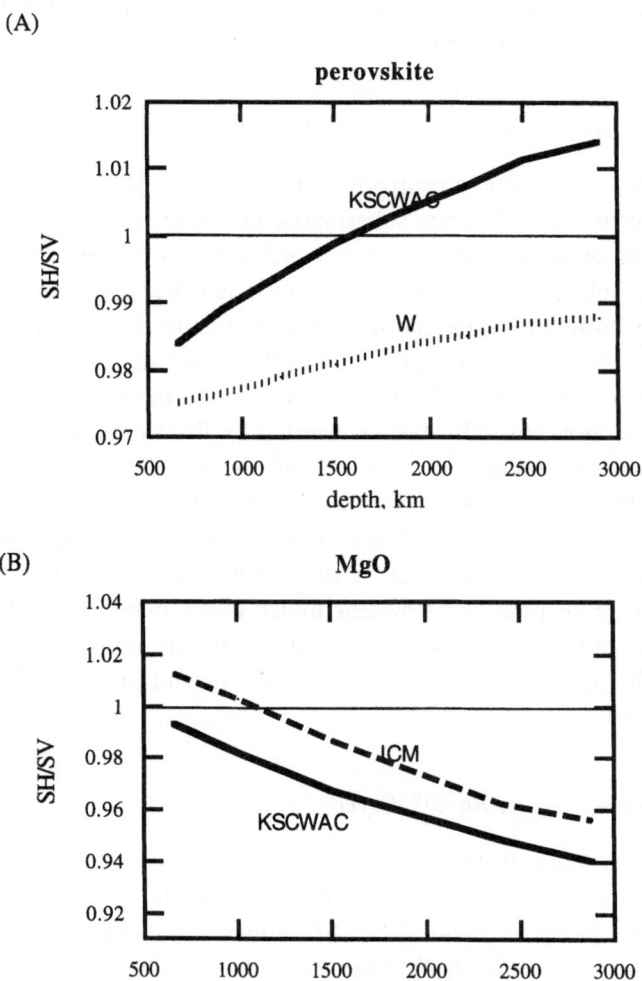

Figure 4

SH/SV anisotropy for horizontal flow in MgO and perovskite as a function of pressure. The lattice preferred orientation (LPO) in MgO and perovskite is assumed to be that corresponding to horizontal shear in dislocation creep regime. The seismic anisotropy is calculated using a theory by MONTAGNER and NATAF (1986). The elastic constants of MgO and for MgSiO$_3$ calculated by theoretical models are used (KARKI *et al.*, 1997 (KSCWC) or ISAAK *et al.*, 1990 (ICM) for MgO: WENTZCOVITCH (W) and STIXRUDE (KSCWC) personal communication for MgSiO$_3$ perovskite).

SH/SV polarization anisotropy. On the other hand, the analyses of shear-wave splitting of SKS or ScS waves investigate the dependence of velocities of vertically propagating shear waves on the direction of polarization. In fact, the analyses of anisotropic structures due to LPO and SPO (or laminated structures) show that although significant polarization anisotropy (*SH/SV* anisotropy) will occur when a strong LPO is developed in this depth region, no significant shear-wave splitting is expected in the bottom of the transition zone. Similarly, SPO or laminated structure will result in *SH/SV* anisotropy but not shear-wave splitting. Therefore, these two sets of observations appear mutually compatible.

2. Absence of Anisotropy and the Strength of the Subducting Slabs

The absence of seismic anisotropy in the lower mantle (near subduction zones) has been interpreted to indicate superplastic flow there (KARATO *et al.*, 1995). The present results on MgO reinforces this conclusion since MgO (or (Mg, Fe)O) also would cause substantial seismic anisotropy when deformed in dislocation creep regime. Thus both perovskite and magnesiowüstite must be deformed by superplastic flow near subducting slabs. This implies that grain size there must be reasonably small and hence slabs must be weak. Weak slabs in the transition zone due to grain-size reduction have also been suggested by Riedel and KARATO (1997). High resolution seismic tomography reveals significant deformation of slabs in the transition zone and in the lower mantle (VAN DER HILST *et al.*, 1991; FUKAO *et al.*, 1992; CADEK *et al.*, 1997), which is consistent with the notion that slabs are weak in this depth range. YAMAZAKI *et al.* (1996) studied grain-growth kinetics in $MgSiO_3$ perovskite plus MgO system under lower mantle conditions and found very sluggish grain-growth kinetics. Their results are consistent with the present results which suggest superplastic deformation in and around subducting slabs in the deep mantle.

3. Seismic Anisotropy in the D″ Layer

The interpretation of the $SH > SV$ polarization anisotropy in the D'' layer is not straightforward. Seismic anisotropy due to LPO of perovskite is either weak $SH > SV$ anisotropy or weak $SV > SH$ anisotropy, depending on the pressure dependence of C_{ij} (see Fig. 4(a)). Anisotropy due to LPO of magnesiowüstite at the D'' layer is strong $SV > SH$ (Fig. 4(b)). Therefore, anisotropies due to LPO of perovskite and magnesiowüstite tend to cancel out and the net effect will be very weak and is difficult to explain a significant $SH > SV$ anisotropy (Table 1). Likewise, the shear-wave splitting from the D'' layer due to LPO of magnesiowüstite and perovskite tend to cancel each other (Table 1). KENDALL and SILVER (1996) proposed that the $SH > SV$ anisotropy is due to the presence of aligned melt. However, I consider that this is unlikely because the $SH > SV$

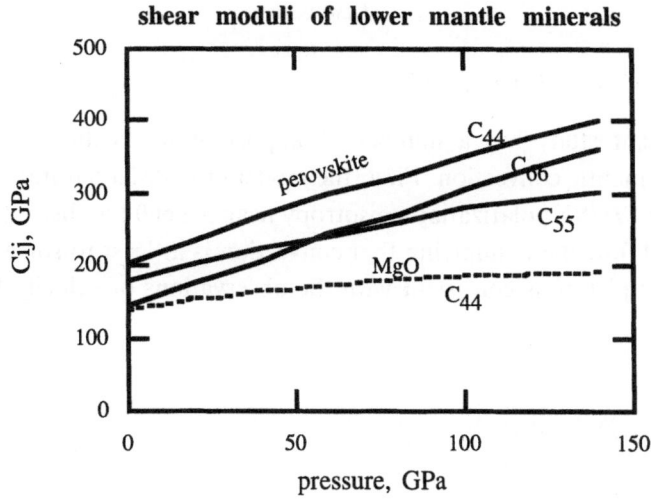

Figure 5

Some shear moduli (C_{ij}) of MgSiO$_3$ and MgO under lower mantle pressures (KARKI *et al.*, 1997; STIXRUDE, personal communication). Bulk modulus and shear modulus of the two phases are similar, but the differences in individual component of shear moduli are very large at deep lower mantle pressures although not at shallow lower mantle conditions. This is due mostly to the anomalously small pressure dependence of C_{44} in MgO.

anisotropy is found in the D'' layer beneath the Caribbean Sea (KENDALL and SILVER, 1996) and beneath the northern part of North America (MATZEL *et al.*, 1996; GARNERO and LAY, 1997) where average velocity is higher than other portions of the D'' layer.

I propose an alternative model that is consistent with mineral physics and seismological observations. In this model, the $SH > SV$ anisotropy is due to SPO of mineral aggregates at the D'' layer caused by dislocation creep (Fig. 1(b)). Although the two major constituents of the D'' layer (perovskite and magnesiowüstite) have similar elastic moduli on average, there is a significant contrast in individual shear modulus because of the very large anisotropy in shear moduli in magnesiowüstite (Fig. 5). Therefore significant scattering of shear waves occurs at grain boundaries which will cause the $SH > SV$ polarization anisotropy when grain boundaries are aligned horizontally (BACKUS, 1962). BACKUS (1962) demonstrated that the magnitude of SH/SV polarization anisotropy at very extensive elongation of grains is given by $SH/SV = 1/\sqrt{1 - (\delta\mu/\mu)^2}$ where $\delta\mu/\mu$ is the contrast in shear moduli (see also GEE and JORDAN, 1988). To explain the observed $\sim 2\%$ $SH > SV$ anisotropy, one needs $\delta\mu/\mu \sim 0.2$ which is compatible with the large constants in shear moduli between magnesiowüstite and perovskite at the deep lower mantle conditions (Fig. 5). Note, however, that the contrast in elastic constants at the shallow lower mantle conditions is too small to cause significant anisotropy due to SPO.

Discussion

1. Is Mantle Convection Layered?

The present study has a number of important implications for the possible geometry of mantle convection. First, the model provides a natural explanation for the observed *SH/SV* polarization anisotropy near the 660 km discontinuity in terms of horizontal flow there, implying that convection is at least to some extent layered (Fig. 6). This picture is consistent with the observations of velocity heterogeneities

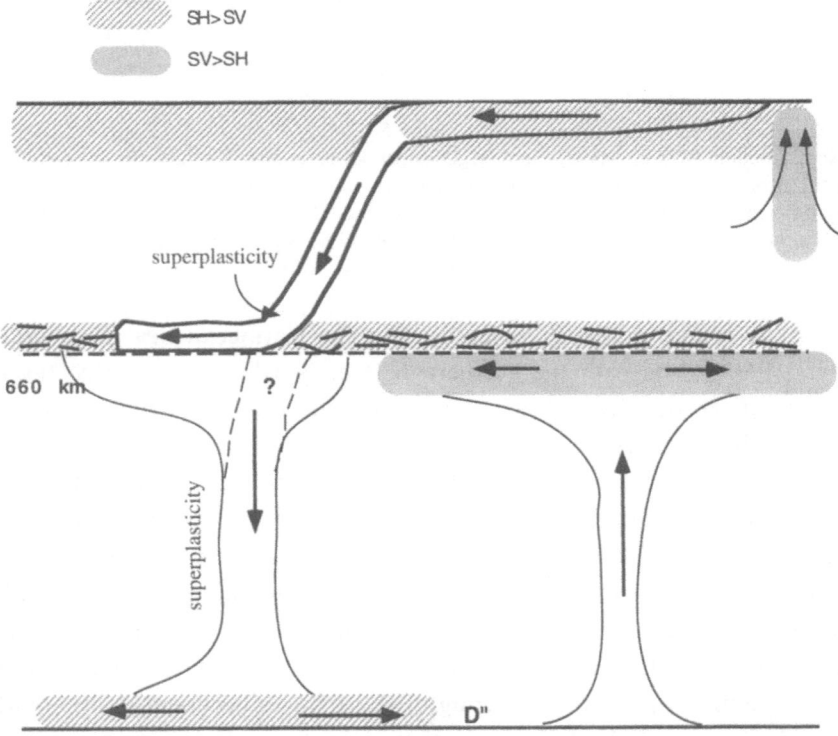

Figure 6
A cartoon of mantle convection model and associated seismic anisotropy (*SH/SV* polarization anisotropy) consistent with mineral physics and seismological observations. Observed significant *SH > SV* anisotropy in cold portions of the *D″* layer is likely to be due to strong shear caused by the sinking of cold materials. The *SH/SV* anisotropy near the 660 km discontinuity can be attributed to shear motion there, implying (partially) layered mantle convection. Most of other portions of the deep mantle, including regions near or inside of subducting slabs, do not display anisotropy, presumably because of deformation by diffusion or superplastic creep. Solid state mechanisms can explain most of the seismological observations, including the radial and lateral variation in anisotropy, although lattice preferred orientation alone cannot explain some of the observations. Seismic anisotropy in the bottom transition zone and in the cold portions of the *D″* layer may be due to laminated structures.

which indicate significant deflection of subducted slabs in the western Pacific (VAN DER HILST et al., 1991; FUKAO et al., 1992; CADEK et al., 1997). Thus the observations of both velocity heterogeneity and velocity anisotropy support a model that mantle convection is layered to some extent.

The present study also provides a clue to understand why layered convection occurs in some regions. According to DAVIES (1995), the role of density anomalies near the 660 km discontinuity which cause layered convection is highly dependent on slab rheology. When slabs are strong, they will penetrate into the lower mantle with little deformation and hence whole mantle convection will occur. It is only when slabs are very weak that the 660 km discontinuity causes (partially) layered convection. An important parameter to control slab rheology is temperature and grain size, and KARATO (1997a) and RIEDEL and KARATO (1997) showed that grain-size reduction associated with mantle phase changes could significantly reduce the strength of slabs. The observed absence of SKS splitting in the deep mantle near subduction zones implies that mantle materials there deform by superplasticity and hence their creep strength is significantly smaller than expected from dislocation creep models (KARATO et al., 1995). Therefore it is conceivable that (some) slabs become very weak due to phase transformations and hence the 660 km discontinuity effectively prevents the penetration of these weak slabs. Evidence of weak slabs is also presented by MORESI and GURNIS (1996). The recent experimental study by YAMAZAKI et al. (1996) showed very sluggish grain-growth kinetics in perovskite plus magnesiowüstite aggregates which suggests that superplastic creep is important in a large portion of the lower mantle.

The nature of seismic anisotropy associated with layered convection depends on the details of convection current. If, for example, subducting slabs lie on the 660 km discontinuity as a result of trench roll-back (VAN DER HILST and SENO, 1993), then not much shear is expected in the top lower mantle, and hence anisotropy will occur around the subduction zone only in the bottom transition zone as a result of the separation of garnetite (KARATO, 1997b) and the resultant laminated structure (ALLÉGRE and TURCOTTE; 1986). If on the other hand, the deflecting slabs push into the transition zone, then shear deformation and resultant anisotropy will develop in both the transition zone and in the lower mantle. The observed absence of shear-wave splitting in the seismic records near subduction zones suggests that horizontal shear is not strong in the shallow lower mantle in these regions. This suggests that the former mode of slab deformation occurs in these regions or that the stress in the topmost lower mantle caused by slab deflection is low' and deformation mechanism is superplastic.

Deformation due to the collision of upwelling plumes with the 660 km discontinuity (e.g., NAKAKUKI et al., 1994; STEINBACH and YUEN, 1995) is also a likely mechanism for seismic anisotropy. In this case one expects significant $SV > SH$ anisotropy at the top lower mantle above the hot D'' regions such as Africa and the South Pacific (Fig. 6). The $SV > SH$ polarization anisotropy inferred from the global inversion of free oscillation data (MONTAGNER and KENNETT, 1996) may be mostly from these regions.

2. Anisotropy and Dynamics of the D″ Layer

Although the present model prefers a solid state mechanism for the $SH > SV$ anisotropy in the D'' layer in the regions where it is detected, this does not imply that partial melting is unlikely in any area of the D'' layer. In fact, the observed velocity reduction may require partial melting in some regions of the D'' layer (WILLIAMS and GARNERO, 1996). However, shear-wave polarization anisotropy is not observed in these low velocity and thin regions of the D'' layer, but rather in high velocity and thick regions of the D'' layer (KENDALL and SILVER, 1996; MATZEL et al., 1996; GARNERO and LAY, 1997). This suggests that the observed SH/SV anisotropy in the D'' layer is due to solid state mechanisms.

Given some evidence for partial melting (e.g., WILLIAMS and GARNERO, 1996), one might ask why regions of the D'' layer where partial melting is likely to occur, do not show anisotropy. There are several possible explanations. First, the regions where strong anisotropy is not observed correspond to the areas of lower than average velocities. Therefore these regions are likely to be hotter and softer than average. Thus, stress level is expected to be low and the deformation mechanism might be superplastic creep, in which case, no mechanism is present to cause seismic anisotropy by solid state processes. In addition, ZHANG et al. (1995) demonstrated that the preferred alignment of melt occurs only at relatively high stress levels. Thus a low stress level in these regions is a possible reason for the absence of anisotropy. Second, and an alternative hypothesis is that the anisotropic structure may also be formed in the hot regions of the D'' layer by the same mechanism (i.e., SPO), although the geometry of deformation in these hot regions is different. It is likely that these hotter than average regions in the D'' layer are regions from which upwelling mantle plumes (hotspots or megaplumes) are generated (e.g., DZIEWON-SKI and WOODHOUSE, 1987; FUKAO et al., 1992). The seismic anisotropy associated with vertical motion caused by SPO would be $SV > SH$, but since strain is not large in the source regions of plumes, its amplitude will be small. This hypothesis implies that one would expect $SV > SH$ anisotropy in an ascending plume far from the D'' layer but its amplitude in the shallower portions of the lower mantle will be small because of less contrast in elastic moduli there.

In summary, the present model for anisotropy in the D'' suggests that the regions of the D'' layer where significant anisotropy occurs, are the regions where strong shear deformation occurs or has occurred at high stress level. These regions coincide with the regions where subducted slabs are likely to have been accumulated during the recent geological history. It is noted also that if anisotropy is due to SPO rather than LPO as the present model suggests, then anisotropy will appear only when deformation is currently on-going or has recently occurred and not when deformation stopped in the distant past. This is because the recovery time scale of SPO is considerably less than that of LPO (e.g., ZHANG *et al.*, 1995). Therefore, the presence of anisotropy in some portions of the D'' layer is evidence of active shear deformation at the present time and/or in the recent past (Fig. 6).

In this model, the top boundary of the D'' layer corresponds to the boundary between isotropic and anisotropic layers (see also MATZEL *et al.*, 1996) similar to the model proposed for the Lehmann discontinuity (REVENAUGH and JORDAN, 1991; KARATO, 1992). The transition is considered to be due to the presence of strong horizontal flow and hence the layer will be thick where a large volume of materials flows due to the accumulation of subducted materials, which is consistent with the seismological observations (GARNERO and LAY, 1997).

Summary and Concluding Remarks (Open Questions)

Mineral physical interpretation of deep mantle seismic anisotropy provides important constraints on mantle dynamics. This paper presents a first attempt in such an endeavor. Although this synthesis appears to provide a sensible interpretation of seismic anisotropy in terms of deep mantle dynamics, a number of problems remain which must be further investigated to clarify the dynamics of the earth's deep mantle.

(1) The LPO assumed in this paper is mostly based on the results in analogue materials due to dislocation glide. The LPO of high-pressure minerals must ultimately be determined under deep mantle conditions (high P and T). Experimental technology to allow this has recently been developed by KARATO and RUBIE (1997). Application of this shear deformation technique under deep conditions will provide the first direct experimental data on LPO to interpret seismic anisotropy without relying on analogue materials.

(2) The elastic constants in deep mantle minerals must be evaluated at high P and T. The present study used elastic constant at high P and low T. The temperature effects must also be evaluated. Effects of Fe also require investigation.

(3) The stability of orthorhombic perovskite must be better understood. If orthorhombic perovskite becomes unstable and transformed to other structures or phases in the deep lower mantle, as suggested by some of the recent studies (e.g., MEADE *et al.*, 1995b), then all of the present discussions on these depths will need revision.

(4) More seismological observations are obviously needed. This is particularly the case for the mid-mantle boundary layer. This boundary layer and the flow associated with it are likely to be regional and studies of anisotropy at regional scales will provide us with further important constraints on the nature of mantle convection.

Acknowledgments

The present study was initiated in Bayerisches Geoinstitut where the author spent one year during his sabbatical year supported by the Alexander von Humboldt Stiftung. Much of the present study was motivated by a lively discussion with Alex Forte, Jean-Paul Montagner and Lev Vinnik in Paris. Shuqing Zhang is thanked for the calculation of elastic constants of deformed perovskites, Renata Wentzcovitch and Lars Stixrude for discussions on the pressure dependence of elastic constants, and Daisuke Yamazaki and Ed Garnero for sending me the preprints of their respective papers. Renata Wentzcovitch and Lars Stixrude kindly provided me with the permission to use their unpublished results of the theoretical calculation of elastic moduli in MgSiO$_3$ perovskite. Dave Yuen is thanked for the discussions on the interaction of plumes and the phase transition boundaries and Bob Liebermann and Ganglin Chen on elasticity in mantle minerals. This research was supported by the grant EAR-9505451 by NSF.

REFERENCES

ALLÈGRE, C., and TURCOTTE, D. L. (1986), *Implications of a Two-component Marble-cake Mantle*, Nature *323*, 123–127.

BACKUS, G. E. (1962), *Long-wave Elastic Anisotropy Produced by Horizontal Layering*, J. Geophys. Res. *67*, 4427–4440.

BASS, J. D., *Elasticity of minerals, glasses and melts*. In *Mineral Physics and Crystallography: A Handbook of Physical Constants* (ed. Ahrens, T. J.) (American Geophysical Union, Washington DC 1995) pp. 45–63.

BOEHLER, R. (1996), *Melting of Mantle and Core Materials at Very High Pressures*, Phil. Trans. R. Soc. Lond. *A354*, 1265–1278.

BREUER, D., ZHOU, H., YUEN, D. A., and SPOHN, T. (1996), *Phase Transition in the Martian Mantle: Implications for the Planet's Volcanic History*, J. Geophys. Res. *101*, 75731–75742.

BUSSE, F. H., *Fundamentals of thermal convection*. In *Mantle Convection* (ed. Peltier, W. R.) (Gordon and Breach, New York 1989), pp. 23–95.

CADEK, O., YUEN, D. A., CIKOVA, H., KIDO, M., ZHOU, H-W., BRUNET, D., and MACHETEL, P. (1997), *New Perspectives on Mantle Dynamics from High-resolution Seismic Tomographic Model P1200*, Pure appl. geophys. *151*, 503–525.

CHEN, G., LI, B., and LIEBERMANN, R. C. (1996), *Selected Elastic Moduli of Single-crystal Olivines from Ultrasonic Experiments to Mantle Pressures*, Science *272*, 979–980.

CHIN, G. Y., and MAMMEL, W. L. (1973), *A Theoretical Examination of the Plastic Deformation of Ionic Crystals: II. Analysis of uniaxial Deformation and Axisymmetric Flow for Slip on {110}⟨110⟩ and {100}⟨110⟩ Systems*, Metall. Trans. *4*, 335–340.

DAVIES, G. F. (1995), *Penetration of Plates and Plumes through the Mantle Transition Zone*, Earth Planet. Sci. Lett. *133*, 507–516.

DAVIES, G. F., and RICHARDS, M. A. (1992), *Mantle Convection*, J. Geol. *100*, 151–206.

DUPAS, C., GREEN, H. W., II., DOUKHAN, N., DOUKHAN, J.-C., and TINGLE, T. N. (1997), *The Rheology of Olivine and Spinel Magnesium Germanate (Mg$_2$GeO$_4$): TEM Study of the Defect Microstructure*, Phys. Chem. Mineral, submitted.

DZIEWONSKI, A. M., and WOODHOUSE, H. J. (1987), *Global Images of the Earth's Interior*, Science *236*, 37–48.

EDINGTON, J. W., MELTON, K. N., and CUTLER, C. P. (1976), *Superplasticity*, Prog. Mater. Sci. *21*, 63–170.

FISCHER, K. M., and WIENS, D. A. (1996), *The Depth Distribution of Mantle Anisotropy beneath the Tonga Subduction Zone*, Earth Planet. Sci. Lett. *142*, 253–260.

FOUCH, M. J., and FISCHER, K. M. (1996), *Mantle Anisotropy beneath Northwest Pacific Subduction Zones*, J. Geophys. Res. *101*, 15987–16002.

FRANSSEN, R. (1993), *Rheology of Synthetic Rocksalt*, Ph.D. Thesis, University of Utrecht, 221 pp.

FROST, H. J., and ASHBY, M. F., *Deformation Mechanism Maps* (Pergamon Press, Oxford 1982) 167 pp.

FUJIMURA, A. (1984), *Preferred Orientation of Silicate Spinel Inferred from Experimentally Deformed Aggregates of Trevorite*, J. Phys. Earth *32*, 273–297.

FUKAO, Y., OBAYASHI, M., INOUE, H., and NENBAI, M. (1992), *Subducting Slabs Stagnant in the Transition Zone*, J. Geophys. Res. *97*, 4809–4822.

GARNERO, E. J., and LAY, T. (1997), *Lateral Variation in Lowermost Mantle Shear-wave Anisotropy beneath the North Pacific and Alaska*, J. Geophys. Res. *102*, 8121–8135.

GEE, L. S., and JORDAN, T. H. (1988), *Polarization Anisotropy and Fine-scale Structure of the Eurasian Upper Mantle*, Geophys. Res. Lett. *15*, 824–827.

HEMLEY, R. J., and COHEN, R. E. (1996), *Structure and Bonding in the Deep Mantle and Core*, Phil. Trans. R. Soc. Lond. *A354*, 1461–1479.

ISAAK, D. G., COHEN, R. E., and MEHL, M. J. (1990), *Calculated Elastic and Thermal Properties of MgO at High Pressures and Temperatures*, J. Geophys. Res. *95*, 7055–7067.

ITO, E., and SATO, H. (1990), *Aseismicity in the Lower Mantle by Superplasticity in a Subducting Slab*, Nature *351*, 140–141.

JARVIS, G. T., and PELTIER, W. R., *Convection models and geophysical observations*. In *Mantle Convection* (ed. Peltier, W. R.) (Gordon and Breach, New York, 1989), pp. 480–593.

JORDAN, T. H., LERNER-LAM, A. L., and CREAGER, K. C., *Seismic imaging of boundary layers and deep mantle convection*. In *Mantle Convection* (ed. Peltier, W.R.) (Gordon and Breach, New York 1989) pp. 97–201.

JORDAN, T. H., PUSTER, P., GLAZMAIER, G. A., and TACKLEY, P. J. (1993), *Comparisons Between Seismic Earth Structures and Mantle Flow Models Based on Radial Correlation Functions*, Science *261*, 1427–1431.

KANESHIMA, S., and SILVER, P. G. (1995), *Anisotropic Loci in the Mantle beneath Central Peru*, Phys. Earth Planet. Inter. *88*, 257–272.

KARATO, S. (1988), *The Role of Recrystallization in the Preferred Orientation in Olivine*, Phys. Earth Planet. Inter. *51*, 107–122.

KARATO, S., *Seismic anisotropy: mechanisms and tectonic implications*. In *Rheology of Solids and of the Earth* (eds. Karato, S. and Toriumi, M.) (Oxford University Press, Oxford, 1989), pp. 393–422.

KARATO, S. (1992), *On the Lehmann Discontinuity*, Geophys. Res. Lett. *19*, 2255–2258.

KARATO, S., *Phase transformations and rheological properties of mantle minerals*. In *Earth's Deep Interior* (eds. Crossley, D. and Soward, A. M.) (Gordon and Breach, New York, 1997a) pp. 223–272.

KARATO, S. (1997b), *On the Separation of Crustal Component from Oceanic Lithosphere near the 660 km Discontinuity*, Phys. Earth Planet. Inter. *99*, 103–111.

KARATO, S., and WU, P. (1993), *Rheology of the Upper Mantle: A Synthesis*, Science *260*, 771–778.

KARATO, S., and RUBIE, D. C. (1997), *Towards an Experimental Study of Deep Mantle Rheology: A New Multi-anvil Specimen Assembly for Deformation Experiments under High Pressures and Temperatures*, J. Geophys. Res. in press.

KARATO, S., ZHANG, S., and WENK, H.-R. (1995), *Superplasticity in Earth's Lower Mantle: Evidence from Seismic Anisotropy and Rock Physics*, Science *270*, 458–461.

KARKI, B. B., STIXRUDE, L., CLARK, S. J., WARREN, M. C., ACKLAND, G. J., and CRAIN, J. (1997), *Structure and Elasticity of MgO at High Pressure*, Am. Mineral. *82*, 51–60.

KENDALL, J.-M., and SILVER, P. G. (1996), *Constraints from Seismic Anisotropy on the Nature of the Lowermost Mantle*, Nature *381*, 409–412.

LI, P., KARATO, S., and WANG, Z. (1996), *High Temperature Creep in Fine-grained Aggregates of CaTiO₃*, Phys. Earth Planet. Inter. *95*, 19–36.

MADON, M., and POIRIER, J.-P. (1980), *Dislocations in Spinel and Garnet High Pressure Polymorphs of Olivine and Pyroxene: Implications for Mantle Rheology*, Science *207*, 66–68.

MATZEL, E., SEN, M. K., and GRAND, S. P. (1996), *Evidence for Anisotropy in the Deep Mantle beneath Alaska*, Geophys. Res. Lett. *23*, 2417–2420.

MAUPIN, V. (1994), *On the Possibility of Anisotropy in the D″ Layer as Inferred from the Polarization of Diffracted S Waves*, Phys. Earth Planet. Inter. *87*, 1–32.

MCKENZIE, D. (1979), *Finite Deformation during Fluid Flow*, Geophys. J. R. Astr. Soc. *58*, 689–715.

MEADE, C., SILVER, P. G., and KANESHIMA, S. (1995a), *Laboratory and Seismological Observations of Lower Mantle Isotropy*, Geophys. Res. Lett. *22*, 1293–1296.

MEADE, C., MAO, H.-K., and HU, J. (1995b), *High-temperature Phase Transition and Dissociation of (Mg, Fe)SiO₃ perovskite at Lower Mantle Pressures*, Science *268*, 1743–1745.

MITCHELL, T. E., HWANG, L., and HEUER, A. H. (1976), *Dislocations in Spinel*, J. Material. Sci. *11*, 264–272.

MONTAGNER, J.-P. (1994), *Can Seismology Tell us Anything About Convection in the Mantle?* Rev. Geophys. *32*, 115–137.

MONTAGNER, J.-P., and KENNETT, B. L. N. (1996), *How to Reconcile Body-wave and Normal-mode Reference Earth Model*, Geophys. J. Int. *125*, 229–248.

MONTAGNER, J.-P., and NATAF, H.-C. (1986), *A Simple Method for Inverting the Azimuthal Anisotropy of Surface Waves*, J. Geophys. Res. *91*, 511–520.

MONTAGNER, J.-P., and TANIMOTO, T. (1991), *Global Upper Mantle Tomography of Seismic Velocities and Anisotropies*, J. Geophys. Res. *96*, 20337–20351.

MORESI, L., and GURNIS, M. (1996), *Constraints on the Lateral Strength of Slabs from Three-dimensional Dynamic Flow Models*, Earth Planet. Sci. Lett. *138*, 15–28.

NAKAKUKI, T., SATO, H., and FUJIMOTO, H. (1994), *Interaction of the Upwelling Plumes with the Phase and Chemical boundary at the 660 km Discontinuity: Effects of Temperature-dependent Viscosity*, Earth Planet. Sci. Lett. *121*, 369–384.

NATAF, H.-C., NAKANISHI, I., and ANDERSON, D. L. (1986), *Measurement of Mantle Wave Velocities and Inversion for Lateral Heterogeneities and Anisotropy, 3. Inversion*, J. Geophys. Res. *91*, 7261–7307.

NICOLAS, A., and CHRISTENSEN, N. I., *Formation of anisotropy in upper mantle peridotites: A review. In Composition, Structure and Dynamics of the Lithosphere/Asthenosphere System* (eds. Fuchs, K. and Froidevaux, C) (American Geophysical Union, Washington, D.C., 1987) pp. 111–123.

PELTIER, W. R. (1996), *Phase-transition Modulated Mixing in the Mantle of the Earth*, Phil. Trans. R. Soc., Lond. *A354*, 1425–1447.

REVENAUGH, J., and JORDAN, T. H. (1991), *Mantle Layering from ScS Reverberations 3. The Upper Mantle*, J. Geophys. Res. *96*, 19781–19810.

RICE, R. W., *Hot-working of oxides. In High Temperature Oxides, Part III* (ed. Alper, A. M.) (Academic Press, New York, 1970) pp. 235–281.

RIEDEL, M. R., and KARATO, S. (1997), *Grain-size Evolution in Subducted Oceanic Lithosphere Associated with the Olivine-spinel Transformation and its Effects on rheology*, Earth Planet. Sci. Lett. *148*, 27–44.

RIGDEN, S. M., GWANMESIA, G. D., FITZGERALD, J. D., JACKSON, I., and LIEBERMANN, R. C. (1991), *Spinel Elasticity and Seismic Structure of the Transition Zone of the Mantle*, Nature *354*, 143–145.

RINGWOOD, A. E. (1991), *Phase Transformations and their Bearings on the Constitution and Dynamics of the Mantle*, Geochim. Cosmochim. Acta *55*, 2083–2110.

RINGWOOD, A. E., and IRIFUNE, T. (1988), *Nature of the 650 km Seismic Discontinuity: Implications for Mantle Dynamics and Differentiation*, Nature *331*, 131–136.

SHARP, T. G., BUSSOD, G. Y. A., and KATSURA, T. (1994), *Microstructures in β-Mg$_{1.8}$Fe$_{0.2}$SiO$_4$ Experimentally Deformed at Transition Zone Conditions*, Phys. Earth Planet. Inter. *86*, 69–83.

SILVER, P. G. (1996), *Seismic Anisotropy beneath the Continents: Probing the Depths of Geology*, Ann. Rev. Earth Planet. Sci. *24*, 385–432.

SILVER, P. G., CARLSON, R. W., and OLSON, P. (1988), *Deep Slabs, Geochemical Heterogeneity, and the Large-scale Structure of Mantle Convection: Investigation of an Enduring Paradox*, Ann. Rev. Earth Planet. Sci. *16*, 477–541.

STEINBACH, V., and YUEN, D. A. (1995), *The Effects of Temperature-dependent Viscosity on Mantle Circulation with Two Major Phase Transition*, Phys. Earth Planet. Inter. *90*, 13–36.

SYLVANDER, M., and SOURIAU, A. (1996), *Mapping S-velocity Heterogeneities in the D″ Region, from SmKS Differential Travel Times*, Phys. Earth Planet. Inter. *94*, 1–21.

TANIMOTO, T., and ANDERSON, D. L. (1984), *Mapping Mantle Convection*, Geophys. Res. Lett. *11*, 287–290.

TONG, C., GUDMUNDSSON, O., and KENNETT, B. L. N. (1994), *Shear-wave Splitting in Refracted Waves Returned from the Upper Mantle Transition Zone beneath Northern Australia*, J. Geophys. Res. *99*, 15783–15797.

VAN DER HILST, R., and SENO, T. (1993), *Effects of Relative Plate Motion on the Deep Structure and Penetration Depth of Slabs below the Izu-Bonin and Mariana Island Arcs*, Earth Planet. Sci. Lett. *120*, 395–407.

VAN DER HILST, R., ENGDAHL, R., SPAKMAN, W., and NOLET, G. (1991), *Tomographic Imaging of Subducted Lithosphere below Northwest Pacific Island Arcs*, Nature *353*, 37–43.

VAN HOUTTE, P., and WAGNER, F., *Development of textures by slip and twinning*. In *Preferred Orientation in Deformed Metals and Rocks* (ed. Wenk, H-R.) (Academic Press, Orlando 1985) pp. 233–258.

VINNIK, L., and MONTAGNER, J.-P. (1996), *Shear-wave Splitting in the Mantle Ps Phases*, Geophys. Res. Lett. *23*, 2449–2452.

VINNIK, L., ROMANOWICZ, B., LE STUNFF, Y., and MAKEYEVA, L. (1995), *Seismic Anisotropy in the D″ Layer*, Geophys. Res. Lett. *22*, 1657–1660.

WEIDNER, D. J., SAWAMOTO, H., SASAKI, S., and KUMAZAWA, M. (1984), *Single-crystal Elastic Properties of the Spinel Phase of Mg_2SiO_4*, J. Geophys. Res. *89*, 7852–7860.

WILLIAMS, R. O. (1962), *Shear Textures in Copper, Brass, Aluminum, Iron, and Zirconium*, Trans. Metall. Soc. AIME *224*, 129–140.

WILLIAMS, Q., and GARNERO, E. J. (1996), *Seismic Evidence for Partial Melt at the Base of Earth's Mantle*, Science *273*, 1528–1530.

WYSESSION, M. E. (1996), *Large-scale Structure at the Core-mantle Boundary from Diffracted Waves*, Nature *382*, 244–248.

YAMAZAKI, D., KATO, T., OHTANI, E., and TORIUMI, M. (1996), *Grain Growth Kinetics of Mg_2SiO_4, under Lower Mantle Condition*, Science *274*, 2052–2054.

YOUNG, C. J., and LAY, T. (1987), *The Core-mantle Boundary*, Ann. Rev. Earth Planet. Sci. *15*, 25–46.

ZHANG, S., and KARATO, S. (1997), *Simple Shear Deformation of Polycrystalline Orthorhombic Perovskite $CaTiO_3$*, Tectonophysics, in press.

ZHANG, S., ZIMMERMAN, M. R., DAINES, M. J., KARATO, S., and KOHLSTEDT, D. L. (1995), *Lattice Preferred Orientation and Melt Distribution in Experimentally Sheared Olivine-basalt Rocks*, EOS, Trans. AGU *75*, S281.

(Received November 5, 1996, revised April 18, 1997, accepted May 22, 1997)

To access this journal online:
http://www.birkhauser.ch

Pure appl. geophys. 151 (1998) 589–603
0033–4553/98/040589–15 $ 1.50 + 0.20/0

▌Pure and Applied Geophysics

Experimental Studies of Shear Deformation of Mantle Materials: Towards Structural Geology of the Mantle

S. KARATO,[1] S. ZHANG,[1,2] M. E. ZIMMERMAN,[1] M. J. DAINES,[1] and D. L. KOHLSTEDT[1]

Abstract—A brief outline is given on experimental studies carried out in the Minnesota Mineral and Rock Physics Laboratory of microstructural evolution and rheology of mantle mineral aggregates or their analogues, using a simple shear deformation geometry. A simple shear deformation geometry allows us to unambiguously identify controlling factors of microstructural evolution and to obtain large strains at high pressures and temperatures, and thus provides a unique opportunity to investigate the "structural geology of the mantle." We have developed a simple shear deformation technique for use at high pressures and temperatures (pressure up to 16 GPa and temperature up to 2000 K) in both gas-medium and solid-medium apparati. This technique has been applied to the following mineral systems: (i) olivine aggregates, (ii) olivine basaltic melt, (iii) $CaTiO_3$ perovskite aggregates. The results have provided important data with which to understand the dynamics of the earth's mantle, including the geometry of mantle convection, mechanisms of melt distribution and migration beneath mid-ocean ridges, and the mechanisms of shear localization. Limitations of laboratory studies and future directions are also discussed.

Key words: Simple shear deformation, structural geology, seismic anisotropy, partial melting, lattice preferred orientation, shear localization.

Introduction

Understanding microstructural development during rock deformation is critical to a number of geological and geophysical problems, including the inference of mantle convection patterns from seismic anisotropy (e.g., NATAF *et al.*, 1986; NICOLAS and CHRISTENSEN, 1987; KARATO, 1989a, this issue), mechanisms of melt migration beneath mid-ocean ridges (e.g., PHIPPS MORGAN, 1987; FORSYTH, 1992; KOHLSTEDT, 1992), and deformation geometry in the mantle such as the flow direction and/or the sense of shear from deformation microstructures (e.g., NICOLAS and CHRISTENSEN, 1987; NICOLAS, 1989). These geological or geophysical problems are related to various questions in mineral and rock deformation,

---

[1] Department of Geology and Geophysics, University of Minnesota, Minneapolis, MN 55455, U.S.A.

[2] Currently at Research School of Earth Sciences, the Australian National University, Canberra, Australia.

including the development of lattice preferred orientation (LPO), the geometry of secondary phases (such as melts) during deformation and the mechanisms of shear localization (e.g., DRURY *et al.*, 1991; JIN *et al.*, 1997; see Fig. 1).

Laboratory studies of mineral and rock deformation can provide important constraints on these issues, but the uni-axial (or tri-axial) deformation geometry usually employed in laboratory studies has major limitations. In co-axial deformation such as uni-axial or tri-axial compression, the principal axes of stress are parallel to those of strain. Therefore, it is difficult to distinguish the influence of stress from that of strain. In contrast, in non-coaxial deformation such as simple shear, the strain ellipsoid rotates with respect to the external framework (such as shear plane and shear direction), although the orientation of the principal stresses remains constant (e.g., HOBBS *et al.*, 1976). Thus, various factors that may control deformation microstructures, namely the stress orientation, the strain ellipsoid orientation and the shear plane/shear direction are clearly distinguished in simple

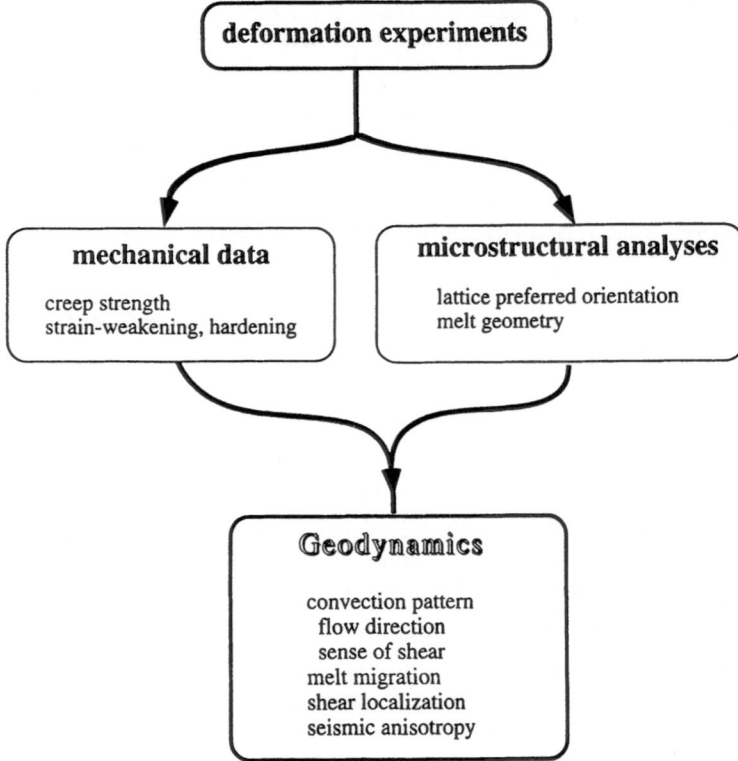

Figure 1
A schematic diagram showing the strategy of laboratory studies of "structural geology of the mantle." For studies on structural geology of the mantle, large strain deformation experiments in which controlling factors for microstructural evolution can be clearly identified are critical. Simple shear deformation geometry is suited for this purpose as compared to more conventional uni-axial (or tri-axial) compression geometry.

shear, which is not the case in uni-axial (or tri-axial) compression. In addition, large strain deformation experiments are readily achieved in a simple shear geometry whereas well-constrained large strain deformation experiments are difficult with uni-axial compression because of barreling effects. Thus, a non-coaxial deformation geometry such as simple shear is better suited for laboratory studies of rock deformation in which microstructural development plays an important role. This paper provides a progress report of our project "structural geology of the mantle" in which microstructural developments and large strain mechanical behavior in mantle minerals or their analogues are studied by simple shear deformation experiments at high pressures and temperatures. (Results from other laboratories are mentioned when they are closely related to ours, although no intention is made to extensively review this area.)

Experimental Procedure

We have developed a simple shear deformation technique for use in a gas medium deformation apparatus (the Paterson apparatus, P (pressure) < 300 MPa, T (temperature) < 1600 K), a piston-cylinder type solid medium deformation apparatus (the Griggs apparatus, P < 3 GPa, T < 1600 K) and for a multi-anvil apparatus (P < 16 GPa, T < 1900 K). The details of experimental techniques are given in ZHANG et al. (1997) for a gas medium apparatus and in KARATO and RUBIE (1997) for a multi-anvil apparatus. The basic design is common to all of the apparatus (Fig. 2). A thin specimen, sandwiched between saw-cut pistons (at 45° with respect to the compression axis), is squeezed by uni-axial movement of the pistons. Pistons are allowed to move laterally with little resistance so that the uni-axial movement is transformed to nearly simple shear deformation of a specimen.

To help determine the deformation geometry of a specimen after shearing, we put a strain marker made of a thin Ni foil or a layer of vacuum coated Pt that is initially perpendicular to the specimen and piston interface or shear boundary and shear direction. The rotation of this strain marker, as well as other measurements of finite strain in the sample such as change in thickness, provide critical information for determining the kinematics of deformation.

The choice of piston materials and the roughness of the piston/specimen interfaces are critical. Piston materials must be significantly stronger than specimens and should be chemically inert. Thoriated tungsten and alumina work reasonably well for olivine, although neither of them is perfect. Deformation of tungsten becomes significant when large stresses are applied, and chemical reaction between tungsten or alumina and olivine becomes significant at high temperatures. The surface roughness is also critical at relatively low pressures to prevent sliding between specimen and piston.

Mechanical tests are mostly conducted either at a constant displacement rate (~constant strain-rate) or constant load (~constant stress). However, in experi-

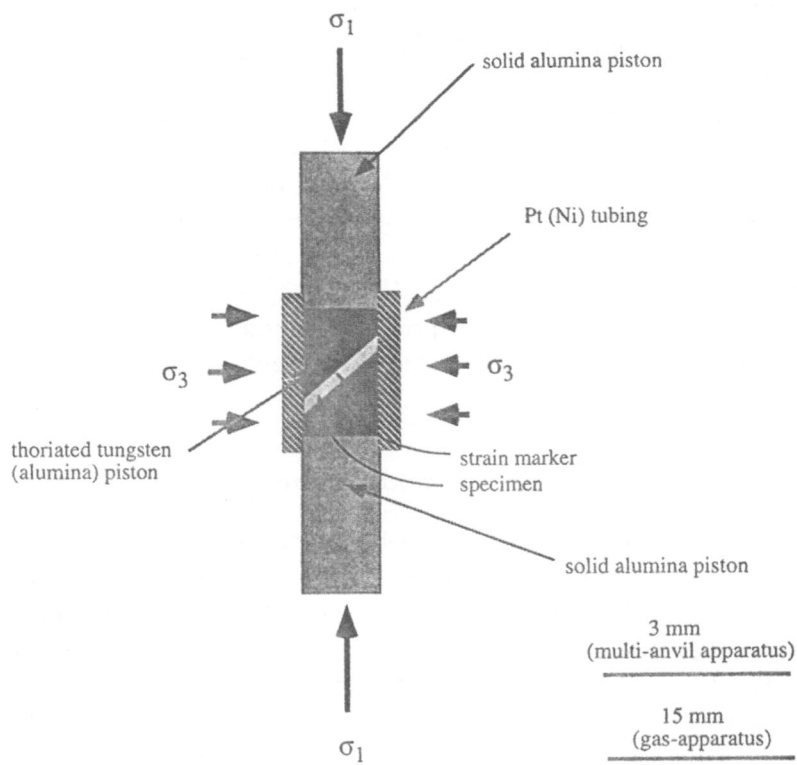

Figure 2

A schematic drawing of a specimen assembly for simple shear deformation at high pressures and temperatures. The whole assembly is surrounded by a pressure medium (either Ar gas, polycrystalline MgO or NaCl) and a furnace. The uni-axial motion of pistons is transformed to simple shear deformation of a thin slice of specimen. The relative lateral motion of pistons is accommodated by the deformation of a soft jacket material (Pt or Ni). Deformation geometry can be characterized by the rotation of a strain marker and the change in specimen thickness.

ments using a multi-anvil apparatus, the dominant mode of deformation turns out to be stress relaxation (KARATO and RUBIE, 1997). Samples are quenched rapidly ($>1°$/sec) at pressure in order to preserve the high strain microstructure. Sample sections are then examined at various magnifications, using both optical and electron microscopy. Digital image analysis has also been incorporated to quantify the microstructure of specimens.

Major Results

1. Olivine

Most of the previous experimental studies on LPO in olivine or olivine-rich rocks were carried out with uniaxial compression (AVÉ LALLEMANT and CARTER,

1970; NICOLAS et al., 1973; KARATO, 1987). Therefore, the relative importance of stress and finite strain on LPO could not be distinguished. In addition, the geometry of LPO from these experiments is different from most of those observed in naturally deformed peridotites which are presumably deformed in nearly simple shear (e.g., NICOLAS, 1989). Consequently these experimental studies cannot be directly applied to interpret the LPO of naturally deformed peridotites or seismic anisotropy caused by LPO.

Most geological or geophysical applications of LPO have used the information obtained by the microstructural analysis of naturally deformed peridotites (e.g., NICOLAS and CHRISTENSEN, 1987; NICOLAS, 1989; MAINPRICE and SILVER, 1993) or the results of computer modeling (e.g., ETCHECOPAR and VASSEUR, 1987; WENK et al., 1991). Although some consensus has been reached regarding the gross features of LPO and flow geometry such as the olivine [100] axis parallel to the flow direction, several key issues remain unclear. These include:

 (i) the relation between the sense of shear and LPO,
 (ii) the role of dynamic recrystallization in LPO,
(iii) the role of deformation mechanisms on LPO and
(iv) the role of water.

To resolve these issues, we have conducted simple shear deformation experiments of olivine aggregates at high pressures (mostly at 300 MPa) and temperatures (up to 1573 K). The starting materials are synthetic olivine aggregates with various mean grain-sizes ($\sim 7-35$ μm). Deformation experiments were conducted in both dislocation and diffusional creep regimes. The samples contain only a small amount of water (~ 100 ppm H/Si or less). The results are summarized in a series of papers (ZHANG and KARATO, 1995, 1997), and the main observations include the following:

 (i) Significant LPO develops when an olivine aggregate is deformed in the dislocation creep regime (Fig. 3).
 (ii) In the dislocation creep regime, LPO is stronger at lower stresses, close to the diffusional creep regime, than at higher stresses. The stronger LPO is also characteristic of single slip while the weaker LPO indicates activation of multiple slip systems.
(iii) In the dislocation creep regime, the LPO rotates with respect to the shear plane with progressive simple shear as opposed to the model by ETCHECOPAR and VASSEUR (1987) and NICOLAS (1989). However, the LPO rotates much faster than the model by WENK et al. (1991) predicts, and the pattern of LPO becomes indistinguishable from those predicted by ETCHECOPAR and VASSEUR (1987) and NICOLAS (1989) above $\sim 100\%$ shear strain.
(iv) Dynamic recrystallization leads to significant rheological weakening.
 (v) No appreciable LPO develops in the diffusional creep regime, although LPO is observed in specimens deformed in the transition region between the dislocation and diffusional creep regimes.

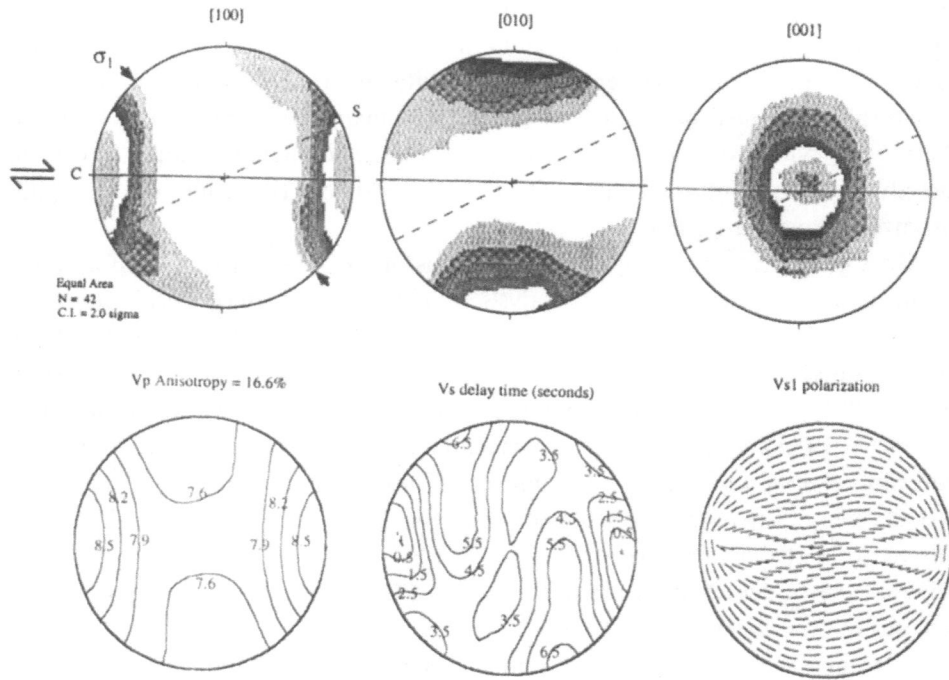

Figure 3

Lattice preferred orientation and calculated seismic anisotropy in an olivine aggregate sheared to 150% strain (strain rate $\sim 1 \times 10^{-4}\ s^{-1}$) at 300 MPa confining pressure at 1573 K under water poor conditions (from ZHANG and KARATO, 1995). Equal area stereographic projection is used. V_p: P-wave velocity, V_s: S-wave velocity. The delay times are the differences in travel times of two shear waves with different polarization through a 200-km thick layer. V_{s1}: the faster shear wave. C: shear plane, S: pole of maximum strain ellipsoid.

Prominent LPO of olivine, the dominant upper mantle mineral, will produce considerable seismic anisotropy due to the anisotropic elastic properties of olivine. Therefore these results can be applied to interpret seismic anisotropy in the upper mantle and to infer the flow geometry from LPO in naturally deformed peridotites. In addition to the well known relation of olivine [100] axis being parallel to flow direction (e.g., NICOLAS and CHRISTENSEN, 1987), our studies provide the first experimental constraints on the relation between LPO and the sense of shear; the orientation of the olivine [100] axis in simple shear deviates from the flow direction and is nearly parallel to the orientation of maximum elongation. However, this difference becomes negligibly small above $\sim 70–100\%$ strain where dynamic recrystallization becomes significant.

The pronounced LPO produced in the dislocation creep regime close to the boundary of the diffusional creep regime and the absence of anisotropy in the diffusional creep regime may explain the observed sharp transition from anisotropic to isotropic structure at around 200–250 km depths in the upper mantle (e.g.,

REVENAUGH and JORDAN, 1991; GAHERTY and JORDAN, 1995), if a change in deformation mechanism occurs at this depth (KARATO, 1992; KARATO and WU, 1993). However, our observation of significant LPO in diffusion creep near the boundary of dislocation creep raises a question as to how sharp this transition (from anisotropic to isotropic structure) can be. Further studies are needed to assess this point. Localized shear around this depth may be needed to explain the observed sharp boundary.

The strain-weakening associated with dynamic recrystallization provides a possible mechanism of shear localization. In fact, JIN et al. (1997) found evidence in a naturally deformed peridotite from Ivrea zone (northwestern Italy) that grain-size reduction promoted shear localization to finally cause shear melting.

2. Olivine Plus Basaltic Melt

The geometry of melt in partially molten rocks has important influence on melt transport and other physical properties. Observations of naturally deformed rocks from ophiolites provide some clues on melt geometry at a large scale ($\sim$cm to $\sim$km), presumably corresponding to the latest stage of melt segregation (e.g., NICOLAS, 1989). However, before melt is segregated in larger scale features, it is also likely to be transported by a porous flow at smaller scales (AHERN and TURCOTTE, 1979). Laboratory deformation of partially molten peridotite may provide important constraints on melt distribution at the grain scale and contribute to our understanding of melt migration under mid-ocean ridges.

Previous experiments (WAFF and BULAU, 1979; COOPER and KOHLSTEDT, 1982) have demonstrated the importance of surface tension in controlling the hydrostatic distribution of melt at the grain scale. At small melt fractions, melt is primarily distributed in triple junction tubules and along some low index boundaries (WAFF and FAUL, 1992) under hydrostatic conditions. However, the dynamic distribution of melt during deformation remains controversial. Compression experiments have shown that melt is redistributed along grain boundaries in a preferred orientation (BUSSOD and CHRISTIE, 1991; DAINES and KOHLSTEDT, 1997) or as a thin film along grain boundaries (JIN et al., 1994). The role of stress and strain have been considered in these experiments but are still not well understood. To help clarify these issues, we have started large strain simple shear deformation experiments of partially molten peridotite.

We have deformed dry (<30 ppm H/Si) fine-grained ($\sim$10–20 μm) olivine aggregates with 3% MORB and partially molten (3–5% melt) spinel Iherzolite samples at $T = 1473 - 1523$ K, $P = 300$ MPa and shear stresses at 15–100 MPa. Preliminary results have been reported in three abstracts (ZHANG et al., 1995; ZIMMERMAN, et al., 1995, 1996). We observed several features (Fig. 4):

(i) Olivine grains in samples deformed to greater than 100% strain in dislocation creep show a strong LPO consistent with the previous experiments on olivine aggregates without melt.

(ii) Olivine grains are elongated in the principal stretching direction although the shape preferred orientation (SPO) is less pronounced than in samples without melt.

(iii) The orientation of the long axis of melt pockets (MPO) shifts from the random distribution found in hydrostatic experiments to a conjugate set of MPOs about 20 degrees from the applied stress. The single MPO about 25 degrees from the shear boundary grows with increasing strain to form prominent interconnected melt channels.

(iv) When the differential stress is removed, MPO becomes subparallel to the shear boundary, preferentially wetting the (010) plane of olivine, consistent with the anisotropic wetting of olivine in hydrostatic experiments.

(v) 3-D reconstructions of serial optical micrograph images of deformed samples reveal the interconnected melt channels as planar features.

(vi) The dynamic distribution of melt is controlled by the orientation of the stress and not by strain or crystallography as indicated by the difference in MPO versus SPO and LPO in sheared samples. This notion is inferred from the observation that the MPO does not change with progressive deformation.

The results suggest that deformed partially molten peridotites have anisotropic physical properties including melt permeability, electrical conductivity and seismic wave velocities, and provide a basis to interpret geophysical observations and modeling of dynamic processes beneath mid-ocean ridges. The contribution from MPO to seismic anisotropy can be quite large but is highly dependent upon the melt fraction and, in particular, on the aspect ratio of melt pockets (e.g., SCHMEL-ING, 1985) and hence is difficult to assess. However, for a typical case of $\sim 1\%$ melt fraction and $\sim 1:10$ aspect ratio, for example, the magnitude of anisotropy due to MPO is comparable to that due to LPO and one can expect a few percent polarization anisotropy of shear waves. It is interesting to note that in a simple mid-ocean ridge spreading model with olivine [100] axis aligned parallel to the vertical flow direction, splitting of vertically traveling shear waves due to LPO will be small because polarization anisotropy of shear waves propagating along shear direction is small (Fig. 3; the delay time of shear waves (V_s) is nearly zero for waves propagating along the shear direction). Under these conditions, shear-wave splitting will primarily reflect anisotropy of melt geometry (MPO) and hence provide valuable information as to the geometry of melt flow beneath mid-ocean ridges.

3. Perovskite

Although seismic anisotropy in the deep mantle can potentially provide valuable information as to the pattern and dynamics of mantle convection (e.g., KARATO, this issue), deformation fabrics in deep mantle minerals are largely unknown. This is primarily because of the difficulties in conducting deformation experiments on deep mantle minerals under the conditions in which they are stable. Although a new

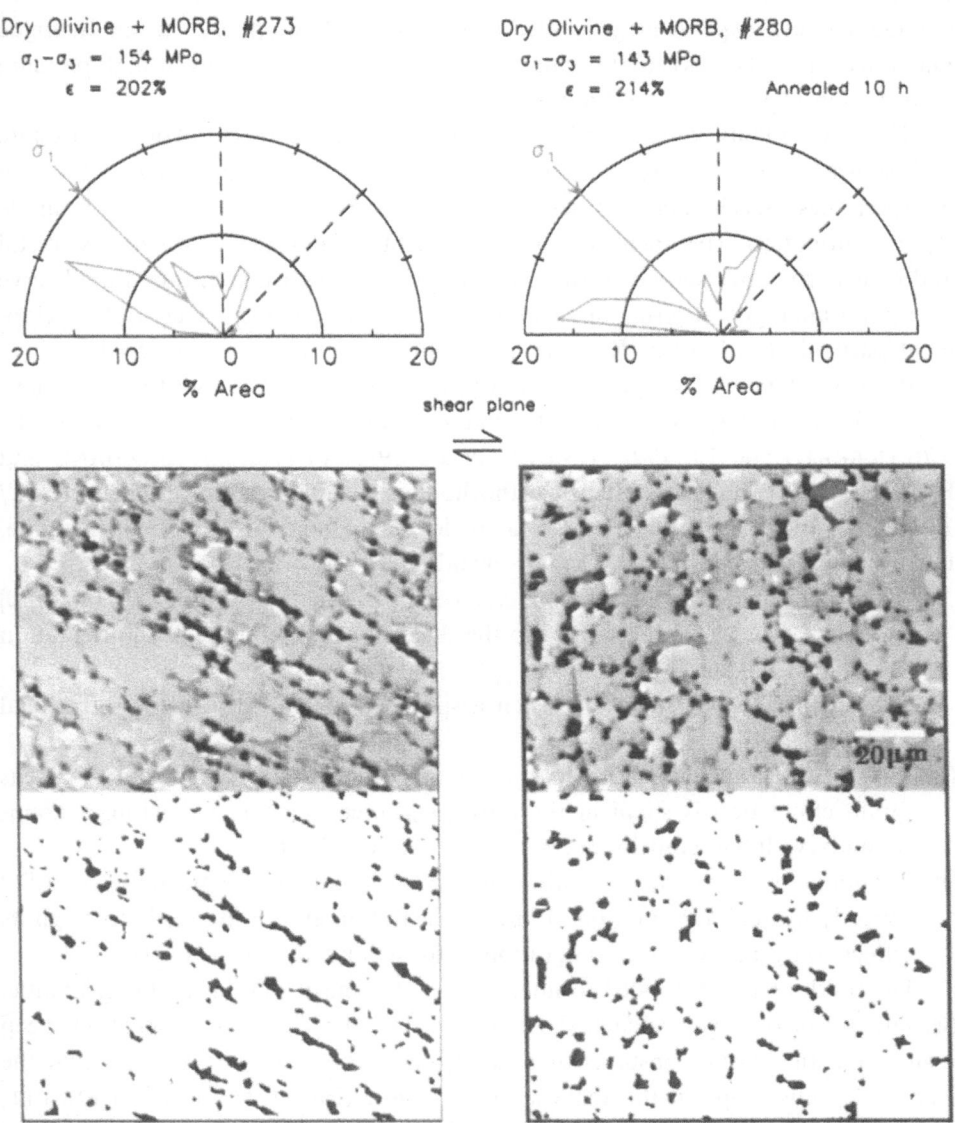

Figure 4
The geometry of melt in sheared and annealed partially molten peridotite (from KOHLSTEDT and
ZIMMERMAN, 1996). (Top) Polar diagrams showing the distribution of melt pockets. (Bottom) Binary
images of optical micrographs. #273: sheared at 154 MPa differential stress to 203% strain at P = 300
MPa and T = 1523 K, #280: sheared at 143 MPa differential stress to 213% strain at P = 300 MPa and
T = 1523 K, and then annealed for 10 hours at the same P, T condition without differential stress.
Arrows indicate shear direction. Melt is aligned at ∼20° with respect to the shear plane and to the
orientation of the maximum compressive stress under differential stress, but after annealing melt
distribution markedly changes and melt pockets are oriented preferentially along the (010) plane of
olivine that is subparallel to the shear plane (see also Fig. 3).

technical development (KARATO and RUBIE, 1997) has opened up a possibility of conducting such experiments, detailed studies of the basic physics using analogue materials will remain essential because the experimental conditions of high pressure deformation experiments are limited.

The lower mantle of the earth is mainly composed of $(Mg, Fe)SiO_3$ perovskite and shows unique features of anisotropy (e.g., KARATO, this issue). To help interpret these seismological observations, we have conducted a series of simple shear deformation experiments on the analogue material, $CaTiO_3$, which has crystal and defect structures similar to those of $(Mg, Fe)SiO_3$ perovskite, thus would serve as a good analogue in terms of rheological properties or microstructural development particularly LPO (see KARATO *et al.*, 1995, for further discussions).

Polycrystalline $CaTiO_3$ perovskites with mean grain-sizes of ~ 7 to $70\ \mu m$ have been deformed to large strains ($< 300\%$) at T/T_m (T_m:melting temperature) $= 0.64$–0.76 (KARATO and LI, 1992; KARATO *et al.*, 1995; LI *et al.*, 1996; ZHANG and KARATO, 1997). $CaTiO_3$ assumes orthorhombic symmetry below ~ 1515 K ($T/T_m = 0.69$) and may serve as a good analogue material of $MgSiO_3$ perovskite. Experimental observations of $CaTiO_3$ include the following:

 (i) Deformation in the dislocation creep regime causes strong LPO, the [100] orientation becomes subparallel to the shear direction and the [010] orientation normal to the shear plane.

 (ii) No significant LPO develops when a specimen is deformed in the diffusional creep regime.

 (iii) Significant grain elongation occurs in the dislocation creep regime, whereas grain elongation is small in the diffusional creep regime, indicating a grain-boundary sliding contribution (RAJ and ASHBY, 1971).

 (iv) Twinning is extensive in both regimes. Twinning may contribute to LPO directly through the rotation of lattice orientation and/or indirectly through its effects on grain-boundary migration (ZHANG and KARATO, 1997).

The results may be applied to interpret seismic anisotropy in the lower mantle. Seismic anisotropy was calculated from the LPO of $CaTiO_3$ in dislocation creep regime and the elastic constants of $MgSiO_3$ perovskite. The results show that the $V_{SV} > V_{SH}$ anisotropy in the shallow lower mantle (MONTAGNER and KENNETT, 1996) indicates a horizontal flow there, implying that mantle convection is (at least partially) layered (KARATO, this issue). The absence of anisotropy in other portions of the lower mantle implies deformation by diffusional creep or superplasticity (KARATO *et al.*, 1995). This latter point suggests that subducted oceanic lithosphere in the lower mantle is likely to be weak because of grain-size reduction due to the transformation to perovskite magnesiowüstite (e.g., ITO and SATO, 1991; KARATO and LI, 1992; LI *et al.*, 1996). In fact, the recent high resolution seismic tomography shows significant thickening of slabs in the lower mantle (VAN DER HILST *et al.*, 1997), a result which is consistent with softening of slabs in the lower mantle.

Future Directions

Our results so far have shown that a rich variety of information pertaining to microstructural development and mechanical behavior at high strains can be obtained by simple shear deformation experiments. We suggest the following new directions to further improve our understanding of the dynamics of the earth's mantle.

(i) Extension to a wider range of thermodynamic conditions:

a. High pressures (LPO in high pressure minerals). Although knowledge of the development of LPO in high pressure minerals is critical for the interpretation of seismic anisotropy in terms of mantle convection (KARATO, this issue), no direct experimental studies have been performed on the development of LPO in high pressure minerals such as silicate spinel or perovskite (except a study at room temperature; MEADE *et al.*, 1995). The new technique of KARATO and RUBIE (1997) has made it possible to investigate deformation microstructures such as LPO in deep mantle minerals directly through laboratory experiments. The first successful application of this technique to the beta (modified spinel) phase of $(Mg, Fe)_2SiO_4$ has recently been made by one of the authors (KARATO, unpublished data). Extension of this technique to the gamma (spinel) phase and $(Mg, Fe)SiO_3$ perovskite will provide key data to interpret seismic anisotropy in terms of the geometry of mantle convection (KARATO, this issue).

b. High water fugacity. Water is known to enhance dislocation creep and dislocation mobility, however the effects appear highly anisotropic (MACKWELL *et al.*, 1985; YAN, 1992). In addition, water enhances dynamic recrystallization (CHOPRA and PATERSON, 1984) and grain-boundary migration (KARATO, 1989b). Therefore addition of a significant amount of water is expected to affect the nature of LPO through the possible change in slip systems and/or through the enhancement of dynamic recrystallization (KARATO, 1995). Since the solubility of water in olivine and other silicates increases strongly with pressure under water-saturated conditions (e.g., KOHLSTEDT *et al.*, 1996), LPO under high water fugacities might be very different from those at low water fugacities thus far investigated. These effects would have important influence on the interpretation of seismic anisotropy in the upper mantle where water fugacity is expected to be high such as the upper mantle in back-arc regions. Shear experiments at high water fugacity will be important to test this hypothesis.

(ii) Mechancial behavior, shear localization and plastic anisotropy:

Our efforts so far have been focused on microstructural development. Another important piece of information is the mechanical behavior including shear localization and plastic anisotropy. We have already observed a variety of mechanical behaviors, including strain softening and strain hardening, depending on deformation conditions and/or deformation geometry (ZHANG and KARATO, 1997; ZHANG *et al.*, 1995, 1997; ZIMMERMAN *et al.*, 1996). Further characterization of conditions for strain softening is critical to the better understanding of the mechanisms of shear localization (DRURY *et al.*, 1991; JIN *et al.*, 1997). It is also important to characterize the full viscosity tensor η_{ijkl} by combining the results of simple shear deformation experiments with those of uni- (or tri-) axial deformation experiments to investigate plastic anisotropy of textured rocks.

(iii) Better technique of microstructural analyses:

a. Three-dimensional microstructural analysis in two-phase materials (e.g., partial melt). Traditional microstructural analysis relies on the observation of two-dimensional sections. Some aspects of microstructures are better analyzed by three-dimensional imaging, including the three dimensional network of melt channels in partially molten materials. Characterization of melt morphology through serial sectioning is underway in our laboratory combined with computerized image analysis technique.

b. Electron backscattering pattern analysis (EBSP) for lattice preferred orientation measurements (e.g., DINGLEY and RANDLE, 1992). Lattice preferred orientation has traditionally been measured using mostly an optical microscope with a universal stage or by X-ray pole figure goniometry (e.g., WENK, 1985). A new technique using scanning electron microscope (EBSP) has recently been developed (e.g., DINGLEY and RANDLE, 1992). This technique allows us to determine the crystallographic orientation of individual grains with a spatial resolution of better than ~ 1 μm, and is ideal for LPO measurements in fine-grained specimens that are typical of high pressure experiments. This technique also provides orientation relationships of individual grains that is critical to the understanding of mechanisms of LPO and of dynamic recrystallization.

The combination of these new analytical techniques with simple shear deformation techniques at high pressures and temperatures will provide a rich variety of new experimental data to investigate "structural geology of the mantle."

Finally, a word of caution is in order. These experimental studies are inevitably on small samples (mm size). When applying the results to the mantle, questions of scale must always be evaluated. When LPO is concerned, the question of scales is relatively unimportant to the extent that LPO is homogeneous in the earth. However, such questions as melt geometry or shear localization are highly dependent upon scales, and the applications of laboratory data to geological problems must be made with caution. For example, in our experimental set-up, shear localization or melt channels with a characteristic scale less than ~ 1 mm can be detected, but all other possible large-scale features cannot be detected. Therefore, the conditions for shear localization that can be identified from these types of studies will be the sufficient conditions for localization but are not the necessary conditions. Similarly, melt transport at larger scales cannot be directly investigated from these studies. Likewise, seismic anisotropy caused by laminated structures (ALLÈGRE and TURCOTTE, 1986) cannot be evaluated from these laboratory experiments. With these caveats in mind, laboratory studies of "structural geology of the mantle" will open a new field of interdisciplinary research in which mineral physics research is integrated with structural geology, seismology and geodynamics to provide a new picture of how this planet works.

Acknowledgments

The research summarized in this paper has been supported by grants from NSF (EAR–9220172, 9306871, 9505451, 9526239, OCE–9529744) and from the Alexander von Humboldt Stiftung. The development of simple shear deformation technique with a multi-anvil apparatus was made at Bayerisches Geoinstitut (BGI) collaboratively with Dave Rubie. The comments by A. Nicolas and an anonymous reviewer helped clarify some of the presentations.

REFERENCES

(References are mostly from our laboratory and are not intended to be exhaustive).

AHERN, J. L., and TURCOTTE, D. L. (1979), *Magma Migration beneath an Ocean Ridge*, Earth Planet. Sci. Lett. *45*, 115–122.

ALLÈGRE, C. J., and TURCOTTE, D. L. (1986), *Implications of a Two-component Marble-cake Mantle*, Nature *323*, 123–127.

AVÉ LALLEMANT, H. G., and CARTER, N. L. (1970), *Syntectonic Recrystallization of Olivine and Modes and Flow in the Upper Mantle*, Geol. Soc. Am. Bull. *81*, 2203–2220.

BUSSOD, G. Y., and CHRISTIE, J. M. (1991), *Textural Development and Melt Topology in Spinel Iherzolite Experimentally Deformed at Hypersolidus Conditions*, J. Petrol., Spec. Issue, 17–39.

CHOPRA, P. N., and PATERSON, M. S. (1984), *The Role of Water in the Deformation of Dunite*, J. Geophys. Res. *89*, 7861–7876.

COOPER, R. F., and KOHLSTEDT, D. L., *Interfacial energies in the olivine-basalt system. In High Pressure Research in Geophysics* (eds. Akimoto, S., and Manghnani, M. H.) (Center for Acadeic Publications, Tokyo 1982) pp. 217–228.

DAINES, M. J., and KOHLSTEDT, D. L. (1997), *Influence of Deformation on Melt Topology in Peridotites*, J. Geophys. Res. *102*, 10257–10271.

DINGLEY, D. L., and RANDLE, V. (1992), *Microtexture Determination by Electron Back-scatter Diffraction*, J. Mater. Sci. *27*, 4545–4566.

DRURY, M. R., VISSERS, R. L. M., VAN DER WAL, D., and HOOGERDUIJN STRATING, E. H. (1991), *Shear Localization in Upper Mantle Peridotites*, Pure appl. geophys *137*, 439–460.

ETCHECOPAR, A., and VASSEUR, G. (1987), *A 3-D Kinematic Model of Fabric Development in Polycrystalline Aggregates: Comparisons with Experimental and Natural Examples*, J. Struct. Geol. *9*, 705–717.

FORSYTH, D. W., *Geophysical constraints on mantle flow and melt generation beneath mid-ocean ridges.* In *Mantle Flow and Melt Generation and Mid-Ocean Ridges* (eds. Phipps Morgan, J., Blackman, D. K., and Sinton, J. M.) (Am. Geophys. Union, Washington DC 1992) pp. 1–65.

GAHERTY, J. B., and JORDAN, T. H. (1995), *Lehmann Discontinuity as the Base of an Anisotropic Layer beneath Continents*, Science *268*, 1468–1471.

HOBBS, B. E., MEANS, W. D., and WILLIAMS, P. F., *An Outline of Structural Geology* (*Wiley International, New York* 1976) pp. 571.

ITO, E., and SATO, H. (1991), *Aseismicity in the Lower Mantle by Superplasticity of the Descending Slab*, Nature *351*, 140–141.

Jin, Z. M., Green, H. W., and Zhou, Y. (1994), *Melt Topology during Dynamic Partial Melting of Mantle Peridotite*, Nature *372*, 164–167.

Jin, D., Karato, S., and Obata, M. (1997), *Mechanisms of Shear Localization in the Continental Lithosphere: Inference from the Deformation Microstructures of Peridotite from the Ivrea Zone, Northern Italy*, J. Struct. Geol. (in press).

Karato, S., *Seismic anisotropy due to lattice preferred orientation of minerals: kinematic or dynamic?* In *High Pressure Research in Mineral Physics* (eds. Akimoto, S., and Manghnani, M. H.) (Am. Geophys. Union, Washington DC 1987) pp. 455–471.

Karato, S., *Seismic anisotropy: Mechanisms and tectonic implications.* In *Rheology of Solids and of the Earth* (eds. Karato, S., and Toriumi, M.) (Oxford University Press, Oxford 1989a) pp. 393–422.

Karato, S. (1989b), *Grain Growth Kinetics in Olivine Aggregates*, Tectonophysics *168*, 255–273.

Karato, S. (1992), *On the Lehmann Discontinuity*, Geophys. Res. Lett. *19*, 2225–2228.

Karato, S. (1995), *Effects of Water on the Velocities of Seismic Waves in the Earth's Upper Mantle*, Proc. Japan Acad. *70B*, 61–66.

Karato, S. (1997), *Seismic Anisotropy in the Deep Mantle, Boundary Layers and the Geometry of Mantle Convection*, Pure appl. geophys. *151*, 565–587.

Karato, S., and Li, P. (1992), *Diffusion Creep in Perovskite: Implications for the Rheology of the Lower Mantle*, Science *255*, 1238–1240.

Karato, S., and Rubie, D. C. (1997), *Towards an Experimental Study of Deep Mantle Rheology: A New Multi-Anvil Sample Assembly for Deformation Experiments under High Pressures and Temperatures*, J. Geophys. Res. (in press).

Karato, S., and Wu, P. (1993), *Rheology of the Upper Mantle: A Synthesis*, Science *260*, 771–778.

Karato, S., Zhang, S., and Wenk, H.-R. (1995), *Superplasticity in Earth's Lower Mantle: Evidence from Seismic Anisotropy and Rock Physics*, Science *270*, 458–461.

Kohlstedt, D. L. *Structure, rheology and permeability of partially molten rocks at low melt fraction.* In *Mantle Flow and Melt Generation at Mid-Ocean Ridges* (eds. Phipps Morgan, J., Blackman, D. K., and Sinton, J. M.) (Am. Geophys. Union, Washington DC 1992) pp. 103–121.

Kohlstedt, D. L., and Zimmerman, M. E. (1996), *Rheology of Partially Molten Mantle Rocks*, Ann. Rev. Earth Planet. Sci. *24*, 41–62.

Kohlstedt, D. L., Keppler, H., and Rubie, D. C. (1996), *Solubility of Water in the α, β and γ Phases of $(Mg, Fe)_2SiO_4$*, Contrib. Mineral. Petrol. *123*, 345–357.

Li, P., Karato, S., and Wang, Z. (1996), *High-temperature Creep in Fine-grained Polycrystalline $CaTiO_3$ an Analogue Material of $(Mg, Fe)SiO_3$ Perovskite*, Phys. Earth. Planet. Inter. *95*, 19–36.

Mackwell, S. J., Kohlstedt, D. T., and Paterson, M. S. (1985), *The Role of Water in the Deformation of Olivine Single Crystals*, J. Geophys. Res. *90*, 11319–11333.

Mainprice, D., and Silver, P. G. (1993), *Interpretation of SKS Waves Using Samples from the Subcontinental Lithosphere*, Phys. Earth Planet. Inter. *78*, 257–280.

Meade, C., Silver, P. G., and Kaneshima, S. (1995), *Laboratory and Seismological Observations of Lower Mantle Isotropy*, Geophys. Res. Lett. *22*, 1293–1296.

Montagner, J.-P., and Kennett, B. L. N. (1996), *How to Reconcile Body-wave and Normal-mode Reference Earth Model*, Geophys. J. Int. *125*, 229–248.

Nataf, H.-C., Nakanishi, I., and Anderson, D. L. (1986), *Measurement of Mantle Wave Velocities and Inversion for Lateral Heterogeneities and Anisotropy, 3. Inversion*, J. Geophys. Res. *91*, 7261–7307.

Nicolas, A., *Structure of Ophiolites and Dynamics of Oceanic Lithosphere* (Kluwer, New York 1989) 367 pp.

Nicolas, A., and Christensen, N. I., *Formation of anisotropy in upper mantle peridotites—a review.* In *Composition, Structure and Dynamics of the Lithosphere-Asthenosphere System* (eds. Fuchs, K., and Froidevaux, C.) (Am. Geophys. Union 1987) pp. 111–123.

Nicolas, A., Boudier, F., and Boullier, A. M. (1973), *Mechanisms of Flow in Naturally and Experimentally Deformed Peridotites*, Am. J. Sci. *273*, 853–876.

Phipps Morgan, J. (1987), *Melt Migration Beneath Mid-ocean Spreading Centers*, Geophys. Res. Lett. *14*, 1238–1241.

RAJ, R., and ASHBY, M. F. (1971), *On Grain Boundary Sliding and Diffusional Creep*, Metall. Trans. 2, 1113–1127.

REVENAUGH, J. S., and JORDAN, T. H. (1991), *Mantle Layering from ScS Reverberations, 3. The Upper Mantle*, J. Geophys. Res. 96, 19781–19810.

SCHMELING, H. (1985), *Numerical Models on the Influence of Partial Melt on Elastic, Anelastic and Electric Properties of Rocks. Part I: Elasticity and Anelasticity*, Phys. Earth Planet. Inter. 41, 34–57.

VAN DER HILST, R. D., WIDIYANTORO, S., and ENGDAHL, E. R. (1997), *Evidence for Deep Mantle Circulation from Global Tomography*, Nature 386, 578–584.

WAFF, H. S., and BULAU, J. R. (1979), *Equilibrium Fluid Distribution in an Ultramafic Partial Melt under Hydrostatic Stress Conditions*, J. Geophys. Res. 84, 6109–6114.

WAFF, H. S., and FAUL, U. H. (1992), *Effects of Crystalline Anisotropy on Fluid Distribution in Ultramafic Partial Melts*, J. Geophys. Res. 97, 9003–9014.

WENK, H.-R., *Measurement of pole figures. In Preferred Orientation in Deformed Metals and Rocks: An Introduction to Modern Texture Analysis* (ed. Wenk, H.-R.) (Academic Press, Orlando, Florida 1985) pp. 11–47.

WENK, H.-R., BENNETT, K., CANOVA, G. R., and MOLINARI, A. (1991), *Modelling Plastic Deformation of Peridotite with the Self-consistent Theory*, J. Geophys. Res. 96, 8337–8349.

YAN, H. (1992), *Dislocation Recovery in Olivine*, MSc. Thesis, University of Minnesota, 98 pp.

ZHANG, S., and KARATO, S. (1995), *Lattice Preferred Orientation of Olivine Aggregates Deformed in Simple Shear*, Nature 375, 774–777.

ZHANG, S., and KARATO, S. (1997), *Simple Shear Deformation of Polycrystalline CaTiO₃ Perovskite*, Phys. Earth Planet. Inter. (submitted).

ZHANG, S., KARATO, S., and ZHOU, Y. (1997), *Simple Shear Deformation of Olivine Aggregates*, Tectonophysics (submitted).

ZHANG, S., ZIMMERMAN, M. E., DAINES, M. J., KARATO, S., and KOHLSTEDT, D. L. (1995), *Lattice Preferred Orientation and Melt Distribution in Experimentally Sheared Olivine-basalt Rocks*, EOS, Trans. Amer. Geophys. Union 76, 281.

ZIMMERMAN, M. E., KOHLSTEDT, D. L., and KARATO, S. (1995), *Shear Localization in Deformed Olivine-basalt Aggregates*, EOS, Trans. Am. Geophys. Union 76, F559–F560.

ZIMMERMAN, M. E., KOHLSTEDT, D. L., and KARATO, S. (1996), *Melt Channels in Peridotites Deformed in Shear*, Abstract, An International Conference on Structure and Properties of High Strain Zones in Rocks.

(Received December 12, 1996, revised May 5, 1997, accepted May 27, 1997)

 To access this journal online:
http://www.birkhauser.ch

Pure appl. geophys. 151 (1998) 605–618
0033–4553/98/040605–14 $ 1.50 + 0.20/0

❘ **Pure and Applied Geophysics**

Free Energy Minimization by Simulated Annealing with Applications to Lithospheric Slabs and Mantle Plumes

CRAIG R. BINA[1]

Abstract—An optimization algorithm based upon the method of simulated annealing is of utility in calculating equilibrium phase assemblages as functions of pressure, temperature, and chemical composition. Operating by analogy to the statistical mechanics of the chemical system, it is applicable both to problems of strict chemical equilibrium and to problems involving metastability. The method reproduces known phase diagrams and illustrates the expected thermal deflection of phase transitions in thermal models of subducting lithospheric slabs and buoyant mantle plumes. It reveals temperature-induced changes in phase transition sharpness and the stability of Fe-rich γ phase within an $\alpha + \gamma$ field in cold slab thermal models, and it suggests that transitions such as the possible breakdown of silicate perovskite to mixed oxides can amplify velocity anomalies.

Key words: Phase relations, simulated annealing, subduction zones, mantle plumes.

Introduction

Determining stable mineral phase assemblages as functions of pressure, temperature, and composition is an important part of modeling the structure and composition of earth's interior. For systems with few chemical components, such problems are typically approached (e.g., HELGESON *et al.*, 1970; BINA and WOOD, 1987; POWELL and HOLLAND, 1988) by numerically solving sets of equilibrium equations, $\mu_i^\phi = \mu_i^\psi$, for all phases ϕ and ψ for each component i, where the chemical potential of component i in phase ϕ (μ_i^ϕ) is a function of pressure (P), temperature (T), and composition (as expressed by the mole fractions X_i^ϕ). For more complex systems with larger numbers of chemical components, such problems are commonly approached (e.g., VAN ZEGGEREN and STOREY, 1970; SMITH and MISSEN, 1982) by minimizing the Gibbs free energy $G = \Sigma_{i,\phi} X_i^\phi \mu_i^\phi$, so as to find the equilibrium X_i^ϕ as functions of P and T. This latter approach amounts to the minimization of a highly nonlinear objective function (G) of many variables (P, T and the X_i^ϕ) subject to several linear equality constraints (mass balance conditions) and linear inequality constraints (mole fractions between zero and one).

Many methods have been employed to solve such free energy minimization problems, including linear (e.g., "steepest-descent") algorithms (STOREY and VAN

[1] Department of Geological Sciences, Northwestern University, 1847 Sheridan Road, Evanston, IL 60208-2150, U.S.A. E-mail: craig@earth.nwu.edu

ZEGGEREN, 1964; ERIKSSON, 1975; SAXENA, 1982; WOOD and HOLLOWAY, 1984; BINA and WOOD, 1987; WOOD, 1987; SAXENA, 1996), quadratic (e.g., "quasi-Newton," "conjugate-gradient") algorithms (SUNDMAN et al., 1985; GHIORSO, 1985; DE CAPITANI and BROWN, 1987; HARVIE et al., 1987; GHIORSO and SACK, 1995), or hybrid linear-quadratic methods (BINA, 1987). Such techniques, which explore the free energy hypersurface by accumulating information on local slope and curvature, work well for the study of gradual phase transitions, in which equilibrium phase proportions and compositions vary slowly and smoothly. However, they often encounter difficulties when applied to more abrupt phase transitions in non-ideal multicomponent systems, in which elemental partitioning may shift dramatically or individual phases may appear or disappear over small ranges of pressure, temperature, or composition. Such problems can be partially alleviated by the use of active set methods (GILL et al., 1981), but this does not change the fact that the properties of the free energy hypersurface can become unstable for small values of X_i^ϕ (given the logarithmic dependence of μ_i^ϕ upon X_i^ϕ).

In response to such issues, as well as to the more general problem of avoiding entrapment in local minima (ERIKSSON and HACK, 1990), stochastic optimization methods such as Simulated Annealing (KIRKPATRICK et al., 1983; KOREN et al., 1991), Genetic Algorithm (GOLDBERG, 1989; SAMBRIDGE and DRIJKONINGEN, 1992), and Taboo Search (GLOVER, 1989, 1990; CVIJOVIĆ and KLINOWSKI, 1995) have been applied to certain classes of optimization problems. Here I present a free energy minimization technique based upon the method of simulated annealing (KIRKPATRICK et al., 1983). Because it is a stochastic method which does not rely upon estimation of local hypersurface slope and curvature, it works well for both gradual and abrupt phase transitions, and avoids entrapment in local minima. Because it operates by analogy with the statistical mechanics of the chemical system, it is also ideal for the study of metastability. I illustrate the method using simple compositional systems subject to various thermal structures which represent subduction zones and mantle plumes. Upon solving for the stable phase assemblages by free energy minimization, I also calculate corresponding geophysical properties, such as density, buoyancy, seismic wave speed structures, and potential elastic anisotropy contrasts.

Method

KIRKPATRICK et al. (1983) identify four key elements of a simulated annealing algorithm: a description of system configuration, a random generator of rearrangements, a quantitative objective function, and an annealing schedule. The system configuration here is described in terms discrete units (thus allowing use of efficient integer arithmetic) of chemical components, in this case corresponding to moles of constituent oxides (e.g., MgO). These components "occupy" various phases, according to the stoichiometry appropriate to their structural formulae (e.g., n moles of Mg_2SiO_4 forsterite will consist of n units of SiO_2 and $2n$ units of MgO).

Random rearrangements of the system configuration are generated by a procedure analogous to reaction mechanisms (Fig. 1). The components occupying a random number of phases are transferred to an (amorphous) "excited state." All components are then allowed to "precipitate" from the excited state into randomly selected phases. (Note that, in order to ensure full evacuation of the excited state for each rearrangement, the constituent oxides must be part of the set of allowable phases.)

For each such rearrangement, the objective function G is evaluated, and the Metropolis algorithm (METROPOLIS *et al.*, 1953) is employed to determine whether or not the rearrangement should be adopted as the new system configuration. The Metropolis algorithm mimics a Boltzmann distribution function, always accepting the rearrangement if it results in an energy decrease ($\Delta G < 0$), otherwise ($\Delta G \geq 0$) randomly accepting the rearrangement with probability $\exp(-\Delta G/\theta)$, where θ is the annealing "temperature" characterizing this stage of the minimization. This procedure allows rearrangements resulting in an energy increase to occur with high probability early in a minimization but with falling probability as θ decreases later in the minimization, thus permitting the algorithm to escape from local minima. A random number of rearrangements are tested at each θ.

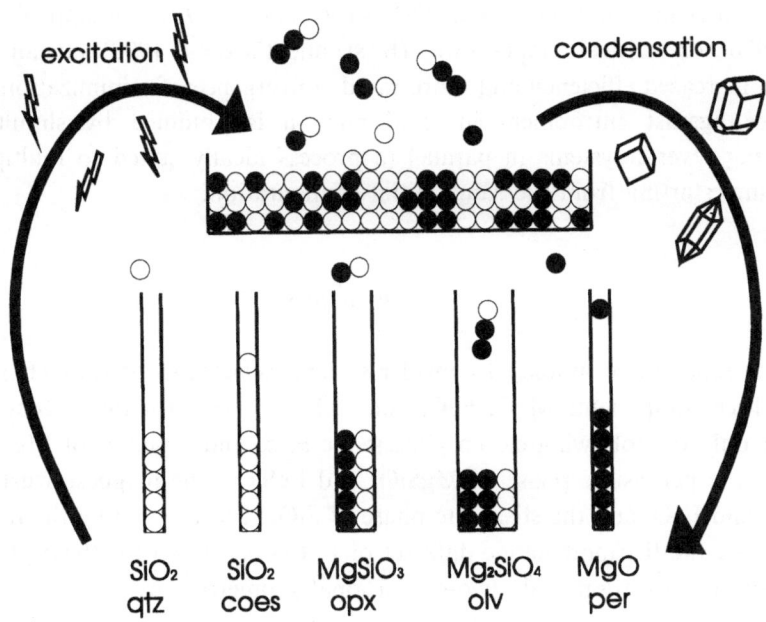

Figure 1
Cartoon illustrating method by which random rearrangements of system configuration are generated. In a procedure analogous to reaction mechanisms, components occupying random phases are transferred to an "excited state" and then allowed to "precipitate" into random phases.

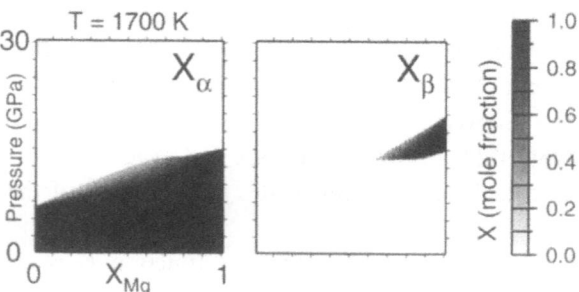

Figure 2

Test phase diagram for olivine polymorphs. Temperature (T) is fixed at 1700 K. Bulk composition (X_{Mg}) varies from 0 (Fe_2SiO_4) to 1 (Mg_2SiO_4). Pressure varies from 0 to 30 GPa. Equilibrium phase distributions show $\alpha \rightarrow \alpha + \gamma \rightarrow \gamma$ and (at high X_{Mg}) $\alpha \rightarrow \alpha + \beta \rightarrow \beta$ transitions.

The annealing temperature θ is gradually lowered according to an annealing schedule. Convergence is assumed when many random rearrangements yield no successful reductions in free energy at a given θ or at several successive θs. In this case, I employ the simple empirical expedient of reducing θ by a constant fraction (1/2) at each iteration. While this does result in convergence, adoption of more sophisticated schedules which evolve dynamically in response to the nature of energy changes in the problem, such as those of "constant thermodynamic speed" (MOSEGAARD and VESTERGAARD, 1991; KOREN et al., 1991), "adaptive" (INGBER, 1989, 1996), or "critical temperature" (BASU and FRAZER, 1990) annealing, would result in increased efficiency and more rapid convergence of minimization. Further insurance against entrapment in local minima is obtained by simultaneously minimizing several systems in parallel (a process ideally suited to multiprocessor computing) starting from the same initial configuration.

Some Examples

In the following examples, I consider systems which can be fully characterized by the three components MgO, FeO, and SiO_2. As hosts for these components, I consider only the following eleven phases: the α, β, and γ phases of Mg_2SiO_4 and Fe_2SiO_4, the perovskite phase of $MgSiO_3$ and $FeSiO_3$, the magnesiowüstite phase of MgO and FeO, and the stishovite phase of SiO_2. For calculating the free energy G, I employ the thermodynamic data set of FEI et al. (1991) for these phases. For simplicity of illustration, only two-dimensional geometries are shown.

I first test the algorithm by computing a simple phase diagram for the olivine polymorphs. Fixing temperature at 1700 K, a P-X section is constructed by allowing bulk composition to vary from Fe_2SiO_4 to Mg_2SiO_4 along the x-axis and pressure to increase from 0 to 30 GPa along the y-axis. The resulting equilibrium distributions of the phases α and β (Fig. 2), for example, clearly depict the $\alpha \rightarrow \alpha + \gamma \rightarrow \gamma$ and $\alpha \rightarrow \alpha + \beta \rightarrow \beta$ transitions at high pressures (KATSURA and ITO, 1989). Henceforth, I illustrate cases in which the bulk composition is fixed globally at $(Mg_{0.9}Fe_{0.1})_2SiO_4$ (i.e., a pure forsterite-90 olivine mantle model). Further complexity can be introduced by including such components as CaO and Al_2O_3 in pyroxene and garnet phases and by allowing bulk composition to vary with spatial position, thus allowing, for example, investigation of compositional layering in the lithosphere (HELFFRICH et al., 1989).

I next investigate a subduction zone thermal model, computed on a 120×90 grid using N. H. Sleep's finite difference algorithm (TOKSÖZ et al., 1973). The model (Fig. 3a) is constructed in a 890×686.63 km box for 140 Ma lithosphere with an initial GDH1 (STEIN and STEIN, 1992) thermal structure, subducting at a dip angle of 60 degrees with a velocity of 8 cm/yr (KIRBY et al., 1996; LEFFLER, pers. comm.), Pressures are determined by vertical integration of a reference density profile (DZIEWONSKI et al., 1975). Upon computing the equilibrium phase distribu-

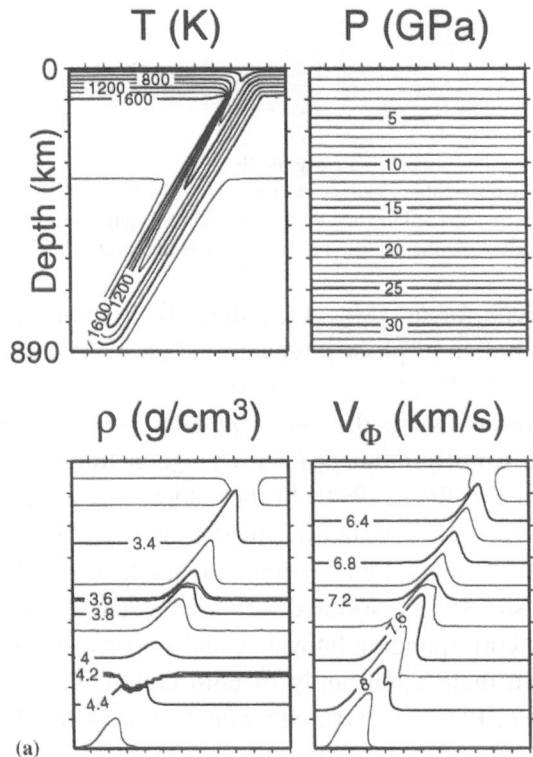

(a)

Figure 3a

Top: Temperature (T) and pressure (P) for subduction zone thermal model from finite difference algorithm (KIRBY et al., 1996; LEFFLER, pers. comm.). Bottom: Density (ρ) and bulk sound velocity (V_ϕ) structures attending consequent equilibrium phase relations in $(Mg_{0.9}Fe_{0.1})_2SiO_4$ composition.

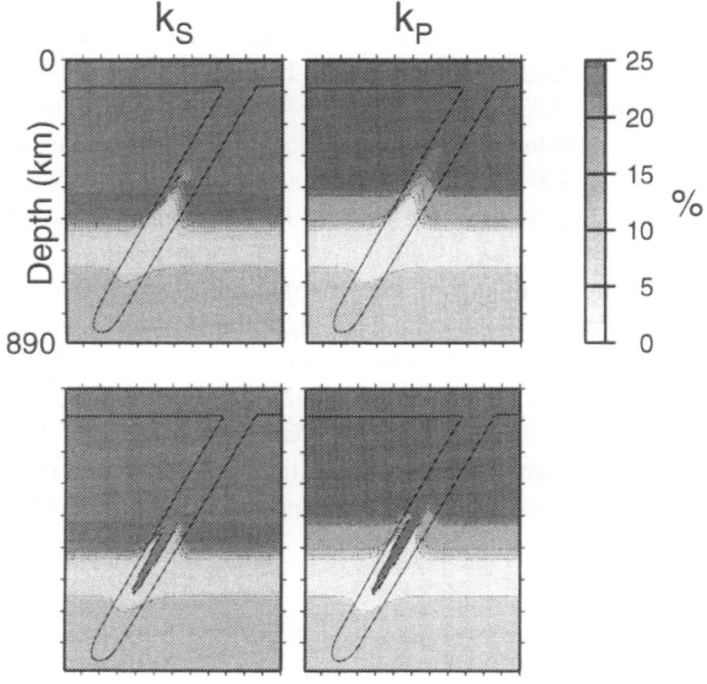

Figure 3b
Top: Maximum potentially observable elastic anisotropy, at room P and T, in V_S (k_S) and V_P (k_P) for equilibrium phase relations in subduction zone thermal model of Figure 3a. Bottom: Same quantities for case in which α-olivine persists metastably at temperatures below 1000 K.

tions and compositions for an $(Mg_{0.9}Fe_{0.1})_2SiO_4$ bulk composition (Plate 1), the expected thermal deflection of phase boundaries (TURCOTTE and SCHUBERT, 1971, 1972; SCHUBERT et al., 1975) is evident—the $\alpha \rightarrow \alpha + \beta \rightarrow \beta$ and $\beta \rightarrow \beta + \gamma \rightarrow \gamma$ transitions deflect upwards while the $\gamma \rightarrow \gamma + pv + mw \rightarrow pv + mw$ transition deflects downwards—as is a temperature-dependent change in transition sharpness within the slab (HELFFRICH and BINA, 1994). In the coldest core of the slab, the uplifted $\alpha + \beta$ region gives way to a shallower region of $\alpha + \gamma$ stability (GREEN and HOUSTON, 1995). Furthermore, Fe-rich γ phase is stable within this $\alpha + \gamma$ zone. The density and velocity structures associated with these equilibrium phase assemblages (Fig. 3a) give rise to corresponding buoyancy and velocity anomalies (Plate 1), the former of which contribute significantly to both convective dynamics (RICHTER, 1973; SCHUBERT et al., 1975; CHRISTENSEN and YUEN, 1984, 1985; BINA and LIU, 1995) and seismicity-related stresses (ISACKS and MOLNAR, 1971; TURCOTTE and SCHUBERT, 1971, 1972; ITO and SATO, 1992; BINA, 1996). The velocity anomalies may be further examined by comparing these results for the equilibrium case to those obtained for the case of metastably persisting α-olivine, in which the higher pressure olivine polymorphs are not permitted to precipitate at temperatures below

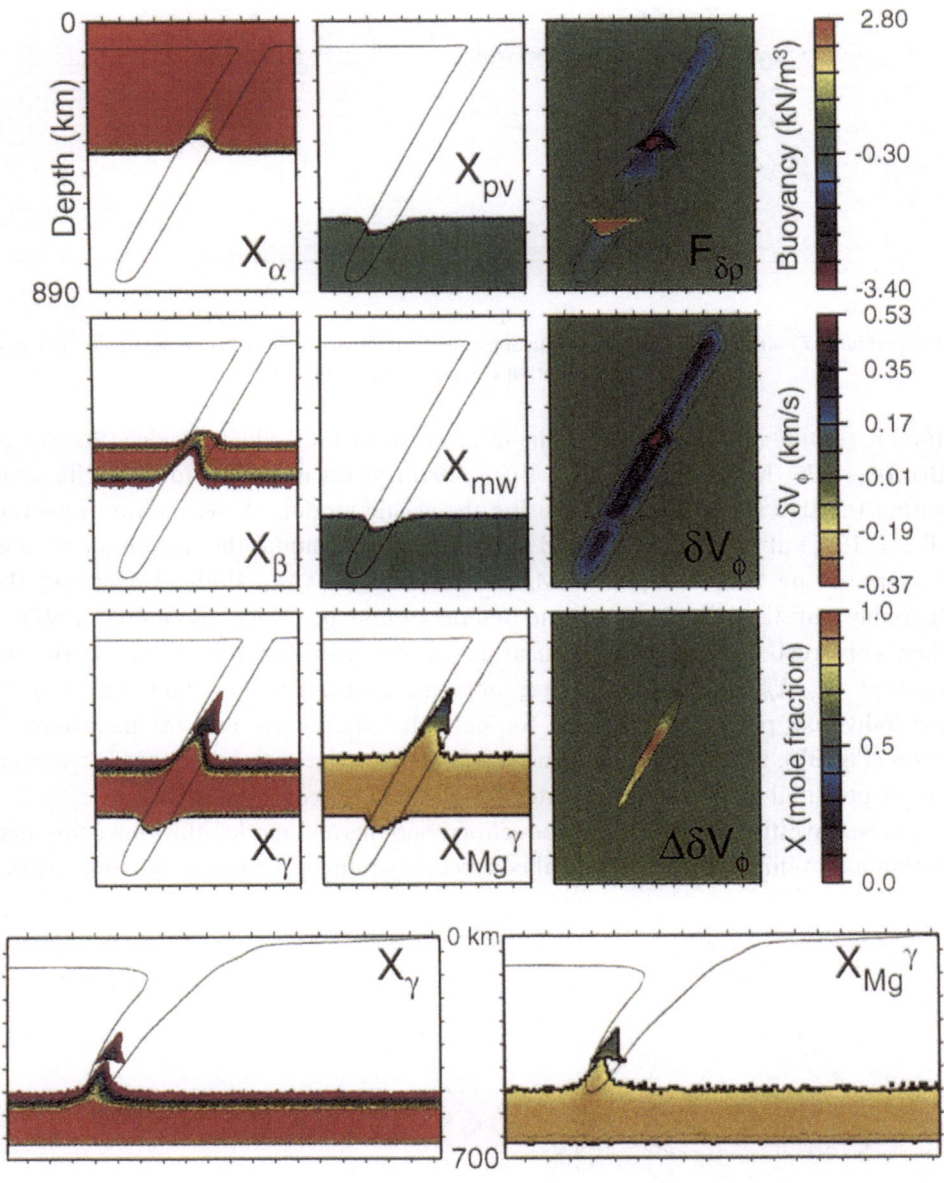

Plate 1 [Bina, 1997]

Plate 1
Top: Equilibrium distribution of phases (X_α, X_β, X_γ, X_{pv}, X_{mw}) for $(Mg_{0.9}Fe_{0.1})_2SiO_4$ bulk composition in slab thermal model of Figure 3. Corresponding equilibrium composition (X^γ_{Mg}) of phase γ. Buoyancy ($F_{\delta\rho}$) and bulk sound velocity (δV_ϕ) anomalies corresponding to equilibrium density and velocity structures of Figure 3a. Residual bulk sound velocity anomalies ($\Delta\delta V_\phi$) for case of metastable α-olivine, after subtraction of anomalies for equilibrium case. Bottom: Equilibrium distribution (X_γ) and composition (X^γ_{Mg}) of phase γ for $(Mg_{0.9}Fe_{0.1})_2SiO_4$ composition in slab thermal model of Figure 4.

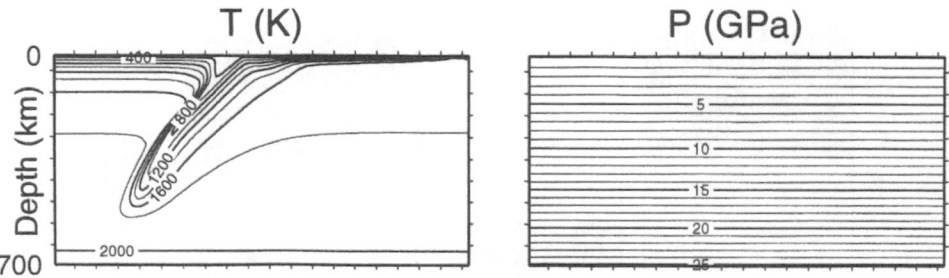

Figure 4
Temperature (*T*) and pressure (*P*) for subduction zone thermal model from two-dimensional numerical convection simulation (KINCAID and SACKS, 1997).

1000 K (RUBIE and ROSS, 1994). The resulting metastable olivine wedge (SUNG and BURNS, 1976; KIRBY *et al.*, 1996; BINA, 1996) yields a narrow low-velocity zone within the slab (Plate 1). Finally, using the elastic moduli of the relevant minerals (BASS, 1995) at room pressure and temperature, I compute the maximum V_S and V_P anisotropy for each phase (MAINPRICE and SILVER, 1993). Neglecting the (possibly substantial) pressure- and temperature-dependence (KARATO, 1997), I then construct an approximate measure of the potential maximum elastic anisotropy (k_S, k_P) of each assemblage, applying a simple mole-weighted average of the individual phases. The results, for both the equilibrium and the metastable α cases (Fig. 3b), illustrate how a metastable olivine wedge may be expected to exhibit larger potential anisotropy to greater depths than an equilibrium slab.

I also investigate a different subduction zone thermal model, this one computed within a two-dimensional numerical convection simulation employing temperature-

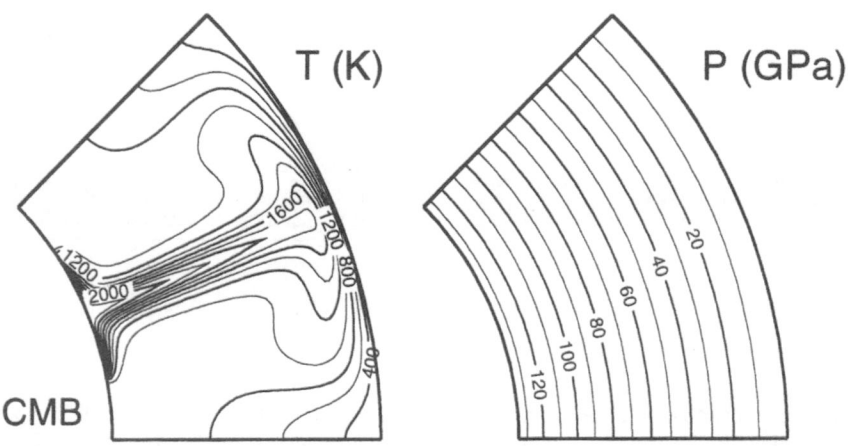

Figure 5
Temperature (*T*) and pressure (*P*) for quasi-steady state, axisymmetric, mantle plume thermal model, with dimensionless basal temperature maximum (KELLOGG and KING, 1997) scaled to 2500°C.

dependent viscosity (Rayleigh number $\sim 8 \times 10^5$). The model (Fig. 4) is constructed in a 700×1400 km box in which temperature varies over 0–1800°C (KINCAID and SACKS, 1997). Again, pressures are determined by vertical integration of a reference density profile (DZIEWONSKI et al., 1975). The computed equilibrium phase assemblages for an $(Mg_{0.9}Fe_{0.1})_2SiO_4$ bulk composition are broadly similar to the previous case but with a more marked asymmetry to the isotherms. Again (Plate 1), the expected thermal deflection of phase boundaries is evident, although deflection of the $\gamma \rightarrow \gamma + pv + mw \rightarrow pv + mw$ transition is not observed since the relatively young slab has not yet penetrated to so great a depth. The temperature-dependence of transition sharpness is again detected, along with the stability of Fe-rich γ phase in a cold $\alpha + \gamma$ field. While the oceanic lithosphere thins dramatically at the edge of this dynamical model, the major features within the slab itself remain robust.

Finally, I investigate a thermal model for a quasi-steady state, axisymmetric, mantle plume. The model (Fig. 5) is constructed within the framework of a strongly temperature dependent viscosity (reference viscosity Rayleigh number $\sim 10^7$) by heating a patch at the base of the mantle, without accounting for any additional effects of internal heating or latent heat of phase change (KELLOGG and KING, 1997). I have scaled the dimensionless temperature of the heated patch to 2500°C. Pressures are determined by radial integration of a reference density profile (DZIEWONSKI et al., 1975). Again, the computed equilibrium phase assemblages for an $(Mg_{0.9}Fe_{0.1})_2SiO_4$ bulk composition reveal (Plate 2) thermal deflection of phase boundaries, although in the opposite sense to that observed for a cold slab. The associated density anomalies generate buoyancy forces (Plate 2) important to convective dynamics, forces which are reversed relative to those associated with slabs. It is important to note that this scaled dynamical model, representing a large thermal upwelling and its counterflow, features unrealistically low temperatures at the edges of the model which are unlikely to be found in earth's mantle. Nonetheless, it is interesting to note (Plate 2) that the thermodynamic parameters of FEI et al. (1991) predict the breakdown of silicate perovskite to mixed oxides in the cold regions at the edge of this plume model. While the thermodynamic description of SiO_2 stishovite in the FEI et al. (1991) data set, derived from the silica polymorph parameterizations of FEI et al. (1990), yields a narrow region of perovskite breakdown, more recent analyses of the coesite-stishovite transition (AKAOGI et al., 1995; LIU et al., 1996) yield a greater thermodynamic stability for stishovite and

---

Plate 2
Left: Equilibrium distribution of phases (X_γ, X_{st}) for $(Mg_{0.9}Fe_{0.1})_2SiO_4$ bulk composition in plume thermal model of Figure 5. Stishovite parameters from F90 (FEI et al., 1990) and from L96 (LIU et al., 1996). Right: Buoyancy $(F_{\delta\rho})$ and bulk sound velocity (δV_ϕ) anomalies in upper half of model mantle arising from corresponding density and velocity structures for L96. Total (left) velocity anomalies (δV_ϕ) contain component (right) due only to $pv \rightarrow mw + st$ reaction, obtained by removing thermal signature from total.

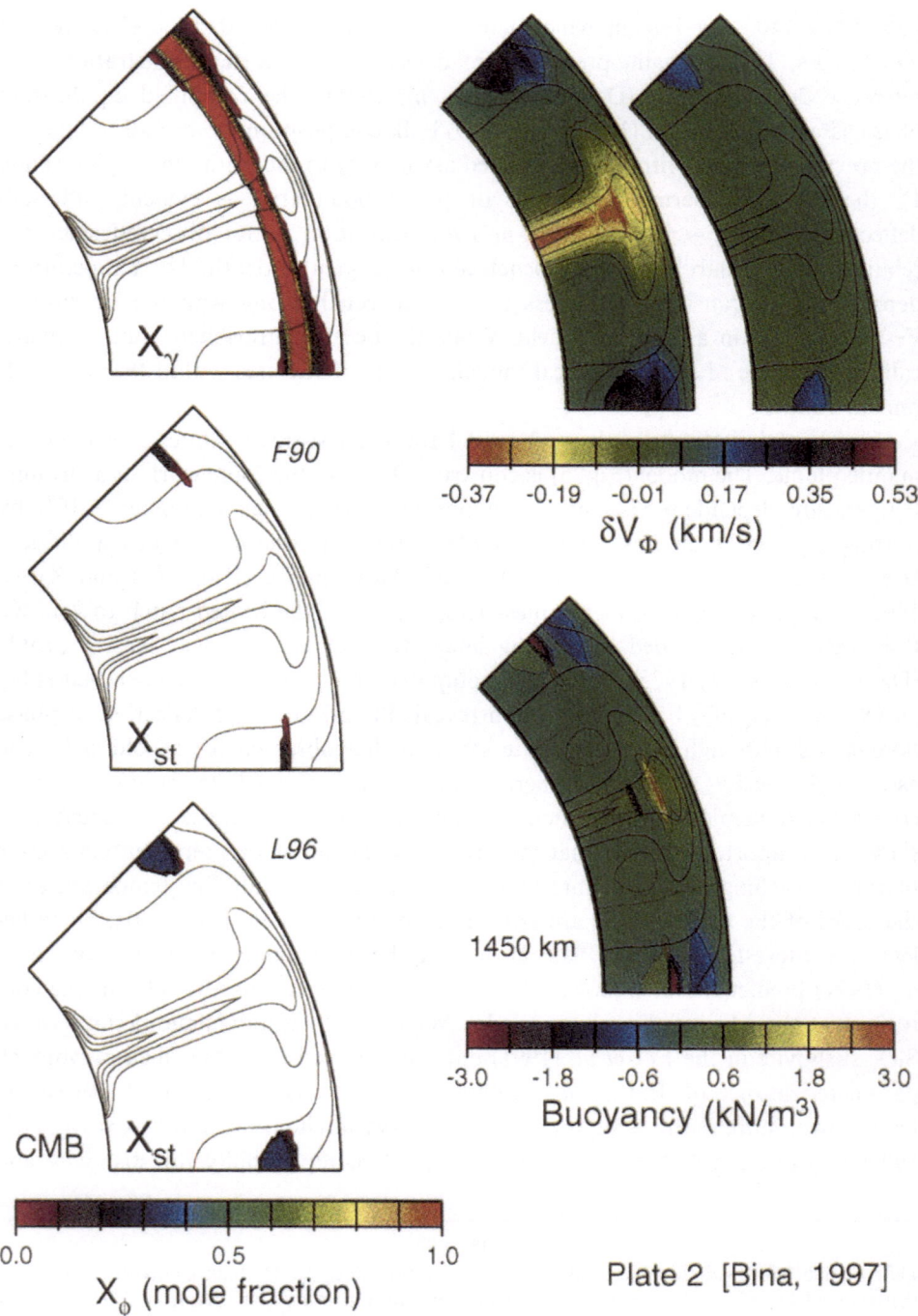

δV_Φ (km/s)

-0.37   -0.19   -0.01   0.17   0.35   0.53

Buoyancy (kN/m³)

-3.0   -1.8   -0.6   0.6   1.8   3.0

X_ϕ (mole fraction)

0.0   0.5   1.0

Plate 2 [Bina, 1997]

thus a broader region of perovskite breakdown (Plate 2). Because the possibility of such perovskite breakdown has been proposed at higher temperatures in the mantle (HEMLEY and COHEN, 1992; MEADE et al., 1995; SAXENA et al., 1996), it is instructive to note that this reaction ($pv \rightarrow mw + st$) is accompanied by a corresponding fast velocity anomaly. Thus, in this plume model, the disproportionation of perovskite to mixed oxides serves to amplify the velocity anomaly arising simply from the low temperatures (Plate 2). Despite the low model temperatures, the continuing evolution of the thermodynamic parameterization of silicate perovskite (FABRICHNAYA, 1995), and ambiguity about post-stishovite silica phases (BELONOSHKO et al., 1996), these calculations illustrate the principle that interpretation of seismic velocity anomalies solely in terms of thermal structure can be complicated by superposition of phase stability boundaries.

Concluding Remarks

An optimization algorithm based upon the method of simulated annealing can be of utility in calculating equilibrium phase assemblages as functions of pressure, temperature, and chemical composition in complex multicomponent systems. The method reproduces known phase diagrams, and it illustrates the expected thermal deflection of phase transitions in thermal models of subducting lithospheric slabs and buoyant mantle plumes. Application of the method to two different slab thermal models demonstrates temperature-induced changes in phase transition sharpness and reveals the stability of Fe-rich γ phase in an $\alpha + \gamma$ field within a cold slab. Application of the method to a plume thermal model illustrates that reactions such as the breakdown of silicate perovskite to mixed oxides can amplify thermal velocity anomalies. Finally, it is important to note that this method operates by analogy to the statistical mechanics of the chemical system. Thus, it is not restricted to problems of strict chemical equilibrium. It is also ideal for the investigation of metastability effects (BINA, 1996).

Acknowledgements

I thank L. Kellogg, C. Kincaid, S. Stein, and L. Leffler for providing useful thermal models. H.-C. Nataf and S. Saxena for helpful reviews, and J. Linton for producing Figure 1. All other figures were produced using GMT software (WESSEL and SMITH, 1995). I acknowledge the support of the National Science Foundation (EAR-9158594).

REFERENCES

AKAOGI, M., YUSA, H., SHIRAISHI, K., and SUZUKI, T. (1995), *Thermodynamic Properties of α-Quartz, Coesite, and Stishovite and Equilibrium Phase Relations at High Pressures and High Temperatures*, J. Geophys. Res. *100*, 22,337–22,347.

BASS, J. D., *Elasticity of minerals, glasses, and melts*. In *Mineral Physics and Crystallography, AGU Ref. Shelf 2* (ed. T. J. Ahrens) (Amer. Geophys. U., Washington, D.C. 1995) pp. 45–63.

BASU A., and FRAZER, L. N. (1990), *Rapid Determination of the Critical Temperature in Simulated Annealing Inversion*, Science *249*, 1409–1412.

BELONOSHKO, A. B., DUBROVINSKY, L. S., and DUBROVINSKY, N. A. (1996), *A New High-pressure Silica Phase Obtained by Molecular Dynamics*, Am. Mineral. *81*, 785–788.

BINA, C. R., and WOOD, B. J. (1987), *The Olivine-spinel Transitions: Experimental and Thermodynamic Constraints and Implications for the Nature of the 400 km Seismic Discontinuity*, J. Geophys. Res. *92*, 4853–4866.

BINA, C. R. (1987), *Mineralogic Transformations and Seismic Velocity Variations in the Upper Mantle of the Earth*, Ph.D. Dissertation, Northwestern Univ., Evanston, Illinois.

BINA, C. R. (1996), *Phase Transition Buoyancy Contributions to Stresses in Subducting Lithosphere*, Geophys. Res. Lett. *23*, 3563–3566.

BINA, C. R., and LIU, M. (1995), *A Note on the Sensitivity of Mantle Convection Models to Composition-dependent Phase Relations*, Geophys. Res. Lett. *22*, 2565–2568.

CHRISTENSEN, U. R., and YUEN, D. A. (1984), *The Interaction of a Subducting Lithospheric Slab with a Chemical or Phase Boundary*, J. Geophys. Res. *89*, 4389–4402.

CHRISTENSEN, U. R., and YUEN, D. A. (1985), *Layered Convection Induced by Phase Transitions*, J. Geophys. Res. *90*, 10,291–10,300.

CVIJOVIĆ, D., and KLINOWSKI, J. (1995), *Taboo Search: An Approach to the Multiple Minima Problem*, Science *267*, 664–666.

DE CAPITANI, C., and BROWN, T. H. (1987), *The Computation of Chemical Equilibrium in Complex Systems Containing Non-ideal Solutions*, Geochim. Cosmochim. Acta *51*, 2639–2652.

DZIEWONSKI, A. M., HALES, A. L., and LAPWOOD, E. R. (1975), *Parametrically Simple Earth Models Consistent with Geophysical Data*, Phys. Earth Planet. Inter. *10*, 12–48.

ERIKSSON, G. (1975), *Thermodynamic Studies of High Temperature Equilibria. XII. SOLGASMIX, A Computer Program for Calculation of Equilibrium Compositions in Multiphase Systems*, Chem. Scr. *8*, 100–103.

ERIKSSON, G., and HACK, K. (1990), *ChemSage—A Computer Program for the Calculation of Complex Chemical Equilibria*, Metall. Trans. B *21B*, 1013–1023.

FABRICHNAYA, O. B. (1995), *Thermodynamic Data for Phases in the FeO–MgO–SiO$_2$ System and Phase Relations in the Mantle Transition Zone*, Phys. Chem. Minerals *22*, 323–332.

FEI, Y., MAO, H.-K., and MYSEN, B. O. (1991), *Experimental Determination of Element Partitioning and Calculation of Phase Relations in the MgO-FeO-SiO$_2$ System at High Pressure and High Temperature*, J. Geophys. Res. *96*, 2157–2169.

FEI, Y., SAXENA, S. K., and NAVROTSKY, A. (1990), *Internally Consistent Thermodynamic Data and Equilibrium Phase Relations for Compounds in the System MgO-SiO$_2$ at High Pressure and High Temperature*, J. Geophys. Res. *95*, 6915–6928.

GILL, P. E., MURRAY, W., and WRIGHT, M. H., *Practical Optimization* (Academic, London 1981) 401 pp.

GHIORSO, M. S. (1985), *Chemical Mass Transfer in Magmatic Processes. I. Thermodynamic Relations and Numerical Algorithms*, Contrib. Mineral. Petrol. *90*, 107–120.

GHIORSO, M. S., and SACK, R. O. (1995), *Chemical Mass Transfer in Magmatic Processes. IV. A Revised and Internally Consistent Thermodynamic Model for the Interpolation and Extrapolation of Liquid-Solid Equilibria in Magmatic Systems at Elevated Temperatures and Pressures*, Contrib. Mineral. Petrol. *119*, 197–212.

GLOVER, F. (1989), *Tabu Search. Part I*, ORSA J. Comput. *1*, 190–206.

GLOVER, F. (1990), *Tabu Search. Part II*, ORSA J. Comput. *2*, 4–32.

GOLDBERG, D. E., *Genetic Algorithms in Search, Optimization, and Machine Learning* (Addison-Wesley, Reading, Massachusetts 1989) 412 pp.

GREEN, H. W., II, and HOUSTON, H. (1995), *The Mechanics of Deep Earthquakes*, Ann. Rev. Earth Planet. Sci *23*, 169–213.

HARVIE, C. E., GREENBERG, J. P., and WEARE, J. H. (1987), *A Chemical Equilibrium Algorithm for Highly Non-ideal Multiphase Systems: Free Energy Minimization*, Geochim. Cosmochim. Acta *51*, 1045–1057.

HELFFRICH, G., and BINA, C. R. (1994), *Frequency Dependence of the Visibility and Depths of Mantle Seismic Discontinuities*, Geophys. Res. Lett. *21*, 2613–2616.

HELFFRICH, G., STEIN, S., and WOOD, B. J. (1989), *Subduction Zone Thermal Structure and Mineralogy and their Relationship to Seismic Wave Reflections and Conversions at the Slab/Mantle Interface*, J. Geophys. Res. *94*, 753–763.

HELGESON, H. C., BROWN, T. H., NIGRINI, A., and JONES, T. A (1970), *Calculation of Mass Transfer in Geochemical Processes Involving Aqueous Solutions*, Geochim. Cosmochim. Acta *34*, 569–592.

HEMLEY, R. J., and COHEN, R. E. (1992), *Silicate Perovskite*, Annu. Rev. Earth Planet. Sci. *20*, 553–600.

INGBER, L. (1989), *Very Fast Simulated Re-annealing*, Math. Comput. Model. *12*, 967–973.

INGBER, L. (1996), *Adaptive Simulated Annealing (ASA): Lessons Learned*, Control Cybern. *25*, 33–54.

ISACKS, B., and MOLNAR, P. (1971), *Distribution of Stresses in the Descending Lithosphere from a Global Survey of Focal-mechanism Solutions of Mantle Earthquakes*, Rev. Geophys. Space Phys. *9*, 103–174.

ITO, E., and SATO, H., *Effect of Phase Transformations on the Dynamics of the Descending Slab*. In *High-pressure Research: Application to Earth and Planetary Sciences* (eds. Y. Syono, and M. H. Manghnani) (Amer. Geophys. U., Washington, D.C. 1992) pp. 257–262.

KARATO, S. (1997), *Seismic Anisotropy in the Deep Mantle, Boundary Layers and the Geometry of Mantle Convection*, Pure appl. geophys. *151*, 565–587.

KATSURA, T., and ITO, E. (1989), *The System Mg_2SiO_4-Fe_2SiO_4 at High Pressures and Temperatures: Precise Determination of Stabilities of Olivine, Modified Spinel, and Spinel*, J. Geophys. Res. *94*, 15,663–15,670.

KELLOGG, L. H., and KING, S. D. (1997), *The Effect of Temperature Dependent Viscosity on the Structure of New Plumes in the Mantle: Results of a Finite Element Model in a Spherical, Axisymmetric Shell*, Earth Planet. Sci. Lett. *148*, 13–26.

KINCAID, C., and SACKS, I. S. (1997), *Thermal and Dynamical Evolution of the Upper Mantle in Subduction Zones*, J. Geophys. Res. *101*, 12,295–12,315.

KIRBY, S. H., STEIN, S., OKAL, E. A., and RUBIE, D. C. (1996), *Metastable Mantle Phase Transformations and Deep Earthquakes in Subducting Oceanic Lithosphere*, Rev. Geophys. *34*, 261–306.

KIRKPATRICK, S., GELATT, C. D., Jr., and VECCHI, M. P. (1983), *Optimization by Simulated Annealing*, Science *220*, 671–680.

KOREN, Z., MOSEGAARD, K., LANDA, E., THORE, P., and TARANTOLA, A. (1991), *Monte Carlo Estimation and Resolution Analysis of Seismic Background Velocities*, J. Geophys. Res. *96*, 20,289–20,299.

LIU, J., TOPOR, L., ZHANG, J., NAVROTSKY, A., and LIEBERMANN, R. C. (1996), *Calorimetric Study of the Coesite-stishovite Transformation and Calculation of the Phase Boundary*, Phys. Chem. Miner. *23*, 11–16.

MAINPRICE, D., and SILVER, P. G. (1993), *Interpretation of SKS-Waves Using Samples from the Subcontinental Lithosphere*, Phys. Earth Planet. Inter. *78*, 257–280.

MEADE, C., MAO, H. K., and HU, J. (1995), *High-temperature Phase Transition and Dissociation of $(Mg, Fe)SiO_3$ Perovskite at Lower Mantle Pressures*, Science *268*, 1743–1745.

METROPOLIS, N., ROSENBLUTH, A. W., ROSENBLUTH, M. N., TELLER, A., and TELLER, E. (1953), *Equation of State Calculations by Fast Computing Machines*, J. Chem. Phys. *21*, 1087–1092.

MOSEGAARD, K., and VESTERGAARD, P. D. (1991), *A Simulated Annealing Approach to Seismic Model Optimization with Sparse Prior Information*, Geophys. Prospect. *39*, 599–611.

POWELL, R., and HOLLAND, T. J. B. (1988), *An Internally Consistent Thermodynamic Dataset with Uncertainties and Correlations: 3. Applications to Geobarometry, Worked Examples and a Computer Program*, J. Meta. Pet. *6*, 173–204.

RICHTER, F. M. (1973), *Finite Amplitude Convection Through a Phase Boundary*, Geophys. J. R. Astron. Soc. *35*, 265–287.

RUBIE, D. C., and ROSS, C. R., II (1994), *Kinetics of the Olivine-Spinel Transformation in Subducting Lithosphere: Experimental Constraints and Implications for Deep Slab Processes*, Phys. Earth Planet. Inter. *86*, 223–241.

SAMBRIDGE, M., and DRIJKONINGEN, G. (1992), *Genetic Algorithms in Seismic Waveform Inversion*, Geophys. J. Int. *109*, 323–342.

SAXENA, S. K., *Computation of multicomponent phase equilibria*. In *Advances in Physical Geochemistry* (ed. S. K. Saxena) (Springer, New York 1982) pp. 225–242.

SAXENA, S. K. (1996), *Earth Mineralogical Model: Gibbs Free Energy Minimization Computation for the System MgO-FeO-SiO_2*, Geochim. Cosmochim. Acta *60*, 2379–2395.

SAXENA, S. K., DUBROVINSKY, L. S., LAZOR, P., CERENIUS, Y., HÄGGKVIST, P., HANFLAND, M., and HU, J. (1996), *Stability of Perovskite ($MgSiO_3$) in the Earth's Mantle*, Science *274*, 1357–1359.

SCHUBERT, G., YUEN, D. A., and TURCOTTE, D. L. (1975), *Role of Phase Transitions in a Dynamic Mantle*, Geophys. J. R. Astr. Soc. *42*, 705–735.

SMITH, W. R., and MISSEN, R. W., *Chemical Reaction Equilibrium Analysis* (Wiley, New York 1982) 364 pp.

STEIN, C. A., and STEIN, S. (1992), *A Model for the Global Variation in Oceanic Depth and Heat Flow with Lithospheric Age*, Nature *359*, 123–129.

STOREY, S. H., and VAN ZEGGEREN, F. (1964), *Computation of Chemical Equilibrium Compositions*, Can. J. Chem. Eng. *42*, 54–55.

SUNDMAN, B., JANSSON, B., and ANDERSON, J. O. (1985), *The Thermo-Calc Databank System*, Calphad *9*, 153–190.

SUNG, C.-m., and BURNS, R. G. (1976), *Kinetics of the High-pressure Phase Transformations: Implications to the Evolution of the Olivine-spinel Phase Transition in the Downgoing Lithosphere and its Consequences on the Dynamics of the Mantle*, Tectonophys. *31*, 1–32.

TOKSÖZ, M. N., SLEEP, N. H., and SMITH, A. T. (1973), *Evolution of the Downgoing Lithosphere and the Mechanisms of Deep Focus Earthquakes*, Geophys. J. R. Astr. Soc. *35*, 285–310.

TURCOTTE, D. L., and SCHUBERT, G. (1971), *Structure of the Olivine-spinel Phase Boundary in the Descending Lithosphere*, J. Geophys. Res. *76*, 7980–7987.

TURCOTTE, D. L., and SCHUBERT, G. (1972), *Correction*, J. Geophys. Res. *77*, 2146.

VAN ZEGGEREN, F., and STOREY, S. H., *The Computation of Chemical Equilibrium* (Cambridge Univ., Cambridge, England 1970) 176 pp.

WESSEL, P., and SMITH, W. H. F. (1995), *New Version of the Generic Mapping Tools Released*, Eos Trans. Amer. Geophys. U. *76*, 329.

WOOD, B. J., and HOLLOWAY, J. R. (1984), *A Thermodynamic Model for Subsolidus Equilibria in the System CaO-MgO-Al_2O_3-SiO_2*, Geochim. Cosmochim. Acta *48*, 159–176.

WOOD, B. J. (1987), *Thermodynamics of Multicomponent Systems Containing Several Solid Solutions*, Rev. Mineral. *17*, 71–95.

(Received October 13, 1996, revised April 30, 1997, accepted May 22, 1997)

To access this journal online:
http://www.birkhauser.ch

Pure appl. geophys. 151 (1998) 619–629
0033–4553/98/040619–11 $ 1.50 + 0.20/0

❙Pure and Applied Geophysics

Laboratory Approach to the Study of Elastic Anisotropy on Rock Samples

ZDENĚK PROS,[1] TOMÁŠ LOKAJÍČEK,[1] and KAREL KLÍMA[1]

Abstract—The experimental approach (hardware and software) to the study of the elastic anisotropy of rocks on spherical samples under hydrostatic pressure up to 400 MPa is discussed. A substantial innovation of the existing measuring system and processing methods enabled us to make a detailed investigation and evaluation of the kinematic as well as dynamic parameters of elastic waves propagating through anisotropic media. The innovation is based on digital recording of the wave pattern with a high sampling density of both time and amplitude. Several options and results obtained with the innovated laboratory equipment are presented.

Key words: Laboratory measurements, elastic anisotropy, kinematic and dynamic parameters of elastic waves, hydrostatic pressure.

Introduction

The elastic anisotropy of rocks is an important geophysical parameter which can be used to classify rocks in petrophysical studies, to interpret seismic field measurements, and also to study the structure of the crust and the upper mantle of the earth. The physical parameters of rocks are determined mainly by their modal composition, properties of grains, grain-size distribution, contact conditions between grains, pore space, population of microcracks and by the combined action of the above-mentioned parameters affected by external conditions, mainly by stress-state conditions.

The anisotropy of *P* waves was studied by BIRCH (1960, 1961); BAYUK *et al.* (1967); GIESEL (1963); BABUŠKA (1966); PROS and BABUŠKA (1967, 1968) and others. The laboratory measuring equipment, developed in the Geophysical Institute, enables elastic anisotropy of *P* waves to be studied by means of ultrasonic sounding under conditions of high hydrostatic pressure reaching 400 MPa (PROS and PODROUŽKOVÁ, 1974; PROS, 1977). In this paper the actual state of the laboratory equipment and the methodical approaches used to study elastic anisotropy are presented.

---

[1] Geophysical Institute, Academy of Sciences of the Czech Republic, Boční II/1401, 14131 Praha 4—Spořilov, Czech Republic.

Experimental Setup

During recent years laboratory measuring equipment has been improved considerably, mainly the ultrasonic data recording and the processing equipment has been modified. Now, not only are the travel times of *P*-wave ultrasonic signals stored, but the whole wavelet of the recorded ultrasonic wave is stored and used for further analysis. A schematic diagram of the improved measuring system is shown in Figure 1. The entire measuring system consists of a pressure vessel, pressure generator, ultrasonic transducers and a sample-positioning control unit, a device for generating and recording of ultrasonic signals. The digitizing scope HP 54540C

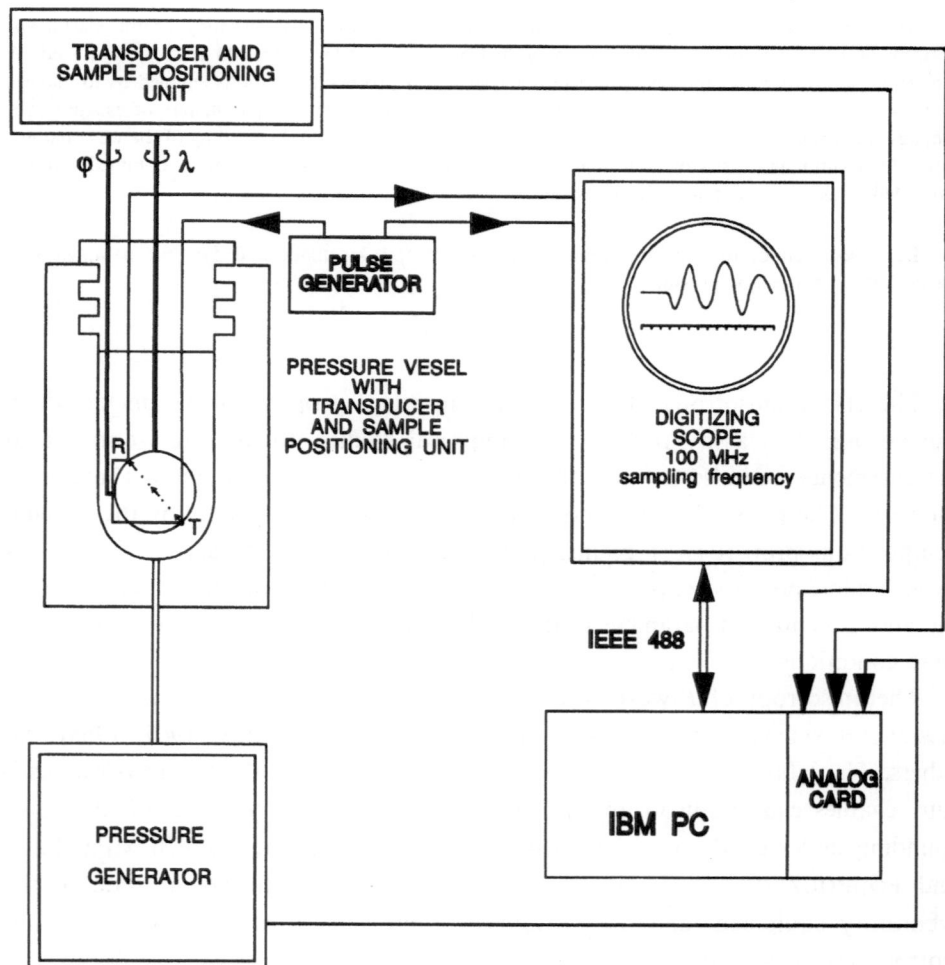

Figure 1

Equipment for the laboratory study of *P*-wave anisotropy on spherical samples under hydrostatic pressure of up to 400 MPa. *T*—transmitter of ultrasonic signal, *R*—receiver of ultrasonic signal. Transducers are located on opposite sides of the sample.

with a 100 MHz sampling frequency is used to record and store the form of the exciting pulse and ultrasonic signal in digital form. A PC card with an A/D converter is used to measure the rock sample position, the ultrasonic transducers positions and the acting pressure. All the measuring equipment is controlled by an IBM PC via an IEEE 488 interface. The measuring software is divided into two independent parts. First, ultrasonic data recording and storing, when the transmitted signal (40 μs, time duration) is recorded with a sampling frequency of 100 MHz and stored on the computer HD. The spherical sample can be measured in 132 independent directions (with a step of 15° in both spherical coordinates) at each pressure level; or it can be measured in a particular direction under variable acting pressure reaching 400 MPa. Second, the stored data are analysed by specialized software packages; the parameters to be studied being selected semi-automatically (time of signal arrival, value of the first amplitude, etc.). The recorded data are stored in the file on the computer HD only after being corrected or confirmed by an operator. These files are used for further analysis of the P-wave anisotropy of rock samples.

Experiments

The following experiments only show some examples of methodical approaches to anisotropy study used in our laboratory. As an experimental approach the pulse-transmission method for the velocity measurement was used. The couple of piezoceramic transducers was used; the transmitter excited by a high voltage pulse and receiver located on the opposite side of the spherical sample. The receiver picked up the ultrasonic signal propagating through the sample. The same transducers were used for transmitting and receiving the ultrasonic signal. The transducers are equipped with an acoustic transformer with a spherical surface tapered to the sample. The active area of the piezoelectric element is 3 mm in diameter. The natural frequency of the element is 2.5 MHz. The shape of the samples in our experiments is a sphere with a diameter of 50 mm. Before the measurements the samples are vacuum dried to a constant weight at a temperature of 50°C during 12 hours. Then the samples are covered by an epoxy resin film (thickness 0.05 mm) to protect the sample against pressure media penetration.

An example of the wave pattern of the ultrasonic signal picked up by a receiver at different pressure levels is depicted in Figure 2. The dashed line shows the transmitter excitation time. Significant changes can be seen in this figure, as not only the travel times of the ultrasonic signal vary, but also their amplitude and frequency contents.

The measuring system enables the travel-time dependence, or of P-wave velocity to be investigated in selected directions. An example of P-wave velocity dependence on hydrostatic stress from atmospheric pressure to 400 MPa for pair of ultrasonic transducers positioned in the maximum and then in the minimum velocity direction

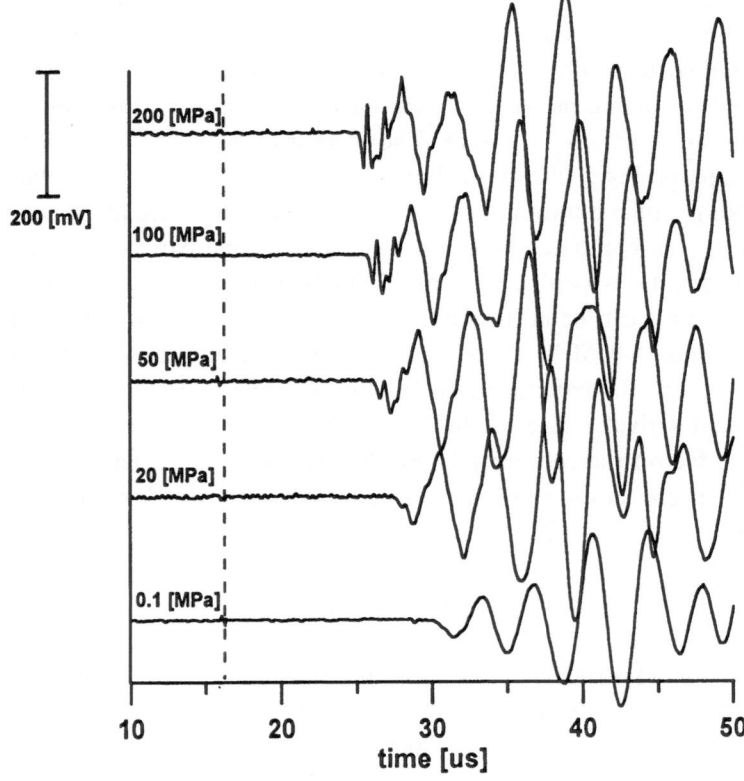

Figure 2
Example of *P*-wave signals (sample RP3, orthogneiss) recorded at different values of hydrostatic
pressure.

is depicted in Figure 3. The directions were determined at 400 MPa at first loading
cycle. This particular rock sample—granite from the West Bohemia region with
signature ZC19—displays a very fast change of *P*-wave velocity at a low value of
applied hydrostatic stress. This fast change is mainly due to the closing of
microcracks and will be discussed in more detail later. For higher stress values the
changes of *P*-wave velocity are less pronounced. It also can be seen that nearly the
same anisotropy is preserved in this sample for the whole range of acting pressure.
Slight scatter shows the limits of measurement accuracy due to the resolution of the
A/D convertor.

As the measuring system records the wavelet of the ultrasonic signal as a whole,
it enables us to investigate the amplitude changes of the ultrasonic signal passing
through the spherical rock sample with direction. A comparison of the *P*-wave
velocity dependence and amplitude dependence A_{12} (A_{12} is the difference between
the first local minimum and first local maximum of the recorded signal) on
hydrostatic pressure up to 400 MPa is shown in Figure 4. The *P*-wave velocity

varies from 5250 m/sec. to 6500 m/sec., which represents a change of about 24%. On the contrary, the A_{12} amplitude varies from 8 mV to 380 mV, which is nearly 50 times higher value. The variations described above indicate that additionally the amplitude dependence of ultrasonic waves on hydrostatic pressure can also be considered as a very promising tool for describing and studying the elastic anisotropy of rocks. The acting force to sliding contact between the sample and the transducer does not vary substantially during the change of hydrostatic pressure. Only the elastic properties of oil film between them change with pressure. Due to this fact, the drastic change of A_{12} amplitude with an increasing load is mainly caused by the consolidation of contact conditions between mineral grains and closing microcracks.

The experimental setup also enables a detailed study of the "velocity-confining pressure equation," which describes the v_P (or v_S) velocity dependence on applied hydrostatic "confining" pressure. WEPFER and CHRISTENSEN (1991) proposed an empirical formula (1) relating velocity to confining pressure in the form:

$$V(P) = A(P/100 \text{ MPa})^a + B(1 - e^{-bP}),\tag{1}$$

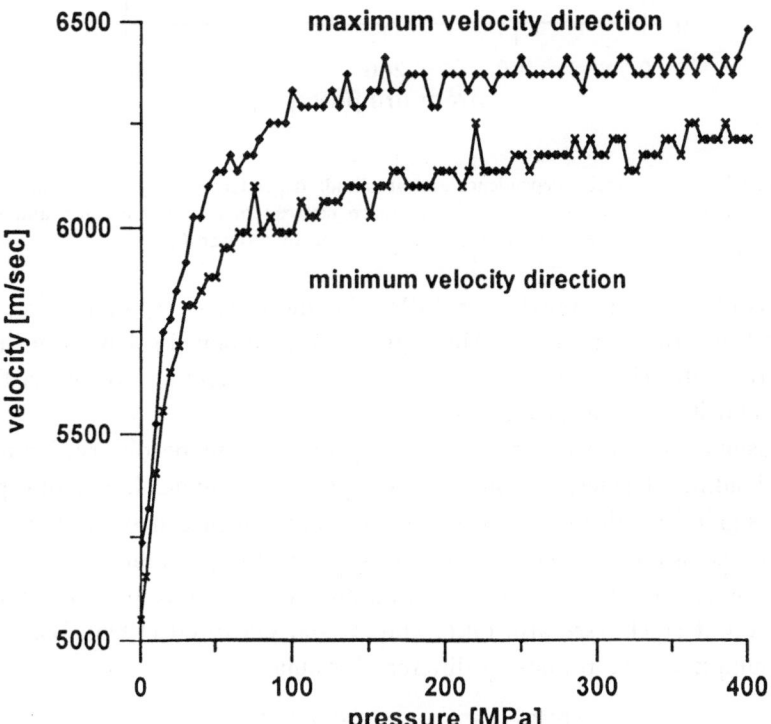

Figure 3

Dependence of P-wave velocity on hydrostatic pressure in the maximum and minimum velocity direction (sample ZC19, granite).

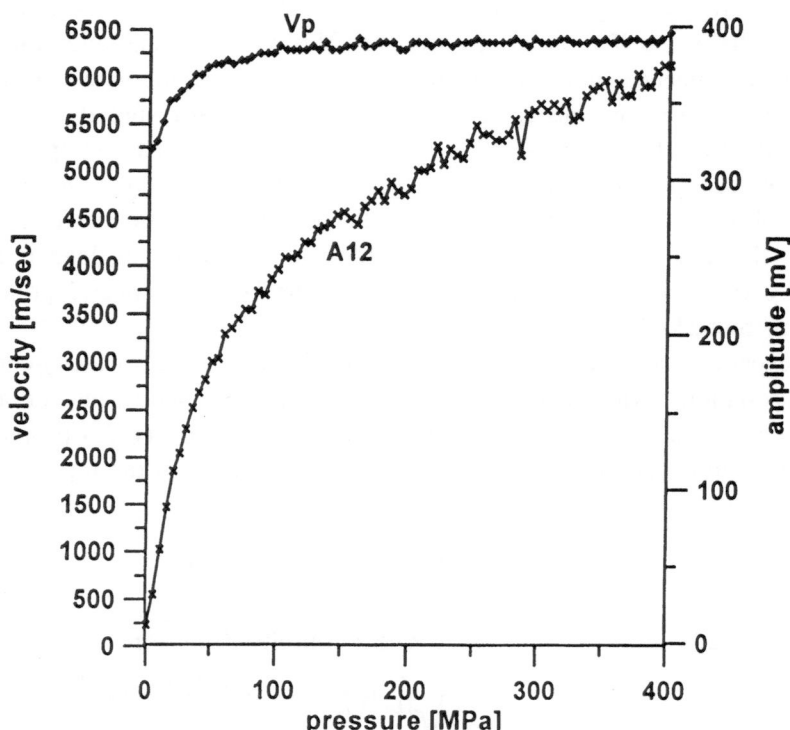

Figure 4

Comparison of *P*-wave velocity dependence and amplitude dependence on hydrostatic pressure up to 400 MPa (sample ZC19, granite). A_{12} is a difference between first local minimum and first local maximum of the signal recorded by receiver *R*.

where *P* is the confining pressure in MPa, *V* is the velocity (v_P or v_S) and *A*, *a*, *B* and *b* are four fitted parameters. This equation was obtained solely from the shape of measured data. However the values of fitted parameters do not reflect the rock structure and its physical properties.

We assume that two different physical processes can be observed during the confined loading of heterogeneous rock samples. First, under low acting pressure the most significant phenomenon is crack closing, which can be described by an exponential equation. Second, under high pressure the prevailing phenomenon is the linear increase of elastic parameters with pressure, as can be expected in an ideal sample without cracks, see also GREENFIELD and GRAHAM (1996). Based on the above assumptions, we propose a different formula:

$$V(P) = Vo + kP - dV10^{(-P/Po)}, \tag{2}$$

where *Vo*—is the velocity under a pressure of 0.1 MPa (in an ideal sample without cracks), *k*—increasing velocity coefficient of an ideal sample without cracks,

P—confining pressure in MPa, dV—crack influence at the 0.1 MPa pressure level, Po—pressure at which the crack influence dV decreases to 10% (see Fig. 5). For the given example $Vo = 6297$, $k = 0.2771$, $dV = 954.16$ and $Po = 75.82$.

This formula was compared with the Wepfer and Christensen empirical formula (1). The comparison of the formulas indicates that formula (2) proposed in this paper yields better results in more cases than formula (1). A superior advantage of our approach is that the parameters we obtained during the fitting process provide for a better description of the physical properties and fabric of rock sample under study. The Po parameter denotes the pressure level at which most of low aspect ratio microcracks appear to be closed.

The digital recording of the entire wavelet of the ultrasonic signal enables us to investigate its changes not only in the time domain, but also in the frequency domain. For this study only, the very beginning (6 μs) of the ultrasonic signal in time domain was used. It is assumed that mainly the very beginning of the recorded ultrasonic signal (direct P wave) reflects the changes in the rock fabric due to the

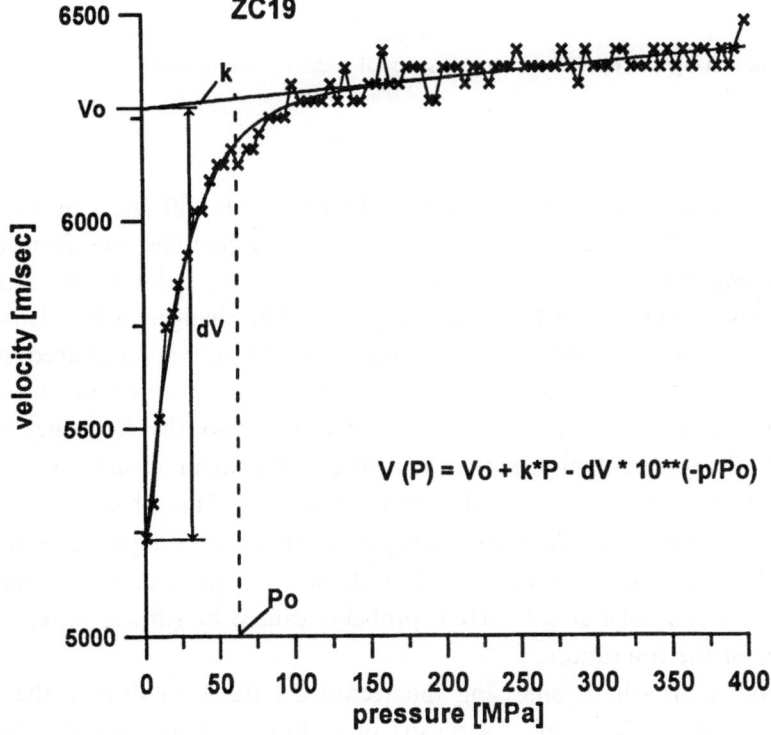

Figure 5

Dependence of P-wave velocity on hydrostatic pressure (sample ZC19, granite) and its fitted approximation. Vo—velocity of an ideal sample without defects at 0.1 MPa, k—increasing velocity coefficient of an ideal sample without defects, dV—crack influence at the 0.1 MPa pressure level, Po—pressure at which the crack influence dV decreases to 10%.

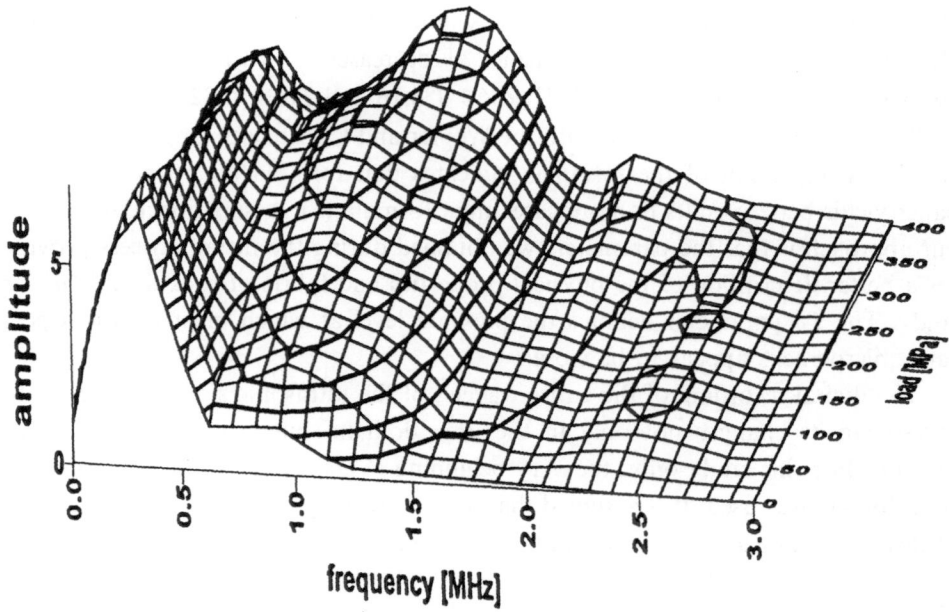

Figure 6
The frequency changes of the *P*-wave ultrasonic signal under increasing hydrostatic pressure up to 400 MPa.

increasing pressure. Also after this time reflected signals will arrive at the receiver and the rest of the wavelet is superpositioned by reflected and transformed waves (PROS *et al.*, 1969). The signal in the time domain is windowed by a Gaussian function after which an FFT analysis is applied. The changes of the signal in the frequency domain data with increasing hydrostatic pressure are displayed in Figure 6. This figure indicates that, due to the increasing hydrostatic pressure, not only the amplitude of the signal changes significantly, but also the frequency content. Although the pressure affects the transducers, the main change in amplitude observed is caused by changes of sample properties. This phenomenon can be explained by changes in the transmitting function of the sample. The amplitude spectra show two maxima, one at 1.1 MHz which represents the wavelet of *P* waves. The second one at 0.3 MHz is probably caused by surface waves or radial resonance of the transducer.

A good approach to analysing the recorded data is to display the *P*-wave velocities in the rock sample by means of isolines and to calculate the elastic constants characterising the material. The comparison of *P*-wave velocities calculated from the elastic constants, which are in principle smooth values, with the measured velocities, provides information mainly about the samples homogeneity and also about the quality of the velocity measurement.

The calculation of elastic constants based on *P*-wave velocities is described in more detail in KLÍMA (1973). It consists in the solution of cubic equations with 21 unknown elastic parameters. Only 15 independent equations are available and six conditions must be added to make the problem solvable. These conditions can vary within relatively wide limits, but during reverse calculations of *P*-wave velocities they are not affected by them. On the contrary, the *S*-wave velocities, which we are not dealing with, are greatly affected by the selected limits. Nevertheless, the *P*-wave velocities calculated from these elastic constants are fully sufficient to describe the quality of the measurement and samples heterogeneity.

In depicting the *P*-wave velocity distribution in terms of isolines, it is necessary to solve three separate problems in turn. Velocity approximation by a continuous function, calculation of the isolines on a spherical surface and, finally, projection of the spherical surface onto a plane.

To approximate the velocity of a spline function, the 2-D harmonic analysis of velocities in a regular grid of 12×24 points was used, due to the measuring step of 15° over the entire sphere. This 2-D analysis enables a simple calculation of functional values, as well as partial derivatives at every point. It has been proved that the calculations can be limited to 5×7 harmonics.

The isoline shape is calculated in two steps. In the first one the approximate position of the next isoline point is determined, based on the knowledge of partial derivatives at a selected distance. The second step consists in the correction of the isoline point position in the perpendicular direction on the basis of the difference between the calculated function value at a given point and a given isoline value.

Equal-area projection of calculated isolines to the lower hemisphere is used. The algorithm used to display isolines enables us to apply arbitrary transformation before the display. An example of the isolines of measured *P*-wave velocities is pictured in Figure 7. The isolines are displayed for all measured loading and unloading pressure levels of granite sample from the West Bohemia region (sample G6).

Conclusions

The experimental tests have proved that the improved measuring equipment can be used to study the elastic anisotropy of *P* waves under high hydrostatic pressure; mainly to investigate the effect of fabric change on the directional dependence of elastic wave velocities, its amplitude changes due to the transmitting function of the sample.

It was demonstrated that increasing hydrostatic pressure changes not only the ultrasonic wave velocity, but also significantly changes the amplitude and frequency contents of the transmitted signal due to the closing microcracks and consolidation of contact conditions between mineral grains.

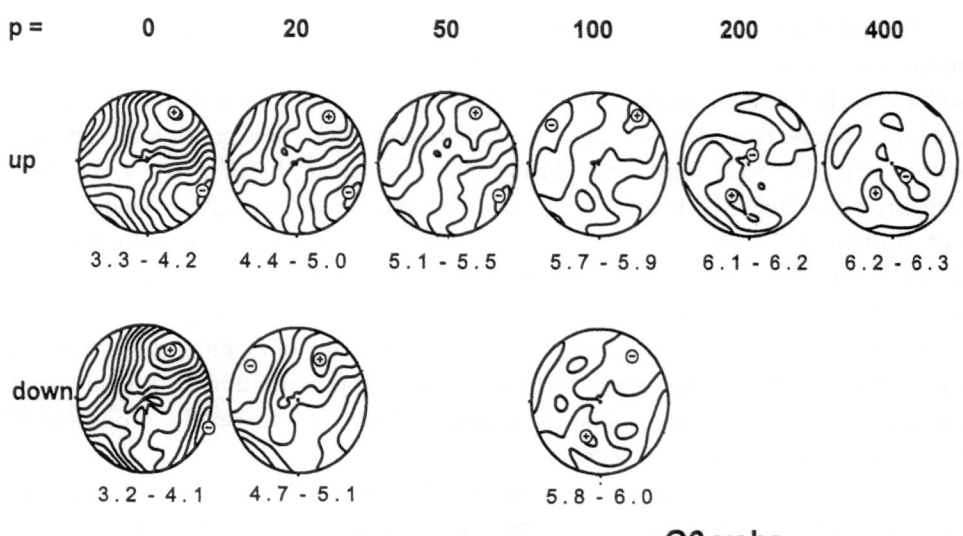

p =   0   20   50   100   200   400

up

3.3 - 4.2   4.4 - 5.0   5.1 - 5.5   5.7 - 5.9   6.1 - 6.2   6.2 - 6.3

down

3.2 - 4.1   4.7 - 5.1   5.8 - 6.0

G6-vobs

Figure 7
Isolines of *P*-wave velocities calculated from elastic constants (sample G6, granite). The isolines are displayed for the complete loading and unloading cycle. Upper line denotes measured pressure values *p* (in MPa). Numbers under the individual graphs denote appropriate velocity limits of isolines (in km/s).

The velocity-confining pressure equation, based on the superposition of the linear and exponential dependence, fits the experimental data very well. The calculated coefficients can be used to describe rock behaviour under hydrostatic pressure. The number of pressure values (at least 6) for which the *P*-wave velocity is determined, does not change the fitted coefficients significantly.

The comparison of Vp and Ap dependencies indicated that the amplitude changes of the ultrasonic signal should be used to study rock fabric due to their high sensitivity to the hydrostatic pressure applied. For hydrostatic pressures extending 400 MPa, velocities were observed to change by as much as 24%. On the contrary, amplitude changes were nearly 50 times.

We believe that quantified elastic parameters obtained by the above-described measuring approach can be used as a powerful tool for realistic seismic modelling of anisotropic crustal environments.

Acknowledgements

We thank Prof. H. Kern and an anonymous reviewer for their constructive comments which helped us improve the manuscript. Part of this project was supported by the Grant Agency of the Academy of Sciences of the Czech Republic, Grant No. A3012603 and also by the Grant Agency of the Czech Republic, Grant Nos. 205/95/0263 and 205/97/0905.

REFERENCES

BABUŠKA, V. (1966), *Velocity of Compressional Waves and Anisotropy of Some Igneous and Metamorphic Rocks*, Travaux Inst. Geophys. Acad. Tchecosl. Sci. *223*, 275–292.

BAYUK, E. I., VOLAROVICH, M. P., KLÍMA, K., PROS, Z., and VANĚK, J. (1967), *Velocity of Longitudinal Waves in Eclogite and Ultrabasic Rocks under Pressures to 4 Kilobars*, Studia Geoph. et Geod. *11*, 271–280.

BIRCH, F. (1960), *The Velocity of Compressional Waves in Rocks to 10 Kilobars*, Part I, J. Geophys. Res. *65*, 1083–1102.

BIRCH, F. (1961), *The Velocity of Compressional Waves in Rocks to 10 Kilobars*, Part II, J. Geophys. Res. *66*, 2199–2224.

GIESEL, W. (1963), *Elastische Anisotropie in tektonisch verformten Sedimentgesteinen*, Geophys. Prosp. *11*, 423–458.

GREENFIELD, R. J., and GRAHAM, E. K. (1996), *Application of a Simple Relation for Describing Wave Velocity as a Function of Pressure in Rocks Containing Microcracks*, J. Geophys. Res. *101*, 5643–5652.

KLÍMA, K. (1973), *The Computation of the Elastic Constants of an Anisotropic Medium from the Velocities of Body Waves*, Studia Geoph. et Geod. *17*, 115–122.

PROS, Z. (1977), *Investigation of anisotropy of elastic properties of rocks on spherical samples at high hydrostatic pressure*. In *High Pressure and Temperature Studies of Physical Properties of Rocks and Minerals* (eds. Volarovich, M. P. and Stiller, H.) (Naukova Dumka, Kijev 1977) pp. 56–67 (in Russian).

PROS, Z., and BABUŠKA, V. (1967), *A Method for Investigating the Elastic Anisotropy on Spherical Rock Samples*, Zeitschrift für Geophysik *33*, 289–291.

PROS, Z., and BABUŠKA, V. (1968), *An Apparatus for Investigating the Elastic Anisotropy on Spherical Samples*, Studia Geoph. et Geod. *12*, 192–198.

PROS, Z., and PODROUŽKOVÁ, Z. (1974), *Apparatus for Investigating the Elastic Anisotropy on Spherical Rock Samples at High Pressure*, Veröff. Zentralinst. Physik d. Erde *22*, 42–47.

PROS, Z., VANĚK, J., KLÍMA, K., and BABUŠKA, V. (1969), *Experimentelle Untersuchung des Wellenbildes bei der Ultraschall-Durchstrahlung einer Kugel*, Zeitschrift für Geophysik *35*, 287–296.

WEPFER, W. W., and CHRISTENSEN, N. I. (1991), *A Seismic Velocity-Confining Pressure Relation, With Applications*, Int. J. Rock. Mech. Min. Sci. and Geomech. Abstr. *28* (5), 451–456.

(Received December 6, 1996, revised June 15, 1997, accepted August 10, 1997)

Pure appl. geophys. 151 (1998) 631–646
0033–4553/98/040631–16 $ 1.50 + 0.20/0

© Birkhäuser Verlag, Basel, 1998

❙ Pure and Applied Geophysics

Elastic Parameters of West Bohemian Granites under Hydrostatic Pressure

Z. Pros,[1] T. Lokajíček,[1] R. Přikryl,[2] A. Špičák,[1] V. Vajdová[1]
and K. Klíma[1]

Abstract—The West Bohemian seismoactive region is situated near the contact of the Moldanubian, Bohemian and Saxothuringian units in which a large volume is occupied by granitoid massifs. The spatial distribution of P-wave velocities and the rock fabric of five representative samples from these massifs were studied. The P-wave velocities were measured on spherical samples in 132 independent directions under hydrostatic pressure up to 400 MPa, using the pulse-transmission method. The pressure of 400 MPa corresponds to a depth of about 15 km in the area under study. The changes of P-wave velocity were correlated with the preferred orientations of the main rock fabric elements, i.e., rock forming minerals and microcracks. The values of the P-wave velocity from laboratory measurements on granite samples fit the velocity model used by seismologists in the West Bohemian seismoactive region.

Key words: Granite, Bohemian Massif, P-wave velocity, velocity anisotropy, fabric elements, hydrostatic pressure.

Introduction

The western part of the Bohemian Massif occupies an extraordinary position in Central Europe from the point of view of recent geodynamic activity. It is characteristic namely in the periodic occurrence of intraplate earthquake swarms. Other uncommon phenomena are the occurrence of juvenile carbon dioxide waters, mineral springs, mofettes, young quaternary volcanism (age 0.8–2.8 Ma), steep gravity gradients, and increased heat flow.

The Hercynian granitoid massifs form a substantial part of the upper crust in the region under study. Seismic rays travelling from earthquake foci to local seismic stations propagate over most of their path through granitic rocks. Aware of the importance of granitic rocks in the region under study, we collected several oriented samples from those massifs.

For a better understanding and interpretation of existing seismic data, we have measured the P-wave velocity under hydrostatic pressure up to 400 MPa. For the

[1] Geophysical Institute, Academy of Sciences of the Czech Republic, Boční II/1401, 141 31 Prague 4, Czech Republic, E-mail: tl@ig.cas.cz
[2] Institute of Geochemistry, Mineralogy and Mineral Resources, Faculty of Science, Charles University, Albertov 6, 128 43 Prague 2, Czech Republic, E-mail: prikryl@mail.natur.cuni.cz

region under study, this value corresponds to the mean lithostatic pressure at a depth of about 15 km, which is the maximum focal depth observed in the region (HORÁLEK *et al.*, 1996).

The key question is why the West Bohemian region of intraplate seismicity responds so exclusively to a relatively uniform tectonic stress which dominates in this part of Europe? Can rock properties also contribute to the anomalous behaviour of the region? This major problem, together with the unending demand of seismologists for a reliable 3-D velocity model, has raised the need for knowledge of the physical properties of rocks composing the dominant geological units of the region.

Geology of the Region

The region is situated at the intersection of principal deep fault systems of the Ohře Graben (Litoměřice and Krušné hory faults running WSW–ENE) and the Tachov-Domažlice Graben (Tachov and Mariánské Lázně faults running NNW–SSE). These systems separate the main geological units of the Moldanubian, Saxothuringian, and Bohemian. Most faults, running both NE–SW and NW–SE, originated during Cadomian orogeny and were reprinted later on the platform cover of Cretaceous, Tertiary and even Quaternary sediments (DUDEK, 1987). The faults served as intrusion channels for the Hercynian granitoid massifs of Smrčiny and Karlovy Vary, and during Alpine orogeny their reactivation resulted in the formation of Tertiary basins accompanied by extensive alkaline volcanism.

Since recent denudation exposed Variscan granitoids as separated bodies (Fig. 1) (the largest are the Karlovy Vary and Smrčiny massifs), they are assumed to form one major differentiated granitic batholit, the so-called Krušné hory pluton (ZOUBEK, 1963, 1978). The analysis of geological and geophysical data indicates that this body extends to great depths (POLANSKÝ, 1977). POLANSKÝ and ŠKVOR (1975) consider this pluton to be a 10-km-thick flat body, resting on the crystalline rocks of the Saxothuringian unit. According to radiometric data, the Krušné hory pluton intruded into anticlinorial structures along tectonically predetermined lineaments between the Westfalian and Permian period (BERNARD and KLOMÍNSKÝ, 1975).

Rocks Studied

Five samples, representing all dominant types of granitic rocks of the region (3 from the Karlovy Vary massif and 2 from the Smrčiny massif), were chosen for the analysis (for location of sampling points see Fig. 1).

The Karlovy Vary massif represents the eastern part of the Krušné hory pluton (Fig. 1). Among several intrusive phases, two dominant events could be distinguished, based on petrological and geochronological criteria (e.g., ZOUBEK, 1963, 1978; ŠTEMPROK, 1986). The members of the older intrusive complex are medium- to coarse-grained biotite granites with porphyry of K-feldspars, regionally called mountain or normal granite (samples G5 and ZC19 in this study). The younger intrusive complex is represented by so-called autometamorphic two-mica and muscovite granites (sample G9).

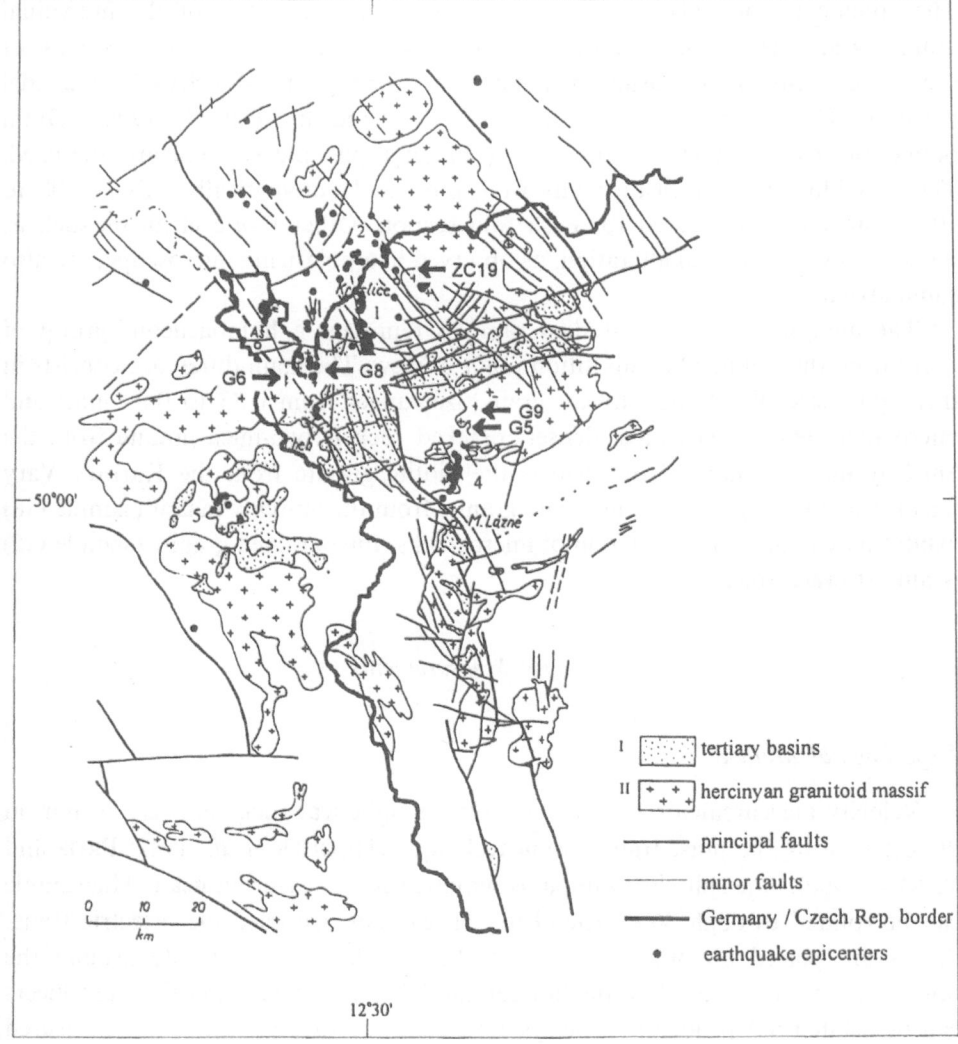

Figure 1

Geological sketch of the studied area. The geology has been compiled from (DUDEK, 1987), seismic foci (black dots) from HORÁLEK et al. (1996). Arrows indicate sample location.

The Smrčiny massif forms the western part of the Krušné hory pluton. Three granite types can be distinguished there. The porphyritic biotite granite represents the older part of the massif. Medium-grained two-mica granite is a younger member of the intrusion (sample G6). This rock type gradually passes to muscovite granite which exhibits signs of metasomatic alteration (sample G8).

Petrographic Characteristics

Model analyses and other characteristics of the samples are given in Table 1. The volume of minerals, as well as the geometric parameters of the individual grains (grain size, aspect ratio) were determined on oriented thin sections of samples G5, G6, G8, G9 using the image analysis program (SIGMASCAN, Jandel Scientific Corp.) and a measuring system devised by Přikryl (1998). Grain boundaries were manually outlined from photographs and subsequently digitized. The individual grain measurements were processed automatically. About 500 to 1000 grains were measured for each thin section. Other fabric elements such as microcracks and mineral alteration, as observed in a polarizing microscope, are also summarized in Table 1.

The analyses show that studied samples represent a homogeneous group of rocks from their mineral composition point of view. The main difference consists in their grain-size distribution (mean grain size ranges from 0.33 to 0.89 mm) and microcrack density. The most densely cracked rock is two-mica granite from the Smrčiny massif (sample G6) whereas the biotite granite from the Karlovy Vary massif (sample G9) and the muscovite granite from the Smrčiny massif (sample G8) exhibit only a medium population of microcracks. Fine-grained granite (sample G5) is almost crack-free.

Velocity Measurements

Experimental Method

Velocity measurements were carried out on spherical rock samples, 50 mm in diameter, using the pulse-transmission technique (HUGHES *et al.*, 1949; PROS and VANĚK, 1960). The spherical sample is mounted in a measuring head. The sample can be rotated through 360° around its vertical axis. Coupled piezoelectric transducers are placed on opposite sides of the sample and can rotate around the horizontal axis to $\pm 75°$ from the horizontal plane. The sphere and the transducers can be rotated independently. The system enables P-wave velocity to be measured in any direction (except in the area near the vertical axis of rotation) with the same accuracy. The net of measuring directions is usually established by dividing both coordinates in steps of 15°, thus defining 132 independent measuring directions.

Table 1

Description of samples under study (petrographic, geometric and physical characteristics)

Lithology	Unit	Sample number	Density [g·cm^{-3}]*	Porosity [%]	Grain size [mm]**	Grain shapes (shape ratios)	Modal composition [%]	Microstructures
granite	KVP	G5	2.651	n.d.	mean 0.36, range of 50% = 0.19–0.45, max. 2.49 (pl), 1.9 (Kf)	mean 2, max. 11.6 (Kf, pl), range of 50% = 1.42–2.2	35 (q), 24 (Kf), 33 (pl), 5 (mu), 3 (bi), +cl, ap	magmatic and heterogranular fabric q—equigranular grains with straight-forward boundaries, no microcracks; pl—subeuhedral grains, albite twinning, medium cracked (intragranular short cracks), all grains partly decomposed and replaced by very fine-grained micas (up to 90% of single grain volume); Kf—almost crack-free orthoclase; micas—platy and irregular grains
granite	SP	G6	2.627	2.89	mean 0.33, range of 50% = 0.16–0.39, max. 3 (pl), 2.1 (Kf)	mean 2.1, max. 10 (mu, pl), range of 50% = 1.5–2.4	36 (q), 20 (Kf), 32 (pl), 7 (mu), 5 (bi), +ap	magmatic and heterogranular fabric q—densely cracked (intra- and transgranular), undulatory extinction, lobate grain boundaries; pl—subeuhedral porphyry grains, larger grains partly decomposed (up to 20% of mineral volume) and replaced by very fine-grained mica, few intragranular cracks; Kf—very few microcracks in microcline; micas—platy and irregular grains
granite	SP	G8	2.621	2.98	mean 0.89, range of 50% = 0.37–1.1, max. 4.7 (pl), 4.1 (q)	mean 2.0, max. 11.3 (mu, pl), range of 50% = 1.4–2.3	42 (q), 20 (Kf), 30 (pl), 8 (mu), + bi, ap	magmatic and heterogranular fabric q—anhedral grains, long transgranular microcrack zones, healed microcracks, undulatory extinction, some lobate grain boundaries;

Table 1 (continued)

Lithology	Unit	Sample number	Density [g · cm⁻³]*	Porosity [%]	Grain size [mm]**	Grain shapes (shape ratios)	Modal composition [%]	Microstructures
								pl—subeuhedral porphyry grains, albite twinning, common intergrowths of tiny muscovite, few intragranular cracks; Kf—anhedral porphyry grains, plagioclas intergrowths, almost no microcracks; micas—platy and irregular grains
granite	KVP	G9	2.608	4.14	mean 0.64, range of 50% = 0.31–0.83, max. 4.41 (q), 4.0	mean 1.9, max. 8.4 (mu, pl), range of 50% = 1.4–2.2	40 (q), 22 (Kf), 30 (pl), 1 (mu), 7 (bi), + ap, cl	magmatic and heterogranular fabric q—anhedral grains, medium density of intra- and transgranular cracks, some grains with irregular grain boundaries; pl—sub- to euhedral grains, albite twinning, total alteration (sericite), few intragranular cracks; Kf—sub- to anhedral porphyry grains, no microcracks in orthoclase; micas—platy and irregular grains; densely cracked apatite grains
granite	KVP	ZC19	2.578	n.d.	n.d.	n.d.	40 (q), 20 (Kf), 30 (pl), 10 (bi)	n.d.

Explanation: KVP—Karlovy Vary pluton, SP—Smrčiny pluton, n.d.—not determined, q—quartz, pl—plagioclase, Kf—potassium feldspar, mu—muscovite, bi—biotite, cl—chlorite, ap—apatite;
* volume density of dry spherical samples, ** grain size expressed as the Feret diameter (diameter of a circle having the same areas as the measured grain), Modal composition, grain size and grain shapes of samples G5, G6, G8, and G9 measured by image analysis system of thin sections (using SIGMASCAN and SIGMASTAT software, Jandel Scientific Corp.) developed by R. Přikryl (1998). The modal composition of sample ZC19 determined using the standard point-counting method by J. Kotková (Czech Geological Survey, Prague).

The complete measuring system consists of a pressure vessel with transducers and sample positioning control, a pressure generator, a device for generating ultrasonic pulses, a travel-time measuring device and data acquisition unit. Several steps in the development of this system were described elsewhere (PROS and PODROUŽKOVÁ, 1974; PROS, 1977; PROS et al., 1998, this issue).

The P-wave velocity was measured at 0.1, 10, 20, 50, 100, 200 and 400 MPa while increasing pressure, and at 200, 100, 20, 0.1 MPa while reducing pressure. The measurement was conducted on vacuum dried samples.

Data Handling

At each pressure we obtained a P-wave velocity distribution in spherical coordinates λ and ϕ, related to the position of the sample in the measuring device. The coordinate system was carefully transferred to the spherical sample.

This so-called "basic source data set" enables us to construct the P-wave velocity distribution in the form of a map of isolines (PŠENČÍK, 1975; KLÍMA and PŠENČÍK, 1977), to determine real extreme velocities and the mean P-wave velocity as a weighted average of all independent measured directions for a given pressure. Another possibility is to construct the P-wave velocity versus pressure curve for any chosen direction.

Velocity-pressure Relationship

With regard to their behaviour under pressure, the studied granites can be divided into two groups: the first represented by samples G5 and ZC19 (both from the Karlovy Vary massif) and the second represented by samples G6, G8 (both from the Smrčiny massif) and G9 (Karlovy Vary massif).

Sample G5 displays a faint orthorhombic configuration of isolines at atmospheric pressure, but this pattern seems to disappear with increasing pressure (Fig. 2A). By eliminating data scattering we can detect this pattern even under high pressures.

The curves of minimum, mean and maximum P-wave velocities display a steep slope up to 50 MPa, and a gradual slope up to 400 MPa. The variation of velocity with pressure in this rock type, as well as the hysteresis observed, are the lowest of all the rocks studied. The highest coefficient of velocity anisotropy (defined as $k = [(v_{P_{max}} - v_{P_{min}})/v_{P_{mean}}]100[\%]$, modified after BIRCH (1961)) is observed at atmospheric pressure, and drops slightly with increasing pressure (compare Table 2). The difference between $v_{P_{max}}$ and $v_{P_{min}}$ remains nearly constant at about 300 m/s over the whole pressure range. The hysteresis hv_P (closure of microcracks that do not partly reopen during decrease of confining pressure) is computed by the following equation

$$hv_P = \{[v_P(0.1_a \text{ MPa}) - v_P(0.1_b \text{ MPa})/v_P(0.1_b \text{ MPa})\}100\%$$

$$v_P(0.1_b \text{ MPa}) - P\text{-wave velocity before sample loading}$$

$$v_P(0.1_a \text{ MPa}) - P\text{-wave velocity after sample unloading.}$$

The behaviour of sample ZC19 is very similar to G5. The only difference is at low pressure (less than 20 MPa) where the velocity isolines exhibit more pro-

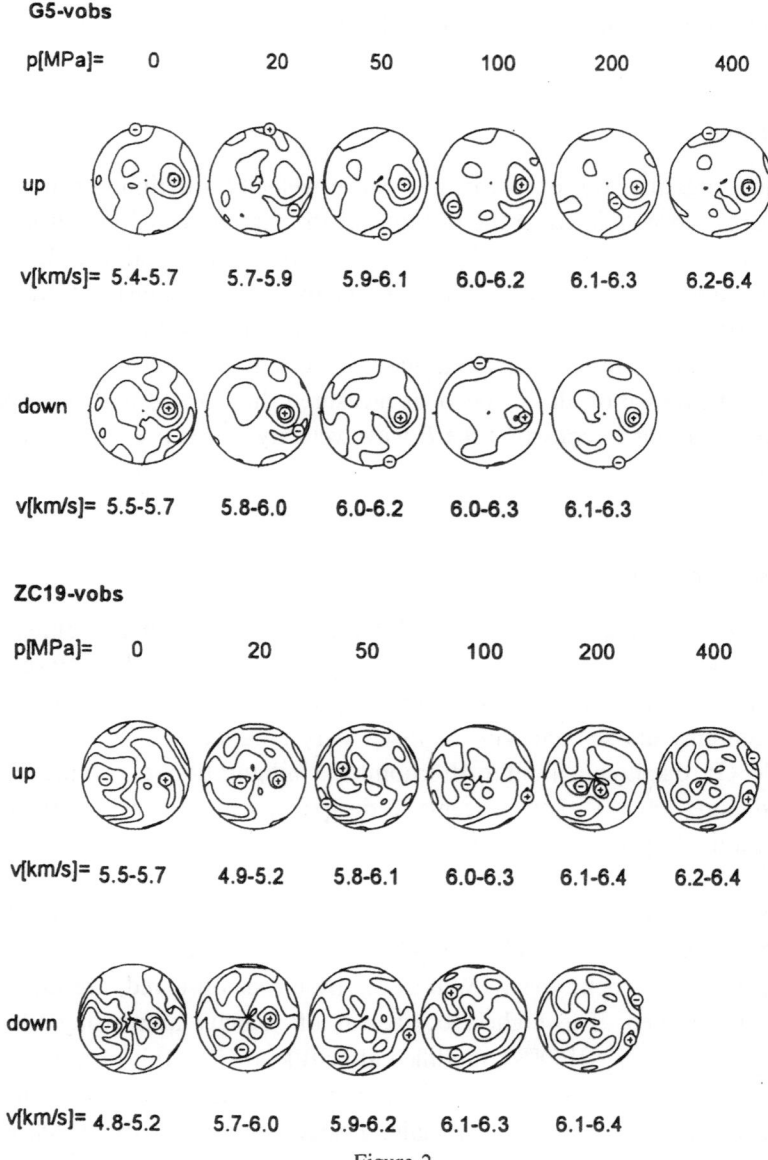

Figure 2

A, B: Behaviour of the spatial distribution of *P*-wave velocities at different hydrostatic pressures. A—sample G5, B—sample ZC19. Lower hemisphere, equal-area projection. Plus and minus signs indicate position of maximum and minimum observed *P*-wave velocity, respectively.

nounced orthorhombic symmetry (Fig. 2B). The increase of P-wave velocity with pressure up to 100 MPa is also higher than for G5. Once the microcracks close, the minimum, mean and maximum P-wave velocity versus pressure curve are nearly equidistant. The difference between $v_{P_{max}}$ and $v_{P_{min}}$ is about 400 m/s for the whole pressure range. The pronounced orthorhombic configuration of isolines based on the observed data disappears between hydrostatic pressure levels 20 and 50 MPa. By smoothing the isolines, the orthorhombic symmetry transforms to transverse isotropy, which persists up to maximum pressure, 400 MPa. After unloading the sample, the isoline pattern is restored to the orthorhombic configuration.

The second group of samples (G6, G8 and G9) exhibits a larger increase of velocities with increasing hydrostatic pressure, as well as a remarkable decrease in the coefficient of anisotropy (Table 2), mainly in the pressure range from 0.1 MPa to 200 MPa. All samples exhibit orthorhombic symmetry and high anisotropy at atmospheric pressure (Figs. 2C,D,E). The least pronounced configuration was observed in sample G8. The orthorhombic pattern of velocity distribution disappears at maximum hydrostatic pressure (400 MPa) and no further symmetry is observed (Fig. 2D). The original orthorhombic pattern in sample G6 that was observed at atmospheric pressure changes to the transverse isotropy symmetry at low confining pressure—between 50 and 100 MPa (Fig. 2C). Further increase of confinement disrupts the ordered pattern of P-wave velocities. While unloading, the symmetry of P-wave velocities develops in a counter manner.

Table 2

Numerical notations of samples under pressure

Sample	G5	ZC19	G6	G8	G9
$\Delta v_{P_{min}}$ [%]	12.2	24.1	53.3	77.3	61.0
$\Delta v_{P_{mean}}$ [%]	13.1	25.0	67.7	94.7	84.6
$\Delta v_{P_{max}}$ [%]	13.8	28.9	86.1	111.7	131.4
$hv_{P_{mean}}$ [%]	1.05	−0.26	−3.10	17.06	5.60
$k_{0.1}$ [%]	6.14	9.41	23.73	24.55	28.67
k_{400} [%]	4.74	5.71	4.40	6.76	6.01

Explanation of symbols used:

$$\Delta v_{P_{min}} = \{[v_{P_{min}}(400 \text{ MPa}) - v_{P_{min}}(0.1 \text{ MPa})]/v_{P_{min}}(0.1 \text{ MPa})\}100\%$$

$$\Delta v_{P_{max}} = \{[v_{P_{max}}(400 \text{ MPa}) - v_{P_{max}}(0.1 \text{ MPa})]/v_{P_{max}}(0.1 \text{ MPa})\}100\%$$

$$\Delta v_{P_{mean}} = \{[v_{P_{mean}}(400 \text{ MPa}) - v_{P_{mean}}(0.1 \text{ MPa})]/v_{P_{mean}}(0.1 \text{ MPa})\}100\%$$

$$hv_{P_{mean}} = \{[v_{P_{mean}}(0.12_a \text{ MPa}) - v_{P_{mean}}(0.1_b \text{ MPa})]/v_{P_{mean}}(0.1_b \text{ MPa})\}100\%$$

$$v_{P_{mean}}(0.1_b \text{ MPa}) - \text{mean velocity before sample loading}$$

$$v_{P_{mean}}(0.1_a \text{ MPa}) - \text{mean velocity after sample unloading}$$

$$k_i = [(v_{P_{max}} - v_{P_{min}})/v_{P_{mean}}]100\%, \quad \text{the index } i \text{ denotes pressure}$$

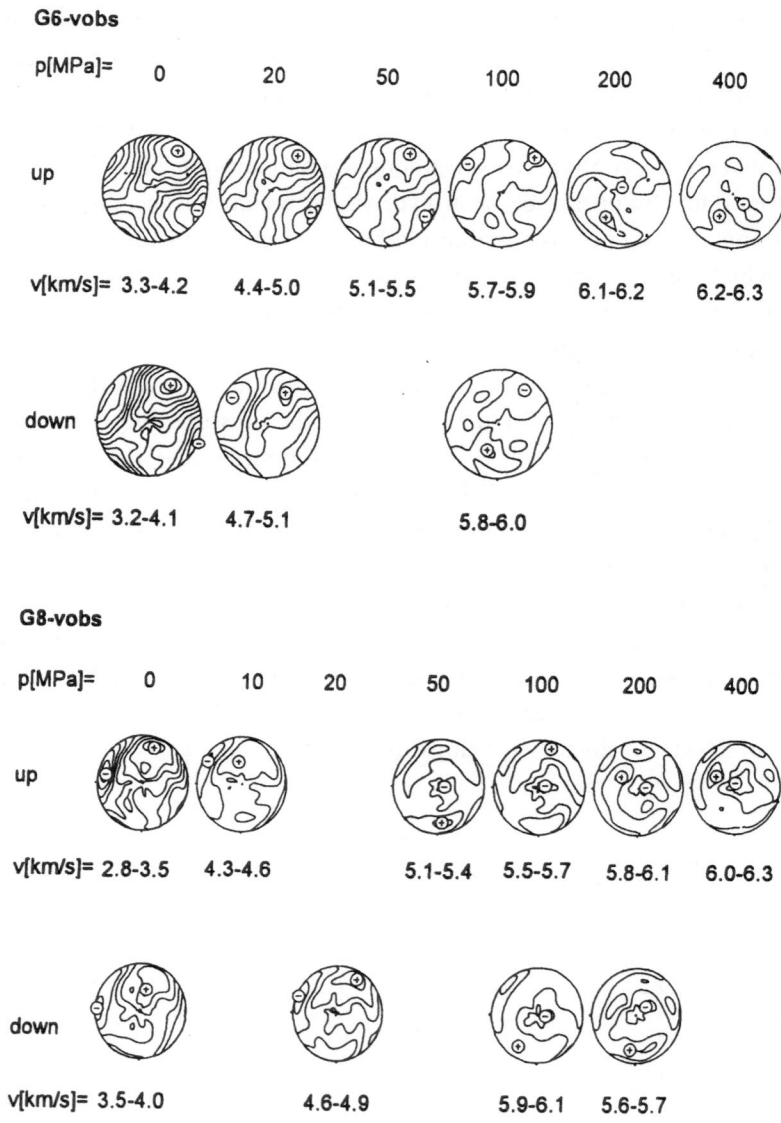

Figure 2

C, D: Behaviour of the spatial distribution of *P*-wave velocities at different hydrostatic pressures. C—sample G6, D—sample G8. Lower hemisphere, equal-area projection. Plus and minus signs indicate position of maximum and minimum observed *P*-wave velocity, respectively.

Discussion

The *P*-wave velocities of West Bohemian granites increase with increasing confining pressures (Table 3). The measured values are compared with published data in other studies (NUR and SIMMONS, 1969).

G9-vobs

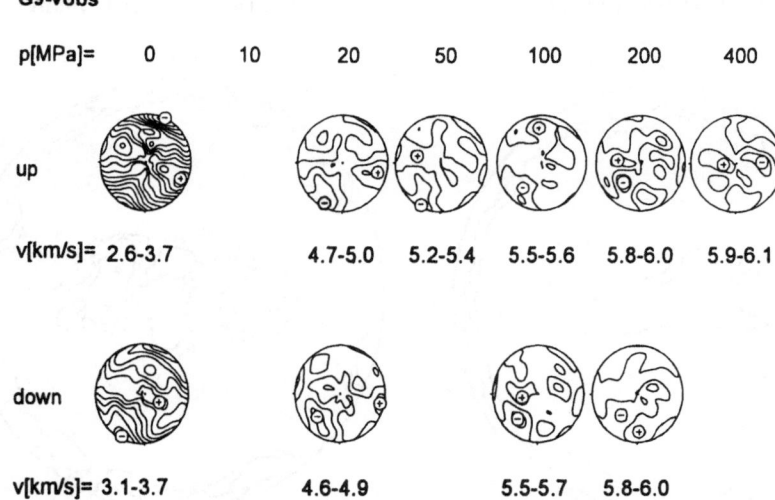

Figure 2

E: Behaviour of the spatial distribution of P-wave velocities at different hydrostatic pressures. E—sample G9. Lower hemisphere, equal-area projection. Plus and minus signs indicate position of maximum and minimum observed P-wave velocity, respectively.

The behaviour of P-wave velocities at low pressures could be explained by the existence of two populations of microcracks, e.g., sample G6. The first one, more pronounced, is perpendicular to the direction of $v_{P_{min}}$. The second one lies

Table 3

Changes of P-wave velocities versus pressure for West Bohemian granites (this study) and Casco, Westerly and Troy granites (NUR and SIMMONS, 1969)

Pressure [MPa]		0	20	100	200
	Density [g/cm^3]	Velocity [km/s]			
Casco granite	2.626	3.30	5.69	6.46	6.55
Westerly granite	2.646	3.80	5.31	5.85	6.06
Troy granite	2.670	4.40	6.17	6.45	—
G5-max	2.651	5.73	5.93	6.22	6.30
G5-mean	2.651	5.52	5.76	6.05	6.14
G6-max	2.627	4.19	5.09	5.97	6.27
G6-mean	2.627	3.74	4.74	5.78	6.10
G8-max	2.621	3.59	—	5.83	6.16
G8-mean	2.621	3.21	—	5.66	5.98
G9-max	2.608	3.87	5.09	5.79	6.11
G9-mean	2.608	3.29	4.83	5.61	5.90
ZC19-max	2.578	5.24	5.79	6.34	6.46
ZC19-mean	2.578	5.05	5.62	6.15	6.23

The mean values are computed from 132 directions, whereas the maximum values and the Nur and Simmons data are measured in one direction.

microcracks (poles)   micas (poles to (001) plane)

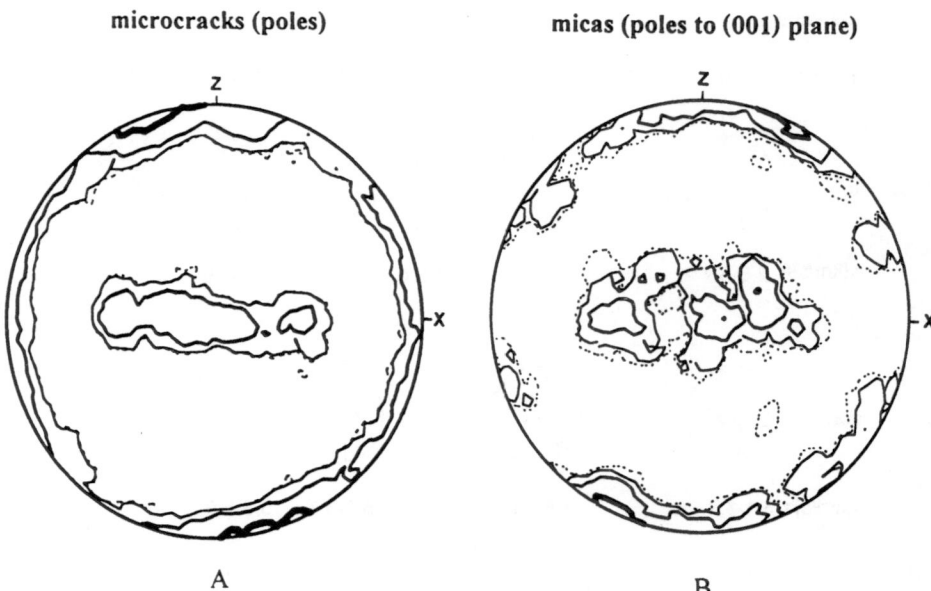

A   B

P-wave velocity

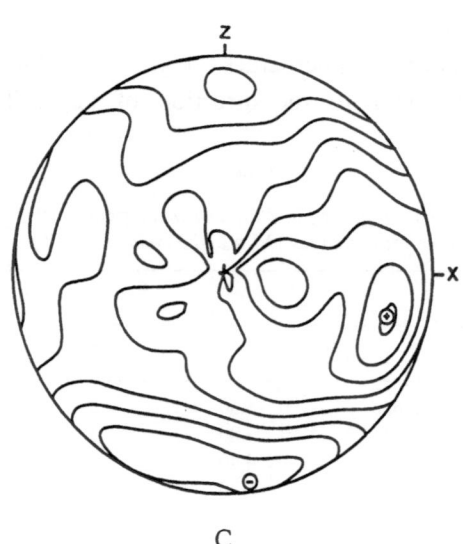

C

Figure 3

Preferred orientation of intragranular microcracks (A, 312 data, contours at 0.881, 1, 2.42 and 6.66%), poles to basal planes of micas (B, 139 data, contours at 1, 1.48, 3.06 and 6.33%) and *P*-wave velocity pattern for sample G8 at atmospheric pressure (C, 132 data, velocity range 2.7–3.5 km/s, steps of 0.1 km/s). All diagrams projected on the lower hemisphere, equal-area projection.

perpendicularly to the first one. Their intersection is parallel with the direction of $v_{P_{max}}$. Figure 2C shows that those microcrack populations are closed at different pressure levels.

To evaluate the effect of rock fabric and microcracks on the P-wave velocity distribution, we have measured the orientation of micas (poles to basal planes), plagioclase (poles to the plane of twinning (010)) and intragranular microcracks that are easily visible. The orientation was determined by light-polarizing microscopes equipped with a standard universal stage. The extremes of the distribution of micas and microcracks display good correlation with the P-wave velocity extremes at atmospheric pressure (see Fig. 3). It is obvious that micas also contribute to the elastic anisotropy of granites; however, it is not possible to evaluate the influence of micas and microcracks to the apparent elastic anisotropy separately. The content of micas in granites is significantly lower than that in metapelites and mylonites which exhibit high elastic anisotropy (BURLINI and FOUNTAIN, 1993; JI et al., 1993).

The fabric of plagioclase is not so well developed as the fabric of micas and microcracks. The measured c axes of quartz exhibit random distribution and thus do not contribute to the observed anisotropy of P-wave velocities in granites.

We assume that other fabric elements exhibit less substantial effect on the elastic anisotropy of granites than the presence of microcracks. A similar observation has been reported by BABUŠKA and PROS (1981).

The effect of hysteresis in high-pressure velocity measurements was observed and discussed in numerous studies (e.g., GARDNER et al., 1965; JI et al., 1993).

The comparison of P-wave velocities versus density proves the well-known fact that rocks of higher density exhibit higher elastic wave velocities. Even such closely related rocks as West Bohemian granites exhibit differences of maximum and mean P-wave velocities due to the diverse rock density (Fig. 4).

The laboratory velocities were compared with two seismic models proposed by NOVOTNÝ (1996). The laboratory velocities fit both models satisfactorily (Fig. 5). The granites with higher velocities (G5 and ZC19) display better correlation with Ore Mts. model, whereas the second group (granite samples G6, G8, and G9) correlates with the WB95 model (NOVOTNÝ, 1996). The data scatter at low pressures (see Fig. 5) could be attributed to the different populations of fractures and cracks. Increasing geostatic (hydrostatic under laboratory conditions) pressure thus weakens the effect of defects.

Conclusion

The seemingly isotropic granites under study exhibit distinct anisotropy of P-wave velocity under atmospheric and low hydrostatic pressure conditions. The P-wave velocity increases with pressure. The anisotropy of the granites under study decreases with increasing pressure by several percent; however, the differences between $v_{P_{max}}$ and $v_{P_{min}}$ remain almost unchanged. The ordered pattern of spatial distribution of elastic waves disappears.

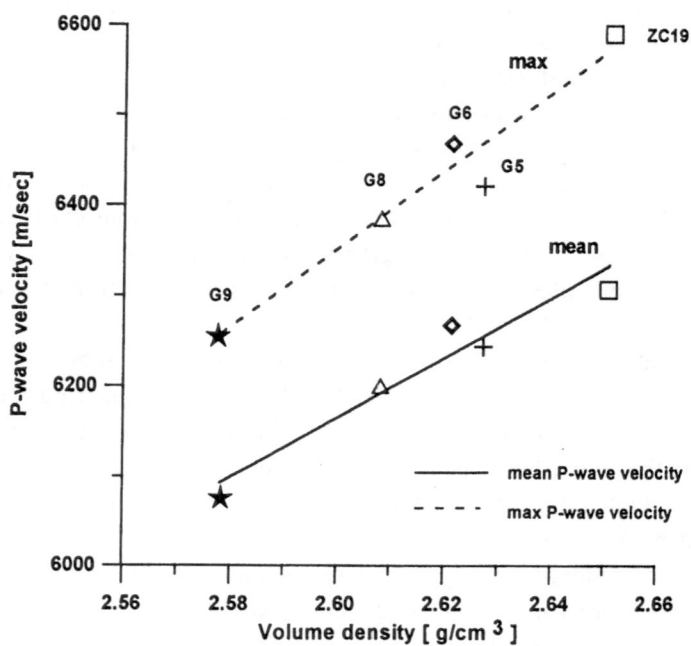

Figure 4

Relationship between rock density and mean and maximum *P*-wave velocities under hydrostatic pressure of 400 MPa.

From the observed behaviour of granite elastic properties and rock fabric, we deduce that the elastic anisotropy at atmospheric and low hydrostatic pressures is governed by the preferred orientation of microcracks. The ordered distribution of microcrack systems probably originated after denudation of the massif. The other systems of microcracks could have developed after the cooling of the massifs, or due to tectonic stress.

The laboratory measurements of *P*-wave velocities show good correlation with the seismic velocity profiles proposed for West Bohemia.

Acknowledgements

We are grateful to Dr. J. Kotková (Czech Geological Survey, Prague) for the petrographic analysis of sample ZC19, and to Mr. J. Vyhnal and Mr. J. Štís for preparing the samples. We thank two anonymous reviewers for the constructive comments that helped improve the manuscript.

We wish to acknowledge the financial support of the Grant Agency of the Czech Republic, Grant Nos. 205/94/1556 and 205/97/0905, of the Grant Agency of the

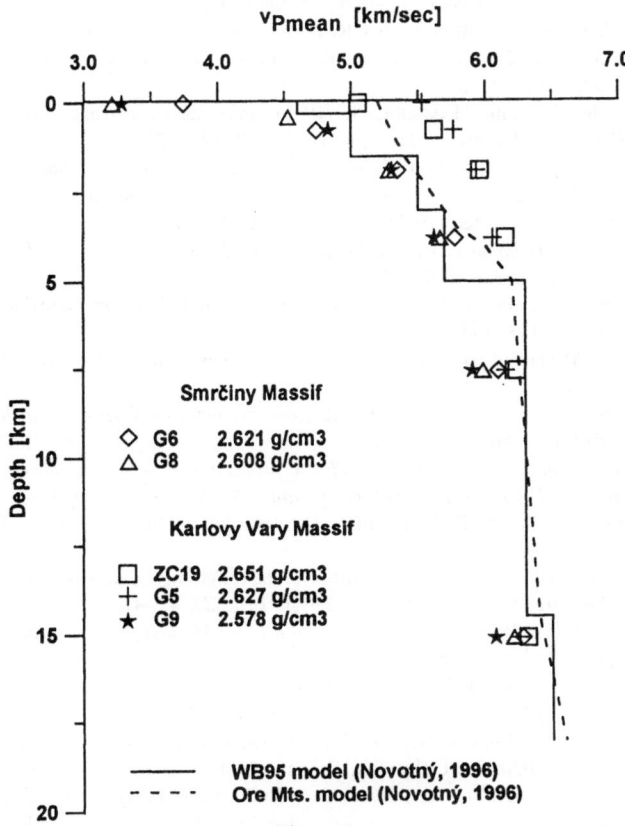

Figure 5

Comparison between seismic velocity models (NOVOTNÝ, 1996) and laboratory mean *P*-wave velocities of West Bohemian granites.

Academy of Sciences of the Czech Republic, Grant No. A3012603, and of the Grant Agency of the Charles University, Grant No. 20130.

REFERENCES

BABUŠKA, V., and PROS, Z. (1981), *Velocity Anisotropy in Granodiorite and Quartzite due to the Distribution of Microcracks*, Geophys. J. R. Astr. Soc. *76*, 121–127.

BERNARD, J., and KLOMÍNSKÝ, J. (1975), *Geochronology of the Variscan Plutonism and Mineralization in the Bohemian Massif*, Věst. Ústř. Úst. Geol. *50* (2), 71–82.

BIRCH, F. (1961), *The Velocity of Compressional Waves in Rocks to 10 Kilobars*, J. Geophys. Res. *66*, 2199–2224.

BURLINI, L., and FOUNTAIN, D. M. (1993), *Seismic Anisotropy of Metapelites from the Ivrea-Verbano Zone and Serie dei Laghi (Northern Italy)*, Phys. Earth Planet. Inter. *78*, 301–317.

DUDEK, A., *Geology and tectonic pattern of the Western Bohemian seismic area*. In *Earthquake Swarm 1985/86 in Western Bohemia (Proceedings of Workshop in Mariánské Lázně, December 1–5, 1986)* (ed. Procházková, D.) (Geophys. Inst. Czechosl. Acad. Sci., Praha 1987) pp. 34–37.

GARDNER, H. F., WYLLIE, M. R. J., and DROSCHAK, D. M. (1965), *Hysteresis in the Velocity-pressure Characteristics of Rocks*, Geophysics *30*, 111–116.

HORÁLEK, J., RUDAJEV, V., NOVOTNÝ, O., BOUŠKOVÁ, A., HAMPL, F., ŠÍLENÝ, J., BROŽ, M., FISCHER, T., JÍRA, T., and JANSKÝ, J. (1996), *Seismic Regime of the West-Bohemian Earthquake Swarms Region*, Acta Montana IRSM AS CR, Series AB, No. 2 (99), 59–69.

HUGHES, D. S., PONDROM, W. L., and MIMS, R. L. (1949), *Transmission of Elastic Pulses in Metal Rods*, Phys. Rev. *75*, 1552–1556.

JI, S., SALISBURY, M. H., and HAMMER, S. (1993), *Petrofabric, P-wave Anisotropy and Seismic Reflectivity of High-grade Tectonites*, Tectonophysics *222*, 195–226.

KLÍMA, K., and PŠENČÍK, I., *Processing of the velocity anisotropy of rocks and minerals*. In *High Pressure and Temperature Studies of Physical Properties of Rocks and Minerals* (eds. Volarovich, M. P. and Stiller, H.) (Naukova Dumka, Kiev 1977) pp. 78–87 (in Russian).

NOVOTNÝ, O. (1996), *A Preliminary Seismic Model for the Regions of the West-Bohemian Earthquake Swarms*, Studia Geophys. et Geod. *40*, 353–366.

NUR, A., and SIMMONS, G. (1969), *The Effect of Saturation on Velocity in Low Porosity Rocks*, Earth and Planet. Sci. Lett. *7*, 183–193.

POLANSKÝ, J. (1977), *Structural-tectonic Position of the Smrčiny Massif* (in Czech), Geol. Průzkum *19* (8), 227–229.

POLANSKÝ, J., and ŠKVOR, V. (1975), *Structural-tectonic Problems of Northwestern Bohemia* (in Czech), Sbor. geol. Věd.-užitá geofyzika *13*, 47–64.

PROS, Z. (1977), *Investigation of anisotropy of elastic properties of rocks on spherical samples at high hydrostatic pressure*. In *High Pressure and Temperature Studies of Physical Properties of Rocks and Minerals* (eds. Volarovich, M. P. and Stiller, H.) (Naukova Dumka, Kiev 1977) pp. 56–67 (in Russian).

PROS, Z., and PODROUŽKOVÁ, Z. (1974), *Apparatus for Investigating the Elastic Anisotropy on Spherical Samples at High Pressure*, Veröff. Zentralinst. Physic Erde *22*, 42–47.

PROS, Z., and VANĚK, J. (1960), *Experimental Study of a Pulse Method for Measuring Elastic Parameters of Rocks on Samples*, Studia geophys. et geodet. *4*, 338–349.

PROS, Z., LOKAJÍČEK, T., and KLÍMA, K. (1998), *Laboratory Approach to the Study of Elastic Anisotropy on Rock Samples*, Pure appl. geophys. *151*, 619–629.

PŘIKRYL, R., *The Effect of Rock Fabric on Some Mechanical Properties of Rocks: An Example of Granites*, Doctoral thesis (Charles University, Prague 1998).

PŠENČÍK, I. (1975), *Continuous Computer Contouring*, Studia geophys. et geodet. *19*, 184–187.

ŠTEMPROK, M. (1986), *Petrology and Geochemistry of the Czechoslovak Part of the Krušné Hory Mts. Granite Pluton*, Sbor. geol. Věd-ložisk. geol., mineral. *27*, 111–156.

ZOUBEK, V., *Tectonic control and structural evidence of the development of the Krušné hory (Erzgebirge) tin-bearing pluton*. In *Metallization Associated with Acid Magmatism* (eds. Štemprok, M., Burnol, L. and Tischendorf, G.) (Geological Survey, Praha 1978) pp. 57–76.

ZOUBEK, V., ed., *Explanation to General Geological Map of Czechoslovakia 1:200 000, M-33-XIII Karlovy Vary* (in Czech) (Geological Survey, Praha 1963).

(Received February 11, 1997, revised June 15, 1997, accepted August 10, 1997)

To access this journal online:
http://www.birkhauser.ch

IV. Mathematical Aspects of Complex Wave Propagation and their Applications

Pure appl. geophys. 151 (1998) 649–667
0033–4553/98/040649–19 $ 1.50 + 0.20/0

Inferring the Orientation-distribution Function from Observed Seismic Anisotropy: General Considerations and an Inversion of Surface-wave Dispersion Curves

WOLFGANG FRIEDERICH[1]

Abstract—A general relation linking the elasticity tensor of an anisotropic medium with that of the constituting single crystals and the function describing the orientation distribution of the crystals is derived. By expanding the orientation distribution function (ODF) into tensor spherical harmonics and using canonical components of the elasticity tensors, it is shown that the elastic tensor of the medium is completely determined by a finite number of expansion coefficients, namely those with harmonic degree $l \leq 4$. The number of expansion coefficients actually needed to determine the elastic constants of the medium depends on the symmetry of the single crystals. For hexagonal symmetry of the single crystals it is shown that only 8 real numbers are required to fix the 13 elastic constants which are for example needed to determine the azimuthal dependence of surface wave velocities. Thus, inversions of observations of seismic anisotropy are feasible which do not make any *a priori* assumptions on the orientation of the crystals. As a byproduct of the derivation, a formula is given which allows the easy calculation of the elastic constants of a medium composed of hexagonal crystals obeying an arbitrary ODF.

An application of the theoretical results to the inversion of surface wave dispersion curves for an anisotropic 1D-mantle model is presented. For the *S*-wave velocities the results are similar to those of previous inversions but the new approach also yields *P*-wave velocities consistent with the assumption of oriented olivine. Moreover it provides a hint of the orientation distribution of the crystals.

Key words: Anisotropy, elastic constants, inversion, ODF, surface waves.

Introduction

One of the major causes of seismic anisotropy in the earth's upper mantle is the preferential orientation of mantle minerals which in turn is attributed to finite strain (ZHANG and KARATO, 1995; RIBE, 1992). Thus, seismic anisotropy can be regarded as a diagnostic for past and current dynamic processes in the mantle (SILVER and CHAN, 1991). A wealth of evidence for the existence of seismic anisotropy in the upper mantle has been accumulated. The principal observations are shear-wave splitting (e.g., VINNIK *et al.*, 1994; SILVER and CHAN, 1991; BORMANN *et al.*, 1993; TONG *et al.*, 1994), azimuthal variation of P_n and surface wave velocities (e.g.,

[1] Institute of Geophysics, Stuttgart University, Richard-Wagner-Str. 44, 70184 Stuttgart, Germany.
Phone: +49-711-1213424, Fax: +49-711-2361218, E-Mail: wolle@geophys.uni-stuttgart.de

BAMFORD, 1977; RAITT *et al.*, 1969; KAWASAKI and KON'NO, 1984; MONTAGNER, 1990; MONTAGNER and TANIMOTO, 1991), discrepancies between dispersion curves of Love and Rayleigh waves (MAUPIN and CARA, 1992; CARA and LÉVÊQUE, 1988) and Love-Rayleigh coupling (YU and PARK, 1994).

The interpretation of the observations in terms of the elastic constants of the mantle, however, is hampered by a severe non-uniqueness caused by the limited sensitivity of the data to the elastic constants of the medium. Restriction of the observations to certain wave types or to special phenomena caused by seismic anisotropy further reduces the number of elastic constants that can be resolved.

On the other hand, much is known about the composition of typical mantle material from the analysis of xenolithes and natural peridotite massifs (PESELNICK and NICOLAS, 1978; MAINPRICE and SILVER, 1993). There is also substantial knowledge of the elastic constants of the constituting minerals from laboratory measurements on single and polycrystals (KUMAZAWA and ANDERSON, 1969; KERN, 1993). However, including this additional information on mantle materials within an interpretation of seismic anisotropy introduces additional difficulties: first, the composition of the upper mantle varies from place to place and can therefore only be regarded as approximately known. Second, the elastic constants of the single crystals depend on pressure and temperature, and temperature in the mantle is not known with great precision. Third, the elastic constants of a medium not only depend on the elastic constants of the constituting minerals but also on their orientation in space. In general, the orientation of the minerals will not be uniform but will be described by an orientation-distribution function (ODF). Introducing composition and ODF as additional parameters into an interpretation will increase the non-uniqueness instead of decreasing it.

In order to keep the number of unknowns small although still taking into account the information on mantle composition and single crystal elastic constants, the following *a priori* assumptions regarding the medium are now commonly made when interpreting observations of seismic anisotropy: the composition is either kept fixed or tight constraints are imposed on it. Olivine is assumed as the major cause of the observed anisotropy. In addition, it is assumed that a certain fraction of the olivine is perfectly oriented while the rest behaves isotropically.

In this paper I propose the use of a more realistic orientation model when interpreting seismic anisotropy. Can the inversion of seismic data for the anisotropic properties of the medium be made more general by working with the ODF without unnecessarily increasing the number of parameters? To answer this question I first derive the relation between the elastic tensor of the single crystal, the ODF and the elastic tensor of the polycrystalline assemblage. Using an expansion of the ODF into tensor spherical harmonics I show that the elastic constants of the assemblage only depend on a finite number of expansion coefficients of the ODF. Hence, working with the ODF only introduces a finite number of parameters into the inversion process. How many parameters there are depends on the symmetry of the single crystals. For hexagonal symmetry of the single crystals I demonstrate that 8 real numbers are sufficient to evaluate the 13 elastic constants required to determine the azimuthal dependence of seismic velocities. Thus, it is indeed possible to reduce the number of unknowns without making any assumptions about the form of the ODF. I propose a method built on the above ideas to invert Love and Rayleigh wave dispersion curves for depth-dependent expansion coefficients of the ODF and apply the method to surface wave dispersion curves representative of Southern Germany.

1. *Polycrystalline Elastic Constants and the ODF*

When dealing with anisotropy, generally two coordinate systems are involved: the global frame attached to the observer and the crystal frame attached to an individual crystal of the medium. Let g_i and k_m be orthonormal basis vectors in the global and crystal frame of reference, respectively. Then, the elasticity tensor can be written in two alternative forms: Either

$$\mathbf{E} = E_{Obs}^{ijkl} \sum_{ijkl} \mathbf{g}_i \mathbf{g}_j \mathbf{g}_k \mathbf{g}_l \tag{1}$$

in the global frame or

$$\mathbf{E} = E_{Cr}^{mnrs} \sum_{mnrs} \mathbf{k}_m \mathbf{k}_n \mathbf{k}_r \mathbf{k}_s \tag{2}$$

in the crystal frame. To describe the relative orientation of the observer and crystal coordinate systems I use Euler angles as defined in Figure 1. The three Euler angles parameterize three elementary rotations by which the crystal frame can be rotated into the observer frame. First the crystal frame is rotated by an angle ψ around the basis vector k_3 aligning the k_1 basis vector with the meridian. A rotation by an angle θ around the new k_2 basis vector brings the 3-axes of both coordinate systems into coincidence. Finally, there remains a rotation by an angle ϕ around the new 3-axis. By multiplying the transformation matrices of each elementary rotation one obtains the following relation between the basis vectors in the observer and crystal coordinate system:

$$\mathbf{g}_i = \sum_n R_{in} \mathbf{k}_n \tag{3}$$

with

$(R_{in}) =$

$$\begin{pmatrix} \cos\psi\,\cos\theta\,\cos\phi - \sin\psi\,\sin\phi & -\sin\psi\,\cos\theta\,\cos\phi - \cos\psi\,\sin\phi & \sin\theta\,\cos\phi \\ \cos\psi\,\cos\theta\,\sin\phi + \sin\psi\,\cos\phi & -\sin\psi\,\cos\theta\,\sin\phi + \cos\psi\,\cos\phi & \sin\theta\,\sin\phi \\ -\sin\theta\,\cos\psi & \sin\theta\,\sin\psi & \cos\theta \end{pmatrix}.$$

$$\tag{4}$$

If all crystals in a polycrystalline material have a common orientation the relation between the components of the elasticity tensor in both frames can be expressed as

$$E_{Obs}^{ijkl} = \sum_{mnrs} E_{Cr}^{mnrs} R_{im} R_{jn} R_{kr} R_{ls}. \tag{5}$$

Since the above relation with the explicit form of the matrix R is inaccessible to further analytical treatment it is favorable to switch to a new set of basis vectors—the canonical basis (GELFAND and SHAPIRO (1956); BURRIDGE (1969)):

$$\mathbf{f}_{-1} = \frac{1}{\sqrt{2}} (\mathbf{g}_1 - i\mathbf{g}_2)$$

$$\mathbf{f}_0 = \mathbf{g}_3$$

$$\mathbf{f}_1 = \frac{1}{\sqrt{2}} (-\mathbf{g}_1 - i\mathbf{g}_2). \tag{6}$$

Let the canonical components of the elasticity tensor in the global frame be denoted by C_{Obs}^{abcd} and in the crystal frame by C_{Cr}^{pqrs} where the indices now assume values between -1 and 1. Using these components instead of the Cartesian ones the transformation relation between the components simplifies dramatically:

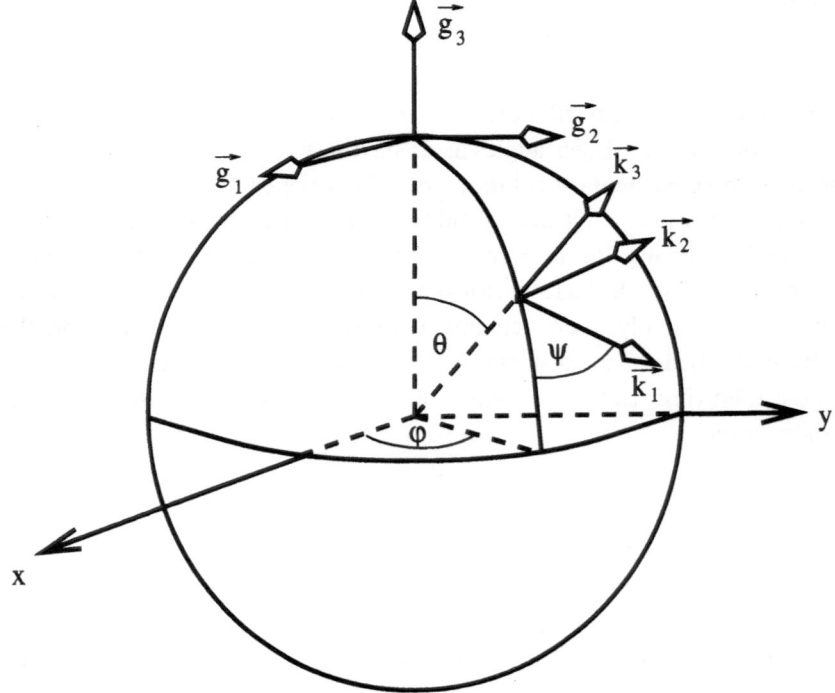

Figure 1
Sketch showing the definition of the three Euler angles parameterizing three elementary rotations by which the basis vectors $\mathbf{k}_i$ are rotated into the basis vectors $\mathbf{g}_i$.

$$C_{Obs}^{abcd} = \sum_{pqrs} C_{Cr}^{pqrs}$$

$$= e^{-i(p+q+r+s)\psi} P_1^{pa}(\theta) P_1^{qb}(\theta) P_1^{rc}(\theta) P_1^{sd}(\theta)\, e^{-i(a+b+c+d)\phi},$$

where the $P_l^{Nm}(\cos\theta)$ are generalized Legendre functions as defined by PHINNEY and BURRIDGE (1973). We write this relation more compactly by introducing the tensor spherical harmonics (GELFAND and SHAPIRO, 1956; BURRIDGE, 1969)

$$T_l^{Nm}(\phi, \theta, \psi) = e^{im\phi} P_l^{Nm}(\theta)\, e^{iN\psi} \tag{7}$$

and obtain with the asterisk denoting complex conjugation

$$C_{Obs}^{abcd} = \sum_{pqrs} C_{Cr}^{pqrs}\, (T_1^{pa} T_1^{qb} T_1^{rc} T_1^{sd})^*. \tag{8}$$

According to GELFAND and SHAPIRO (1956) a product of two tensor spherical harmonics of degree l and degree 1 can be written as a sum of tensor spherical harmonics of degree $l-1$, l and $l+1$:

$$T_1^{pa} T_l^{qb} = \sum_{j=-1}^{1} B_{aj}^{l,a+b} B_{pj}^{l,p+q} T_{l-j}^{p+q,a+b}, \tag{9}$$

where the coefficients B_{aj}^{lm} are given in GELFAND and SHAPIRO (1956) in eq. (9) on page 253 and are reproduced by BURRIDGE (1969) in eq. (A7). In both publications there is a printing error in the $(-1, 1)$-element (first row, third column) of B. The correct expression is

$$B_{-1,1}^{lm} = \sqrt{\frac{(l+m)(l+m+1)}{2l(2l+1)}}. \tag{10}$$

By repeated application of eq. (9) we can write the product of the four tensor spherical harmonics as follows

$$T_1^{pa} T_1^{qb} T_1^{rc} T_1^{sd} = \sum_{l=0}^{4} A_{l,pqrs}^{abcd}\, T_l^{p+q+r+s,a+b+c+d}(\phi, \theta, \psi) \tag{11}$$

and obtain for the transformation relation

$$C_{Obs}^{abcd} = \sum_{pqrs} C_{Cr}^{pqrs} \sum_{l=0}^{4} A_{l,pqrs}^{abcd}\,(T_l^{p+q+r+s,a+b+c+d}(\phi, \theta, \psi))^*. \tag{12}$$

Now, assume that the orientation of the single crystals is described by an ODF $f(\phi, \theta, \psi)$. Then, using the Voigt average the effective elastic constants of the medium are the average of the single crystal elastic constants weighted with the ODF:

$$C_{Obs}^{abcd} = \sum_{pqrs} C_{Cr}^{pqrs} \sum_{l=0}^{4} A_{l,pqrs}^{abcd} \int_0^{\pi} \int_0^{2\pi} \int_0^{2\pi} \sin\theta\, d\theta\, d\phi\, d\psi\, f(\phi, \theta, \psi)$$

$$\times (T_l^{p+q+r+s,a+b+c+d}(\phi, \theta, \psi))^*. \tag{13}$$

Since the T_l^{Nm} form a complete set of functions in orientation space (GELFAND and SHAPIRO, 1956), the ODF can be expanded into tensor spherical harmonics (see also BUNGE, 1982)

$$f(\phi, \theta, \psi) = \frac{1}{8\pi^2} \sum_{l=0}^{\infty} \sum_{N=-l}^{l} \sum_{m=-l}^{l} z_l^{Nm} T_l^{Nm}(\phi, \theta, \psi) \tag{14}$$

where the z_l^{Nm} are complex numbers with $z_0^{00} = 1$ in order to ensure that the ODF integrates to 1. Moreover, since the ODF is real we have the symmetry relations

$$z_l^{-p,-q} = (-1)^{p+q}(z_l^{pq})^*. \tag{15}$$

Inserting the expansion of the ODF into eq. (13) and applying the orthogonality relation for the tensor spherical harmonics

$$\int_0^\pi \int_0^{2\pi} \int_0^{2\pi} \sin\theta \, d\theta \, d\phi \, d\psi \, (T_l^{N,m})^* T_{l'}^{N'm'} = \frac{8\pi^2}{2l+1} \delta_{l,l'} \, \delta_{N,N'} \, \delta_{m,m'} \tag{16}$$

one obtains

$$C_{Obs}^{abcd} = \sum_{pqrs} C_{Cr}^{pqrs} \sum_{l=0}^{4} A_{l,pqrs}^{abcd} \frac{1}{2l+1} z_l^{p+q+r+s,a+b+c+d}. \tag{17}$$

From this result several conclusions can be drawn: the effective elastic constants of the medium are completely determined by a finite number of expansion coefficients, namely those with harmonic degrees $l \le 4$. A negative consequence of this fact is that with macroscopic seismology we will never be able to determine the orientation of mantle minerals. A positive consequence is that we may use eq. (17) to determine the elastic constants of the medium from observations of seismic anisotropy without any *a priori* assumptions pertaining to the ODF. Moreover, it is evident from eq. (17) that the number of expansion coefficients involved depends on the symmetry of the single crystals leading to a further reduction of parameters for high symmetries.

2. Hexagonal Symmetry of the Single Crystals

As one example of geophysical interest, I treat here the case of hexagonal symmetry of the single crystals. In this case, only the 19 single crystal elastic constants with $p + q + r + s = 0$ contribute to the sum in eq. (17) of which the following five are independent:

$$c_1^{Cr} = C_{Cr}^{0000} = C$$

$$c_2^{Cr} = C_{Cr}^{+0-0} = -L$$

$$c_3^{Cr} = C_{Cr}^{+,-00} = -F$$

$$c_4^{Cr} = C_{Cr}^{+,-+-} = A - N$$

$$c_5^{Cr} = C_{Cr}^{+,+--} = 2N. \tag{18}$$

Equation (17) simplifies to

$$C_{Obs}^{abcd} = \sum_{l=0}^{4} z_l^{0,a+b+c+d} \frac{1}{2l+1} \sum_{i=1}^{5} c_i^{Cr} F_l^{abcd,i}, \tag{19}$$

where the $F_l^{abcd,i}$ are derived from the $A_{l,pqrs}^{abcd}$ by collecting all terms in the sum over $pqrs$ belonging to one of the five independent elastic constants c_i^{Cr}. From observations of the azimuthal dependence of the velocities of seismic waves we have access to at most 13 of the 21 elastic constants. The reason is that the velocity for azimuth ϕ must be the same as for azimuth $\phi + \pi$. Hence, the azimuthal dependence of seismic velocities must be independent of the 8 elastic constants transforming according to $\exp(\pm im\phi)$, where m is odd. The remaining 13 elastic constants fall into three groups:

$a+b+c+d=0$: Into this group fall five constants which depend on the expansion coefficients z_0^{00}, z_2^{00} and z_4^{00}. Since $z_0^{00} = 1$ and the z_l^{00} are real due to the symmetry relations (15), only 2 real numbers are required to determine the 5 elastic constants.

$a+b+c+d=\pm2$: Into this group fall six elastic constants which depend on the expansion coefficients z_2^{02}, $z_2^{0,-2}$, z_4^{02}, $z_4^{0,-2}$. Since the z_l^{0m} are complex numbers, it seems at first sight that 8 real numbers are needed to determine 6 elastic constants. But again due to the symmetry relation (15) this is reduced to 4 independent real numbers.

$a+b+c+d=\pm4$: Into this group fall 2 elastic constants determined by the expansion coefficients z_4^{04}, $z_4^{0,-4}$. Again due to eq. (15) only 2 real numbers are required to determine 2 elastic constants.

The odd degree expansion coefficients are not involved because the $F_l^{abcd,i}$ vanish for odd l if $a+b+c+d$ is even.

In the appendix, I give values of the $F_l^{abcd,i}$ calculated with a computer algebra system for all the 13 elastic constants mentioned above. With these values it is possible by using eq. (19) to easily compute elastic constants for a medium consisting of crystals with hexagonal symmetry whose orientation is described by an arbitrary ODF. The simplest application of eq. (19) is the calculation of the equivalent isotropic elastic constants of a hexagonal crystal by setting all expansion coefficients to zero except z_0^{00}.

In summary, it is now clear that in the case of hexagonal symmetry of the single crystals the use of eq. (19) in an interpretation of seismic anisotropy reduces the number of parameters to only 8 real numbers which completely determine 13 elastic constants. Moreover, there exist simple relations between the elastic constants of the medium and the expansion coefficients. For example, the 5 elastic constants with $a + b + c + d = 0$ only depend on the two real expansion coefficients z_2^{00} and z_4^{00}.

3. Inversion of Surface Wave Dispersion Curves

In this section I propose a method to invert dispersion curves of Love and Rayleigh waves which is built on the general ideas outlined above. If the dispersion curves have been obtained by averaging over profiles of various azimuths, there is only information relative to the elastic constants with $a + b + c + d = 0$ which are invariant under a rotation of the observer coordinate system around the 3-axis. Of course, realistic upper mantle material does not consist of one mineral species only. However since the anisotropic properties of upper mantle material are likely to be dominated by olivine, I will neglect in the following the contributions of other minerals to seismic anisotropy. Let r denote the volume fraction of olivine in the upper mantle. Thereafter, again using the Voigt average which is not optimal but is used here for simplicity, the elastic constants of the medium can be written

$$C_{Obs}^{abcd} = (1 - r)C_R^{abcd} + rC_{Ol}^{abcd}, \tag{20}$$

where the subscript Ol indicates the olivine part and the subscript R denotes the contribution of the other minerals. Olivine has orthorhombic symmetry and therefore does not fit into the theory of the previous section. Replacing its elasticity tensor by one with hexagonal symmetry by taking girdles around the a-axis, however, produces elastic constants which are very similar to those of olivine itself. If the 3-axis of the coordinate system is taken parallel to the a-axis of olivine, there are only slight deviations in E_{1111}, E_{2222}, E_{1133} and E_{2233}, which should be acceptable for an inversion of surface wave dispersion curves. Assuming then that olivine can be described by five elastic constants c_i^{Ol}, eq. (19) can be applied with $a + b + c + d = 0$ yielding

$$c_k^{Obs} = (1 - r)c_k^R + r \sum_{l=0}^{4} \frac{z_l^{00}}{2l+1} \sum_{i=5}^{5} c_i^{Ol} F_l^{ki}, \tag{21}$$

where the F_l^{ki} are obtained from the $F_l^{abcd,i}$ by inserting the appropriate indices given in eq. (18). The F_l^{ki} form the first five rows of the tables in appendix A.

In a second step I decompose the elastic constants c_i^{Ol} and c_i^R into an isotropic and anisotropic part:

$$c_i = I_i + A_i, \tag{22}$$

where $I_1 = \lambda + 2\mu$, $I_2 = -\mu$, $I_3 = -\lambda$, $I_4 = \lambda + u$ and $I_5 = 2\mu$. The Lamé parameters are obtained from the c_i by the relations

$$\mu = \frac{1}{2} \sum_{i=1}^{5} c_i F_0^{5i} = \tfrac{1}{15}(A + C - 2F + 6L + 5N) \tag{23}$$

and

$$\lambda + 2\mu = \sum_{i=1}^{5} c_i F_0^{1i} = \tfrac{1}{15}(3C + 8L + 4F + 8A). \tag{24}$$

Neglecting the contribution to the seismic anisotropy from the other minerals, eq. (21) reduces to

$$c_k^{Obs} = (1 - r)I_k^R + r \sum_{l=0}^{4} \frac{z_l^{00}}{2l + 1} \sum_{i=1}^{5} (I_i^{Ol} + A_i^{Ol})F_l^{ki}. \tag{25}$$

Since the average of an isotropic elastic tensor over orientation space must reproduce itself regardless of the ODF it follows

$$\sum_i I_i^{Ol}F_2^{ki} = \sum_i I_i^{Ol}F_4^{ki} = 0$$

$$\sum_i I_i^{Ol}F_0^{ki} = I_k^{Ol}. \tag{26}$$

And because the purely anisotropic part of an elastic tensor averaged with a constant ODF vanishes it also follows that

$$\sum_{i=1}^{5} A_i^{Ol}F_0^{ki} = 0. \tag{27}$$

Thus, eq. (25) reduces to

$$c_k^{Obs} = (1 - r)I_k^R + rI_k^{Ol} + r \sum_{i=1}^{5} A_i^{Ol}(\tfrac{1}{5}F_2^{ki}z_2^{00} + \tfrac{1}{9}F_4^{ki}z_4^{00}). \tag{28}$$

At this stage there are several possibilities to formulate an inversion. Here, I combine the isotropic parts into effective constants λ_{eff} and μ_{eff} and fix the fraction of olivine to 0.7, a value which was measured by MAINPRICE and SILVER (1993) in South African kimberlite xenoliths and which is also representative for xenoliths found in southern Germany (ENDERELE et al., 1996) where the olivine content ranges from 56% to 90%. Moreover, I write for the anisotropic part

$$A_i^{Ol} = f_i I_{i,\text{ref}}, \tag{29}$$

where $I_{i,\text{ref}}$ are the elastic constants of an isotropic reference mantle model and the f_i give the relative amount of anisotropy. This locking of the anisotropic part to the reference model is done in order to avoid accounting for pressure and temperature dependence of the elastic constants of olivine. Values of the f_i are derived from experimental data by KUMAZAWA and ANDERSON (1969) after averaging the orthorhombic elastic tensor of olivine by taking girdles around the a-axis. Choosing the a-axis as the 3-axis of the crystal coordinate system, I obtain the following values for the f_i:

$$f_1 = \frac{C - (\lambda + 2\mu)}{\lambda + 2\mu} = 0.366$$

$$f_2 = \frac{L - \mu}{\mu} = -0.035$$

$$f_3 = \frac{F - \lambda}{\lambda} = -0.125$$

$$f_4 = \frac{A - N - (\lambda + \mu)}{\lambda + \mu} = -0.074$$

$$f_5 = \frac{N - \mu}{\mu} = -0.166.$$

4. Results

Dispersion curves of Love and Rayleigh waves for southern Germany (see FRIEDERICH and HUANG, 1996, for more detail) are inverted for the depth dependence of the five elastic constants c_i applying the method outlined above. Starting model is the isotropic model IASP91 (KENNETT and ENGDAHL, 1991) with density taken from PREM (DZIEWONSKI and ANDERSON, 1981). Parameters are the functions $\mu_{\text{eff}}(z)$, $z_2^{00}(z)$ and $z_4^{00}(z)$ which themselves are expanded into even-order Hermite-Gaussian functions below the Moho. For each function I use 9 parameters below the Moho plus one in the lower and upper crust, respectively. Density and $\lambda_{\text{eff}}(z)$ are kept fixed. Below the Moho a roughness constraint is imposed on all three functions and in addition I look for solutions $\delta\mu_{\text{eff}}(z)$, $z_2^{00}(z)$ and $z_4^{00}(z)$ with minimum L_2 norm. The results are expressed in terms of horizontal and vertical P-wave velocities, $v_{ph} = \sqrt{A/\rho}$, $v_{pv} = \sqrt{C/\rho}$, velocities of horizontally and vertically polarized S waves, $v_{sh} = \sqrt{N/\rho}$, $v_{sv} = \sqrt{L/\rho}$, and the parameter $\eta = F/(A - 2L)$ which has no obvious interpretation.

In previous work by FRIEDERICH and HUANG (1996) it was shown that polarization anisotropy is required in a depth range of 70 km to 250 km to explain both Love and Rayleigh dispersion curves. However, the data do not exclude the presence of anisotropy in the crust or in the deeper mantle. In the inversion discussed here anisotropy is suppressed in the crust and is forced to zero for depths greater than 350 km. In Figure 2 I present the results of the inversion. The top panels show the 1D models and the bottom panels the predicted and observed phase velocities with error bars. The left panels display the starting model and the right panels the inverted one. The parameter η which takes values close to 1 is not

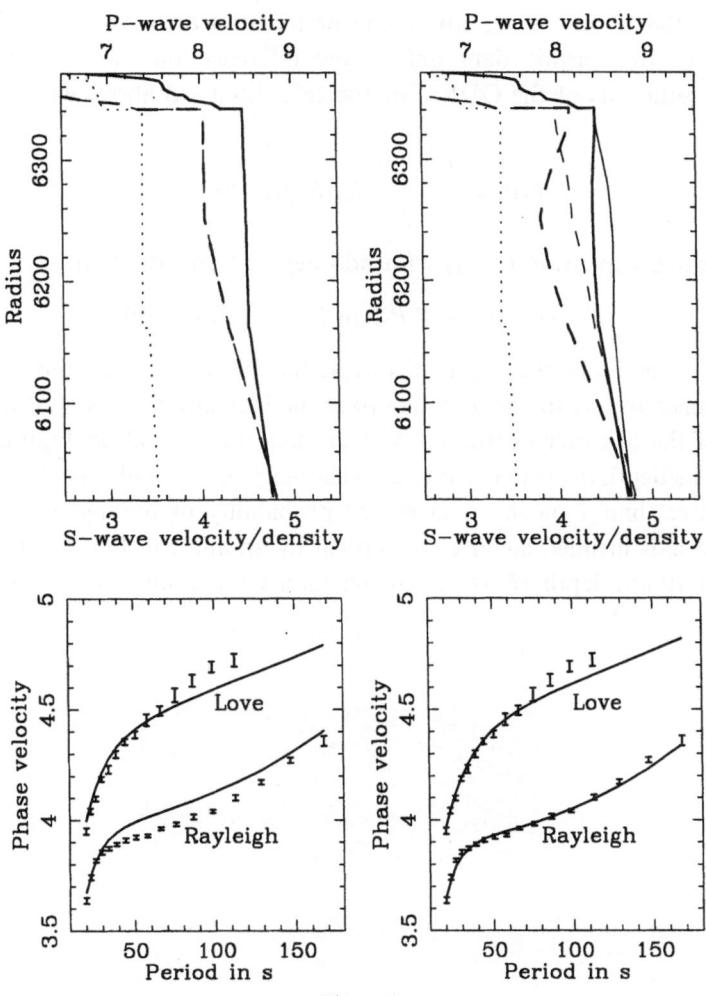

Figure 2

Results of the inversion of the dispersion curves for a 1D-mantle model. Top panels: Model parameters as a function of depth: density (dotted) in g/cm³, PV-wave velocity (thick dashed line), PH-wave velocity (thin dashed line), SH-wave velocity (thin solid line) and SV-wave velocity (thick solid line) in km/s. The parameter η which assumes values near 1 is not plotted. Bottom panels: Observed phase velocities with error bars in km/s together with the dispersion curve predicted by the corresponding mantle model. From left to right: isotropic initial model, inverted anisotropic model.

shown. Regarding the shear-wave velocities the result is quite similar to that obtained by FRIEDERICH and HUANG (1996) (FH96): a shear-wave anisotropy of up to 4 percent, mainly located in a depth range between 70 and 250 km. However in contrast to FH96 I now also get a considerable P-wave anisotropy. Since surface wave dispersion data are mainly sensitive to S-wave velocities the P-wave anisotropy may partially be a consequence of the chosen parameterization. Systematic studies might show how far the P-wave anisotropy can be reduced without abandoning the concept of preferentially oriented olivine.

What do the results tell us about the preferential orientation of olivine? Not very much as the seismic data only allow inference on the low order ($l \leq 4$) expansion coefficients of the ODF. For the case discussed above we can construct a function

$$F(\theta) = \int_0^{2\pi} \int_0^{2\pi} d\phi \, d\psi \, f(\phi, \theta, \psi), \tag{30}$$

whose truncated expansion ($l \leq 4$) with odd degree terms omitted,

$$F_T(\theta) = \tfrac{1}{2}(1 + z_2^{00} P_2(\cos \theta) + z_4^{00} P_4(\cos \theta)), \tag{31}$$

is available to us. Since the a-axis of olivine has been chosen as the 3-axis of the crystal reference frame, the angle θ describes the inclination of the fast axis against the 3-axis of the observer coordinate system, here the vertical. In Figure 3, I show $2F_T(\theta)$ and its depth dependence in a 2D-shading plot. Roughly speaking, values of $2F_T(\theta)$ greater than 1 mean an increased probability of finding olivine minerals with the fast axis inclined against the vertical by an angle θ. Below the Moho and above about 70 km depth $2F_T(\theta)$ is greater than 1 for small values of θ, indicating

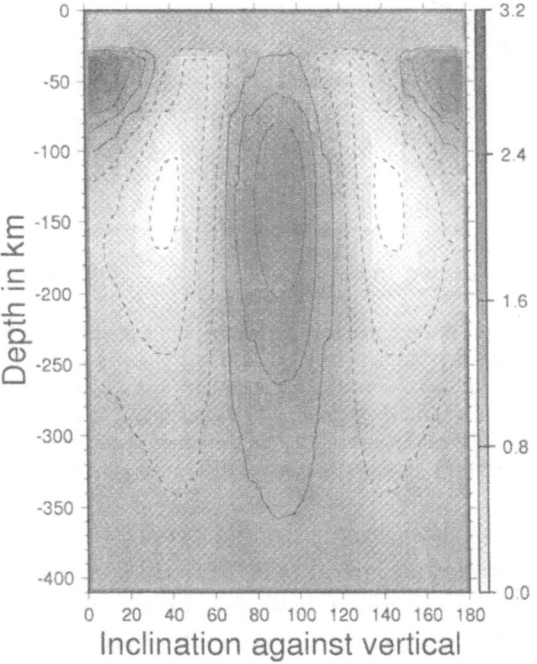

Figure 3
2D-shading plot with contour lines of the function $2F(\theta)$ introduced in eq. (31) and its dependence on depth. Values greater than 1 indicate an enhanced probability to find olivine minerals with an inclination against the vertical by the given angle.

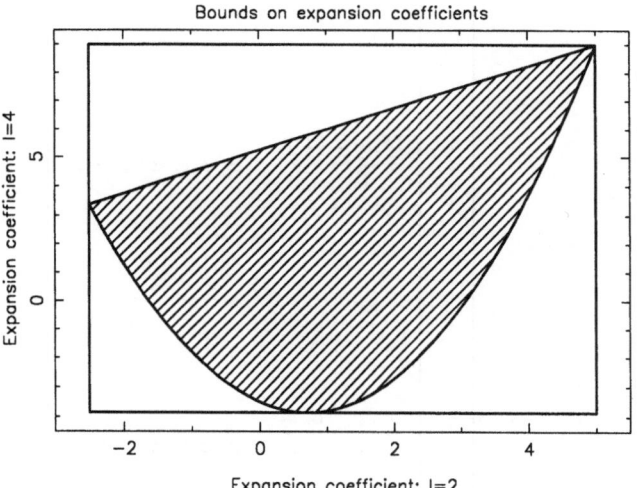

Figure 4
Sketch visualizing the necessary (rectangular box) and sufficient but probably not necessary (hatched area) conditions for any pair of real numbers (z_2^{00}, z_4^{00}) to be valid expansion coefficients of an ODF.

preferential orientation along the vertical. Below 70 km down to 250 km depth the orientation of the fast axis is preferentially horizontal ($\theta = 90°$).

The results shown in Figure 3 are hard to reconcile with the strong azimuthal dependence of P_n velocities found by BAMFORD (1977). If it is true that the a-axis of the olivine crystals just below the Moho is preferentially oriented along the vertical it appears to be very difficult to generate a significant azimuthal anisotropy of P waves due to the small difference between velocities along the b- and the c-axes of olivine.

One final remark concerning the interpretation of the truncated ODF is in order. Obviously, $F_T(\theta)$ is not a genuine ODF as it can assume negative values. It may even happen that an inversion yields values for z_2^{00} and z_4^{00} that cannot be obtained from a genuine ODF. What is needed are necessary and sufficient conditions for z_2^{00} and z_4^{00} to be valid expansion coefficients of an ODF. These conditions would define a region in the two-dimensional real vector space $\mathbb{R}^2$ within which a vector (z_2^{00}, z_4^{00}) has to lie in order to represent valid expansion coefficients of an ODF. Although I am not able to specify this region, it is at least possible to give necessary conditions for the expansion coefficients by using the positivity property of the ODF. In Appendix B, I show that the following bounds apply:

$$l = 2: \quad -\tfrac{5}{2} \le z_2^{00} \le 5$$
$$l = 4: \quad -\tfrac{27}{7} \le z_4^{00} \le 9. \tag{32}$$

They describe the rectangular region in $\mathbb{R}^2$ shown in Figure 4. The lower bounds are attained with a delta-like ODF of the form

$$F(\theta) = \frac{\delta(\theta - \theta_0)}{\sin \theta} \tag{33}$$

Table 1

Values of $F_0^{abcd,i}$

$l=0$	C_{Cr}^{0000}	C_{Cr}^{+0-0}	C_{Cr}^{+-00}	C_{Cr}^{+-+-}	C_{Cr}^{+-+--}
C_{Obs}^{0000}	1/5	−8/15	−4/15	8/15	4/15
C_{Obs}^{+0-0}	−1/15	2/5	−2/15	−1/15	−1/5
C_{Obs}^{+-00}	−1/15	−4/15	8/15	−2/5	2/15
C_{Obs}^{+-+-}	2/15	−2/15	−2/5	7/15	1/15
C_{Obs}^{+---}	2/15	−4/5	4/15	2/15	2/5
C_{Obs}^{++00}	0	0	0	0	0
C_{Obs}^{+0+0}	0	0	0	0	0
C_{Obs}^{+++-}	0	0	0	0	0
C_{Obs}^{--00}	0	0	0	0	0
C_{Obs}^{-0-0}	0	0	0	0	0
C_{Obs}^{---+}	0	0	0	0	0
C_{Obs}^{++++}	0	0	0	0	0
C_{Obs}^{----}	0	0	0	0	0

Table 2

Values of $F_2^{abcd.i}$

$l=2$	C_{Cr}^{0000}	C_{Cr}^{+0-0}	C_{Cr}^{+-00}	C_{Cr}^{+-+-}	C_{Cr}^{++--}
C_{Obs}^{0000}	$4/7$	$-8/21$	$-4/21$	$-16/21$	$-8/21$
C_{Obs}^{+0-0}	$-1/21$	$1/7$	$-2/21$	$-1/21$	$1/7$
C_{Obs}^{+-00}	$-1/21$	$-4/21$	$5/21$	$2/7$	$-4/21$
C_{Obs}^{+-+-}	$-4/21$	$-2/21$	$2/7$	$10/21$	$-2/21$
C_{Obs}^{++--}	$-4/21$	$4/7$	$-8/21$	$-4/21$	$4/7$
C_{Obs}^{+-+00}	$(1/21)\sqrt{2}\sqrt{3}$	$(4/21)\sqrt{2}\sqrt{3}$	$-(5/21)\sqrt{2}\sqrt{3}$	$-(2/7)\sqrt{2}\sqrt{3}$	$(4/21)\sqrt{2}\sqrt{3}$
C_{Obs}^{+0+0}	$(1/21)\sqrt{2}\sqrt{3}$	$-(1/7)\sqrt{2}\sqrt{3}$	$(2/21)\sqrt{2}\sqrt{3}$	$(1/21)\sqrt{2}\sqrt{3}$	$-(1/7)\sqrt{2}\sqrt{3}$
C_{Obs}^{+++-}	$-(1/7)\sqrt{2}\sqrt{3}$	$(2/21)\sqrt{2}\sqrt{3}$	$(1/21)\sqrt{2}\sqrt{3}$	$(4/21)\sqrt{2}\sqrt{3}$	$(2/21)\sqrt{2}\sqrt{3}$
C_{Obs}^{-00}	$(1/21)\sqrt{2}\sqrt{3}$	$(4/21)\sqrt{2}\sqrt{3}$	$-(5/21)\sqrt{2}\sqrt{3}$	$-(2/7)\sqrt{2}\sqrt{3}$	$(4/21)\sqrt{2}\sqrt{3}$
C_{Obs}^{-0-0}	$(1/21)\sqrt{2}\sqrt{3}$	$-(1/7)\sqrt{2}\sqrt{3}$	$(2/21)\sqrt{2}\sqrt{3}$	$(1/21)\sqrt{2}\sqrt{3}$	$-(1/7)\sqrt{2}\sqrt{3}$
C_{Obs}^{--+}	$-(1/7)\sqrt{2}\sqrt{3}$	$(2/21)\sqrt{2}\sqrt{3}$	$(1/21)\sqrt{2}\sqrt{3}$	$(4/21)\sqrt{2}\sqrt{3}$	$(2/21)\sqrt{2}\sqrt{3}$
C_{Obs}^{+++}	0	0	0	0	0
C_{Obs}^{---}	0	0	0	0	0

Table 3

Values of $F_4^{abcd,i}$

$l=4$	C_{Cr}^{0000}	C_{Cr}^{+0-0}	C_{Cr}^{+-00}	C_{Cr}^{+-+-}	C_{Cr}^{+---}
C_{Obs}^{0000}	8/35	32/35	16/35	8/35	4/35
C_{Obs}^{+0-0}	4/35	16/35	8/35	4/35	2/35
C_{Obs}^{+-00}	4/35	16/35	8/35	4/35	2/35
C_{Obs}^{+-+-}	2/35	8/35	4/35	2/35	1/35
C_{Obs}^{+---}	2/35	8/35	4/35	2/35	1/35
C_{Obs}^{++00}	$(2/35)\sqrt{2}\sqrt{5}$	$(8/35)\sqrt{2}\sqrt{5}$	$(4/35)\sqrt{2}\sqrt{5}$	$(2/35)\sqrt{2}\sqrt{5}$	$(1/35)\sqrt{2}\sqrt{5}$
C_{Obs}^{+0+0}	$(2/35)\sqrt{2}\sqrt{5}$	$(8/35)\sqrt{2}\sqrt{5}$	$(4/35)\sqrt{2}\sqrt{5}$	$(2/35)\sqrt{2}\sqrt{5}$	$(1/35)\sqrt{2}\sqrt{5}$
C_{Obs}^{+++-}	$(1/35)\sqrt{2}\sqrt{5}$	$(4/35)\sqrt{2}\sqrt{5}$	$(2/35)\sqrt{2}\sqrt{5}$	$(1/35)\sqrt{2}\sqrt{5}$	$(1/70)\sqrt{2}\sqrt{5}$
C_{Obs}^{--00}	$(2/35)\sqrt{2}\sqrt{5}$	$(8/35)\sqrt{2}\sqrt{5}$	$(4/35)\sqrt{2}\sqrt{5}$	$(2/35)\sqrt{2}\sqrt{5}$	$(1/35)\sqrt{2}\sqrt{5}$
C_{Obs}^{-0-0}	$(2/35)\sqrt{2}\sqrt{5}$	$(8/35)\sqrt{2}\sqrt{5}$	$(4/35)\sqrt{2}\sqrt{5}$	$(2/35)\sqrt{2}\sqrt{5}$	$(1/35)\sqrt{2}\sqrt{5}$
C_{Obs}^{---+}	$(1/35)\sqrt{2}\sqrt{5}$	$(4/35)\sqrt{2}\sqrt{5}$	$(2/35)\sqrt{2}\sqrt{5}$	$(1/35)\sqrt{2}\sqrt{5}$	$(1/70)\sqrt{2}\sqrt{5}$
C_{Obs}^{++++}	1	$(8/35)\sqrt{2}\sqrt{5}\sqrt{7}$	$(4/35)\sqrt{2}\sqrt{5}\sqrt{7}$	1	1
C_{Obs}^{----}	1	$(8/35)\sqrt{2}\sqrt{5}\sqrt{7}$	$(4/35)\sqrt{2}\sqrt{5}\sqrt{7}$		

where $\cos \theta_0 = 0$ for $l = 2$ and $\cos \theta_0 = \pm \sqrt{3/7}$ for $l = 4$. The upper bounds are attained with the same ODF and $\cos \theta_0 = \pm 1$.

Furthermore, it is also possible to specify sufficient conditions on the expansion coefficients which are however not necessary. The region in $\mathbb{R}^2$ formed by these conditions is the hatched area shown in Figure 4. The lower curve limiting this area is a parabola described by

$$z_2^{00} = 5P_2(\cos \theta_0)$$

$$z_4^{00} = 9P_4(\cos \theta_0), \tag{34}$$

where θ_0 runs from 0 to $\pi/2$. Pairs of coefficients lying on this parabola are produced by a delta-like ODF of the form (33). The upper limiting curve is a straight line connecting the end points of the parabola. Points on this curve result from an ODF of the form

$$F(\theta) = \frac{1}{\sin \theta} (\alpha \delta(\theta - \pi/2) + \beta \delta(\theta)) \tag{35}$$

with $\alpha + \beta = 1$. For all points within the hatched area there exist at least ODFs formed by a linear combination of two delta functions.

5. Conclusions

It is shown that an interpretation of seismic anisotropy in terms of an arbitrary orientation distribution function is possible and also feasible for high symmetry of the single crystals. For hexagonal symmetry in particular, an inversion for the 13 elastic constants describing the azimuthal dependence of seismic velocities can be reduced to an inversion for 8 real numbers which are linked to the expansion coefficients of the ODF into tensor spherical harmonics. No special assumptions on the orientation of the crystals are further needed. Unfortunately, it is impossible to infer the true ODF since only the low-order expansion coefficients of the ODF are involved in computing the elastic constants. Thus, the method presented in this paper should be seen as a way to parameterize an inversion for the elastic constants of an anisotropic medium and not as a way to determine the ODF. If the bounds depicted in Figure 4 are incorporated as constraints into an inversion scheme it can also be ensured that the resulting coefficients are actually valid expansion coefficients of a semi-positive ODF.

Appendix A

Here, I list explicit values of the quantity $F_l^{abcd,i}$ introduced in eq. (19) and reproduced here for convenience:

$$C_{Obs}^{abcd} = \sum_{l=0}^{4} z_l^{0,a+b+c+d} \frac{1}{2l+1} \sum_{i=1}^{5} c_i^{Cr} F_l^{abcd,i}.$$

The values are given in Tables 1–3 for $l = 0, 2, 4$, respectively. $F_l^{abcd,i}$ vanishes for $l = 1, 3$ if $a+b+c+d$ is even.

Appendix B

Here I give a proof for the bounds on the expansion coefficients stated in eq. (32). Due to the orthogonality relation of the Legendre polynomials, the expansion coefficients are given by

$$z_l^{00} = (2l+1) \int_0^{\pi} F(\theta) P_l(\cos \theta) \sin \theta \, d\theta. \tag{36}$$

Because $F(\theta)$ is a semi-positive function the mean value theorem of integral calculus states that

$$z_l^{00} = (2l+1)c \int_0^{\pi} F(\theta) \sin \theta \, d\theta \tag{37}$$

where $P_{l,\min} \leq c \leq P_{l,\max}$. Since $F(\theta)$ integrates to 1 it follows

$$(2l+1)P_{l,\min} \leq z_l^{00} \leq (2l+1)P_{l,\max}. \tag{38}$$

With $P_{2,\min} = -1/2$, $P_{4,\min} = -3/7$ and $P_{l,\max} = 1$ one obtains the bounds on z_l^{00} stated in eq. (32).

REFERENCES

BAMFORD, D. (1977), P_n Velocity Anisotropy in a Continental Mantle, Geophys. J. R. Astr. Soc. 49, 29–48.
BORMANN, P., BURGHARDT, P. T., MAKEYEVA, L. I., and VINNIK, L. P. (1993), Teleseismic Shear Wave Splitting and Deformations in Central Europe, Phys. Earth Planet. Int. 78, 157–166.
BUNGE, H. J., Texture Analysis in Material Science (Butterworths, London, 1982).
BURRIDGE, R. (1969), Spherically Symmetric Differential Equations, the Rotation Group, and Tensor Spherical Functions, Proc. Camb. Phil. Soc. 65, 157–175.
CARA, M., and LÉVÊQUE, J. J. (1988), Anisotropy of the Asthenosphere: The Higher Mode Data of the Pacific Revisited, Geophys. Res. Lett. 15, 205–208.
DZIEWONSKI, A. M., and ANDERSON, D. L. (1981), Preliminary Reference Earth Model, Phys. Earth Planet. Int. 25, 297–356.
ENDERLE, U., MECHIE, J., SOBOLEV, S., and FUCHS, K. (1996), Seismic Anisotropy within the Uppermost Mantle of Southern Germany, Geophys. J. Int. 125, 747–767.
FRIEDERICH, W., and HUANG, Z.-X. (1996), Evidence for Upper Mantle Anisotropy beneath Southern Germany from Love and Rayleigh Wave Dispersion, Geophys. Res. Lett. 23, 1135–1138.
GELFAND, I. M., and SHAPIRO, Z. YA (1956), Representations of the Group of Rotations of 3-dimensional Space and their Applications, Am. Math. Soc. Transl. 2, 207–316.

KAWASAKI, I., and KON'NO, F. (1984), *Azimuthal Anisotropy of Surface Waves and the Possible Type of Seismic Anisotropy due to Preferred Orientation of Olivine in the Uppermost Mantle beneath the Pacific Ocean*, J. Phys. Earth *32*, 229–244.

KENNETT, B. L. N., and ENGDAHL, E. R. (1991), *Traveltimes for Global Earthquake Location and Phase Identification*, Geophys. J. Int. *105*, 429–465.

KERN, H. (1993), *P- and S-wave Anisotropy and Shear-wave Splitting at Pressure and Temperature in Possible Mantle Rocks and their Relation to Fabric*, Phys. Earth Planet. Int. *78*, 245–256.

KUMAZAWA, M., and ANDERSON, O. L. (1969), *Elastic Moduli, Pressure Derivatives and Temperature Derivatives of Single-crystal Olivine and Single Crystal Forsterite*, J. Geophys. Res. *74*, 5961–5972.

MAINPRICE, D., and SILVER, P. G. (1993), *Interpretation of SKS-waves Using Samples from the Subcontinental Lithosphere*, Phys. Earth Planet. Int. *78*, 257–280.

MAUPIN, V., and CARA, M. (1992), *Love-Rayleigh Wave Incompatibility and Possible Deep Upper Mantle Anisotropy in the Iberian Peninsula*, Pure appl. geophys. *138*, 429–444.

MONTAGNER, J.-P. (1990), *Global Anisotropy in the Upper Mantle Inferred from the Regionalization of Phase Velocities*, J. Geophys. Res. *95*, 4797–4819.

MONTAGNER, J. P., and TANIMOTO, T. (1991), *Global Upper Mantle Tomography of Seismic Velocities and Anisotropies*, J. Geophys. Res. *96*, 20337–20351.

PESELNICK, L., and NICOLAS, A. (1978), *Seismic Anisotropy in an Ophiolite Periodotite: Application to Oceanic Upper Mantle*, J. Geophys. Res. *83*, 1227–1235.

PHINNEY, R. A., and BURRIDGE, R. (1973), *Representation of the Elastic-gravitational Excitation of a Spherical Earth Model by Generalized Spherical Harmonics*, Geophys. J. R. Astr. Soc. *34*, 451–487.

RAITT, R. W., SHOR, G. G., FRANCIS, T. J. G., and CLOUGH, J. W. (1969), *Anisotropy of the Pacific Upper Mantle*, J. Geophys. Res. *74*, 3095–3109.

RIBE, N. M. (1992), *On the Relationship between Seismic Anisotropy and Finite Strain*, J. Geophys. Res. *97*, 8737–8747.

SILVER, P. G., and CHAN, W. W. (1991), *Shear-wave Splitting and Subcontinental Mantle Deformation*, J. Geophys. Res. *96*, 16429–16454.

TONG, C., GUDMUNDSSON, O., and KENNETT, B. L. N. (1994), *Shear-wave Splitting in Refracted Waves Returned from the Upper Mantle Transition Zone beneath Northern Australia*, J. Geophys. Res. *99*, 15783–15797.

VINNIK, L. P., KRISHNA, V. G., KIND, R., BORMANN, P., and STAMMLER, K. (1994), *Shear-wave Splitting in the Records of the German Regional Seismic Network*, Geophys. Res. Lett. *21*, 457–460.

YU, Y., and PARK, J. (1994), *Hunting for Azimuthal Anisotropy beneath the Pacific Ocean*, J. Geophys. Res. *99*, 15399–15421.

ZHANG, S., and KARATO, S.-I. (1995), *Lattice Preferred Orientation of Olivine Aggregates Deformed in Simple Shear*, Nature *375*, 774–777.

(Received October 14, 1996, revised April 18, 1997, accepted June 9, 1997)

To access this journal online:
http://www.birkhauser.ch

Pure appl. geophys. 151 (1998) 669–697
0033–4553/98/040669–29 $ 1.50 + 0.20/0

Pure and Applied Geophysics

P-SH Conversions in Layered Media with Hexagonally Symmetric Anisotropy: A CookBook

Vadim Levin[1] and Jeffrey Park[1]

Abstract—Reflectivity synthetic seismograms demonstrate that the type, layering and orientation of 1-D anisotropy influences strongly the coda of teleseismic *P* waves at periods $T > 1$ sec, particularly *P-SH* converted waves. We assume the simplest form of anisotropy described by an elastic tensor with a symmetry axis $\hat{w}$ of arbitrary orientation. The resulting phase velocities vary as $\cos 2\xi$ with respect to that axis. Using three families of simple crustal models, we compare the effects of an anisotropic surface layer with reverberations caused by both "thick" and "thin" layers of anisotropy at depth. If anisotropy in the surface layer is significant, the polarization of direct *P* can be distorted to generate a transverse component, followed by *Ps* and a prominent shear reverberation converted from direct *P* at the free surface. If the anisotropic layer is buried, the first, and often the most prominent, arrival on the transverse component is the *P-to-SH* conversion at its upper surface. If the anisotropic layer is sufficiently thin, *P-to-SH* conversions from its boundaries interfere to form a derivative pulse shape on the transverse component, which could be mistaken as the signature of shear-wave splitting. If $\hat{w}$ is horizontal, compressional (*P*) and shear (*S*) anisotropy both produce similar waveform perturbations with four-lobed azimuthal patterns, suggesting that a weighted stack of *P* coda from different back-azimuths would improve signal-to-noise. For $\hat{w}$ tilted between the horizontal and vertical, however, the effects of *P*- and *S*-anisotropy differ greatly. The influence of *P*-anisotropy on *P-to-S* conversion is greatest for a symmetry axis tilted at 45° to the vertical, where its azimuthal pattern has two lobes, rather than four. Combinations of *P*- and *S*-anisotropy typically lead to a composite azimuthal dependence in the *P*-coda reverberations.

Key words: Seismic anisotropy, crustal structure, body waves, layered media, scattered waves, synthetic seismograms.

Introduction

There is a mounting body of evidence suggesting that seismic isotropy—an important and common assumption in seismological studies—may actually be a rarity rather than the rule in the shallow earth. The majority of minerals and rocks that form the crust and upper mantle display seismic anisotropy in laboratory measurements (Babuska and Cara, 1991). Bulk anisotropy in the oceanic crust and lithosphere was established by marine refraction experiments over two decades ago (e.g., Raitt *et al.*, 1969). Shear-wave splitting in broad-band seismic data suggests that the continental lithosphere has significant elastic anisotropy (e.g.,

---

[1] Department of Geology and Geophysics, Yale University, PO Box 208109, New Haven, CT 06520-8109, U.S.A.

VINNIK et al., 1992; SILVER, 1996; BABUSKA et al., 1993; HEARN, 1996; LEVIN et al., 1996), which can be used to reconstruct both active and fossil bulk strain. The upper part of the continental crust appears to have particularly strong anisotropic properties (ZHI et al., 1994; LYNN, 1991).

The variety of mechanisms producing seismic anisotropy in the crust centers on a handful of scenarios. In the upper crust the strongest influence is believed to be that of aligned cracks and/or pore spaces (BABUSKA and PROS, 1984), for which slower velocities are found for waves that propagate normal to the average crack plane. The aspect ratio of pore/cracks and presence of fluid in them determine the extent and proportion of anisotropy (HUDSON, 1981; CRAMPIN, 1984, 1991). Alternating thin isotropic layers of higher and lower velocity can also produce an overall anisotropic effect (BACKUS, 1962; HELBIG, 1994), with the velocities slower normal to bedding than along it. In the lower crust and the uppermost mantle, cracks are assumed to close in response to increasing overburden pressure (BABUSKA and PROS, 1984; KERN et al., 1993), though field exposures of (formerly) deep-crustal fluid-filled cracks can be found (AGUE, 1995). In the absence of cracks and inclusions, the lattice-preferred orientation (LPO) of mineral crystals is taken as the main cause of seismic anisotropy. Most minerals composing the bulk of the crust are anisotropic to some degree (BABUSKA and CARA, 1991), while properties of olivine and, to a lesser extent, orthopyroxene dominate the upper mantle anisotropy. Different deformation mechanisms can lead to the alignment of either the slow or the fast crystallographic direction in olivine grains (NICOLAS et al., 1973; RIBE, 1992), but LPO caused by dislocation creep in the shallow mantle is commonly believed to lead to preferred alignment of the fast axis (ZHANG and KARATO, 1995). GAHERTY and JORDAN (1995) argue, on the basis of mantle shear-wave reverberations, that thin-layering of different rock types also plays a role in the bulk anisotropy of the continental upper mantle.

Although minerals often exhibit more complex behavior, many instances of seismic anisotropy in crystalline basement rocks display hexagonal symmetry to the first order (e.g., MAINPRICE and SILVER, 1993; BURLINI and FOUNTAIN, 1993). There is no simple relationship between the amount of P velocity anisotropy and that of S anisotropy, but it is rare for only one type, P or S, to be present in a rock. Anisotropy in velocity up to 10% is a common feature of crustal and upper mantle rocks, and exceeds 15% in some lithologies, e.g. metapelites (BURLINI and FOUNTAIN, 1993). To influence a teleseismic wave in the frequency range 0.2–2 Hz, crack and/or mineral alignment must be coherent within substantial volumes of crust, so the effective anisotropy is often diminished substantially relative to the anisotropy of either minerals and rock samples.

A common assumption in anisotropy studies is that the symmetry axis is either vertical (i.e., as in the case of alternating layers) or horizontal (as in cases of stress-aligned olivine or vertical cracks). On the other hand, a number of studies indicate that an axis with a tilted orientation is needed to explain observations of teleseismic waves (BABUSKA *et al.*, 1993; LEVIN *et al.*, 1996; GRESILLAUD and CARA, 1996). The possible causes of tilted-axis anisotropy are not exceptional, though one expects only local or regional coherence in the associated details of tectonic deformation. BLACKMAN *et al.* (1996) have modeled tilted alignment of olivine LPO beneath mid-ocean ridge systems. Crustal overthrusting is another likely cause of tilted-axis anisotropy.

Anisotropy induces compressional waves to convert to *SH*-type motion in 1-D velocity structures with horizontal interfaces. KOSAREV *et al.* (1984) and VINNIK and MONTAGNER (1996) invoke this mechanism to explain long-period *SH* phases following teleseismic *P* waves as *P-SH* conversions in horizontally stratified anisotropic upper mantle. *P-SH* conversion resulting from reverberations of a plane wave in a stack of flat anisotropic layers can give rise to *P* coda whose complexity approaches that of observations (LEVIN and PARK, 1997b). Before *P-SH* conversions can be useful in studies of crustal structure, certain questions about the effects of the anisotropic layered medium should be addressed. We need to know which of the possible converted phases will have sufficient energy to be observable in real data. Of these, we must identify phases that may be used to distinguish various types of anisotropic structures, i.e., *P* vs *S* anisotropy, fast versus slow velocity alignment. We also must determine which portions of a layered structure are most promising in terms of generating large *P-SH* phases. Back-azimuth dependence unique to this type of reverberation may distinguish 1-D anisotropic models from the two explanations typically offered for *SH*-motion in the teleseismic *P* wave-train: a) scattering due to velocity heterogeneities (e.g., VISSER and PAULLSEN, 1993; HU, 1993) and b) inclined interfaces beneath the receiver (e.g., OWENS and CROSSON, 1988; ZHU *et al.*, 1995). At any seismic station, more than one of these mechanisms may be important. If the azimuthal pattern of converted phases can be related confidently to a particular type of crustal model, stacking seismic records from different back-azimuths may be useful.

In this paper we investigate the influence of 1-D anisotropy, its type, layering and orientation, on the coda of teleseismic *P* waves, particularly on *P-SH* converted waves. We employ a reflectivity technique to compute the transmission response of a flat-layered medium with arbitrarily oriented hexagonally symmetric anisotropy, as developed by PARK (1996) for surface waves and extended by LEVIN and PARK (1977b) to receiver-function geometry. We describe the main features of synthetic seismograms for a variety of simple models and discuss how to interpret converted phases in observations.

Method

The models we consider consist of homogeneous flat layers atop a homogeneous halfspace. The halfspace is isotropic. Each layer may possess seismic anisotropy with an axis of symmetry $\hat{\mathbf{w}}$. The velocity profiles have Poisson ratio ≈ 0.27, consistent with a somewhat mafic continental crust (CHRISTENSEN, 1996). According to the classification by ZANDT and AMMON (1995), velocity values selected for the crust in our models would place them on the old stable continent. A compressional wave is assumed to propagate upwards from the halfspace into the layered part of the model, where it undergoes refraction and conversion. The combination of pulses arriving at the free surface is the "transmission response" of the media. Once computed, this transmission response can be convolved with the pulse of the original compressional wave, yielding a synthetic seismogram.

To compute the interaction of upgoing and downgoing plane waves, we express the elastic properties as a function of depth as $\Lambda(z)$, where Λ_{ijkl} is the fourth-order stress-strain tensor. In the case where the axis of symmetry is horizontal, BACKUS (1965) derived the azimuthal dependence of P and SV velocities for horizontal propagation, which is appropriate for head waves in marine refraction studies. Expressed in terms of the angle ξ from $\hat{\mathbf{w}}$, these head-wave velocities are

$$\rho\alpha^2(\xi) = A + B\cos 2\xi + C\cos 4\xi$$

$$\rho\beta_{SV}^2(\xi) = D + E\cos 2\xi. \tag{1}$$

The SH velocity for horizontal propagation satisfies $\rho\beta_{SH}^2(\xi) = D + C(1 - \cos 4\xi) + E$.

If density perturbations are neglected, knowledge of A, B, C, D, E is sufficient to determine the stress-strain tensor (SHEARER and ORCUTT, 1986) for "weak" anisotropy. An expression for this tensor, outlined in the appendix, can be used for "strong" anisotropy as well. In an isotropic medium, $B = C = E = 0$ and $A = \lambda + 2\mu$ and $D = \mu$, where λ, μ are the Lamé parameters. PARK (1993) showed how these azimuthal relations generalize to other orientations of $\hat{\mathbf{w}}$. We assume a flat earth, $z = 0$ at the free surface, and z increasing downward.

To compute the reverberation response of a crustal model, we prescribe an upgoing plane-wave of the form $\mathbf{U}(\mathbf{x}, t) = \mathbf{u}_0 \, e^{i(\mathbf{k}\cdot\mathbf{x} - \omega t)}$ in the halfspace. A compressional plane wave in an anisotropic 1-D flat-layered structure suffers conversion to both vertically (SV) and horizontally (SH) polarized shear waves, with two exceptions: no P-SH conversion occurs if the axis of symmetry is everywhere vertical, or if the ray and the symmetry axis are contained in the same vertical plane in each layer.

The phase velocity for P and S waves in hexagonally symmetric media can be represented by smooth surfaces symmetric about the axis in 3-D-space defined by $\hat{\mathbf{w}}$ (Fig. 1). If $B, E > 0$, $\hat{\mathbf{w}}$ defines the 'fast' axis for wave propagation, leading to phase-velocity surfaces that resemble tilted melons. If $B, E < 0$, $\hat{\mathbf{w}}$ defines the 'slow' axis for wave propagation, leading to phase-velocity surfaces that resemble tilted pumpkins. The $\cos 4\xi$ coefficient C would distort these ellipsoidal surfaces. As noted in the appendix, departures from phase-velocity ellipticity can be substantial in measurements made from sedimentary rock samples and for special cases in fine-layering anisotropy. However, the parameter C is small in most seismic refraction estimates (e.g., SHEARER and ORCUTT, 1986; ANDERSON, 1989). We also note in the appendix that the formulas of HUDSON (1981), CRAMPIN (1984) for crack-induced anisotropy imply either $C = 0$ or $C \ll B$. Although real-world anisotropy can be more complex than such observations and theories suggest, we set $C = 0$ in order to examine the large model space respresented by media with elliptical velocity surfaces of varying orientation, parameterized by $\hat{w}$, B, and E. Anisotropy of this simplified type has proven useful in modelling the azimuthal variation in both shear-wave splitting (BABUSKA *et al.*, 1993) and *P-SH* conversion LEVIN and PARK (1997a), so a careful examination of the relative influences of symmetry axis $\hat{w}$, S and P anisotropy is useful for the assessment of seismic data sets.

In a layer with constant anisotropic elastic properties, one can calculate three upgoing and three downgoing plane-wave solutions to the equations of motion, with vertical wavenumbers and polarizations determined by the eigenvectors of a

Anisotropy Parameterization

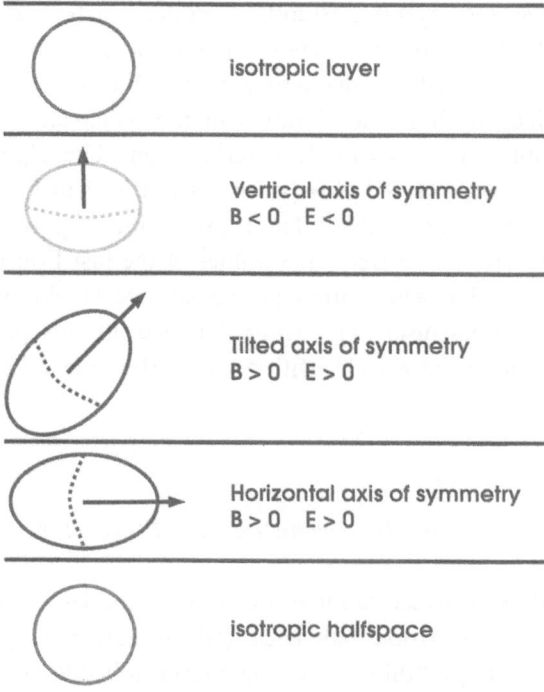

Figure 1
Schematic diagram illustrating possible shapes of velocity distribution for various choices of anisotropic parameters.

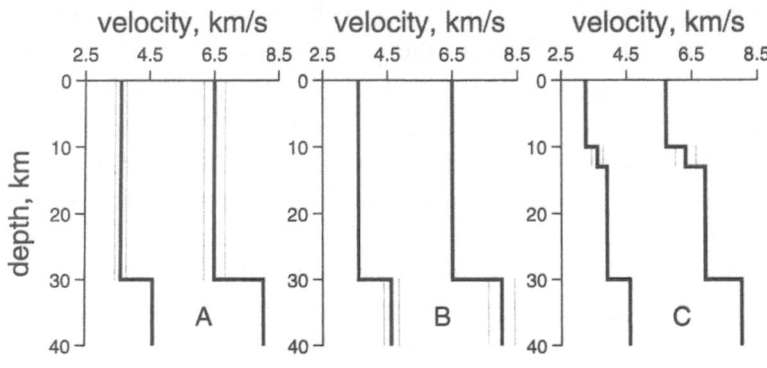

Figure 2

Schematic representation of 1-D velocity models used in simulations: a) anisotropic layer over an isotropic halfspace; b) isotropic layer over an anisotropic halfspace; c) thin anisotropic layer in an isotropic stack. Dashed lines denote range of velocity variation for 5% anisotropy (B, $E = -0.05$).

6×6 matrix eigenvalue problem (GARMANY, 1983; FRYER and FRAZER, 1987; PARK, 1996). Assume K layers over an isotropic halfspace, with interfaces at $z_1, z_2, \ldots z_K$. We compute the generalized transmission response of the layer stack, equivalent to a 3-D receiver function (LANGSTON, 1977), to determine the particle motion at the free surface $z_0 = 0$. A standard propagator formalism determines the response of a stack of anisotropic layers to upgoing wave motion with frequency ω and horizontal phase velocity c (slowness $p = 1/c$). We restrict attention to phase velocities c for which both P and S waves in the halfspace are oscillatory, thus bypassing the problem of leaky-mode reverberation. The algorithm follows the development of KENNETT (1983) closely, and is outlined in more detail in LEVIN and PARK (1997b) and PARK (1996). The transmission response of the medium is calculated at the evenly-spaced frequency values of the fast Fourier Transform of a chosen input wavelet. Particle motion at the surface is obtained by an inverse Fourier Transform. Noncausal 'wraparound' effects in this procedure are minimized by padding the initial wavelet with zeros in the time domain, to interpolate the spectrum.

Models, Ray Geometries and Procedures

We consider three distinct families of velocity models (Fig. 2): (A) an anisotropic layer over an isotropic halfspace; (B) an isotropic layer over a "thick" anisotropic layer; and (C) a "thin" anisotropic layer imbedded in an isotropic stack. In family A, the surface layer is anisotropic. In families B and C, the anisotropic layer is buried. For each model family we consider cases with P anisotropy only (coefficient B in (1)), S anisotropy only (coefficient E in (1)), and equal amounts of

P and *S* anisotropy. While anisotropy in *P* or *S* velocity only is hardly representative of physical reality, simulations with such an assumption provide insight into the relative contributions within "mixed" *P* and *S* anisotropy. The velocity models in family B place anisotropy in a layer below the Moho in the upper mantle, but the general behavior of these synthetic seismograms should carry over to the case of an anisotropic crustal layer overlain by a shallow low-velocity isotropic layer, e.g., a granite pluton atop anisotropic basement gneisses.

All waveforms analyzed in this work arise (through conversion and/or reverberation) from the original compressional plane wave that ascends from the isotropic halfspace beneath the layers. Its pulse shape is prescribed at the bottom of the model, with all converted phases being scaled and/or distorted versions of it. No knowledge of the source of the pulse, or of the propagation effects in the medium outside our model is required for this exercise. In practice, effects of the source and the path outside the receiver region are routinely removed via "source normalization" techniques typical of receiver function analysis (e.g., LANGSTON, 1977).

Most synthetics are computed for 5% peak-to-peak velocity anisotropy (e.g., $B = 0.05$), with systematic variation in the tilt of the symmetry axis $\hat{w}$ and the back-azimuth of the arriving *P* wave. The effects of anisotropy magnitude and velocity contrast across the interface are studied in separate experiments. Both positive (fast symmetry axis—"melon") and negative (slow symmetry axis—"pumpkin") anisotropy are investigated, bringing the number of models examined within each family to 6. In all models the symmetry axis $\hat{w}$ is tilted at an angle η from the vertical towards the north, in 15° increments between 15° and 90°. (At $\eta = 0°$ the axis of symmetry is vertical and the *P-SV* and *SH* equations of motion are uncoupled.) We propagated upgoing plane waves through each of the anisotropic velocity models using a range of back-azimuths, measured clockwise from the north, and incidence angles. Incidence angles vary form 5° to 60° in 5° increments, and back-azimuths vary from 0° to 360° in 15° increments. Computations for 1800 plane waves were performed for each combination of the model family (A, B or C), anisotropy type (*P*, *S* or both) and the anisotropy sign (*B*, $E > 0$ or *B*, $E < 0$). In each simulation, the time-domain waveforms were computed and parameters (timing, amplitude, polarity) of the chosen phases were measured by a guided auto-picking routine.

An identical one-sided pulse waveform was used for the incoming *P* wave in all simulations. Sample synthetic seismograms for models from different families are shown on Figure 3. The converted phases most often have pulse shapes on the horizontal components that resemble either (1) a scaled version of the original pulse (e.g., the direct *P* in model-family A, and most radial phases) or (2) a derivative of the original pulse (e.g., the *Psms* phase in model-family A). In the first case the polarity of a converted phase is defined as "positive" if the pulse is "up," and the amplitude is defined as the maximum absolute value in a chosen time window. In the second case the phase polarity is considered "positive" if the first swing of the pulse is "up." The amplitude of a waveform is then defined as a "peak-to-peak" difference between the smallest and the largest values within a chosen time window. These amplitude and polarity definitions, while not unique, are very helpful for describing how *P-SH* converted phases vary with the back-azimuth of the incident wave. Care was taken to design test models that would prevent an overlap of two phases in time. In real data overlapping phases may be unavoidable, and should be anticipated.

In the following sections we describe general properties of *P-SH* conversion in layered anisotropic media, present results of simulations for each model family and summarize common features. For each synthetic sweep the amplitudes of horizontal components are normalized by the maximum amplitude of corresponding vertical

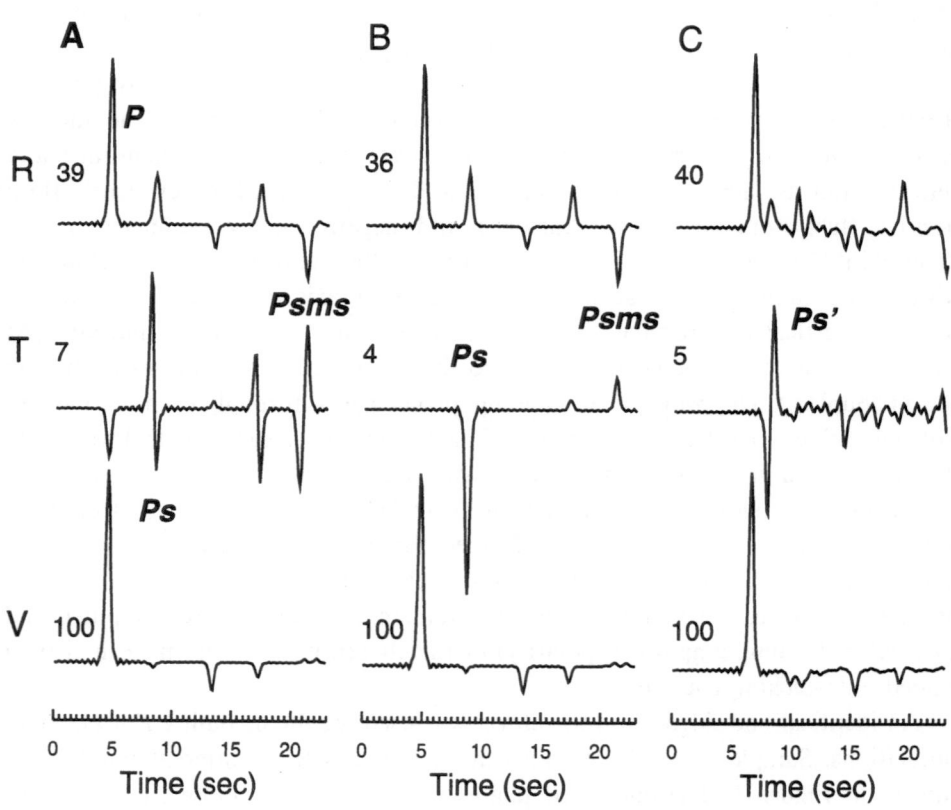

Figure 3
Sample waveforms generated in three velocity structures: a) anisotropic layer over an isotropic halfspace; b) isotropic layer over an anisotropic halfspace; c) thin anisotropic layer in an isotropic stack. Traces are scaled individually, with relative scale within each 3-component seismogram indicated by a number in percent at the beginning of each trace. Parameters of anisotropy used in simulations are: "melon" (positive) anisotropy of 5% in both *P* and *S* velocity, ray incidence angle 25%, back-azimuth 300°, axis tilt 60° from vertical. Converted phases analyzed in this paper are indicated for each model family.

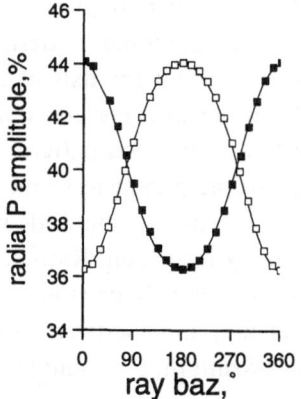

Figure 4

Azimuthal variation in the radial component of the direct *P* wave, model family A, pure *P* anisotropy. Incidence angle 30°, axis tilt 45° from vertical (maximal effect). Positive (melon) anisotropy imposes the pattern shown by open symbols, negative (pumpkin) anisotropy—solid symbols. Patterns are mirror-symmetric and vary as sin ξ with ray back-azimuth.

traces, and expressed in percent. For some converted phases, these amplitude ratios can vary with the period of the initial *P* pulse, especially if the phase is generated by two interfering pulses, as for the transverse component of shear-wave splitting. Therefore the amplitude ratios should be taken as guides to, rather than absolute predictions of, data behavior.

General Observations

The radial amplitudes of converted phases in anisotropic structures vary with incident back-azimuth, with unchanged polarity. The azimuthal pattern of the radial component is controlled by the tilt of the anisotropic symmetry axis, the incidence angle and the velocity contrast. The axis tilt angle η defines the shape of the pattern, whether two-lobed, four-lobed, or a composite. The incidence angle and the velocity contrast affect primarily the amplitude of the converted phase, and less so its azimuthal dependence. Figure 4 shows an amplitude pattern for one incidence angle and a sweep of back-azimuth that is representative of the phase behavior for the combination of model family and anisotropy type. The azimuthal patterns of radial-component converted phases are symmetric about the axis of symmetry $\hat{w}$.

Variations of transverse components with incident back-azimuth are more complex, involving changes in both amplitude and polarity. Variations in incidence angle can lead to different azimuthal patterns independent of changes in other parameters. To describe fully the azimuthal pattern of the transverse-component of the crustal phases, an area plot (Fig. 5) is required, similar in appearance to an earthquake focal mechanism. The azimuthal patterns of transverse-component converted phases are antisymmetric about the axis of symmetry $\hat{w}$. In addition to this polarity switch, azimuthal patterns of transverse phases often have a second set of polarity transitions, leading to a four-lobed pattern. The precise nature of these transitions depends on the anisotropic parameters and incidence angle. While it is usually marked by a moderate depression in amplitude values, pulse amplitude does not typically vanish at the secondary transition. Rather, a polarity transition occurs through a gradual evolution of the pulse shape (Fig. 6). For back-azimuth aligned with $\hat{w}$ there is no P-SH conversion, and the transverse component vanishes.

Systematic changes with back-azimuth ξ of amplitude, polarity and timing of converted phases have some 2-lobed dependence on $\sin \xi$ or $\cos \xi$ in all but special cases. However, secondary polarity and amplitude changes lead typically to asymmetric patterns, depending on the model parameters. Nevertheless, it is usually helpful to describe the azimuthal pattern by the number of lobes (2 or 4) in the complete 360°. For instance, the pattern in Figure 4 is two-lobed, while that in Figure 5 is asymmetrically four-lobed. In the special case of a horizontal symmetry

P&S, 75°

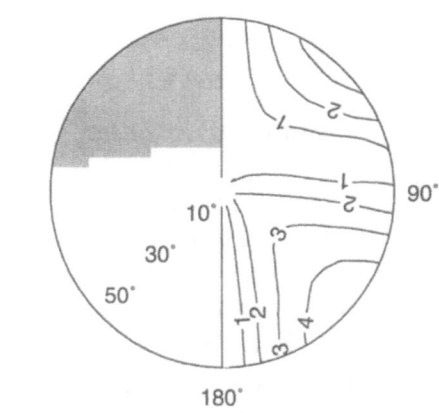

model type A   T component of direct P

Figure 5

A diagram illustrating the azimuthal variation in amplitude and polarity of the transverse component of the direct P wave. Model family A, positive (melon) P and S anisotropy of 5%, axis tilt 75° from vertical. Right side of the plot illustrates variation of the amplitude (in percent of vertical P) as the function of back-azimuth and incidence angle. Back-azimuth varies clockwise from 0° to 180°. Incidence angle increases uniformly from 5° in the center to 60° at the rim. The left side of the plot illustrates polarity (shaded—negative) of the converted pulse for back-azimuths 180°–360°. Two sides of the plot are antisymmetric, since this is the transverse amplitude. In this example, pulses amplitudes for back-azimuths 180°–280° are positive, and for 80°–180° are negative.

model: P&S "melon", family A
axis tilt 60°, incidence angle 30°

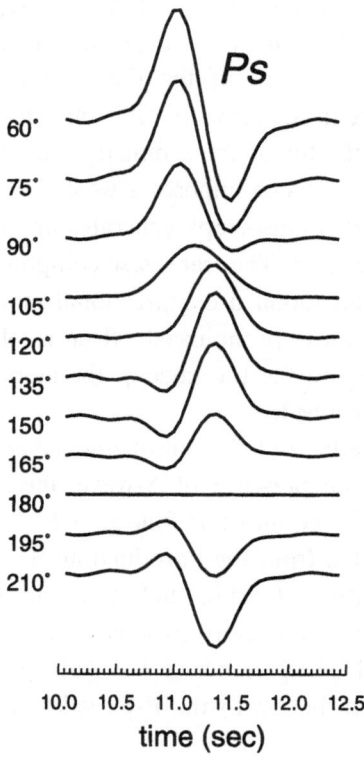

Figure 6

Pulse shape of a transverse *Ps* phase as a function of back-azimuth. Family A model with negative *P* and *S* anisotropy, $\eta = 60°$. Phane wave incidence angle 30°. Traces are plotted on a common scale, and labeled with ray back-azimuth. Change of polarity around BAZ 75° occurs via a gradual evolution of the waveform, while at 180° ($\hat{w}$ direction) the transverse component vanishes.

axis $\hat{w}$, all patterns are four-lobed and symmetric, and may by described by $\sin 2\xi$ or $\cos 2\xi$.

The transverse components of converted phases for opposite signs of anisotropy ("melon" vs. "pumpkin") always have opposite polarity, leading to mirror-image azimuthal patterns. In most cases the radial components of converted phases also display mirror symmetry in models with opposite anisotropy sign. Exceptions from this rule are discussed in following sections.

The effects of *P* and *S* anisotropy differ substantially. The effect of pure *P* anisotropy is typically two-lobed, and is maximized when the symmetry axis $\hat{w}$ is tilted at $\eta = 45°$. The effect of pure *S* anisotropy is more four-lobed, and is strongest for subhorizontal $\hat{w}$. If the anisotropy types are mixed, the azimuthal patterns follow the stronger influence.

Model Family A: Anisotropic Layer over an Isotropic Halfspace

Because the polarization of seismic waves is distorted within the surface layer by its anisotropy, each phase for this family of models typically has a transverse component. Polarization distortion also affects the radial component amplitude significantly. The radial component of the direct *P* wave suffers a two-lobed perturbation to its amplitude in the presence of *P* anisotropy (Fig. 7). The intensity of the pattern depends on the tilt of the symmetry axis. Radial *P* amplitudes are also affected by *S* anisotropy, which imposes a weak four-lobed perturbation. An equal combination of *P* and *S* anisotropy generates amplitude perturbations that resemble those of the pure *P* case. The transverse component of the *P* phase is not a *P-SH* converted phase, but rather a compressional motion deflected out of the source-receiver plane. The most pronounced effect is observed for pure *P* anisotropy (Fig. 8). Depending on the tilt angle η, the amplitude patterns are either two- or asymmetrically four-lobed.

P-to-*S* conversion at the base of the anisotropic layer follows direct *P* on our synthetic seismograms. In the presence of *S*-wave anisotropy its timing on the radial component relative to the direct *P* follows a four-lobed azimuthal pattern (Fig. 9), as would be expected from the introduction of $\cos 2\xi$ variations in shear velocity. Peak-to-peak variation of almost half a second is reached in our models for near-horizontal orientation of anisotropy axis, as the shear wave (*Ps*) traverses the entire crust in the model. The presence of *P* anisotropy in the surface layer adds a smaller, but nonzero, perturbation to the *P-Ps* delay time.

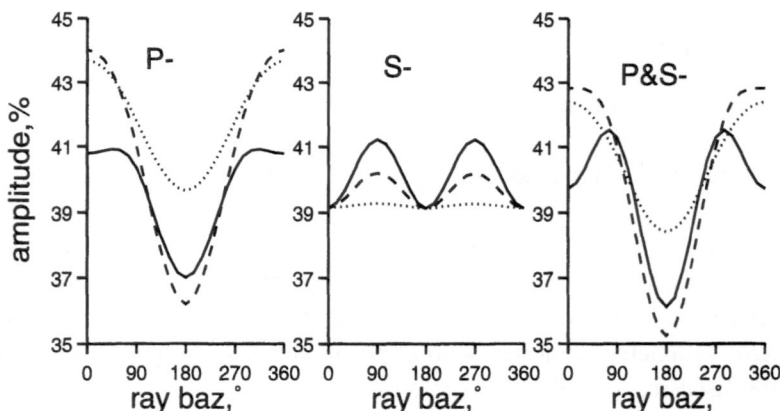

Figure 7

Radial component of the direct *P* wave as a function of back-azimuth in family A models with negative (pumpkin or " $-$ ") anisotropy. Incidence angle of the incoming wave is 25°. The line type indicates the tilt from vertical of the anisotropic symmetry axis: dotted $-15°$, dashed $-45°$, solid $-75°$. Type of anisotropy (*P*-, *S*- or combined) is indicated on the plots.

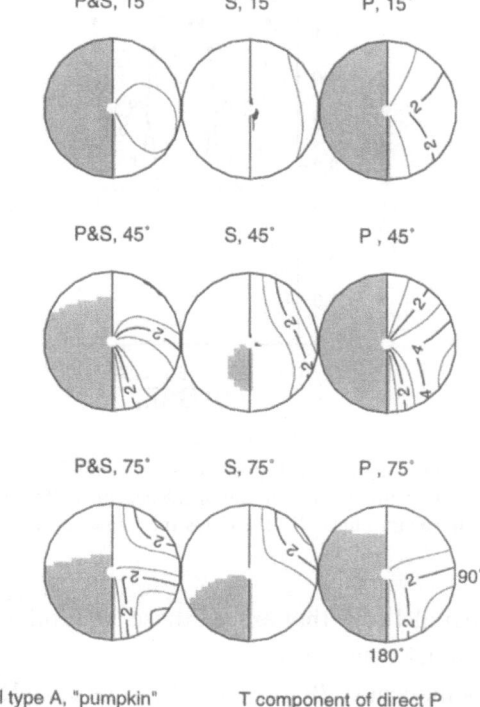

Figure 8
Transverse component of the direct *P* wave in family A models with negative anisotropy as a function of back-azimuth, incidence angle and anisotropic axis tilt from vertical. Parameters of radial plots are as in Figure 5. Anisotropy type and anisotropic axis tilt from vertical are indicated above each plot.

The radial amplitude of the *Ps* phase can change by more than a factor of 2 in the presence of *P* anisotropy (Fig. 10a). The azimuthal pattern is two-lobed. Pure *S* anisotropy causes much smaller variations in the radial *Ps* phase (Fig. 10b). In a deviation from typical behavior, a change from a "pumpkin" to a "melon" anisotropy (i.e., the sign of the anisotropic parameters *B* and *E*) affects the amplitude of the transverse *Ps* phase slightly without altering the distribution of lobes in the azimuthal pattern. This subtle change is not likely to be a useful interpretive tool for observations, however.

A combination of *P* and *S* anisotropy results in azimuthal patterns that depend strongly on the axis tilt (Fig. 10c). Axes inclined no more than 45° result in two-lobed amplitude patterns that are relatively smooth. *Ps* amplitude perturbations for opposite signs of anisotropic parameters *B* and *E* resemble mirror images of each other. If the axis of symmetry $\hat{w}$ is subhorizontal, however, *Ps* amplitude oscillates rapidly with back-azimuth. This pattern is sufficiently asymmetric to cause, in the case of the "pumpkin" (negative) anisotropy, *Ps* amplitudes from

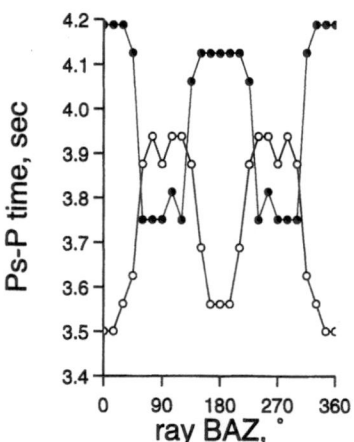

Figure 9

Timing of *Ps* phase in family A models with positive (open symbols) and negative (closed symbols) *P* and *S* anisotropy. Values are computed for the ray incidence angle 25°, and a symmetry axis tilted 75° from vertical. For this incidence angle, 75° tilt yields the largest azimuthal variation of *Ps-P* delay.

back-azimuths subparallel to $\hat{w}$ that exceed those from other directions by almost a factor of 2 on the radial component.

The transverse component of *Ps* phase can be as large as 15% of *P* in our synthetic seismograms (Fig. 11). Near-vertical incidence leads to converted waves with small amplitudes ($\lesssim 5\%$), leaving the largest converted-wave amplitudes to shallow-incidence *P* waves. Azimuthal patterns obtained in models with pure *P* anisotropy are generally two-lobed, with two additional smaller lobes appearing in the pattern for

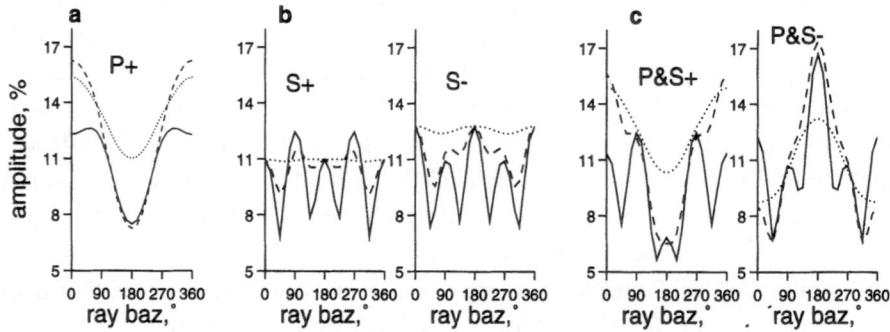

Figure 10

Radial component of the *Ps* phase in family A models. Type and sign of anisotropy is indicated on the plots. The line type indicates the tilt of the anisotropic symmetry axis from vertical: dotted $-15°$, dashed $-45°$, solid $-75°$. a) Pure *P* anisotropy. b) Pure *S* anisotropy. Distribution of lobes in the azimuthal pattern is not sensitive to the sign of anisotropy; c) *P* and *S* anisotropy. Azimuthal patterns strongly depend on η.

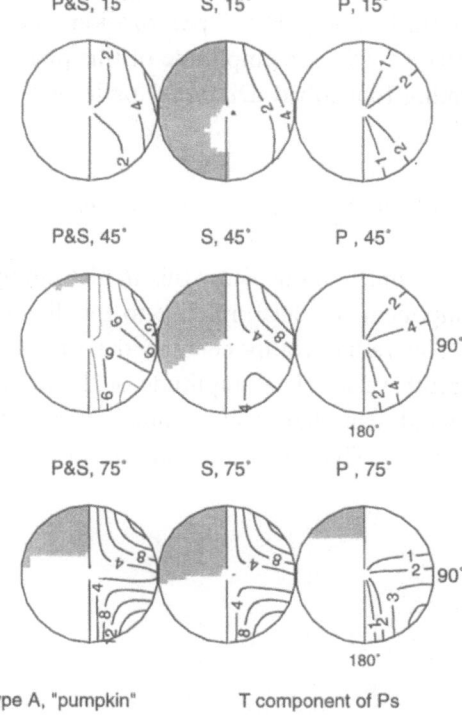

P&S, 15°   S, 15°   P, 15°

P&S, 45°   S, 45°   P , 45°

P&S, 75°   S, 75°   P , 75°

model type A, "pumpkin"   T component of Ps

Figure 11

Transverse component of the *Ps* phase in family A models with negative anisotropy. Parameters of area plots are as in Figure 5. Anisotropy type and anisotropic axis tilt from vertical are indicated above each plot.

near-horizontal axes and relatively shallow incidence. In models with *S* velocity anisotropy the azimuthal pattern of transverse *Ps* amplitudes is always four-lobed, with relative sizes of the lobe pairs controlled by the anisotropic axis tilt. Transverse-component amplitude in the case of *P* anisotropy is relatively small. For *S* anisotropy, derivative-pulse shapes on the transverse component indicate that much of this waveform arises from a shear-wave splitting time delay δt. The relative amplitude of the transverse component therefore depends on the ratio of δt to the period T of the incoming wave i.e., stronger for shorter period waves, weaker for longer-period waves.

The *Psms* phase arrives between 15 and 17 seconds after the direct *P* in our simulations. It is a split shear wave, generated through a substantial *P-S* conversion at the free surface and subsequently reflected back from the top of the halfspace. The *Psms* phase is very prominent on the transverse component if there is *S* wave anisotropy in the model (Fig. 12), as a result of shear-wave splitting. The delay time of the *Psms* arrival, relative to *P*, is only slightly affected by the presence of anisotropy, because both fast and slow *S* polarization contribute to the waveform. For symmetry-axis tilts $\eta > 45°$, the amplitude of this phase can be as large as 15% that of direct *P*, forming four-lobed azimuthal patterns.

Model Family B: Isotropic Layer over Anisotropic Halfspace

In synthetic seismograms computed for this model family, *Ps* is the first arrival on the transverse component. Its delay relative to the direct *P* varies insignificantly with back-azimuth. The radial component of the direct *P* wave is also nearly constant as back-azimuth varies. The amplitude of the radial component of the *Ps* phase, on the other hand, can change by as much as a factor of 2 in a two-lobed azimuthal pattern (Fig. 13). The transverse component of *Ps* is larger (up to 5%) in

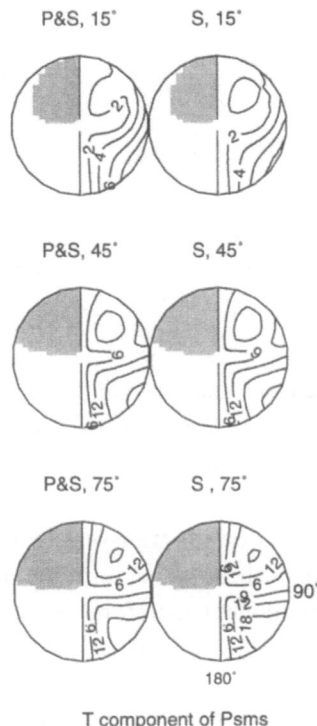

T component of Psms

model type A, "pumpkin"

Figure 12

Transverse component of the *Psms* phase in family A models with negative anisotropy. Parameters of area plots are as in Figure 5. Anisotropy type and anisotropic axis tilt from vertical are indicated above each plot.

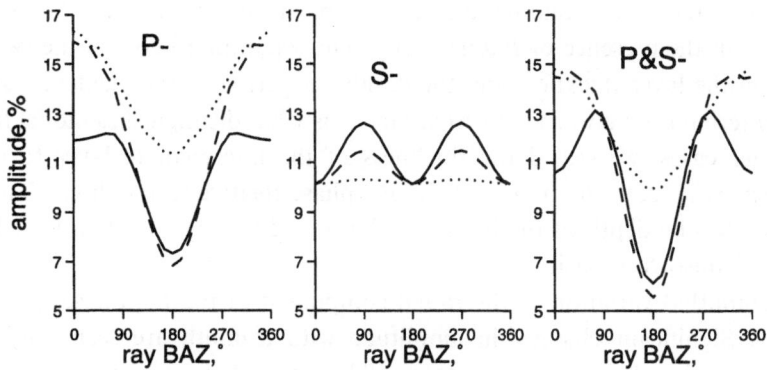

Figure 13

Radial component of the *Ps* phase in family B models. The line type indicates the tilt of the anisotropic symmetry axis from vertical: dotted −15°, dashed −45°, solid −75°. Type of anisotropy is indicated on the plots.

models with *P* anisotropy than in models with pure *S* anisotropy (Fig. 14). Azimuthal patterns of transverse *Ps* amplitude change from two-lobed for subvertical axes to asymmetric four-lobed for subhorizontal axes. The transverse amplitude of the *Psms* phase does not exceed 3% in any of our simulations, with *S* anisotropy models leading to stronger conversions, largely due to the lack of shear-wave splitting in the isotropic surface layer. Azimuthal patterns are mostly two-lobed, with two vestigial lobes present if the symmetry axis $\hat{w}$ is subhorizontal.

Model Type C: Thin Anisotropic Layer in an Isotropic Stack

Many parameters influence the *P* coda generated in this family of models. The ratio of wave period *T* to the travel time through the anisotropic layer is one of the strongest influences, as it determines whether the converted phases generated at the upper and lower boundaries of the anisotropic zone interfere or arrive as separate pulses. We used a velocity-depth profile in which the sense of velocity contrast across all boundaries does not change if 5% anisotropy is introduced into a thin intermediate layer (Fig. 2c). This type of model would be appropriate, for instance, for a shear zone within the crust (COLEMAN, 1996). Other scenarios may be necessitated by particular datasets (e.g., a direction-dependent low-velocity zone), but we defer their investigation.

The radial component of the direct *P* wave does not vary significantly with back-azimuth. The first-arriving energy on the transverse component is the equivalent of the *Ps* phase (labeled *Ps'* to distinguish it from *P*-to-*S* conversion at the base of the "crust"). This phase, composed of two pulses of opposite polarity generated at the two interfaces bounding the anisotropic layer, dominates the transverse component in the presence of *P* anisotropy. The separation between the two pulses depends on the layer thickness and the dominant period of the incident waveform. Because interference between converted phases is what distinguishes model family C from earlier cases, we scaled the thickness of the intermediate layer to examine these effects (Fig. 3c). The timing of the *Ps'* phase relative to the direct *P* arrival is determined by the depth of the imbedded layer and is only weakly affected by the presence of anisotropy in it.

The azimuthal variation of the radial component of the *Ps'* phase is shown on Figure 15. Significant changes in amplitude with azimuth are seen only for the models containing *P* velocity anisotropy. The azimuthal patterns are four-lobed, with the relative size of lobes strongly dependent on the axis tilt. The amplitudes of the transverse component of *Ps'* reach 10% in our simulations, with *P* wave anisotropy leading to considerably stronger converted phases. The azimuthal patterns (Fig. 16) change from two-lobed for subvertical $\hat{w}$ to asymmetric four-

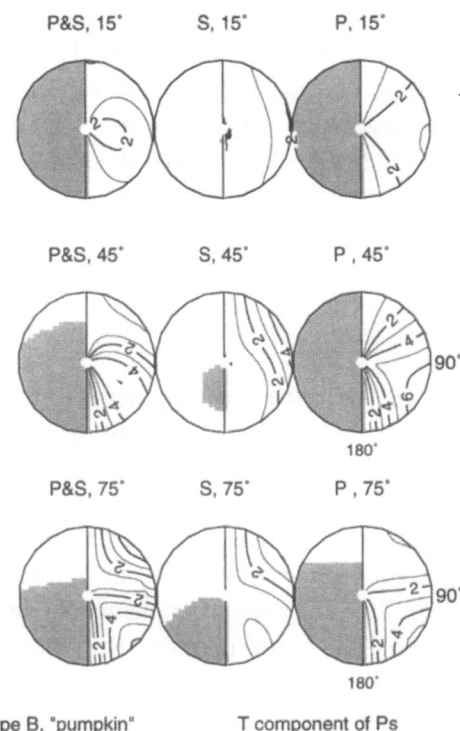

Figure 14

Transverse component of the *Ps* phase in family B models with negative anisotropy. Parameters of area plots are as in Figure 5. Anisotropy type and anisotropic axis tilt from vertical are indicated above each plot.

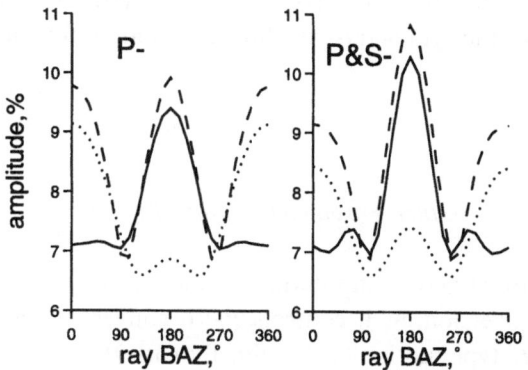

Figure 15

Radial component of the *Ps'* phase in family C models with negative anisotropy. Significant variation of radial amplitude is seen only if *P* anisotropy is present in the model. The line type indicates the tilt of the anisotropic symmetry axis from vertical: dotted $-15°$, dashed $-45°$ solid $-75°$. Type of anisotropy is indicated on the plots.

lobed for subhorizontal $\hat{w}$. The raypath through the anisotropic layer is too short to induce substantial shear-wave splitting, so this patterns depends more on the details of wave conversion at the interfaces. In another deviation from the general tendency, only the radial amplitudes of *Ps'* are affected by a change of sign in the anisotropic parameters *B* and *E*, while the lobe patterns are similar for both "melon" and "pumpkin" models.

Reverberations in a model containing three layers over a halfspace lead to synthetic seismograms sufficiently complex to make later arrivals more difficult to associate with particular interfaces. Although numerous phases comparable in amplitude with *Ps'* appear in synthetics computed for pure *S* anisotropy, their origin and properties are likely to be very model-specific. By extension, multiple-layer models with distinct anisotropic parameters in successive layers will often lead to *P* coda that are difficult to interpret.

Special Case—Horizontal Symmetry Axis

Most studies of shear-wave splitting due to seismic anisotropy assume that the axis of symmetry $\hat{w}$ is horizontal. This assumption is also made by KOSAREV *et al.* (1984), FARRA *et al.* (1991) and VINNIK and MONTAGNER (1996) to analyze *P-SH* conversions from upper-mantle discontinuities. In our synthetics, the azimuthal patterns of converted phases that develop for horizontal axes of anisotropy are qualitatively similar to the "near-horizontal" cases in each model family. The main distinguishing feature of *SH* waveforms in the horizontal axis models is an exact sin 2ξ symmetry in the amplitude patterns. Zero amplitude nodes occur both parallel and at 90° to the symmetry-axis direction, and all four lobes of the pattern are of the same size.

Other Parameters, Other Effects

The velocity contrast across an interface in an anisotropic medium controls the process of *P-SH* conversion, as it controls *P-SV* conversion in the isotropic case: the amplitude of *SH*-type motion scales directly with the velocity jump across the interface (Fig. 17). Converted-phase amplitudes are enhanced with an increase in the incidence angle. An exception is the transverse component of the direct *P*, which depends primarily on polarization distortion in the surface layer, and only weakly on the velocity contrast of the interface.

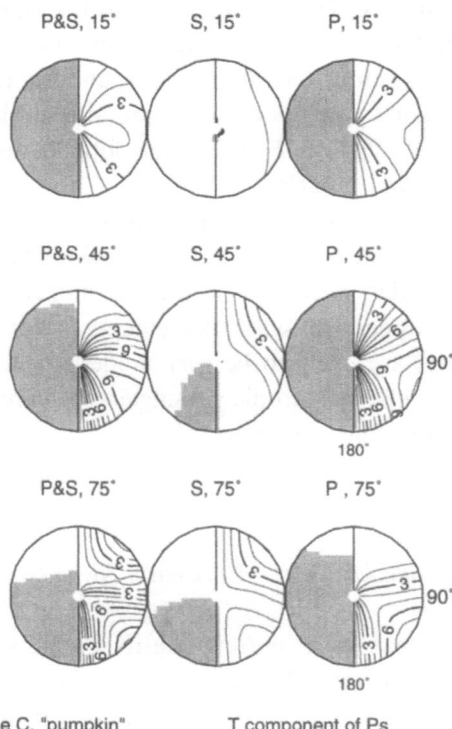

Figure 16

Transverse component of the *Ps'* phase in family C models with negative anisotropy. Parameters of area plots are as in Figure 5. Anisotropy type and anisotropic axis tilt from vertical are indicated above each plot.

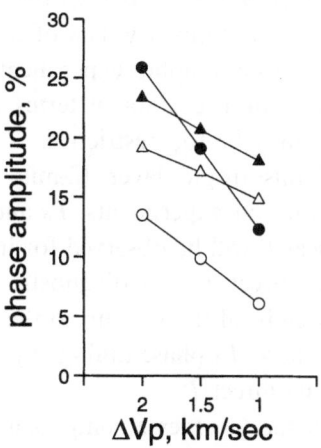

Figure 17

Influence of the velocity contrast across an interface on the *P-SH* conversion. Compressional velocity values of 6.0, 6.5 and 7.0 km/sec were used for the anisotropic layer over an isotropic halfspace with compressional velocity 8.0 km/sec (model family A). Shear velocity values were computed as $Vp/1.8$. A positive (melon) anisotropy of 5% in both *P* and *S* velocities were used in the layer. The plot depicts amplitudes of transverse *Ps* (circles) and *Psms* (triangles) phases for rays incoming from back-azimuth 315° with an incidence angle of 30° (open symbols) and 50° (closed symbols).

The magnitude of anisotropy (i.e., the value of *B* and/or *E* coefficients in (1)) directly scales the amplitude of resulting *P-SH* conversions. Stronger anisotropy leads to azimuthal patterns that are more differentiated but have nodes (polarity changes) in the same places. It should be noted that direct dependence of converted-phase parameters on velocity contrast and anisotropy extent holds only as long as the sense of velocity change across the interface is preserved. These relationships will break down if the velocity difference across the interface is comparable to the anisotropic perturbation. In an extreme case, a directionally-dependent velocity inversion may exist, so that different phases may be generated for different combinations of incidence angle, axis orientation and ray back-azimuth.

Discussion

The effects of anisotropy one may hope to interpret successfully in *P* coda involve both the presence of transverse motions (the phases *Ps* and *Psms* and the transverse component of direct *P*), as well as the azimuthal variation of radial amplitudes (*P* and *Ps*). Synthetic seismograms for a variety of simple 1-D an-isotropic velocity models allow us to make a number of observations, and to propose answers for the questions posed at the onset of this study. The major lesson drawn from this exercise is the importance of P anisotropy in the generation of P-coda from teleseismic body waves, as well as the strong sensitivity of P-to-S converted phases to the tilt of the symmetry axis of anisotropy. Most anisotropy studies, aside from large-scale tomographic experiments, assume a horizontal or vertical axis of symmetry and interpret data in terms of S anisotropy only. Our experiments suggest that this may be too restrictive.

Models with a surface anisotropic layer (family A) are most efficient in generating P-SH conversions in our experiments. Ps and $Psms$ with amplitudes on the order of 10% of the vertical P will be observed for incidence angles over 20°. A transverse component of the direct P is a diagnostic phase of this model type, indicating that anisotropy extends all the way up to the receiver. Also diagnostic is the azimuthal dependence of the P-Ps phase timing, and strong azimuthal variation in the radial component of the direct P.

A delayed first motion on the transverse component is diagnostic of models in families B and C, in which the anisotropic layer is buried. For simple models with a single anisotropic layer, the P-SH converted phase would likely be the most energetic transverse arrival as well. The timing of this arrival relative to direct P may be the best indicator of the depth of the buried anisotropic layer. Since the surface layer is isotropic in both model families B and C, the variation of Ps-P delay times is negligible, as is the variation in radial amplitude of direct P. A possible discriminant between types B and C is the azimuthal variation of the radial Ps phase. If the symmetry axis $\hat{w}$ is tilted 45°, the pattern is two-lobed if the layer is buried and "thick" (family B) and four-lobed if the anisotropic layer is buried and "thin" (family C). It is instructive that both B and C model families yield significant P-SH conversions only if P-wave anisotropy is present. Speculatively, a thin layer of anisotropy within the crust may prove to be more appealing in crustal models than the "anisotropic halfspace" concept. It is also instructive that large transverse-component P coda can be generated with anisotropic layers that are too thin to generate significant shear-wave splitting. The derivative-pulse shape characteristic of the Ps' phase in model family C (Fig. 3) resembles a split shear wave, but actually is generated by the interference of P-to-S conversions at separate interfaces. To distinguish between this type of effect and that of an anisotropic layer above the P-to-S conversion, one might check for a transverse component in direct P due to polarization distortion, and the transverse component of a "$Psms$" phase, generated by P-to-S conversion at the free surface.

It seems that discriminating the "melon" and the "pumpkin" models of anisotropy on the basis of P-SH converted data may be difficult with P coda observations only, as the azimuthal patterns of most phases are similar. In model family A the pattern of the radial Ps phase may help determine the sign of anisotropy. Also a comparison of Ps-P delay pattern with that of the direct P amplitude may be instructive. However, supplementary data may be necessary to distinguish directly whether inferred anisotropy is due to cracks or thin layering ("pumpkin") or LPO of mineral fast-axes ("melon").

Stacking P coda to enhance converted phases is potentially a powerful tool. If the symmetry axis $\hat{w}$ throughout the crust and shallow mantle is horizontal, stacking with $\cos 2\xi$- and $\sin 2\xi$-weighting with back-azimuth ξ should enhance the signal-to-noise ratio and identify the strike of $\hat{w}$. Likewise, the effects of P and S anisotropy are quite similar for horizontal $\hat{w}$. Although this makes distinguishing P from S anisotropy difficult from seismic data alone, the tradeoff between the two is straightforward. However, if the axis is tilted, the converted-phase azimuthal patterns can be four-lobed, two-lobed or a mixture of the two. The two anisotropy types behave differently as $\hat{w}$ varies from horizontal to vertical, making data interpretation more challenging.

Conclusions

Seismic anisotropy in a flat-layered homogeneous medium results in the generation of *P-SH* converted phases and also affects *P-SV* conversions. Since the anisotropy of many rocks can be approximated to possess either a fast or slow axis of symmetry, we have used hexagonal symmetry in our calculations, varying the orientation of the symmetry axis $\hat{w}$. *P-SH* conversions arise in models containing either P and/or S velocity anisotropy, with P anisotropy leading to stronger effects in many of the scenarios we examined. Synthetic P coda from different combinations of anisotropy type, sign and location are substantially distinct, and potentially resolvable in band-limited noisy data.

The strengths of *P-SV* and *P-SH* conversions vary with the back azimuth ξ of the incoming wave relative to the axis of symmetry $\hat{w}$. These patterns can be used to distinguish candidate models of anisotropy. The tilt of the anisotropic symmetry axis and the incidence angle of the incoming P wave control the resulting azimuthal pattern. Perturbative waveforms on the radial component, whether due to converted phases or polarization distortions of direct P, are symmetric to sign changes in ξ. Perturbative waveforms on the transverse component are anti-symmetric to sign changes in ξ. Near-vertical axes of symmetry result in azimuthal patterns that are effectively two-lobed ($\sin \xi$). A transition to asymmetric four-lobed pattern occurs with increasing tilt and is more pronounced for shallow incidence P waves. The perturbative waveforms often do not vanish at the pair of "nodes" that define extra lobes in the azimuthal pattern, but rather distort gradually in a manner that makes the precise location of the polarity transition somewhat subjective. Patterns with exact $\sin 2\xi$ and $\cos 2\xi$ symmetry, and waveforms that vanish at the second set of polarity transitions, occur only when the axis of symmetry $\hat{w}$ is horizontal. The effect of pure P anisotropy is maximized when the symmetry axis $\hat{w}$ is tilted 45° from the vertical, while the effect of pure S anisotropy is strongest for subhorizontal axes. In the case of mixed anisotropy, azimuthal patterns follow the stronger influence. For subvertical axis tilts, models with pure P and pure S anisotropy predict transverse Ps patterns of opposite polarity. Aside from a few cases, changes of anisotropy sign, that is, switching between slow and fast axes of symmetry, results in azimuthal patterns that are mirror images of each other.

Our synthetic P coda suggest that interference between P-to-S converted phases from different interfaces within the crust can create SH-waveforms that resemble the derivative-pulse waveforms diagnostic of shear-wave splitting. Paradoxically, these phases are best generated by thin layers of compressional, not shear, anisotropy with a tilted axis of symmetry.

Acknowledgments

Discussions at the International Workshop on Geodynamics of Lithosphere and Earth's Mantle in Castle Trest in the Czech Republic with V. Farra and N. Girardin to help to motivate this study. This work was supported by Air Force Office of Scientific Research contract F49620–94–1–0043. Computational firepower was supported by NSF equipment grant EAR–9528484. Many figures were generated with the GMT graphics package (WESSEL and SMITH, 1991).

Appendix: Relation to THOMSEN (1986)

Anisotropy with *one* axis of symmetry can be parameterized by seven constants, two of which describe the orientation of the axis of symmetry $\hat{w}$, and five elastic parameters. Several choices for the five elastic parameters exist in the literature (ANDERSON, 1989). The choice we take in (1) was initially derived for weak anisotropy with a horizontal axis of symmetry in the context of marine refraction studies (BACKUS, 1965). In the limit of weak anisotropy, the coefficients in (1) can be related directly to a decomposition of the elastic tensor, each term of which possesses hexagonal symmetry with respect to rotations about $\hat{w}$ (SHEARER and ORCUTT, 1986). We express the elastic tensor in each layer with the following decomposition:

$$\Lambda = A\Lambda_A + B\Lambda_B + C\Lambda_C + D\Lambda_D + E\Lambda_E, \tag{2}$$

where

$$\Lambda_A = \mathbf{I} \otimes \mathbf{I}$$

$$\Lambda_B = \mathbf{W} \otimes \mathbf{I} + \mathbf{I} \otimes \mathbf{W}$$

$$\Lambda_C = 8\mathbf{W} \otimes \mathbf{W} - \mathbf{I} \otimes \mathbf{I} \tag{3}$$

$$\mathbf{\Lambda}_D = {}_{(13)}\mathbf{I} \otimes \mathbf{I} + {}_{(14)}\mathbf{I} \otimes \mathbf{I} - 2\mathbf{I} \otimes \mathbf{I}$$

$$\mathbf{\Lambda}_E = 2[{}_{(13)}\mathbf{\Lambda}_B + {}_{(14)}\mathbf{\Lambda}_B - 2\mathbf{\Lambda}_B] + \mathbf{\Lambda}_D.$$

The '$\otimes$' symbol denotes the tensor product operation, and $\mathbf{W} = \hat{\mathbf{w}} \otimes \hat{\mathbf{w}} - \frac{1}{2}\mathbf{I}$, where $\mathbf{I}$ is the identity tensor. The permutation (ij) indicates the interchange of the ith and jth tensor index e.g., $\{{}_{(13)}\mathbf{I} \otimes \mathbf{I}\}_{ijkl} = \delta_{kj}\delta_{il}$. An isotropic elastic tensor $\mathbf{\Lambda}^{(0)}$ contains only terms proportional to the isotropic tensors $\mathbf{\Lambda}_A$ and $\mathbf{\Lambda}_D$, as neither depends on $\hat{\mathbf{w}}$.

When using expressions (2) and (3), we are neither limited to "weak" anisotropy nor to a horizontal axis of symmetry. For "strong" anisotropy, PARK (1996) shows that the azimuthal phase velocity formulas (1) are the first-order approximations to the P and SV head-wave velocities of a medium with horizontal $\hat{\mathbf{w}}$ and elastic tensor described by (2) and (3). For modeling media with more complexity, it is possible to form a linear combination of anisotropic deviations from an isotropic reference model, each with its own axis of symmetry $\hat{\mathbf{w}}$. This would be useful for media with both oriented cracks and oriented minerals, if the orientations differ, or for media with orthorhombic symmetry.

Another common parameterization for hexagonally-symmetric anisotropy is that derived by THOMSEN (1986) for a vertical axis of symmetry in the context of shallow seismic profiling. Thomsen references phase velocities to the vertical velocity, rather than to the average of the velocity extremes, with three anisotropic parameters γ, ε and δ^*. In applications, Thomsen recommends replacing δ^* with a first-order approximation "δ". To relate these parameters to our anisotropic parameters B, C, and E, we can use the formulas of YU and PARK (1993) to express the elastic tensor 6×6 matrix format $\{C_{jk}\}$ for a vertical axis of symmetry. We adopt the usual conventions, with components 1, 2, 3 corresponding to x, y, z, respectively, and $C_{jk} = \Lambda_{lmnp}$ according to the substitutions $1 \to 11$; $2 \to 22$; $3 \to 33$; $4 \to 23$; $5 \to 13$; and $6 \to 12$. The matrix C is expressed as

$$\begin{bmatrix} A-B+C & A-B+C-2(D-E) & A-3C-2(D+E) & & & \\ A-B+C-2(D-E) & A-B+C & A-3C-2(D+E) & & & \\ A-3C-2(D+E) & A-3C-2(D+E) & A+B+C & & & \\ & & & D+E & & \\ & & & & D+E & \\ & & & & & D-E \end{bmatrix}$$

$$(4)$$

where the blank indices are zero. Using the formulas in PARK (1996), the Christoffel matrix $\mathscr{H}$ for this elastic tensor can be computed for an upgoing plane wave at an angle of incidence θ to the vertical, using wavenumber vector $\mathbf{k} = \hat{\mathbf{x}} \sin\theta - \hat{\mathbf{z}} \cos\theta$. (Note that z increases downward in the coordinate system of our synthetic-seismogram calculations.)

$$\mathscr{K} = \begin{bmatrix} (A-B+C)\sin^2\theta + (D+E)\cos^2\theta & 0 & -(A-3C-D-E)\sin\theta\cos\theta \\ 0 & D+E\cos 2\theta & 0 \\ -(A-3C-D-E)\sin\theta\cos\theta & 0 & (A+B+C)\cos^2\theta + (D+E)\sin^2\theta \end{bmatrix}$$

(5)

The eigenvalues of the Christoffel matrix correspond to the phase velocities of the quasi-P, quasi-SV and quasi-SH polarized waves. These phase velocities correspond to those derived by THOMSEN (1986).

Using (4), we can relate the two sets of anisotropic parameters, using Thomsen's definitions

$$\varepsilon = \frac{C_{11}-C_{33}}{2C_{33}} = -\frac{B}{A+B+C}$$

$$\gamma = \frac{C_{66}-C_{44}}{2C_{44}} = -\frac{E}{D+E}$$

$$\delta = \frac{(C_{13}+C_{44})^2 - (C_{33}-C_{44})^2}{2C_{33}(C_{33}-C_{44})} = \frac{8C^2 - B^2 - 2BC - 2(B+4C)(A-D-E)}{2(A+B+C)(A+B+C-D-E)}.$$

(6)

The first two of Thomsen's parameters relate directly to our parameters B and E for the $\cos 2\xi$ azimuthal variation in phase velocity. The negative sign in the formulas for ε and γ reflects the preponderance of "slow" symmetry axes in crustal environments, corresponding to crack and/or fine-layering anisotropy. The formula for δ is complex, but can be reduced by discarding higher-order terms to obtain

$$\delta \approx \varepsilon - \frac{4C}{A+B+C}$$

(7)

From this formula, one infers that the phase velocity surface for quasi-P waves is elliptical only if $\varepsilon = \delta$. This condition is not satisfied for many of the anisotropy measurements tabulated in THOMSEN (1996), indicated that $C=0$ might be an incorrect assumption. However, extending these measurements to characterize crystalline bedrock may be risky. All elastic properties tabulated in THOMSEN (1986) involve shallow sedimentary facies relevant to oil exploration, with relatively few values measured *in situ*. For the *in situ* measurements of seismic anisotropy tabulated in THOMSEN (1986), $\varepsilon \approx \delta$ tends to be better satisfied. This suggests an enhancement of C by decompression, perhaps via an increase in the porosity of a rock sample.

The case for setting $C=0$ in our calculations is supported by the perturbative expressions for anisotropy in cracked isotropic media (HUDSON, 1981; CRAMPIN, 1984). These expressions estimate that $B=C=0$ for fluid-saturated cracks, and, using (4), one can show that $C=0$ of the first-order perturbation associated with dry cracks. In the second-order dry-crack perturbation in a Poisson solid, $C/B \lesssim 0.1$, suggesting that P-phase velocities are near-elliptical for this case, in Hudson's perturbation theory at least.

A suggestion that C might typically be nonzero arises from a special case in the theory for anisotropy caused by thin-layering of different media (BACKUS, 1962; HELBIG, 1994; THOMSEN, 1986). In this theory, Thomsen's parameter $\delta \approx 0$ if the alternating lithologies share a common Poisson ratio. This is equivalent to $C = -B/4$, a small value, but not zero. However, constant Poisson ratio is not the norm among crustal rocks. A typical intercalation in the crystalline basement might involve felsic and mafic rock types, where the mafic layers have higher seismic velocities α, β *and* higher Poisson ratio (i.e., a higher velocity ratio α/β). Using the averaging formulas in Chapter 7.4 of HELBIG (1994), one can demonstrate that $B < 0$ and $0 \leq (B + 4C)/B \leq 1$ for this case, so that $|C|$ is bounded by $|B/4|$ and can be considerably smaller.

Teleseismic *P*-coda reverberations with periods $T \gtrsim 1$ sec are typically used to investigate the properties of the bulk crust, e.g., its Poisson ratio, or differences between upper and lower crustal layers. Given the above estimates, our choice to neglect the $\cos 4\xi$ azimuthal variation in P velocity seems reasonable as a working hypothesis for the bulk of the crust. However, shallow structures, such as sedimentary basins, may require this parameter for an accurate description of the *P*-coda.

REFERENCES

AGUE, J. J. (1995), *Deep Crustal Growth of Quartz, Kyanite and Garnet into Large Aperture, Fluid-filled Fractures, Northeastern Connecticut, USA*, J. Metam. Geol. *13*, 299–314.

ANDERSON, D. L., *A Theory of the Earth* (Blackwell Scientific, Oxford 1989).

BABUSKA, V., and PROS, Z. (1984), *Velocity Anisotropy in Granodiorite and Quartzite due to the Distribution of Microcracks*, Geophys. J. Roy. Astron. Soc. *76*, 121–127.

BABUSKA, V., and CARA, M., *Seismic Anisotropy in the Earth* (Kluwer Academic, Dordrecht 1991).

BABUSKA, V., PLOMEROVA, J., and SILENY, J. (1993), *Models of Seismic Anisotropy in the Deep Continental Lithosphere*, Phys. Earth and Planet. Int. *78*, 167–191.

BACKUS, G. E. (1962), *Long-wave Elastic Anisotropy Produced by Horizontal Layering*, J. Geophys. Res. *67*, 4427–4440.

BACKUS, G. E. (1965), *Possible Forms of Seismic Anisotropy of the Uppermost Mantle under Oceans*, J. Geophys. Res. *70*, 3429–3439.

BLACKMAN, D. K., KENDALL, J.-M., DAWSON, P. R., WENK, H.-R., BOYCE, D., and JASON PHIPPS MORGAN (1996), *Teleseismic Imaging of Subaxial Flow at Mid-ocean Ridges: Travel-time Effects of Anisotropic Mineral Texture in the Mantle*, Geophys. J. Int. *127*, 415–426.

BURLINI, L., and FOUNTAIN, D. M. (1993), *Seismic Anisotropy of Metapelites from the Ivrea-Verbano Zone and Serie dei Leghi (Northern Italy)*, Phys. Earth and Planet. Int. *78*, 301–317.

CHRISTENSEN, N. I. (1996), *Poisson's Ratio and Crustal Seismology*, J. Geophys. Res. *101*, 3139–3156.

CRAMPIN, S. (1984), *Effective Elastic Constants for Wave Propagation through Cracked Solids*, Geophys. J. Roy. Astron. Soc. *76*, 135–145.

CRAMPIN, S. (1991), *Wave Propagation Through Fluid-filled Inclusions of Various Shapes: Interpretation of Extensive-dilatancy Anisotropy*, Geophys. J. Int. *104*, 611–623.

COLEMAN, M. (1996), *Orogen-parallel and Orogen-perpendicular Extension in the Central Nepalese Himalayas*, GSA Bulletin *108*, 1594–1607.

FARRA, V., VINNIK, L. P., ROMANOWICZ, B., KOSAREV, G. L., and KIND, R. (1991), *Inversion of Teleseismic S Particle Motion for Azimuthal Anisotropy in the Upper Mantle; A Feasibility Study*, Geophys. J. Int. *106*, 421–431.

FRYER, G. J., and FRAZER, L. N. (1987), *Seismic Waves in Stratified Anisotropic Media—II. Elastodynamic Eigensolutions for Some Anisotropic Systems*, Geophys. J. Roy. Astron. Soc. *91*, 73–102.

GAHERTY, J. B., and JORDAN, T. H. (1995), *Lehmann Discontinuity as the Base of an Anisotropic Layer Beneath Continents*, Science *268*, 1468–1471.

GARMANY, J. (1983), *Some Properties of Elastodynamic Eigensolutions in Stratified Media*, Geophys. J. Roy. Astron. Soc. *75*, 565–570.

GRESILLAUD, A., and CARA, M. (1996), *Anisotropy and P-wave Tomography; A New Approach for Inverting Teleseismic Data from a Dense Array of Stations*, Geophys. J. Int. *126*, 77–91.

HEARN, T. M. (1996), *Anisotropic P_n Tomography in the Western United States*, J. Geophys. Res. *101*, 8403–8414.

HELBIG, K., *Foundations of Anisotropy for Exploration Seismics* (Pergamon/Elsevier Science, Oxford 1994).

HU, G. (1993), *Lower Crustal and Moho Structures beneath Beijing, China Determined by Formal Inversion of Single Station Measurements of Teleseismic P-wave Polarization Anomalies*, EOS 74, 426.

HUDSON, J. A. (1981), *Wave Speeds and Attenuation of Elastic Waves in Material Containing Cracks*, Geophys. J. Roy. Astron. Soc. *64*, 133–150.

KENNETT, B. L. N., *Seismic Wave Propagation in Stratified Media* (Cambridge University Press, Cambridge 1983).

KERN, H., WALTHER, Ch., FLUH, E. R., and MARKER, M. (1993), *Seismic Properties of Rocks Exposed in the POLAR Profile Region—Constraints on the Interpretation of the Refraction Data*. Precambrian Research *64*, 169–187.

KOSAREV, G. L., MAKEYEVA, L. I., and VINNIK, L. P. (1984), *Anisotropy of the Mantle Inferred from Observations of P to S Converted Waves*, Geophys. J. Roy. Astron. Soc *76*, 209–220.

LANGSTON, C. A. (1977), *The Effect of Planar Dipping Structure on Source and Receiver Responses for Constant Ray Parameter*, Bull. Seismol. Soc. Am. *67*, 1029–1050.

LEVIN, V., MENKE, W., and LERNER-LAM, A. L. (1996), *Seismic Anisotropy in Northeastern U.S. as a Source of Significant Teleseismic P Travel-time Anomalies*, Geophys. J. Int. *126*, 593–603.

LEVIN, V., and PARK, J. (1977a), *Anisotropy in the Ural Mountains Foredeep from Teleseismic Receiver Functions*, Geophys. Res. Letts. *24*, 1283–1286.

LEVIN, V., and PARK, J. (1977b), *P-SH Conversions in a Flat-layered Medium with Anisotropy of Arbitrary Orientation*, Geophys. J. Int., in press.

LYNN, H. B. (1991), *Field Measurements of Azimuthal Anisotropy: First 60 Meters, San Francisco Bay Area, CA, and Estimation of the Horizontal Stress Ratio from V_{S1}/V_{S2}*, Geophysics 56, 822–832.

MAINPRICE, D., and SILVER, P. G. (1993), *Constraints on the Interpretation of Teleseismic SKS Observations from Kimberlite Nodules from the Subcontinental Mantle*, Phys. Earth and Planet. Int. *78*, 257–280.

NICOLAS, A., BOUDIER, F., and BOULIER, A. M. (1973), *Mechanism of Flow in Naturally and Experimentally Deformed Periodotites*, Am. J. Sci. *10*, 853–876.

OWENS, T. J., and CROSSON, R. S. (1988), *Shallow Structure Effects on Broad-band Teleseismic P Waveforms*, Bull. Seismol. Soc. Am. *77*, 96–108.

PARK, J. (1993), *The Sensitivity of Seismic Free Oscillations to Upper Mantle Anisotropy I. Zonal Structure*, J. Geophys. Res. *98*, 19933–19949.

PARK, J. (1996), *Surface Waves in Layered Anisotropic Structures*, Geophys. J. Int. *126*, 173–183.

RAITT, R. W., SHOR Jr., G. G., FRANCIS, T. J. G., and MORRIS, G. B. (1969), *Anisotropy of the Pacific Upper Mantle*, J. Geophys. Res. *74*, 3095–3109.

RIBE, N. (1992), *On the Relation Between Seismic Anisotropy and Finite Strain*, J. Geophys. Res. *97*, 8737–8747.

SHEARER, P. M., and ORCUTT, J. A. (1986), *Compressional and Shear-wave Anisotropy in the Oceanic Lithosphere—the Ngendie Seismic Refraction Experiment*, Geophys. J. Roy. Astron. Soc. *87*, 967–1003.

SILVER, P. G. (1996), *Seismic Anisotropy Beneath the Continents: Probing the Depths of Geology*, Ann. Rev. Earth Planet. Sci. *24*, 385–432.

SILVER, P. G., and CHAN, W. W. (1991), *Shear-wave Splitting and Subcontinental Mantle Deformation*, J. Geophys. Res. *96*, 16429–16454.

THOMSEN, L. (1986), *Weak Elastic Anisotropy*, Geophysics *51*, 1954–1966.

VINNIK, L. P., MAKEYEVA, L. I., MILEV, A., and USENKO, A. Yu. (1992), *Global Patterns of Azimuthal Anisotropy and Deformations in the Continental Mantle*, Geophys. J. Int. *111*, 433–447.

VINNIK, L. P., and MONTAGNER, J.-P. (1996), *Shear-wave Splitting in the Mantle Ps Phases*, Geophys. Res. Letts. *23*, 2449–2452.

VISSER, J., and PAULLSEN, H. (1993), *The Crustal Structure from Teleseismic P-wave Coda—II. Application to Data of the NARS Array in Western Europe and Comparison with Deep Seismic Sounding Data*, Geophys. J. Int. *112*, 26–38.

WESSEL, P., and SMITH, W. H. F. (1991), *Free Software Helps Map and Display Data*, EOS Trans. AGU *72*, 441.

YU, Y., and PARK, J. (1933), *Upper Mantle Anisotropy and Coupled-mode Long-period Surface Waves*, Geophys. J. Int. *114*, 473–489.

ZANDT, G., and AMMON, C. J. (1995), *Continental Crust Composition Constrained by Measurements of Crustal Poisson's Ratio*, Nature *374*, 152–154.

ZHANG, S., and KARATO, S. (1995), *Lattice Preferred Orientation of Olivine Aggregates Deformed in Simple Shear*, Nature *375*, 774–777.

ZHI, Z., and SCHWARTZ, S. Y. (1994), *Seismic Anisotropy in the Shallow Crust of the Loma Prieta Segment of the San Andreas Fault System*, J. Geophys, Res. *99*, 9651–9661.

ZHU, L., OWENS, T. J., and RANDALL, G. E. (1995), *Lateral Variation in Crustal Structure of the Northern Tibetan Plateau Inferred from Teleseismic Receiver Functions*, Bull. Seismol. Soc. Am. *85*, 1531–1540.

(Received October 30, 1996, revised July 8, 1997, accepted July 17, 1997)

To access this journal online:
http://www.birkhauser.ch

Pure appl. geophys. 151 (1998) 699–718
0033–4553/98/040699–20 $ 1.50 + 0.20/0

Pure and Applied Geophysics

Weak Contrast *PP* Wave Displacement R/T Coefficients in Weakly Anisotropic Elastic Media

Ivan Pšenčík[1] and Václav Vavryčuk[1]

Abstract—Approximate *PP* plane wave displacement coefficients of reflection and transmission for weak contrast interfaces separating weakly but arbitrarily anisotropic elastic media are presented. The *PP* reflection coefficient for such an interface has been derived recently by Vavryčuk and Pšenčík (1997). The *PP* transmission coefficient presented in this paper was derived by the same approach. The coefficients are given as a sum of the coefficient for the weak contrast interface separating two nearby isotropic media and a term depending linearly on contrasts of the so-called weak anisotropy (WA) parameters (parameters specifying deviation of properties of the medium from isotropy), across the interface. While the reflection coefficient depends only on 8 of the complete set of the WA parameters describing *P*-wave phase velocity in weakly anisotropic media, the transmission coefficient depends on their complete set. The *PP* reflection coefficient depends on "shear-wave splitting parameter" γ. Tests of accuracy of the approximate formulae are presented on several models.

Key words: Weak anisotropy, weak contrast interface, plane wave reflection and transmission coefficients.

1. Introduction

Seismic anisotropy is a nearly omnipresent phenomenon, which affects, often considerably, parameters of propagating elastic waves. Displacement coefficients of reflection and transmission (R/T) belong to such parameters. For isotropic media, explicit formulae for the R/T coefficients are well known, see e.g., Aki and Richards (1980). They are, however, relatively complicated and their relation to elastic parameters is often strongly nonlinear. The complexity of the coefficients reduces substantially if the contrast between the two media separated by an interface is weak. The coefficients can be linearized with respect to the contrast in elastic parameters. The linearized formulae become more transparent and they are often very accurate, see again Aki and Richards (1980). In anisotropic media, explicit expressions for the R/T coefficients are available for media with a higher symmetry, whose symmetry planes are especially oriented with respect to an interface, see e.g. Daley and Hron (1977), Keith and Crampin (1977). In the

[1] Geophysical Institute, Academy of Sciences of the Czech Republic, Boční II, Praha 4, Czech Republic; e-mail: ip@ig.cas.cz, vasek@seis.ig.cas.cz.

case of general anisotropy, a common method to determine the coefficients is the numerical solution of the system of equations resulting from the boundary conditions, see e.g. GAJEWSKI and PšENČÍK (1987). As in isotropic media, the problem simplifies considerably if reflection/transmission at a weak contrast interface is considered and if the media surrounding the interface are only weakly anisotropic.

The assumption of weak contrast interface and of weak anisotropy has been used by several authors although higher symmetry anisotropy was always considered. THOMSEN (1993) extended BANIK's (1987) work and derived the PP R/T coefficients for a weak contrast interface separating two weakly transversely isotropic media with axes of symmetry perpendicular to the interface, see also discussion of this formula by TSVANKIN (1996). RUEGER (1996) corrected and generalized Thomsen's results for PP reflections in planes containing symmetry axes of transversely isotropic and orthorhombic media so that media with symmetry axes parallel to the interface could also be considered. HAUGEN and URSIN (1996) derived PP reflection coefficients in the symmetry planes of a model containing an interface separating a TI medium with axis of symmetry perpendicular to the interface from a TI medium with axis of symmetry parallel to the interface.

This paper is an extension of the paper by VAVRYČUK and PšENČÍK (1998) who derived an approximate formula for the PP wave displacement coefficient of reflection for a weak contrast interface separating two arbitrary weakly anisotropic media. In this paper, in addition to the formula for the reflection coefficient, the PP wave displacement coefficient of transmission is presented. Both formulae are obtained by applying the first-order perturbation theory. Continuous isotropic medium with no discontinuity of parameters of the medium across the studied interface is considered as a background medium. The media on both sides of the interface are then perturbed so that the result is a model composed of two slightly different weakly anisotropic halfspaces.

Accuracy of the approximate formulae is tested on models consisting of a homogeneous isotropic halfspace over a halfspace filled by a homogeneous transversely isotropic (TI) material with the horizontal axis of symmetry (the HTI material). Behavior of the PP R/T coefficients at an interface separating two TI halfspaces is also shown. The upper halfspace contains a material with the vertical axis of symmetry (the VTI material), the lower halfspace contains the HTI material or a material with the inclined axis of symmetry (the ITI material).

If not specified differently, the Roman lower-case indices attain values 1, 2 and 3, upper-case Roman indices attain only values 1 and 2. The Greek indices run from 1 to 6. Einstein summation convention is used for the repeated indices.

2. Basic Formulae

VAVRYČUK and PšENČÍK (1998) present a detailed derivation of the formula for the PP reflection coefficient at a weak contrast interface separating two weakly anisotropic media. Here only a short review of basic steps and formulae is made.

We consider a model consisting of two homogeneous weakly anisotropic halfspaces separated by an interface with the unit normal v_i pointing into the halfspace, in which an incident wave propagates. We call it the halfspace 1 and denote its density and the density-normalized elastic parameters $\rho^{(1)}$ and $a_{ijkl}^{(1)}$. The same parameters in the halfspace 2 are denoted $\rho^{(2)}$ and $a_{ijkl}^{(2)}$. The incident and generated waves satisfy the boundary conditions at the interface: continuity of the displacement and the traction vectors. As a consequence of the boundary conditions, we get important relations for the slowness vectors $p_i^{(N)}$ and $p_i^{(0)}$ of the generated and the incident waves and for the reflection/transmission coefficients $U^{(N)}$. The superscripts $N = 1, 2$ and 3 correspond to reflected $S1$, $S2$ and P wave, the superscripts $N = 4, 5$ and 6 correspond to transmitted $S1$, $S2$ and P wave. The relation for the slowness vectors has the form

$$p_i^{(N)} = b_i + \xi^{(N)}v_i = p_i^{(0)} - (p_k^{(0)}v_k)v_i + \xi^{(N)}v_i, \tag{1}$$

where the quantity $\xi^{(N)}$ can be determined from the polynomial equation of the sixth order

$$\det[a_{ijkl}(b_j + \xi v_j)(b_l + \xi v_l) - \delta_{ik}] = 0. \tag{2}$$

The R/T coefficients $U^{(N)}$ are determined by solving the system of six algebraic equations

$$U^{(1)}g_i^{(1)} + U^{(2)}g_i^{(2)} + U^{(3)}g_i^{(3)} - U^{(4)}g_i^{(4)} - U^{(5)}g_i^{(5)} - U^{(6)}g_i^{(6)} = -g_i^{(0)},$$

$$U^{(1)}X_i^{(1)} + U^{(2)}X_i^{(2)} + U^{(3)}X_i^{(3)} - U^{(4)}X_i^{(4)} - U^{(5)}X_i^{(5)} - U^{(6)}X_i^{(6)} = -X_i^{(0)}, \tag{3}$$

where

$$X_i^{(N)} = \rho^{(1)}a_{ijkl}^{(1)}v_j g_k^{(N)}p_l^{(N)}, \quad N = 0, 1, 2, 3,$$

$$X_i^{(N)} = \rho^{(2)}a_{ijkl}^{(2)}v_j g_k^{(N)}p_l^{(N)}, \quad N = 4, 5, 6. \tag{4}$$

The vectors $X_i^{(N)}$ are the amplitude-normalized traction vectors. Equation (3) can be rewritten into the matrix form

$$C_{\alpha\beta}U_\beta = B_\alpha, \tag{5}$$

where $C_{\alpha\beta}$ is the displacement-stress matrix of the R/T waves, U_β is the vector of the R/T coefficients and B_α is the amplitude-normalized displacement-stress vector of the incident wave.

In each halfspace, the density-normalized elastic parameters and the density are considered in the form

$$a_{ijkl}^{(I)} = a_{ijkl}^{(I)0} + \delta a_{ijkl}^{(I)}, \quad \rho^{(I)} = \rho^{(I)0} + \delta\rho^{(I)}, \quad I = 1, 2. \tag{6}$$

The symbols $a_{ijkl}^{(I)0}$ and $\rho^{(I)0}$ denote the elastic parameters and the density of the background isotropic media in both halfspaces. The quantities $\delta a_{ijkl}^{(I)}$ and $\delta\rho^{(I)}$ represent small deviations from the isotropic backgrounds.

We now linearize Eq. (5) with respect to the deviations of elastic parameters $a_{ijkl}^{(I)}$ and the density $\rho^{(I)}$ from the average values $\bar{a}_{ijkl}^0$ and $\bar{\rho}^0$ of the parameters and the density of the isotropic backgrounds. The average value $\bar{w}$ of parameters $w^{(I)}$ is defined as follows

$$\bar{w} = \tfrac{1}{2}(w^{(1)} + w^{(2)}). \tag{7}$$

In this way, we get

$$C_{\alpha\beta}^0 \delta U_\beta = \delta B_\alpha - \delta C_{\alpha\beta} U_\beta^0. \tag{8}$$

The symbols $C_{\alpha\beta}^0$ and U_β^0 denote the matrix $C_{\alpha\beta}$ and the vector U_β, specified for a fictitious interface in a continuous isotropic space characterized by the parameters $\bar{a}_{ijkl}^0$ and the density $\bar{\rho}^0$. For the incident wave with a unit amplitude, the vector δU_α contains perturbations of three reflection and three transmission coefficients from their values U_α^0 in the background isotropic medium. The vector δB_α and the matrix $\delta C_{\alpha\beta}$ are the perturbations of the corresponding vector and matrix in Eq. (5). The linearized reflection/transmission coefficients can be sought in the form $U_\alpha^0 + \delta U_\alpha$, where δU_α is given as

$$\delta U_\alpha = (\mathbf{C}^0)_{\alpha\beta}^{-1} (\delta B_\beta - \delta C_{\beta\gamma} U_\gamma^0). \tag{9}$$

The basic step in making Eq. (9) useful is the inversion of the matrix $C_{\beta\gamma}^0$. VAVRYČUK and PŠENČÍK (1997) found the inverted matrix in the form

$$(\mathbf{C}^0)_{\alpha\beta}^{-1} = \begin{pmatrix} -\dfrac{\bar{\beta}^2 Y p_1^0 \cos\Psi}{Z_S} & \dfrac{\sin\Psi}{2} & -p_1^0\bar{\beta}\cos\Psi & \dfrac{\cos\Psi}{2\bar{\beta}\bar{\rho}^0} & -\dfrac{p_1^0\bar{\beta}\sin\Psi}{Z_S} & \dfrac{\bar{\beta}^2(p_1^0)^2\cos\Psi}{Z_S} \\[4mm] \dfrac{\bar{\beta}^2 Y p_1^0 \sin\Psi}{Z_S} & \dfrac{\cos\Psi}{2} & p_1^0\bar{\beta}\sin\Psi & -\dfrac{\sin\Psi}{2\bar{\rho}^0\bar{\beta}} & \dfrac{\bar{\beta}p_1^0\cos\Psi}{Z_S} & \dfrac{\bar{\beta}^2(p_1^0)^2\sin\Psi}{Z_S} \\[4mm] \dfrac{\bar{\beta}^2 p_1^0}{\bar{\alpha}} & 0 & -\dfrac{p_1^0\bar{\beta}^2 Y}{Z_P} & -\dfrac{\bar{\beta}^2(p_1^0)^2}{Z_P} & 0 & \dfrac{1}{2\bar{\rho}^0\bar{\alpha}} \\[4mm] -\dfrac{\bar{\beta}^2 Y p_1^0 \cos\Phi}{Z_S} & -\dfrac{\sin\Phi}{2} & p_1^0\bar{\beta}\cos\Phi & \dfrac{\cos\Phi}{2\bar{\beta}\rho^0} & -\dfrac{p_1^0\bar{\beta}\sin\Phi}{Z_S} & \dfrac{\bar{\beta}^2(p_1^0)^2\cos\Phi}{Z_S} \\[4mm] \dfrac{\bar{\beta}^2 Y p_1^0 \sin\Phi}{Z_S} & -\dfrac{\cos\Phi}{2} & -p_1^0\bar{\beta}\sin\Phi & \dfrac{\sin\Phi}{2\bar{\rho}^0\bar{\beta}} & \dfrac{\bar{\beta}p_1^0\cos\Phi}{Z_S} & -\dfrac{\bar{\beta}^2(p_1^0)^2\sin\Phi}{Z_S} \\[4mm] -\dfrac{\bar{\beta}^2 p_1^0}{\bar{\alpha}} & 0 & -\dfrac{p_1^0\bar{\beta}^2 Y}{Z_P} & -\dfrac{\bar{\beta}^2(p_1^0)^2}{Z_P} & 0 & -\dfrac{1}{2\bar{\rho}^0\bar{\alpha}} \end{pmatrix}, \tag{10}$$

where

$$Y = \bar{\rho}^0(1 - 2\bar{\beta}^2(p_1^0)^2), \quad Z_P = 2\bar{\alpha}\bar{\rho}^0\bar{\beta}^2 p_1^0 p_3^{0P}, \quad Z_S = 2\bar{\rho}^0\bar{\beta}^3 p_1^0 p_3^{0S}. \tag{11}$$

Here p_1^0, p_3^{0P} and p_3^{0S} denote the x and z components of the slowness vector of the P and S waves in the background isotropic medium. The angles Ψ and Φ are angles of rotation of the polarization vectors of reflected (Ψ) and transmitted (Φ) S waves in the planes perpendicular to their rays. The vectors must be rotated in order to guarantee a small perturbation from the isotropic to the weakly anisotropic medium. The determination of the angles Ψ and Φ, of course, complicates the procedure of the determination of the linearized R/T coefficients. VAVRYČUK and PŠENČÍK (1998) show that this is not the case for the *PP* reflection and transmission coefficients. In the following, we concentrate on these two coefficients. Derivation of the formulae for the converted waves is left for a next study.

3. PP Wave Displacement Coefficients of Reflection and Transmission

The elastic parameters and the density of the isotropic backgrounds in both halfspaces can be chosen arbitrarily but they should not deviate much from the elastic parameters and the density of the weakly anisotropic media in the halfspaces. We choose the elastic parameters in such a way that our results are simply reducible to the results of previous authors. Specifically, we choose the P- and S-wave velocities $\alpha^{(I)}$ and $\beta^{(I)}$ as follows

$$(\alpha^{(I)})^2 = A_{33}^{(I)}, \quad (\beta^{(I)})^2 = A_{55}^{(I)}. \tag{12}$$

In addition to $\alpha^{(I)}$, $\beta^{(I)}$ and $\rho^{(I)}$, we introduce the P-wave impedance $Z^{(I)}$ and the shear modulus $G^{(I)}$,

$$Z^{(I)} = \rho^{(I)}\alpha^{(I)}, \quad G^{(I)} = \rho^{(I)}(\beta^{(I)})^2. \tag{13}$$

The contrast of a parameter w across the interface is denoted by Δw and it is defined as follows

$$\Delta w = w^{(2)} - w^{(1)}. \tag{14}$$

As VAVRYČUK and PŠENČÍK (1998), we specify the R/T coefficients by the direction of the phase normal n_i of the incident plane wave, specifically by the angles of the incidence θ and the azimuth φ

$$n_i \equiv (\cos \varphi \sin \theta, \sin \varphi \sin \theta, \cos \theta)^T. \tag{15}$$

The approximate formula for the *PP* reflection coefficient $R_{PP}(\varphi, \theta)$ at an interface separating two weakly but arbitrarily anisotropic media can be obtained from Eq. (9) and has the form

$$R_{PP}(\varphi, \theta) = R_{PP}^{iso}(\theta) + \frac{1}{2}\left[\Delta\delta_x \cos^2\varphi + \left(\Delta\delta_y - 8\frac{\bar{\beta}^2}{\bar{\alpha}^2}\Delta\gamma\right)\sin^2\varphi \right.$$

$$+ 2\left(\Delta\chi_z - 4\frac{\bar{\beta}^2}{\bar{\alpha}^2}\Delta\left(\frac{A_{45}}{A_{55}}\right)\right)\cos\varphi\,\sin\varphi \left.\right]\sin^2\theta$$

$$+ \tfrac{1}{2}[\Delta\varepsilon_x \cos^4\varphi + \Delta\varepsilon_y \sin^4\varphi + \Delta\delta_z \cos^2\varphi\,\sin^2\varphi$$

$$+ 2(\Delta\varepsilon_{16}\cos^2\varphi + \Delta\varepsilon_{26}\sin^2\varphi)\sin\varphi\,\cos\varphi]\sin^2\theta\,\tan^2\theta. \tag{16}$$

The symbol $R_{PP}^{iso}(\theta)$ in Eq. (16) denotes the weak contrast reflection coefficient at an interface separating two slightly different isotropic media, see e.g. AKI and RICHARDS (1980):

$$R_{PP}^{iso}(\theta) = \frac{1}{2}\frac{\Delta Z}{\bar{Z}} + \frac{1}{2}\left[\frac{\Delta\alpha}{\bar{\alpha}} - 4\left(\frac{\bar{\beta}}{\bar{\alpha}}\right)^2\frac{\Delta G}{\bar{G}}\right]\sin^2\theta + \frac{1}{2}\frac{\Delta\alpha}{\bar{\alpha}}\sin^2\theta\,\tan^2\theta. \tag{17}$$

In addition to averages and differences of the parameters $\rho^{(I)}$, $\alpha^{(I)}$ and $\beta^{(I)}$ of the background media and the angles θ and φ, the reflection coefficient depends on contrasts of eight *weak anisotropy* (WA) parameters (see PŠENČÍK and GAJEWSKI, 1996) describing the P-wave phase velocity and polarization in weakly anisotropic media. The WA parameters characterize deviations of properties of the studied medium from the isotropic background. In isotropic media, the WA parameters become zero. The total number of the WA parameters is 15. For the P-wave velocity α specified by Eq. (12), their number reduces to 14 and they are, see PŠENČÍK and GAJEWSKI (1998):

$$\delta_x = \frac{A_{13} + 2A_{55} - A_{33}}{A_{33}}, \quad \delta_y = \frac{A_{23} + 2A_{44} - A_{33}}{A_{33}}, \quad \delta_z = \frac{A_{12} + 2A_{66} - A_{33}}{A_{33}},$$

$$\chi_x = \frac{A_{14} + 2A_{56}}{A_{33}}, \quad \chi_y = \frac{A_{25} + 2A_{46}}{A_{33}}, \quad \chi_z = \frac{A_{36} + 2A_{45}}{A_{33}},$$

$$\varepsilon_{15} = \frac{A_{15}}{A_{33}}, \quad \varepsilon_{16} = \frac{A_{16}}{A_{33}}, \quad \varepsilon_{24} = \frac{A_{24}}{A_{33}}, \quad \varepsilon_{26} = \frac{A_{26}}{A_{33}}, \quad \varepsilon_{34} = \frac{A_{34}}{A_{33}}, \quad \varepsilon_{35} = \frac{A_{35}}{A_{33}},$$

$$\varepsilon_x = \frac{A_{11} - A_{33}}{2A_{33}}, \quad \varepsilon_y = \frac{A_{22} - A_{33}}{2A_{33}}, \quad \gamma = \frac{A_{44} - A_{55}}{2A_{55}}, \tag{18}$$

Note that in addition to the P-wave WA parameters, the reflection coefficient also depends on the parameters A_{45}/A_{55} and γ. The symbol γ denotes the "shear-wave splitting parameter" introduced by THOMSEN (1986), see also RUEGER (1996). It is also of interest to note that the above reflection coefficient is reciprocal, i.e. it yields the same value for angles θ, φ and θ, $\varphi + \pi$. This indicates that the PP displacement coefficient of reflection is proportional to the PP coefficient of reflection of the square root of the vertical energy flux, see AKI and RICHARDS (1980) and CHAPMAN (1994).

Using the same approach as for the derivation of the reflection coefficient, we obtain the formula for the PP displacement coefficient of transmission. It reads

$$T_{PP}(\varphi, \theta) = T_{PP}^{iso}(\theta) + \tfrac{1}{2}[\Delta\delta_x \cos^2\varphi + \Delta\delta_y \sin^2\varphi + 2\Delta\chi_z \cos\varphi \sin\varphi] \sin^2\theta$$

$$+ 2[\Delta\varepsilon_{35} \cos\varphi + \Delta\varepsilon_{34} \sin\varphi - \Delta\chi_x \cos^2\varphi \sin\varphi - \Delta\chi_y \cos\varphi \sin^2\varphi$$

$$- \Delta\varepsilon_{15} \cos^3\varphi - \Delta\varepsilon_{24} \sin^3\varphi] \sin^3\theta \cos\theta$$

$$+ \tfrac{1}{2}[\Delta\varepsilon_x \cos^4\varphi + \Delta\varepsilon_y \sin^4\varphi + \Delta\delta_z \cos^2\varphi \sin^2\varphi$$

$$+ 2(\Delta\varepsilon_{16} \cos^2\varphi + \Delta\varepsilon_{26} \sin^2\varphi) \sin\varphi \cos\varphi] \sin^2\theta \tan^2\theta$$

$$+ [\Delta\varepsilon_x \cos^4\varphi + \Delta\varepsilon_y \sin^4\varphi + \Delta\delta_z \cos^2\varphi \sin^2\varphi$$

$$+ 2(\Delta\varepsilon_{16} \cos^2\varphi + \Delta\varepsilon_{26} \sin^2\varphi) \sin\varphi \cos\varphi$$

$$- \Delta\delta_x \cos^2\varphi - \Delta\delta_y \sin^2\varphi - 2\Delta\chi_z \cos\varphi \sin\varphi] \sin^4\theta. \qquad (19)$$

The symbol $T_{PP}^{iso}(\theta)$ denotes the weak contrast transmission coefficient at an interface separating two isotropic media, see again AKI and RICHARDS (1980):

$$T_{PP}^{iso}(\theta) = 1 - \frac{1}{2}\frac{\Delta Z}{\bar{Z}} + \frac{1}{2}\frac{\Delta\alpha}{\bar{\alpha}} \tan^2\theta. \qquad (20)$$

The meaning of the other symbols is the same as in (12)–(14) and (18).

By comparing formulae (16) and (19), we can see that the PP reflection coefficient depends on contrast of only 8 P-wave WA parameters while the PP transmission coefficient depends on contrast of the complete set of the WA parameters. The eight WA parameters are the coefficients of the azimuthally symmetric terms of the expressions for the P-wave phase velocity and polarization in weakly anisotropic media, see PŠENČÍK and GAJEWSKI (1998). The remaining WA parameters, which are the coefficients of the azimuthally anti-symmetric terms, disappear in (16) due to the symmetry of the unconverted PP reflection. We can also see that the reflection coefficient contains some information on the vertical shear-wave propagation, see the "shear-wave splitting parameter" γ in (16), while no such information appears in the formula for the transmission coefficient. In contrast to the reflection coefficient, the transmission coefficient is not reciprocal. This indicates that the relation of the displacement coefficient of transmission to the coefficient related to vertical energy flux is more complicated than in the case of reflection. It confirms the well-known fact that the displacement coefficients are generally not reciprocal.

4. PP Wave R/T Coefficients for Transversely Isotropic Media with a Horizontal Axis of Symmetry along x Axis

The R/T coefficients (16) and (19) can be applied to any type of anisotropic media in both halfspaces surrounding the interface. They simplify considerably if higher symmetry anisotropy is considered. Let us, for example, consider that both halfspaces are transversely isotropic with horizontal axes of symmetry along the x axis. This kind of anisotropy is very important since it describes effects of a system of parallel vertical cracks. For simplicity we consider that the axes of symmetry in both media are parallel. For such a case, the non-zero density-normalized elastic parameters satisfy the following relations on both sides of the interface

$$A_{33} = A_{22}, \quad A_{66} = A_{55}, \quad A_{13} = A_{12}, \quad A_{23} = A_{33} - 2A_{44}. \tag{21}$$

For such a specification, the formulae (16) and (19) reduce to

$$R_{PP}(\varphi, \theta) = R_{PP}^{iso}(\theta) + \frac{1}{2}\left[\Delta\delta_x \cos^2\varphi - 8\left(\frac{\bar{\beta}}{\bar{\alpha}}\right)^2 \Delta\gamma \sin^2\varphi\right]\sin^2\theta$$

$$+ \tfrac{1}{2}(\Delta\varepsilon_x \cos^2\varphi + \Delta\delta_x \sin^2\varphi)\cos^2\varphi \sin^2\theta \tan^2\theta \tag{22}$$

and

$$T_{PP}(\varphi, \theta) = T_{PP}^{iso}(\theta) + \tfrac{1}{2}\Delta\delta_x \cos^2\varphi \sin^2\theta$$

$$+ \tfrac{1}{2}(\Delta\varepsilon_x \cos^2\varphi + \Delta\delta_x \sin^2\varphi)\cos^2\varphi \sin^2\theta \tan^2\theta$$

$$+ (\Delta\varepsilon_x - \Delta\delta_x)\cos^4\varphi \sin^4\theta. \tag{23}$$

The symbols δ_x, ε_x and γ are given in Eqs. (18).

5. Test Example

To test the accuracy of the approximate formulae for the R/T coefficients, we first use the same models as in VAVRYČUK and PŠENČÍK (1998). The halfspace, in which the incident wave propagates, is isotropic, the other halfspace is HTI with the axis of symmetry along the x axis. For these models we calculate values of the R/T coefficients $R_{PP}(\varphi, \theta)$ and $T_{PP}(\varphi, \theta)$ using numerical solution of boundary conditions and compare them with values calculated using the approximate formulae (22) and (23). For the isotropic overburden, the formulae (22) and (23) reduce to

$$R_{PP}(\varphi, \theta) = R_{PP}^{iso}(\theta) + \frac{1}{2}\left[\delta_x^{(2)} \cos^2\varphi - 8\left(\frac{\bar{\beta}}{\bar{\alpha}}\right)^2 \gamma^{(2)} \sin^2\varphi\right]\sin^2\theta$$

$$+ \tfrac{1}{2}(\varepsilon_x^{(2)} \cos^4\varphi + \delta_x^{(2)} \cos^2\varphi \sin^2\varphi)\sin^2\theta \tan^2\theta \tag{24}$$

and

$$T_{PP}(\varphi, \theta) = T_{PP}^{iso}(\theta) + \tfrac{1}{2}\delta_x^{(2)} \cos^2 \varphi \sin^2 \theta$$

$$+ \tfrac{1}{2}(\varepsilon_x^{(2)} \cos^2 \varphi + \delta_x^{(2)} \sin^2 \varphi) \cos^2 \varphi \sin^2 \theta \tan^2 \theta$$

$$+ (\varepsilon_x^{(2)} - \delta_x^{(2)}) \cos^4 \varphi \sin^4 \theta. \tag{25}$$

Note that for the models with isotropic overburden, the R/T coefficients depend directly on WA parameters of the anisotropic halfspace.

Two isotropic and two TI halfspaces are considered. The *P*- and *S*-wave velocities in the isotropic halfspaces are (A): $\alpha = 4.0$ km/sec, $\beta = 2.31$ km/sec and $\rho = 2.65$ g/cm³; (B): $\alpha = 3.0$ km/sec, $\beta = 1.73$ km/sec and $\rho = 2.2$ g/cm³. Anisotropy of the anisotropic halfspaces is assumed to be caused by a system of vertical parallel dry cracks, see HUDSON (1981). The *P*- and *S*-wave velocities of the host rock are 4.0 km/sec and 2.31 km/sec and the density is 2.6 g/cm³. The aspect ratio is 10^{-4} and the crack densities are (C): 0.05 and (D): 0.1. The corresponding matrices of the density-normalized elastic parameters (in GPa) with the axis of symmetry along the x axis have the form

$$\begin{pmatrix} 11.96 & 3.99 & 3.99 & 0.00 & 0.00 & 0.00 \\ & 15.55 & 4.88 & 0.00 & 0.00 & 0.00 \\ & & 15.55 & 0.00 & 0.00 & 0.00 \\ & & & 5.33 & 0.00 & 0.00 \\ & & & & 4.76 & 0.00 \\ & & & & & 4.76 \end{pmatrix}$$

in the case C and

$$\begin{pmatrix} 9.43 & 3.14 & 3.14 & 0.00 & 0.00 & 0.00 \\ & 15.27 & 4.60 & 0.00 & 0.00 & 0.00 \\ & & 15.27 & 0.00 & 0.00 & 0.00 \\ & & & 5.33 & 0.00 & 0.00 \\ & & & & 4.25 & 0.00 \\ & & & & & 4.25 \end{pmatrix}$$

in the case D. Sections of the phase velocity surfaces with the vertical plane containing axes of symmetry for the cases C ($e = 0.05$) and D ($e = 0.1$) are shown in Figure 1.

We consider three models a, b and c, see Table 1. In the models a and b, the phase velocity of the halfspace 1 is for all azimuths higher than the phase velocity in the halfspace 2. In the model c, the relation is opposite. In all cases, the values of reflection coefficients start to rise considerably for higher angles of incidence and the approximate formulae of this paper become inapplicable. From this reason, we consider the angles of incidence only in the interval (0°, 42°). The values of the velocities and the density of the background isotropic medium were determined from formulae (7) and (12). For the used values, see the figure captions.

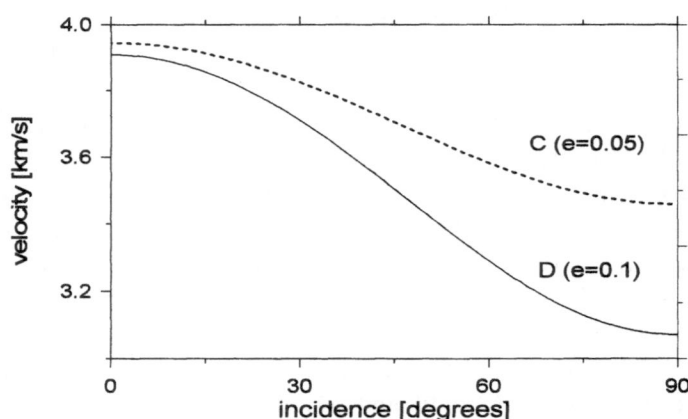

Figure 1

Phase velocity sections with the vertical plane containing the symmetry axes of two "dry crack" models with crack densities $e = 0.05$ (C) and $e = 0.1$ (D).

The test results are displayed in the form of four plots in Figures 2–7. In all the plots the horizontal axis corresponds to the angle of incidence θ, measured in degrees. The vertical axis corresponds to the azimuth φ, also in degrees. Azimuth $\varphi = 0°$ corresponds to the profile along the axis of symmetry, azimuth $\varphi = 90°$ corresponds to the profile in the plane perpendicular to the axis of symmetry, i.e., in the isotropy plane.

The tests of accuracy of the reflection coefficient have been discussed in VAVRYČUK and PŠENČÍK (1998). For completeness, we show in Figures 2–4 only the plots of these results in a different display. Note that in contrast to results discussed in the above-mentioned paper, no shift of values of approximate coefficients is made here. Figures 5–7 contain similar plots as Figures 2–4 but for the transmission coefficients. They are self-explanatory.

A general feature of the presented numerical examples is a higher relative accuracy of the PP transmission coefficients compared to the reflection coefficients.

Table 1

Models used in test examples. I—isotropic, HTI—transversely isotropic (TI) with the horizontal axis of symmetry, VTI—TI with the vertical axis of symmetry, ITI—TI with the inclined axis of symmetry (30° from the horizontal in the (x, z) plane). For the description of parameters of models see the text

	a	b	c	d	e
Halfspace 1	A (I)	A (I)	B (I)	E (VTI)	E (VTI)
Halfspace 2	C (HTI)	D (HTI)	D (HTI)	F (HTI)	F (ITI)

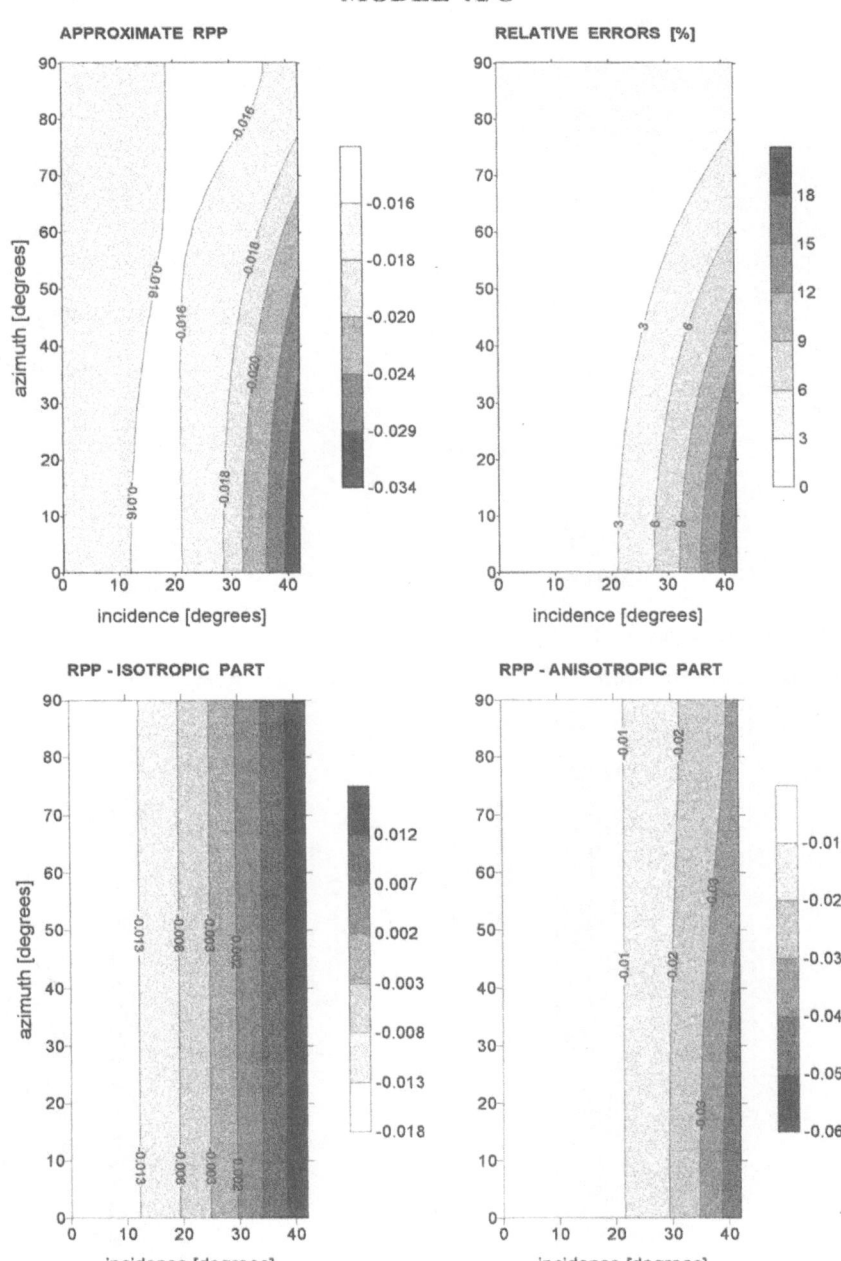

Figure 2
The maps of the approximate R_{PP} and R_{PP}^{iso} reflection coefficients (upper and bottom left), the map of relative errors of the approximate coefficient (upper right) and the map of the difference $R_{PP} - R_{PP}^{iso}$ for the model a of the Table 1. The isotropic halfspace (A): $\alpha = 4.0$ km/sec, $\beta = 2.31$ km/sec, $\rho = 2.65$ g/cm³. The HTI halfspace (C): axis of symmetry along x axis, host rock: $\alpha = 4.0$ km/sec, $\beta = 2.31$ km/sec, $\rho = 2.60$ g/cm³; dry cracks: aspect ratio $a = 0.0001$, crack density $e = 0.05$. Isotropic background: $\bar{\alpha} = 3.97$ km/sec, $\bar{\beta} = 2.25$ km/sec, $\bar{\rho} = 2.63$ g/cm³.

MODEL A/D

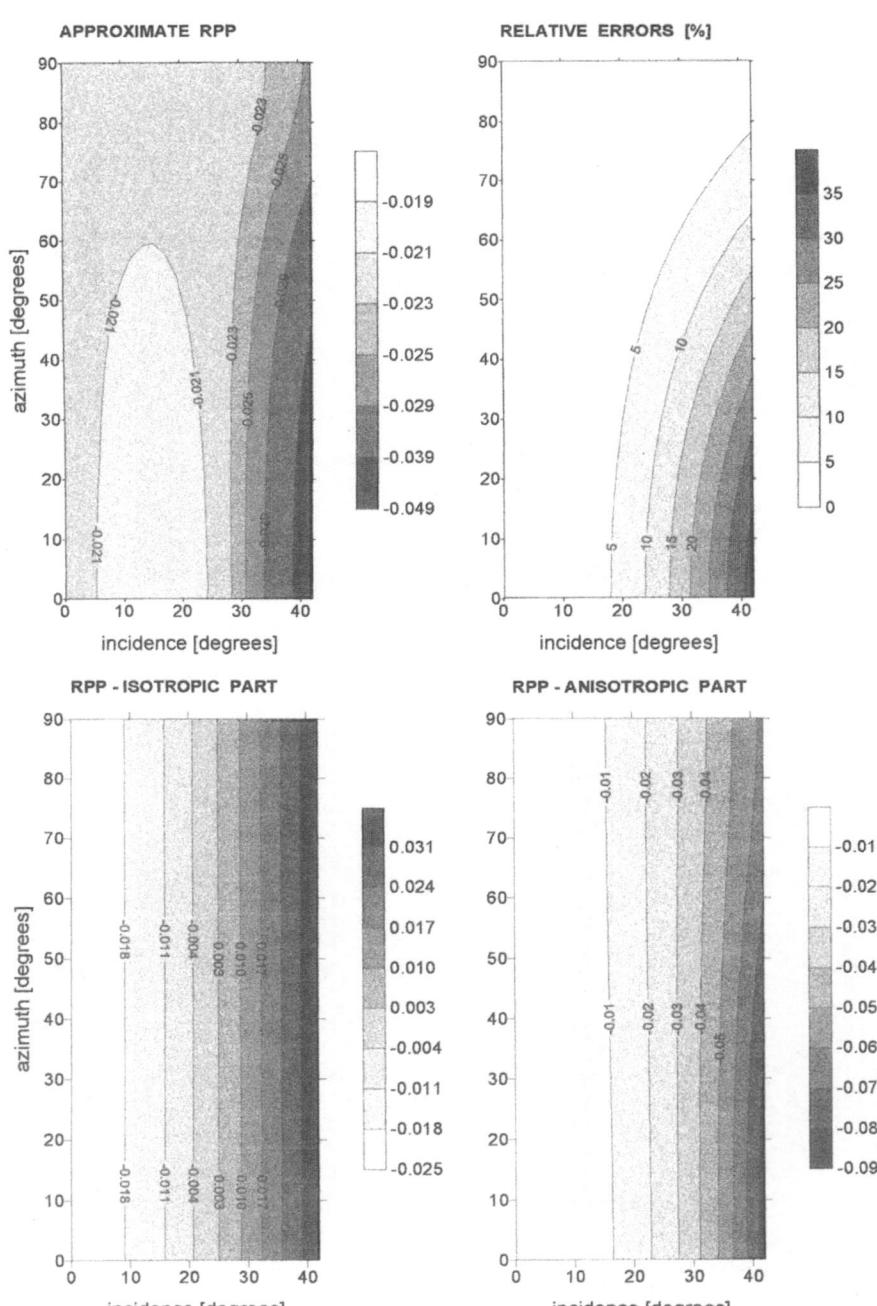

Figure 3
The same as in Figure 2 but for the model b of the Table 1. The isotropic halfspace (A): $\alpha = 4.0$ km/sec, $\beta = 2.31$ km/sec, $\rho = 2.65$ g/cm^3. The HTI halfspace (D): axis of symmetry along x axis, host rock: $\alpha = 4.0$ km/sec, $\beta = 2.31$ km/sec, $\rho = 2.60$ g/cm^3; dry cracks: aspect ratio $a = 0.0001$, crack density $e = 0.01$. Isotropic background: $\bar{\alpha} = 3.95$ km/sec, $\bar{\beta} = 2.19$ km/sec, $\bar{\rho} = 2.63$ g/cm^3.

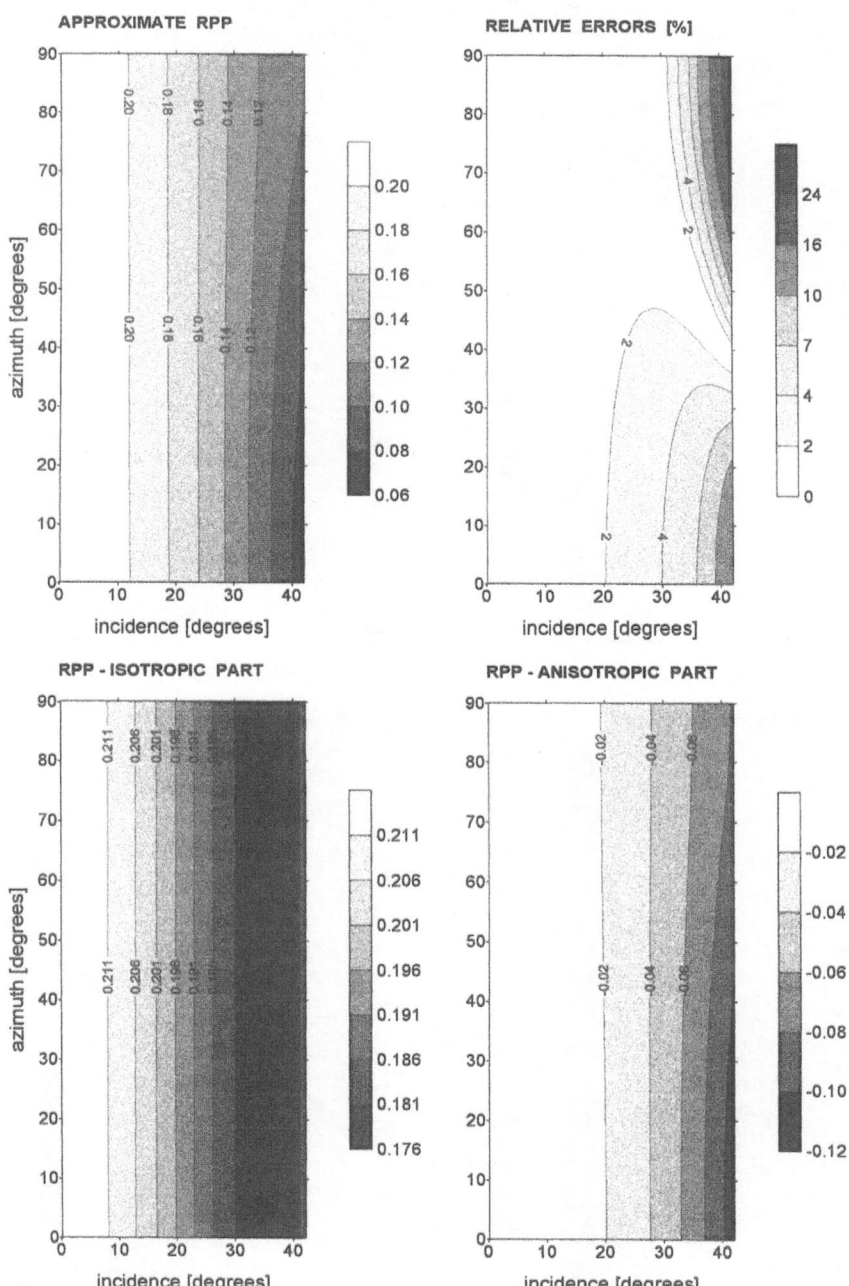

Figure 4
The same as in Figure 2 but for the model c of the Table 1. The isotropic halfspace (B): $\alpha = 3.0$ km/sec, $\beta = 1.73$ km/sec, $\rho = 2.2$ g/cm³. The HTI halfspace (D): axis of symmetry along x axis, host rock: $\alpha = 4.0$ km/sec, $\beta = 2.31$ km/sec, $\rho = 2.60$ g/cm³; dry cracks: aspect ratio $a = 0.0001$, crack density $e = 0.1$. Isotropic background: $\bar{\alpha} = 3.45$ km/sec, $\bar{\beta} = 1.90$ km/sec, $\bar{\rho} = 2.4$ g/cm³.

MODEL A/C

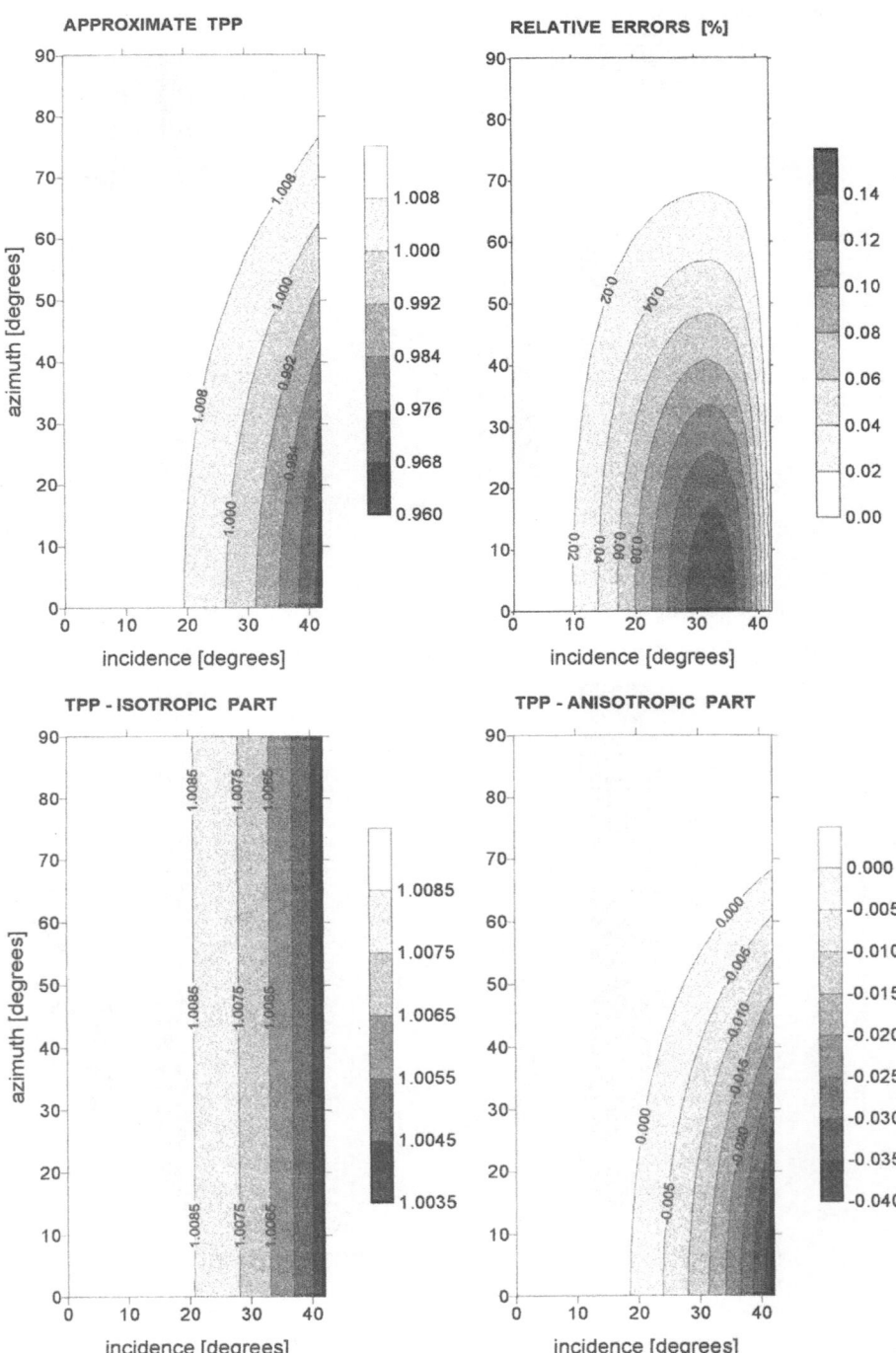

Figure 5
The same as in Figure 2 but for the transmission coefficient T_{PP}.

MODEL A/D

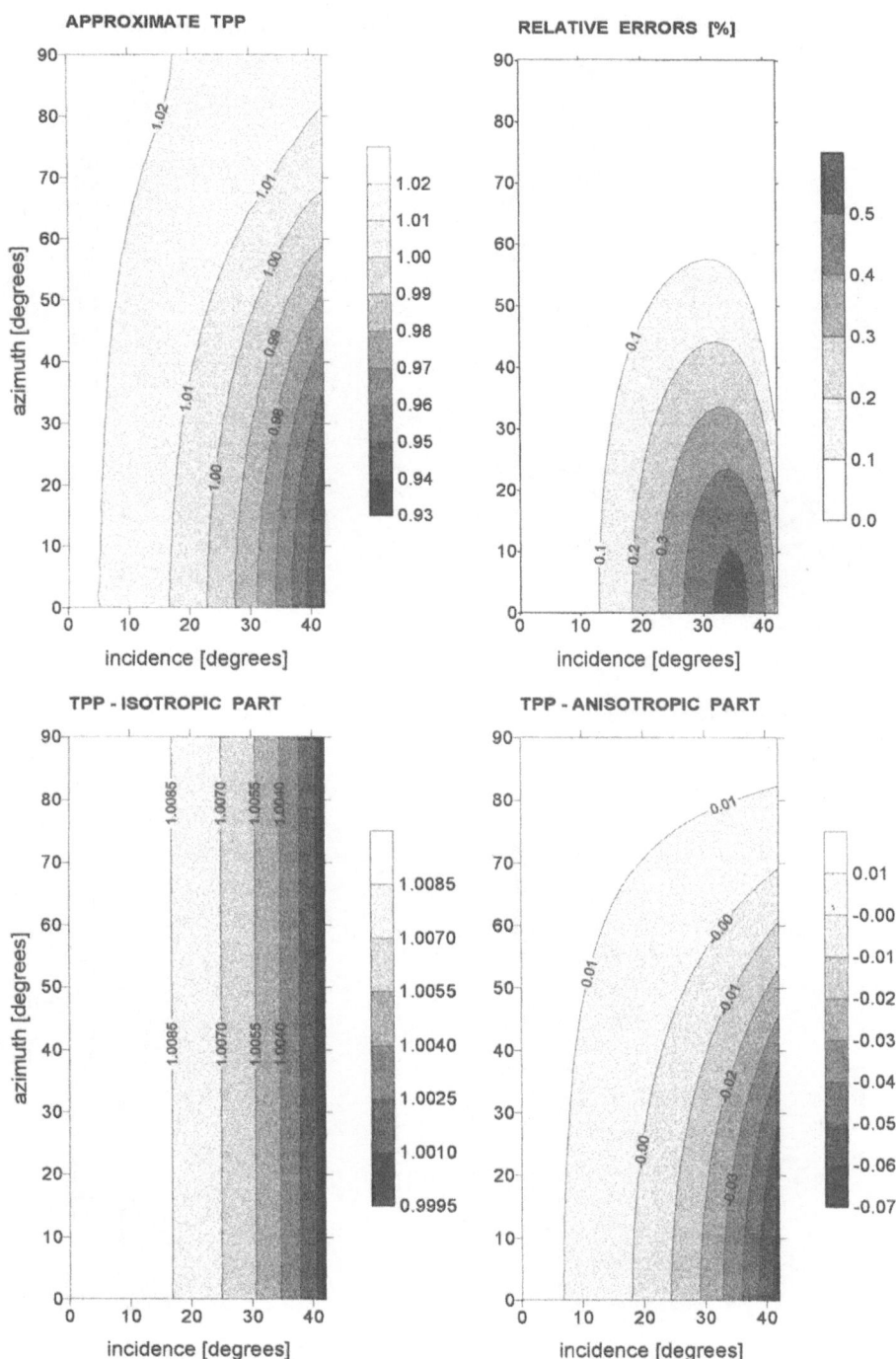

Figure 6
The same as in Figure 3 but for the transmission coefficient T_{PP}.

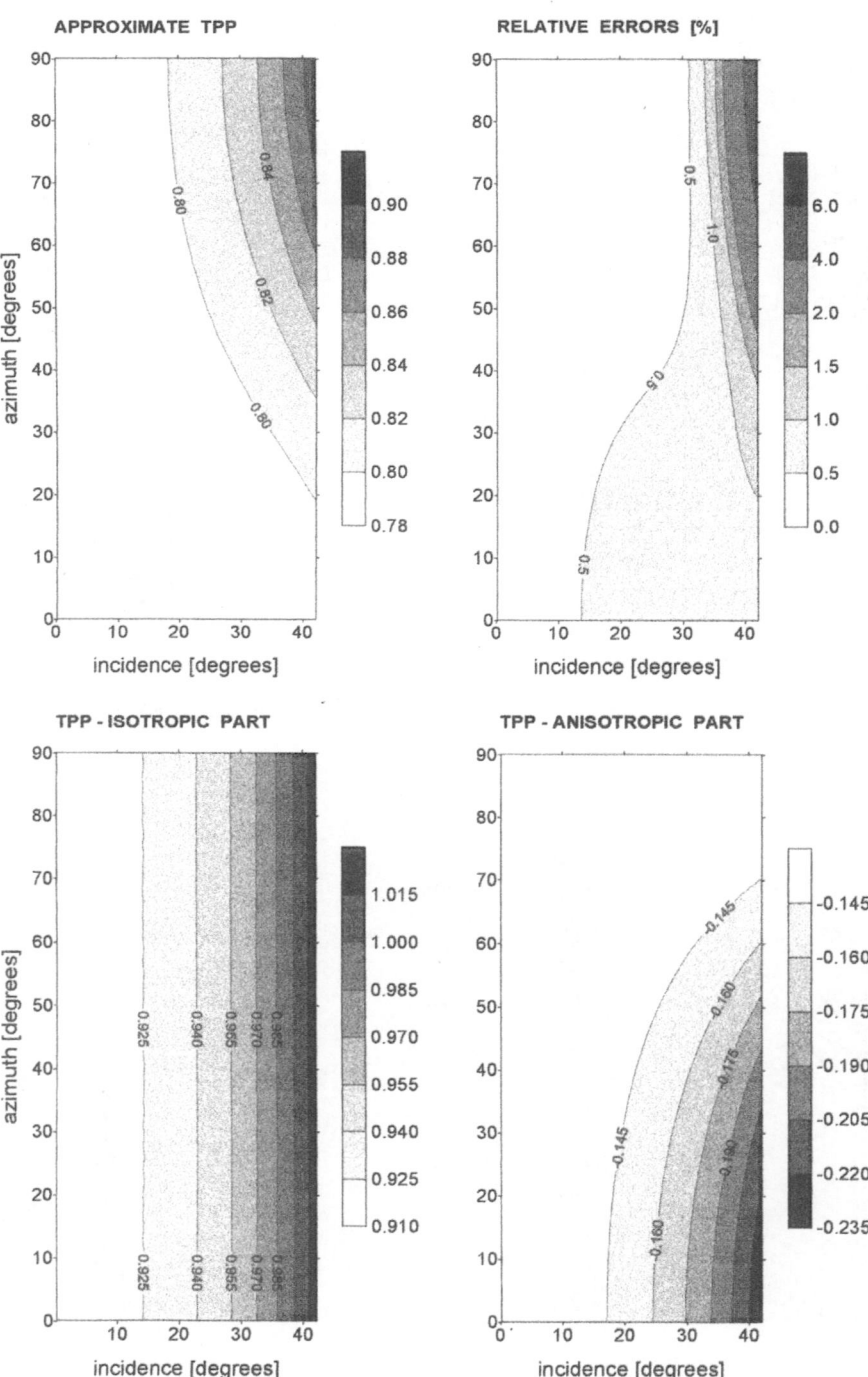

Figure 7
The same as in Figure 4 but for the transmission coefficient T_{PP}.

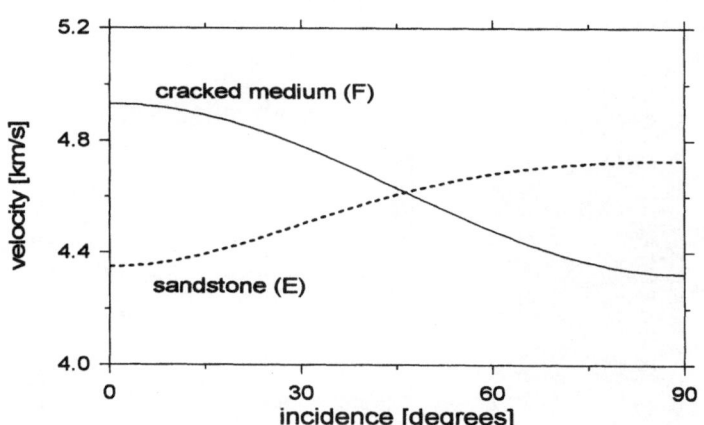

Figure 8
Phase velocity sections of the Mesaverde sandstone ((E)—dashed line) and the "dry crack" model ((F)—full line) with the vertical plane containing the symmetry axes.

It is due to the fact that the reflection coefficients are quite small while the transmission coefficients are close to unity and only vary slightly. We can see that the accuracy of coefficients remains high even in Figures 4 and 7, which correspond to a rather strong contrast (up to 25%) and strong anisotropy (nearly 20%).

Finally, we present the approximate R_{PP} and T_{PP} coefficients for more complicated models than considered before. Figure 8 shows phase velocity sections of two TI materials, E and F, filling the halfspaces in these models. The halfspace 1, in which incident wave propagates, is filled by Mesaverde immature sandstone (E), see THOMSEN (1986), which is VTI. Its matrix of density-normalized elastic parameters (in GPa) has the form

$$
\begin{pmatrix}
22.36 & 6.36 & 8.49 & 0.00 & 0.00 & 0.00 \\
 & 22.36 & 8.49 & 0.00 & 0.00 & 0.00 \\
 & & 18.91 & 0.00 & 0.00 & 0.00 \\
 & & & 6.61 & 0.00 & 0.00 \\
 & & & & 6.61 & 0.00 \\
 & & & & & 8.00
\end{pmatrix}
$$

and the density is $\rho = 2.46$ g/cm³.

The halfspace 2 is filled by a system of parallel dry cracks (F). The cracks are vertical so that the halfspace 2 is HTI with the axis of symmetry parallel to the x axis. The matrix of density-normalized elastic parameters (in GPa) has the form

VERTICAL CRACKS

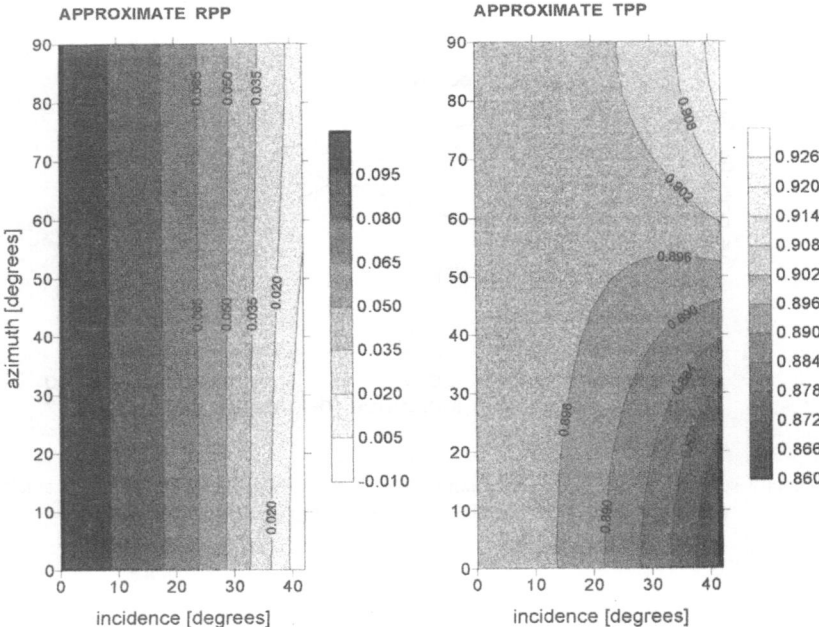

INCLINED CRACKS

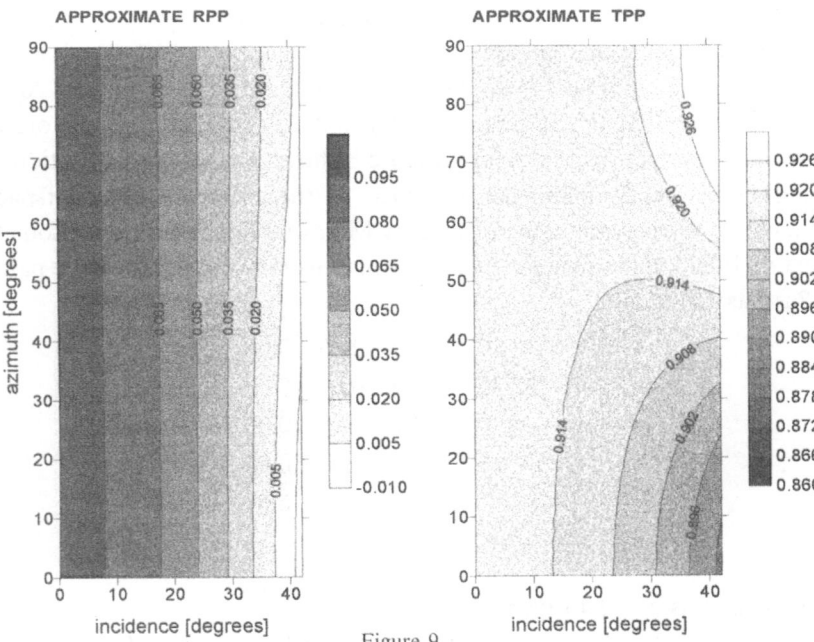

Figure 9

The maps of the approximate R_{PP} and T_{PP} coefficients for the models d and e of Table 1. The upper halfspace (E): the VTI Mesaverde sandstone. The lower halfspace: the dry vertical cracks (model d: upper pictures) exhibiting HTI with the axis of symmetry along the x axis and the dry inclined cracks (model e: bottom pictures) exhibiting ITI with the axis of symmetry making the angle of 30° with the x axis within the (x, z) plane. For parameters of both halfspaces, see the text.

$$\begin{pmatrix} 18.69 & 6.20 & 6.20 & 0.00 & 0.00 & 0.00 \\ & 24.31 & 7.60 & 0.00 & 0.00 & 0.00 \\ & & 24.31 & 0.00 & 0.00 & 0.00 \\ & & & 8.35 & 0.00 & 0.00 \\ & & & & 7.45 & 0.00 \\ & & & & & 7.45 \end{pmatrix}$$

and the density is $\rho = 2.65$ g/cm³. In addition to vertical cracks, cracks inclined by 30° are also considered. This configuration corresponds to the axis of symmetry making the angle of 30° with the x axis in the plane (x, z).

The upper pictures in Figure 9 illustrate the approximate R_{PP} and T_{PP} coefficients in the model d, see Table 1, with vertical cracks in the halfspace 2. The bottom pictures in Figure 9 display the same coefficients for the model e with the inclined cracks in the halfspace 2. We can see that the inclination of cracks has greater effects on the transmission coefficient. The transmission coefficient also behaves in both cases more "anisotropically" than the reflection coefficient.

6. Conclusions

Presented approximate plane wave PP displacement coefficients of reflection and transmission at a weak contrast interface separating two weakly but generally anisotropic media, give a clear insight into the dependence of these coefficients on parameters of media surrounding the interface. The coefficients consist of two parts. The first part is the coefficient for a weak contrast interface separating two slightly different isotropic media. The second part is due to a perturbation of the isotropic background. In addition to the angles θ and φ, the perturbation depends linearly on the contrasts of WA parameters but not on the parameters themselves. The PP reflection coefficient depends on 8; the PP transmission coefficient on 14 P-wave WA parameters (for our choice of the background isotropic medium, 14 WA parameters represent a complete set of the parameters describing the P-wave phase velocity in a weakly anisotropic medium, see PŠENČÍK and GAJEWSKI, 1998). We can conclude that similarly as for the phase velocity and polarization vectors of a P wave propagating in a weakly anisotropic medium, the study of R/T coefficients can yield only limited information on the elastic parameters of the halfspaces surrounding the interface. The PP reflection coefficient contains information on "shear-wave splitting parameter" γ. The reflection coefficient is reciprocal while the transmission coefficient is not.

Presented tests show very good performance of the approximate formulae in the selected region of angles of incidence (0°, 42°) even in cases of rather strong anisotropy and contrast across the interface. Slightly higher relative errors of reflection coefficients are caused by the fact that the coefficients in the studied region are rather small.

Both presented formulae of reflection and transmission coefficients are relatively simple if we take into account that· they describe the case of R/T between two generally anisotropic media. As their simplified forms derived for higher symmetry anisotropic media, they will surely find applications in both forward and inverse seismic modeling.

Acknowledgements

The authors thank Dirk Gajewski and Jeffrey Park for their reviews. Comments of Sláva Červený and Luděk Klimeš are appreciated. This work was supported by the Grant Agency of the Czech Republic, Grant No. 205/96/0968, by the consortium project "Seismic waves in complex 3-D structures" and by the EC, INCO—Copernicus Project IC15 CT96 200. The support of the Brazilian Ministry of Science and Technology under the project Pronex—Engenharia de Petróleo is acknowledged.

REFERENCES

AKI, K., and RICHARDS, P. G., *Quantitative Seismology* (W.H. Freeman, S. Francisco 1980).
BANIK, N. C. (1987), *An Effective Parameter in Transversely Isotropic Media*, Geophysics *52*, 1654–1664.
CHAPMAN, C. H. (1994), *Reflection/Transmission Coefficient Reciprocities in Anisotropic Media*, Geophys. J. Int. *116*, 498–501.
DALEY, P. F., and HRON, F. (1977), *Reflection and Transmission Coefficients for Transversely Isotropic Media*, Bull. Seismol. Soc. Am. *67*, 661–675.
GAJEWSKI, D., and PŠENČÍK, I. (1987), *Computation of High-frequency Seismic Wave Fields in 3-D Laterally Inhomogeneous Anisotropic Media*, Geophys. J. R. Astr. Soc. *91*, 383–411.
HAUGEN, G. U., and URSIN, B. (1996), *AVO-A analysis of vertically fractured reservoir underlaying shale*. In *Expanded Abstracts of the 66th Annual International Meeting of the SEG, Denver*, pp. 1826–1829.
HUDSON, J. A. (1981), *Wave Speeds and Attenuation of Elastic Waves in Material Containing Cracks*, Geophys. J. R. Astr. Soc. *64*, 133–150.
KEITH, C. M., and CRAMPIN, S. (1977), *Seismic Body Waves in Anisotropic Media: Reflection and Refraction at a Plane Interface*, Geophys. J. R. Astr. Soc. *49*, 181–208.
PŠENČÍK, I., and GAJEWSKI, D. (1998), *Polarization, Phase Velocity and NMO Velocity of qP Waves in Arbitrary Weakly Anisotropic Media*, Geophysics, accepted.
RUEGER, A., *Analytic description of reflection coefficients in anisotropic media*. In *EAGE Extended Abstracts, Vol. 1* (Amsterdam, 1996), P026.
THOMSEN, L. (1986), *Weak Elastic Anisotropy*, Geophysics *51*, 1954–1966.
THOMSEN, L., *Weak anisotropic reflections*. In *Offset-dependent Reflectivity—Theory and Practice of AVO Analysis* (eds. Castagna, J. P., and Backus, M. M.) (SEG, Tulsa 1993), pp. 103–111.
TSVANKIN, I. (1996), *P-wave Signatures and Parameterization of Transversely Isotropic Media: An Overview*, Geophysics *61*, 467–483.
VAVRYČUK, V., and PŠENČÍK, I. (1998), *PP Wave Reflection Coefficients in Weakly Anisotropic Elastic Media*, Geophysics, accepted.

(Submitted February 7, 1997, accepted June 23, 1997)

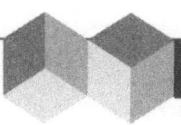

Notes to Authors

PAGEOPH welcomes original contributions in English (and occasionally in French and German) on all aspects of geophysics. All manuscripts should be submitted to the Regular Issues Editor-in-Chief, in triplicate, formatted with double spacing and wide margins. For further details see the following paragraphs.

Format of Manuscripts

Length and Page Charges: A paper should not exceed 16 printed pages including tables and figures. For articles exceeding 16 printed pages the authors will be charged sFr. 80.00 for each additional page. No page charges, except those for color prints, are required for contributors to special issues.

Title Page: This should include the the complete title, full names and addresses of all authors. In addition, corresponding authors should provide their fax number and e-mail address if they are available.

Abbreviated Title: It is necessary to indicate an abbreviated title, which will be used as a running head (no more than 50 characters including spaces).

Abstract: The abstract should be in English, and in the language of the text, if different. It has to be of no more than 10 sentences and should be concise and self-contained.

Keywords: Up to 6 keywords should be listed, suitable for incorporation into information-retrieval systems.

Text: The text must include a citation for each item listed under References; the approximate position of each figure should be indicated in the text. The metric system should be used throughout the text, figures, and tables.

Tables: Tables are to be presented on separate pages, with a brief title for each.

Figures: Figure captions and legends are to be typed on a separate page or pages as the last element in the manuscript. Make sure that line thickness and lettering allow an adequate size reduction. Heliographic or photocopies are not suitable for reproduction. Highquality, glossy, photographic prints must be submitted. Color prints are permitted but authors will be charged for them.

References: They are to be listed in alphabetical order in the following style:
Journal article: Haurwitz, B., and Cowley, A.D. (1973), The Diurnal and Semidiurnal Barometric Oscillations, Global Distribution and Annual Variation, Pure Appl. Geophys. 102, 193-222.

Whole book: Bath, M., Introduction to Seismology (Birkhäuser, Basel 1973).

Article in a book: Haurwitz, B., and Cowley, A.D., Barometric oscillations, In Introduction to Seismology (ed. Bath. M.) (Birkhäuser, Basel 1973) pp. 193-222.

Submission of Manuscripts

Manuscripts must be submitted in triplicate, formatted with double line spacing and wide margins. Copies of the figure should be attached at the end of the manuscript. Original, high quality, glossy figures may be submitted later. All manuscript pages, including references, tables, and captions, should be numbered consecutively, starting with the title page as page o

All manuscripts should be submitted to Regular Issues
Editor-in-Chief
Brian Mitchell
Department of Earth & Atmospheric Sciences
Saint Louis University
3507 Laclede Avenue
St. Louis, MO 63103, USA
e-mail: mitchell@eas.slu.edu

Delivering manuscripts in diskette form may substantially facilitate the publication process provided certain points are taken into consideration:

Texts on diskette should be delivered in either DOS or Macintosh format. They should be saved and delivered in tw separate versions:
• with standard text format as offered by your word process program, and
• in addition in Rich Text Format (RTF) or, as a last resort, an ASCII file.
Numerous word processing programs offer these options wl saving the text.
The final hard copy of the manuscript should be submitted together with the diskette.
The electronic and printed version must be absolutely identic
All pictorial and graphic illustrations should be delivered as hard copy originals and must be 200% of the final printed si
Digital drawings and graphs should be submitted in Encapsulated PostScript (EPS) or Tag Image File Format (TIFF) form. Do not fail to include a hard copy for ready viewing. Back-up copies of the diskettes must be kept. Diske must be adequately protected for transport.

Galley Proofs
Unless indicated otherwise, galley proofs will be sent to the first-named author directly from Birkhäuser Verlag AG and should be returned with the least possible delay. Textual alterations made in the galley proof stage will be charged to the author. One copy of the corrected proof is to be returned immediately to
Editorial Office
Attn. Mrs. Renate D'Arcangelo
Harvard University
233 Pierce Hall, 29 Oxford Street
Cambridge, MA 02138, USA

The editorial office assumes no responsibility for delayed proofs, errors in the original manuscript, or major alteration in proofs for any reason.

Reprints
The authors will receive 50 reprints of each article without charge. Additional reprints may be ordered in lots of 50 wh the final corrected page proofs are returned. Orders submitt thereafter are subject to considerably higher rates.